# Proceedings

Ein stetig steigender Fundus an Informationen ist heute notwendig, um die immer komplexer werdende Technik heutiger Kraftfahrzeuge zu verstehen. Funktionen, Arbeitsweise, Komponenten und Systeme entwickeln sich rasant. In immer schnelleren Zyklen verbreitet sich aktuelles Wissen gerade aus Konferenzen, Tagungen und Symposien in die Fachwelt. Den raschen Zugriff auf diese Informationen bietet diese Reihe Proceedings, die sich zur Aufgabe gestellt hat, das zum Verständnis topaktueller Technik rund um das Automobil erforderliche spezielle Wissen in der Systematik aus Konferenzen und Tagungen zusammen zu stellen und als Buch in Springer.com wie auch elektronisch in Springer Link und Springer Professional bereit zu stellen. Die Reihe wendet sich an Fahrzeug- und Motoreningenieure sowie Studierende, die aktuelles Fachwissen im Zusammenhang mit Fragestellungen ihres Arbeitsfeldes suchen. Professoren und Dozenten an Universitäten und Hochschulen mit Schwerpunkt Kraftfahrzeug- und Motorentechnik finden hier die Zusammenstellung von Veranstaltungen, die sie selber nicht besuchen konnten. Gutachtern, Forschern und Entwicklungsingenieuren in der Automobil- und Zulieferindustrie sowie Dienstleistern können die Proceedings wertvolle Antworten auf topaktuelle Fragen geben.

Today, a steadily growing store of information is called for in order to understand the increasingly complex technologies used in modern automobiles. Functions, modes of operation, components and systems are rapidly evolving, while at the same time the latest expertise is disseminated directly from conferences, congresses and symposia to the professional world in ever-faster cycles. This series of proceedings offers rapid access to this information, gathering the specific knowledge needed to keep up with cutting-edge advances in automotive technologies, employing the same systematic approach used at conferences and congresses and presenting i

t in print (available at Springer.com) and electronic (at Springer Link and Springer Professional) formats. The series addresses the needs of automotive engineers, motor design engineers and students looking for the latest expertise in connection with key questions in their field, while professors and instructors working in the areas of automotive and motor design engineering will also find summaries of industry events they weren't able to attend. The proceedings also offer valuable answers to the topical questions that concern assessors, researchers and developmental engineers in the automotive and supplier industry, as well as service providers.

Alexander Heintzel
Editor

# Automotive Acoustics Conference 2025

## 8. Internationale ATZ-Fachtagung Fahrzeugakustik

Springer Vieweg

*Editor*
Alexander Heintzel
Springer Fachmedien Wiesbaden
Wiesbaden, Germany

ISSN 2198-7432 ISSN 2198-7440 (electronic)
Proceedings
ISBN 978-3-658-51248-4 ISBN 978-3-658-51249-1 (eBook)
https://doi.org/10.1007/978-3-658-51249-1

Die Deutsche Nationalbibliothek verzeichnet diese Publikation in der Deutschen Nationalbibliografie; detaillierte bibliografische Daten sind im Internet über https://portal.dnb.de abrufbar.



This Springer Vieweg imprint is published by the registered company Springer Fachmedien Wiesbaden GmbH, part of Springer Nature.
The registered company address is: Abraham-Lincoln-Str. 46, 65189 Wiesbaden, Germany

# Welcome

As the increasing electrification of mobility continues to reshape the acoustic landscape in the automotive industry, high standards for noise, vibration and harshness (NVH) remain critical to vehicle quality and refinement. In this context, sustainable and smart noise control solutions are key to enhancing the NVH performance of both passenger cars and commercial vehicles while minimizing their environmental impact.

The 8th Automotive Acoustics Conference will cover the latest developments in this area and provide a comprehensive snapshot of global advancements in acoustics and NVH. This year's program features 28 presentations from industry leaders, including keynote lectures by Volvo, Audi, and Stellantis. Topics range from NVH strategies for electric vehicles to cutting-edge materials and simulation techniques.

For the first time, our leading global conference in the field of automotive acoustics will take place at the Bodenseeforum in Constance, Germany. The hybrid format of the 2025 edition once again allows participants worldwide to attend virtually.

We look forward to meeting you in Constance or online.

On behalf of the Scientific Advisory Board

Dr. Davide Caprioli
Dr. Alexander Heintzel

# Contents

# News from the Automotive Acoustics Conference 2025—8th International Conference on Vehicle Acoustics

## The Importance of Vehicle Acoustics is Increasing

Acoustic engineers face challenging tasks due to the electrification of mobility. They'll be talking about the main trends and developments at the Automotive Acoustics Conference 2025.

With the accelerated shift toward electrification of modern powertrain systems, acoustic development work is becoming increasingly important. How to master the new challenges in the field of NVH (noise, vibration, harshness) will be the focus of the Automotive Acoustics Conference, which will take place on July 8 and 9, 2025, in Constance, Germany, with around 180 experts from 18 countries participating. 150 of them will take advantage of the opportunity for intensive exchange on site at the Bodenseeforum, while 30 additional participants will join via online stream. A total of 28 interesting presentations will be available to choose from over the two days. Noise, vibration, and harshness have a significant influence on the perceived quality of a vehicle—a factor that is even more important in electric vehicles. This is because the absence of traditional engine noise means that other interior and exterior noises become more noticeable. The challenge for developers is therefore to optimize not only the powertrain but also other components in a targeted manner.

Davide Caprioli from Autoneum, scientific director of the conference, welcomed all attendees at the start and wished them two interesting and exciting days. For over 50 years, the Automotive Acoustics Conference has been a central forum for this discourse, as Eelco Spoelder, CEO of host Autoneum, emphasized in his opening speech, highlighting the importance of artificial intelligence and sustainability in the development of future vehicle models. For the second time, the conference was held in a hybrid format—on site and digitally. After many years in Rüschlikon, Switzerland, this year's venue was moved to Constance, marking a relocation from Lake Zurich to Lake Constance. Despite the option of participating online, many of the international guests traveled to the conference in person—a sign of the high value placed on direct exchange. Marc Ziegler, editor-in-chief of the ATZ-MTZ Group, welcomed all participants on behalf of the organizer ATZlive.

## Development of the Volvo EX90

In Germany alone, the number of electric vehicles is rapidly approaching 2 million, and there are already an estimated 56 million electric cars on the road worldwide, with China accounting for the largest market share. The number of electric cars has risen by a third in 2024; at the end of 2023, the global fleet stood at just under 42 million vehicles. Once again, China led the growth, followed by the US and Germany.

The central importance of electric drive was directly reflected in the first keynote speech of the conference: Eskil Lindberg from Volvo described how, during the early concept phases of the Volvo EX90, the structural NVH characteristics were addressed in a controlled manner while at the same time creating the conditions for good driving dynamics and a high level of driving comfort. Low noise levels in the interior placed new demands on the NVH of the powertrain, for example in the form of high-frequency whistling noises. In his detailed presentation, he described the importance of body and chassis stiffness in ensuring vibration isolation between the body and chassis and in achieving a design that meets durability targets and strikes the right balance between NVH, ride comfort, steering, and handling.

**Electrification of Commercial Vehicles: Ford Transit Custom**

In the first presentation, Daniel Kierspel from Ford used the Ford Transit Custom as an example to illustrate the challenges involved in electrifying an existing commercial vehicle model. These challenges lie not only in integrating an ICE-HEV-BEV powertrain into a platform, but also in meeting a wide range of requirements for derivatives such as vans, double cab-in-vans, chassis cabs, buses, and the Nugget camper van. As part of the NVH targets, a driving simulator was integrated at Ford to enable the development team to simulate the realistic interior noise levels to be expected. This made it possible to diagnose potential fault conditions even for different road profiles and traffic situations. This method enables an improved target setting and coordination process. NVH performance evaluations are carried out using a customized model that utilizes data from various sources and regularly updates the input data as the program progresses. Thanks to the early model-based evaluations in conjunction with the driving simulator, physical prototype testing could ultimately begin at a high level of maturity. This proved to be extremely helpful in avoiding unpleasant surprises later on.

**Noise Transmission Path Through the Trunk**

After the first refreshment break, the next speaker was Kichang Jo from Hyundai, who gave a presentation entitled “Noise transmission path analysis and sound package optimization for BEVs using numerical simulation”. He explained how, in electric vehicles in particular, the amount of noise entering through the trunk increases, making sound insulation in this area increasingly important. However, since the rear of a vehicle has a complex structure and consists of various parts, it is not easy to identify the path through which noise is transmitted by vehicle testing and to find an efficient method for reducing interior noise. Therefore, Hyundai conducted a panel contribution analysis and a target cascade analysis of interior noise using the Statistical Energy Analysis (SEA) analysis model of the actual sound absorption performance of vehicles to identify the noise transmission path through the trunk of electric vehicles.

**SEA Tool and NVH Analysis of Fully-trimmed Car Bodies**

In their paper entitled "SEA hybrid p/Q correlated modelization process for vehicle transient 'on track' noise assessment", Jean-Baptiste Prunet and Johann Löwen from Stellantis explained that the SEA tool is traditionally used globally to provide a pre-dimensioned design for the acoustic package of the planned vehicle. However, since the insulation performance of the vehicle depends to a large extent on the contributions of smaller components, SEA does not seem to be able to provide a noise prediction for the interior, as SEA works with large assemblies and does not take the details of the vehicle into account. An operating model was therefore created that takes into account detailed component performance and realistic physical parameters. Since the issue of noise prediction at the passengers' ears remains within the scope of the designer's task and the purpose of creating a calculation model for the vehicle, Stellantis also conducted an analysis on the race track, including noise measurements at the noise sources.

Nicolas Driot from Hexagon devoted the last presentation of the first morning to the topic of "Recent advances in efficient fully-trimmed body NVH analysis". He began by presenting a methodology that correctly sequences the modal extraction process for the un-damped in-vacuum body and the rigid wall cavity model, as well as the calculation of the reduced trim impedance matrix (RIM). Care is taken to avoid unnecessary modal data recovery in order to speed up the process without compromising the numerical predictions. He also demonstrated a new air gap coupling condition for trim interfaces that enables a reduction in the size of finite element (FE) models for trim.

# Nachricht von der Automotive Acoustics Conference 2025 – 8. Internationale ATZ-Fachtagung Fahrzeugakustik

**Relevanz der Fahrzeugakustik steigt**

Akustikingenieure stehen aufgrund der Elektrifizierung der Mobilität vor herausfordernden Aufgaben. Welche Trends und Entwicklungen maßgeblich sind, diskutieren sie auf der Automotive Acoustics Conference 2025.

Mit dem beschleunigten Wandel hin zur Elektrifizierung moderner Antriebssysteme steigt auch die Relevanz akustischer Entwicklungsarbeit. Wie sich die neuen Herausforderungen im Bereich NVH (Noise, Vibration, Harshness) meistern lassen, steht im Zentrum der Automotive Acoustics Conference, die am 8. und 9. Juli 2025 in Konstanz stattfindet, an der rund 180 Expertinnen und Experten aus 18 Ländern teilnehmen. 150 davon nutzen die Gelegenheit zum intensiven Austausch vor Ort im Bodenseeforum, während 30 weitere Teilnehmende per Online-Stream zugeschaltet sind. Insgesamt stehen 28 interessante Vorträge in den zwei Tagen zur Auswahl. Geräusche, Vibrationen und Rauigkeit haben maßgeblichen Einfluss auf die wahrgenommene Qualität eines Fahrzeugs – ein Umstand, der insbesondere bei Elektrofahrzeugen noch stärker ins Gewicht fällt. Denn durch den Wegfall des klassischen Motorgeräuschs rücken andere Innen- und Außengeräusche stärker in den Fokus. Die Herausforderung für Entwickler liegt deshalb darin, nicht nur den Antriebsstrang, sondern auch weitere Komponenten gezielt akustisch zu optimieren.

Davide Caprioli von Autoneum, wissenschaftlicher Leiter der Konferenz, begrüßte zum Auftakt alle Anwesenden und wünschte zwei interessante und aufregende Tage. Seit über 50 Jahren ist die Automotive Acoustics Conference ein zentrales Forum für diesen Diskurs – wie auch Eelco Spoelder, CEO des Gastgebers Autoneum, in seiner Eröffnungsrede betonte und dabei die Bedeutung künstlicher Intelligenz und Nachhaltigkeit bei der Entwicklung künftiger Fahrzeugmodelle hervorhob. Bereits zum zweiten Male fand die Konferenz hybrid statt – vor Ort und digital. Nachdem sich in vielen Jahren das schweizerische Rüschlikon als Veranstaltungsort etabliert hatte, fiel in diesem Jahr die Wahl auf Konstanz, es war also ein Umzug vom Zürich- an den Bodensee. Trotz der Möglichkeit zur Online-Teilnahme reisten viele der internationalen Gäste, persönlich an – ein Zeichen für den hohen Stellenwert des direkten Austauschs. Für den Veranstalter ATZlive begrüßte Marc Ziegler, verantwortlicher Redakteur der ATZ-MTZ-Gruppe, alle Teilnehmer.

**Entwicklung des Volvo EX90**

Allein in Deutschland bewegt sich die Zahl der E-Fahrzeuge rasch auf die Zahl von 2 Millionen zu, weltweit sind schätzungsweise bereits 56 Millionen Elektroautos unterwegs, wobei China den größten Marktanteil hat. Die Zahl der Elektroautos ist im Jahr 2024 um ein Drittel gestiegen, Ende 2023 lag der weltweite Bestand noch bei knapp 42 Mio Fahrzeugen. Angeführt wurde das Wachstum erneut deutlich von China, gefolgt von den USA und Deutschland.

Die zentrale Bedeutung des Elektroantriebs spiegelte sich direkt in der ersten Keynote der Konferenz wider: Eskil Lindberg von Volvo beschrieb, wie bei der Entwicklung des Volvo EX90 bereits in den frühen Konzeptphasen die strukturbedingten NVH-Eigenschaften kontrolliert angegangen und gleichzeitig die Voraussetzungen für eine gute Fahrdynamik und einen hohen Fahrkomfort geschaffen wurden. Niedrige Geräuschpegel im Innenraum stellten neue Anforderungen an die NVH des Antriebs, beispielsweise in Form von hochfrequenten Pfeifgeräuschen. In seinen teils sehr detaillierten Ausführungen schilderte er die Bedeutung der Faktoren Steifigkeit der Karosserie und des Fahrwerks, um die Schwingungsisolierung zwischen Karosserie und Fahrwerk zu gewährleisten und eine Konstruktion zu realisieren, die die Haltbarkeitsziele erfüllt und die richtige Balance zwischen NVH, Fahrkomfort sowie Lenkung und Handling ermöglicht.

**Elektrifizierung des Nutzfahrzeugs: Ford Transit Custom**

Im ersten Vortrag verdeutlichte Daniel Kierspel von Ford am Beispiel des Ford Transit Custom die Herausforderungen bei der Elektrifizierung eines bereits vorhandenen Nutzfahrzeugmodells. Diese bestehe nicht allein in der Integration eines ICE-HEV-BEV-Antriebsstrangs in eine Plattform, sondern auch in der Erfüllung eines breiten Anforderungsportfolios für Derivate wie Van, Double Cab-in-Van, Chassis Cab, Bus sowie den Nugget Camper Van. Im Rahmen der NVH-Zielvorgaben wurde bei Ford ein Fahrsimulator integriert, um dem Entwicklungsteam eine Simulation der zu erwartenden realistischen Innenraumgeräusche zu ermöglichen. Potenzielle Fehlerzustände ließen sich so selbst für verschiedene Straßenprofile und Verkehrssituationen diagnostizieren. Diese Methode ermögliche einen verbesserten Zielfestlegungs- und Abstimmungsprozess. Bewertungen der NVH-Performance werden mit einem maßgeschneiderten Modell durchgeführt, das Daten aus verschiedenen Quellen nutzt, und die Eingabedaten je nach Programmfortschritt regelmäßig aktualisiert. Dank der frühzeitigen modellbasierten Bewertungen in Verbindung mit dem Fahrsimulator könnten die physischen Prototypentests schließlich mit einem hohen Reifegrad beginnen. Dies habe sich als äußerst hilfreich erwiesen, um später unliebsame Überraschungen zu vermeiden.

**Geräuschübertragungsweg durch den Kofferraum**

Nach der ersten Erfrischungspause folgten als nächster Referent Kichang Jo von Hyundai und sein Vortrag mit dem Titel „Analyse der Geräuschübertragungswege und Optimierung des Soundpakets für Elektrofahrzeuge unter Verwendung numerischer Simulation“. Er legte dar, wie insbesondere bei Elektrofahrzeugen die

Menge der durch den Kofferraum eindringenden Geräusche zunimmt, wodurch die Schalldämmung in diesem Bereich an Bedeutung gewinnt. Da der hintere Teil eines Fahrzeugs jedoch eine komplexe Struktur aufweist und aus verschiedenen Teilen besteht, ist es nicht einfach, den Weg, über den Geräusche übertragen werden, durch Fahrzeugtests genau zu identifizieren und eine effiziente Methode zur Reduzierung der Innenraumgeräusche zu finden. Daher wurden bei Hyundai eine Panel-Beitragsanalyse und eine Zielkaskadenanalyse der Innenraumgeräusche unter Verwendung des Statistical Energy Analysis (SEA)-Analysemodells der tatsächlichen Schallabsorptionsleistung von Fahrzeugen durchgeführt, um den Geräuschübertragungsweg durch den Kofferraum von Elektrofahrzeugen zu ermitteln.

**SEA-Tool und NVH-Analyse vollständig ausgekleideter Karosserien**
In ihrem Beitrag mit dem Titel „SEA-Hybrid-p/Q-korrelierter Modellierungsprozess zur Bewertung von Fahrzeug-Transientenlärm auf der Strecke“ legten Jean-Baptiste Prunet und Johann Löwen von Stellantis dar, dass traditionell das SEA-Tool global eingesetzt werde, sodass das geplante Fahrzeug einen vorab dimensionierten Entwurf für das Akustikpaket erhält. Da die Dämmleistung des Fahrzeugs in hohem Maße von den Beiträgen kleinerer Komponenten abhängt, scheint SEA aber nicht in der Lage zu sein, eine Geräuschprognose für den Innenraum zu liefern, da SEA mit großen Baugruppen arbeitet und die Details des Fahrzeugs nicht berücksichtigt. Daher wurde ein Betriebsmodell erstellt, das detaillierte Komponentenleistungen und realistische physikalische Parameter berücksichtigt. Da die Frage der Geräuschvorhersage an den Ohren der Passagiere im Aufgabenbereich des Konstrukteurs und im Zweck der Erstellung eines Berechnungsmodells für das Fahrzeug bleibt, hat Stellantis auch eine Analyse auf der Rennstrecke durchgeführt, einschließlich Geräuschmessungen an den Geräuschquellen.

Nicolas Driot von Hexagon widmete sich im letzten Vortrag des ersten Vormittags dem Thema „Jüngste Fortschritte bei der effizienten NVH-Analyse von vollständig ausgekleideten Karosserien“. Er stellte zunächst eine Methodik vor, die den Modalextraktionsprozess für die ungedämpfte In-Vakuum-Karosserie und das starre Wandhohlraummodell sowie die Berechnung der reduzierten Trimmimpedanzmatrix (RIM) richtig sequenziert. Dabei wird darauf geachtet, unnötige Modaldatenwiederherstellungen zu vermeiden, um den Prozess zu beschleunigen, ohne die numerischen Vorhersagen zu beeinträchtigen. Zudem zeigte er eine neue Luftspalt-Kopplungsbedingung für Verkleidungsschnittstellen auf, die eine Verringerung der Größe der Finite-Elemente-Modelle (FE) für Verkleidungen ermögliche.

Frank Jung

# Volvo EX90—The Journey Towards World Class Comfort

Eskil Lindberg(✉), Thomas Englund, Christian Fransson, Mattias Davidsson, David Lennström, Pär Harling, and Per Alenius

Volvo Car Corporation, Gunnar Engellaus Väg 8, 418 78 Gothenburg, Sweden
{eskil.lindberg,thomas.englund,christian.fransson, mattias.davidsson,david.lennstrom,par.harling, per.alenius}@volvocars.com

**Abstract.** Customer expectations on premium E-segment Battery Electric Vehicle (BEV) are high in many aspects, not least when it comes to low levels of Noise, Vibration and Harshness (NVH) along with a high level of Ride Comfort. To create a pleasant riding experience, it is important to consider the balance between different noise sources in the car, i.e. ensuring that they orchestrate well with each other in terms of level and character. In a BEV, the balance between the propulsion NVH, road NVH and wind noise will be very different to the balance in a vehicle with an internal combustion engine and need to be considered from the very start when developing a new BEV.

In this paper the development of the new Volvo EX90 with respect to the forementioned NVH areas along with ride comfort will be presented. A complete vehicle perspective and a common target has been key, approaching a systematic focus on all parts of the vehicle, in combination with persistence in all development stages ranging from early pre-study knowledge building and strategy to industrialization. An efficient high voltage battery integration in combination with comprehensive body structure improvements and improved chassis isolation are some of the key elements in the development, both for reducing NVH levels and increasing ride comfort levels. Furthermore, significant attention was given to designing noise-absorbing and insulating interior trim components to achieve a balance between airborne and structure-borne noise. Finally, the electric drive units were improved throughout the development stages, particularly by refining the active parts affecting e-motor and gear whine.

As a result, significant improvements in NVH and Ride comfort have been achieved and are presented in the paper.

**Keywords:** Road NVH · Ride comfort · World class comfort


A. Heintzel (Ed.): Automotive Acoustics Conference 2025, Proceedings, pp. 1–15, 2026. https://doi.org/10.1007/978-3-658-51249-1_1

## 1.1 Background

The Volvo EX90 development began with initial discussions and conceptual ideas around 2016 after the second-generation Volvo XC90 on the SPA1 platform had been launched a few years earlier.

The vision for the Volvo XC90/SPA1 platform was to create a premium car that had a performance level in line with the best competitors in the world. Realizing comfort levels of the platform has proven a non-trivial task, and a high amount of work was put in to making the Volvo XC90 competitive regarding comfort and noise levels at the launch 2015. A lot of knowledge about platform development was gained during the development of the SPA1 platform, resulting in a big leap into the premium luxury car segment. However, in the subsequent years, it became clear that competitors were not idle, and a subtle 'new premium' began to emerge among the premium brands. The "new premium" meant switched focus for the high-end cars, such as the Volvo XC90, from powerful engine sound towards more refined general NVH and ride comfort levels, to create maximum comfort on the medium to long drives. Volvo Cars have followed this trend, continuously investing in, and updating the SPA1 cars. Moving into the development of the new SPA2 platform, experiences from the SPA1 platform development and updates, in combination with new research, have been utilized to take the next step in NVH and ride comfort levels (see Fig. 1.1).

Focus areas for SPA2 and Volvo EX90 include

- **General customer demands:** There's a growing emphasis on NVH and overall comfort.
- **High-end customer expectations:** Premium customers demand higher standards for NVH and comfort.
- **Chinese market demands:** Chinese consumers have elevated expectations for NVH and comfort.
- **Electric vehicle buyers:** EV buyers generally expect superior NVH performance.
- **Early NVH Consideration:** It's crucial to design the platform with NVH in mind from the start to avoid additional costs later.
- **Balancing attributes early:** It's essential to balance conflicting attributes as early as possible in the concept phase.
- **Unified comfort goals:** All attributes related to comfort must work towards the same high ambition level.
- **Objective communication**: Clear communication based on objective data is vital throughout the project.
- **Holistic approach:** Focus on all parts of the car, starting with dimensions and those affecting multiple attributes.

This paper will outline the efforts that contributed to establishing the EX90 as a benchmark for NVH and ride comfort.

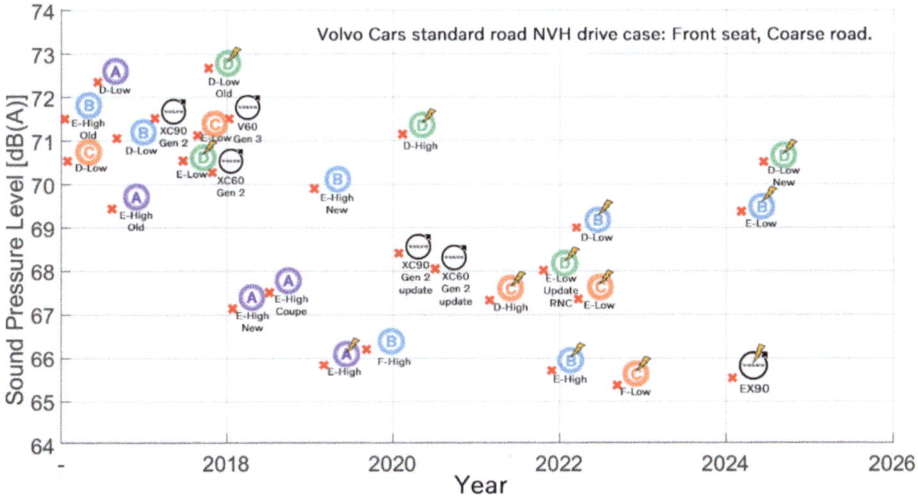


**Fig. 1.1.** Trend towards more refined NVH levels. A, B, C and D are all premium car competitors. The competitor cars are classified as D=large (e.g. V60), E=executive (e.g. XC90) and F=Luxury (e.g. Mercedes- Benz S-Class). "High" means SUV and "low" means non-SUV. The flash indicates that it's a fully electric car.

## 1.2 Attribute Balancing

At Volvo Cars Vehicle Attribute Department, our guiding principle is: "Our consumers enjoy every moment." This means that our cars cannot excel at just one attribute; all attributes and features need to be in harmony. Some attributes, such as road NVH and steering and handling, have conflicting requirements when it comes to chassis design. This well-documented trade-off has inspired extensive research in the field (see [1]).

Achieving top NVH is not possible with late attribute balancing. The key to the Volvo EX90's world-class comfort is the early cross-functional collaboration. Early in the SPA2 project, the road NVH and vehicle dynamics (VDyn) teams work closely together to set the right concept and avoid late attribute balancing. The starting point was that Road NVH identified the need for a certain amount of isolation (soft bushings) between the body and the subframes (front and rear) to meet the overall targets for Road NVH.

The VDyn community was then challenged to find concepts that enabled this level of stiffness. Positively, good isolation is achieved by the difference in stiffness between the bushings and the body structure. One of the countermeasures for the soft bushings for VDyn was increasing body stiffness. A common goal of achieving sufficient body structure stiffness was agreed and worked towards. It is always more compelling to have multiple arguments for why a given design should be a certain way (see Fig. 1.2).

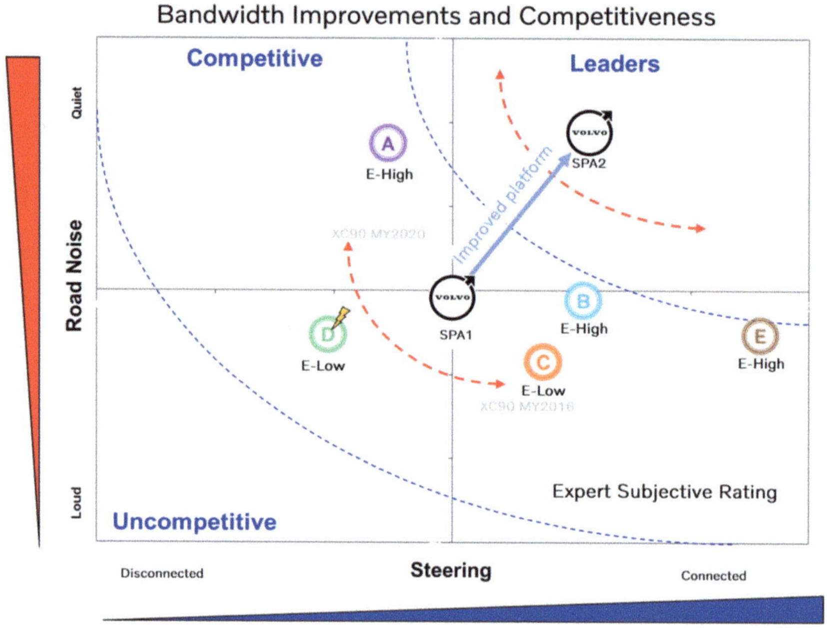


**Fig. 1.2.** Balance Steering & Handling, Road NVH and Ride comfort on a higher level. Body stiffness is one of the key parameters to improve bandwidth of the platform.

### 1.2.1 Road Noise

One key challenge when it comes to developing a car with excellent road noise levels is that nothing can be left to chance. Since road noise is a complete vehicle attribute, it is of outermost importance to always understand what parts of the car that is dimensioning as unfortunately "the weakest link in the chain will be dimensioning". Adding further complexity, road noise contains both structural and airborne noise contributions, where structural borne dominates the interior noise up to 400 Hz [2], (see Figs. 1.3 and 1.4). Taking a systematic approach is therefore critical.

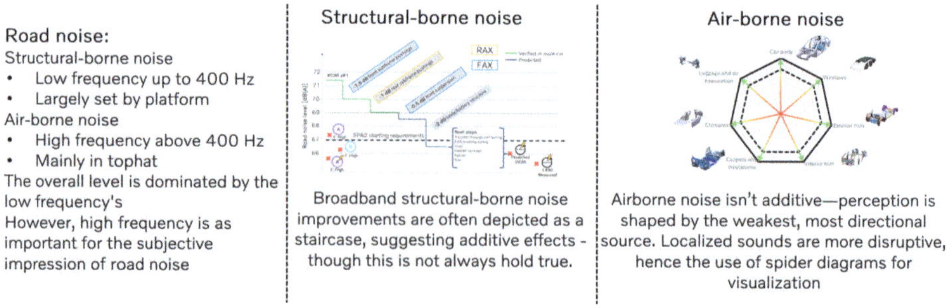


**Fig. 1.3.** Road noise is a complete vehicle attribute with both structural and airborne contributions. The images illustrate how project status has been reported. For larger versions, see Figs. 1.5 and 1.7.

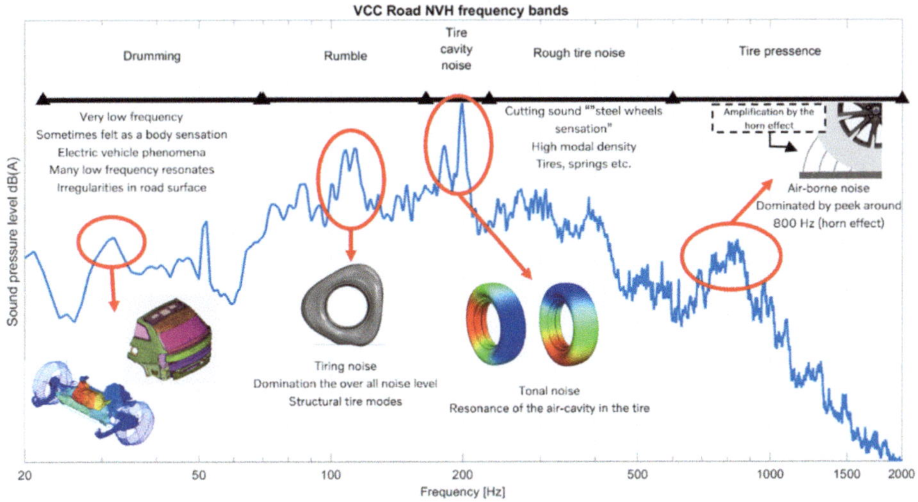


**Fig. 1.4.** Road noise phenomena to consider during development.

Road NVH includes a range of noise phenomena, as outlined in Fig. 1.4. Achieving a premium acoustic environment requires that these elements are well-balanced—none should dominate or disrupt the overall sound character. Since each phenomenon may demand a unique mitigation strategy, a thorough understanding of system interactions is essential to ensure cohesive NVH performance.

**Structural-Borne Road NVH**

The first step was to set the target at an overall level and then detailed analysis of the complete car was done to break down this target into sub targets for different parts of the car. This process, however, is not fixed, it requires ongoing updates throughout development as conditions evolve and optimizations are implemented.

Setting appropriate system targets requires a solid understanding of noise transfer paths in the vehicle structure. To support early-stage analysis, a new complex power Transfer Path Analysis method has been developed [3].

The critical aspect is not to lose focus on the overall setup target. An example of this is illustrated in Fig. 1.5, which shows the glidepath detailing the various broken-down sub targets.

Similar target cascading was done for airborne noise as described in the following section. Although airborne noise has a minor effect on the overall sound pressure level (SPL) it is of outermost importance to find the right balance between high and low frequency noise.

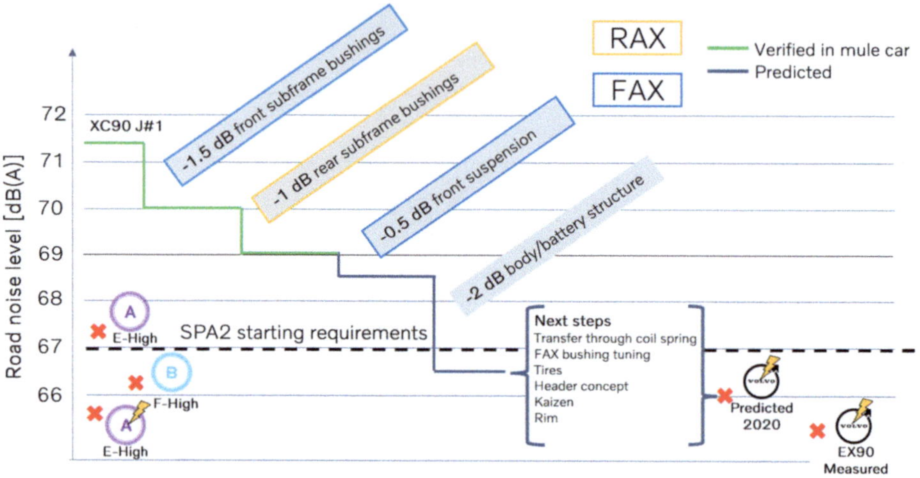


**Fig. 1.5.** Glidepath showing the different broken-down sub targets.

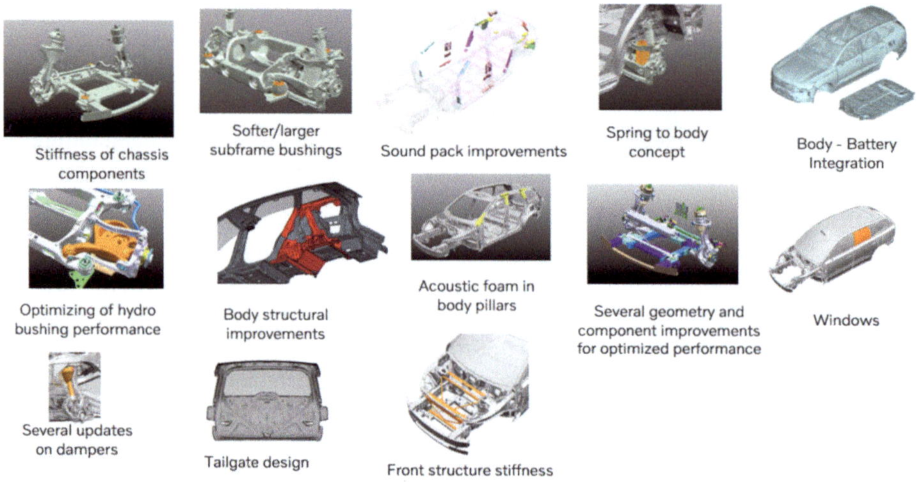


**Fig. 1.6.** Some of the improvements that have been implemented in the EX90 for road NVH.

As seen in Fig. 1.5 above, to reach the road noise targets the whole car design is questioned and improved. For the body structure a huge number of improvements have been made to secure low noise sensitivity. Examples are the integration of casted rear side members and several reinforcements in the front structure. Similar structural improvements have been made to chassis components, such as the subframe and suspension components. Additionally, previous research has shown that even components traditionally considered rigid, such as rims, can significantly influence interior noise levels due to their dynamic behavior [4]. This insight reinforces the importance of a holistic approach to NVH optimization, including wheel design. A key aspect is the integration of the high voltage battery to the rest of the car, really making use of its mass and stiffness. Further, improvements in the body and chassis stiffness are key to secure vibration isolation between the body and the chassis. A lot of effort was spent throughout the development to secure a bushing design that fulfills durability targets and finding the optimum between road noise and steering/handling. To really fine-tune the car's performance, much time was spent on tuning bushings and tires to find the best possible overall solution.

Figure 1.6 highlights several key design enhancements implemented to achieve the targeted road NVH performance in the Volvo EX90.

**Sound Package—Airborne Noise Mitigation**

Solutions for sound absorption, insulation and tightness—often referred to as sound package—were given high attention at the very start of the Volvo EX90 development. Mitigation of airborne noise is largely a matter of balance, both in terms of directivity and frequency and is as important as the measured SPL. To keep track of the status during the program a division into sub areas as described in Fig. 1.7 was used. This reflects the target setting for each sub system and visualizes that all of them need to reach a certain level (rating 8–10, between the dashed and solid lines) for an optimized overall experience. If one of the areas underperforms, it will likely deteriorate the overall performance, and less value is gained from the other systems. The targets were cascaded down to appropriate metrics such as absorption coefficient, sound transmission loss and geometries. Furthermore, the targets were set from collected source strength data, acoustic transfer functions, benchmark data and engineering judgments.

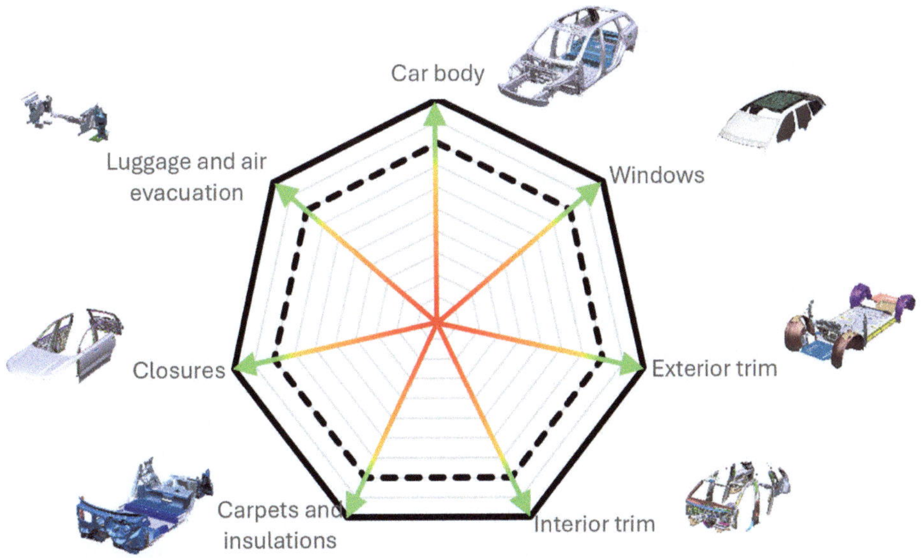


**Fig. 1.7.** Sound pack focus areas. The spider chart reflects target setting for each sub system, rating 1–10, were 8 and above is target.

In the upper car body, polyurethane foam with sound absorbing properties was implemented in distributed sections. The positioning and volume of the foam was investigated in system mules to ensure an optimized solution. This, in combination with sound blocking baffles and tedious work minimizing uncontrolled leakage resulted in a well-insulated car body.

Assuring a tight interface between the high voltage battery and the car body enabled not only efficient noise insulation but also the possibility to save weight on interior insulation. The underbody panels below the electric drive units were designed to meet strict sound absorption requirements. Furthermore, the coverage has been optimized to benefit not only NVH but also aero dynamics and contamination.

The air distributed via the climate system evacuates from the cabin through two outlets in the rear of the car. These controlled openings are often a challenge from an NVH point of view, especially in SUVs and wagons. The concept for air evacuation in the Volvo EX90 is based on pre-studies in collaborations with teams responsible for climate performance, contamination and mechanical integration which gave an understanding how to find a balance fulfilling all attribute requirements.

New standard criteria to virtually evaluate sound insulation status during the development were implemented. These supported the teams to keep track of the effectiveness of the carpets and insulations at each release. A couple of weight optimization loops were done without any negative effect on NVH.

The selection of materials for all sound absorbers, both inside and outside the vehicle, was carefully evaluated to meet the stringent NVH requirements while maintaining a lightweight and compact design. Throughout the program, these sound package solutions were refined to achieve the ambitious full-vehicle NVH standards in an efficient manner regarding cost, weight, and space. Only minimal adjustments were needed during the prototype verification phase.

All in all, a huge number of improvements, leaving nothing to chance and aiming towards a well-defined target were key aspects, securing world-class road NVH levels in the new Volvo EX90.

### 1.2.2 Vehicle Dynamics—Chassis Solutions

To secure a well-balanced car, high efforts have been put into solutions that enable a balance on a high level between steering & handling, ride comfort and road NVH. In the following section some of the chassis technical solutions will be described.

The new Volvo EX90 features an optional dual chamber air suspension system on both front and rear axle. By closing one of the two chambers, the spring stiffness increases by 25% and in the center display, the suspension can be set in two modes: a default comfort setting or a firmer more dynamic mode. The air suspension comes with electronically controlled shock absorbers that constantly adjust to provide a perfect balance between smooth ride comfort and handling performance [5]. This technology monitors the road and driver input and adapts an adequate damping level up to 500 times per second, by in-house designed and developed software. With this dynamic control software, the roll-, pitch- and heave motions of the car is continuously optimized for providing a perfect balance between a superb ride comfort and yet responsive handling characteristics. In addition, the air suspension also comes with a possibility to alter the ride height of the vehicle by –20 to + 40 mm (see Fig. 1.8).

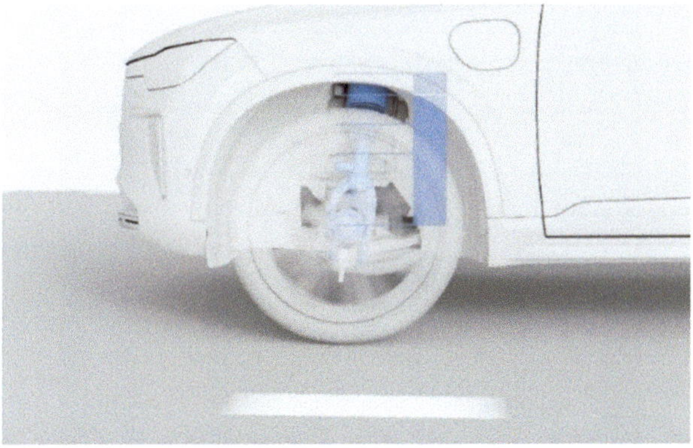

**Fig. 1.8.** Air suspension system to balance ride comfort and handling characteristics.

For customers not opting for the air suspension system, there is a new developed coil spring chassis with the latest state-of-the-art shock absorber technology with frequency selective damping. The tuning of these shock absorbers allows for solving the compromise between offering responsive handling and steering characteristics and yet a plush ride comfort experience on coarse road surfaces.

In all vehicle applications, achieving a vibration free ride experience is key and a focus to offering the best possible ride comfort. In this context, minimizing the sourced excitation is essential, by making sure imbalance from tire and wheel assembly and brake disc installation are minimized. However, in a car driving at different kinds of road types and surfaces, vibration excitation will always exist to some degree. Then it is crucial for the ride experience to introduce damping in the wheel suspension, particularly in the longitudinal direction, since most of the excitation from small road imperfections is translated from the wheel center and through the chassis into the seat and steering wheel. A way to introduce damping in the wheel suspension, is by utilization of hydro bushings, which are liquid filled bushings, where unwanted motion and vibration energy is dissipated. The Volvo EX90 features hydro bushings for both front and rear axles, located in the lower link arm, significantly contributing to the plush ride comfort and isolation of vibrations (see Fig. 1.9).

The damping of the wheel suspension is important to control, for example impact behavior, after shake behavior, damping of imbalance vibrations and brake judder.

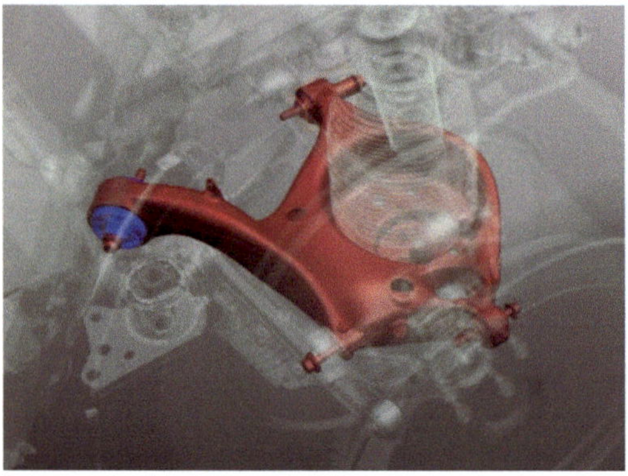

**Fig. 1.9.** Hydro bushings to introduce damping in the wheel suspension.

The air suspension system consists of a compressor, air hoses and an air tank as an accumulator of pressurized air. In the Volvo EX90, the design of the air tank is integrated into the car body (see Fig. 1.10) as a structural crossmember connecting the longitudinal bulkheads. In this way, structural stiffness is provided by utilizing the air tank in a new, inventive and weight efficient way.

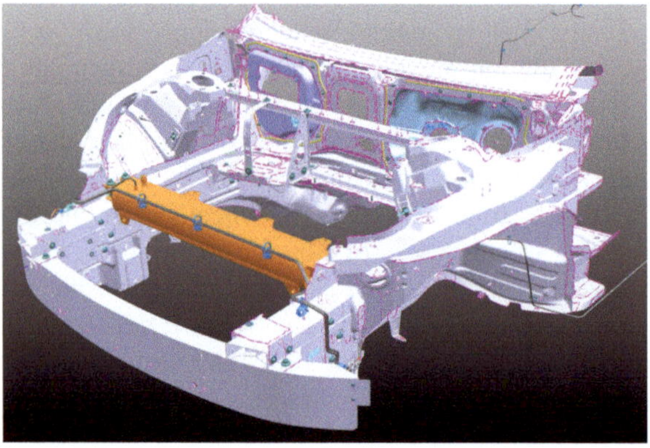

**Fig. 1.10.** The air suspension air tank is integrated as a part of the front structure.

### 1.2.3 Wind Noise

For every attribute of a car, including wind noise, it is essential to begin work at an early stage. Achieving the optimal balance between noise reduction, aerodynamic performance, and maintaining the design language is where it all begins. Using CAE and full-scale clay car tests, various ideas and shapes are tested and refined. For EX90, these included the curvature of the A-pillar, the complex side rear-view mirror system with a camera-equipped foot, and the Lidar dome on the roof, which is positioned in a sensitive area affected by airflow (see Fig. 1.11). Small details that improve wind noise, such as dimples on the bottom side of the rear-view mirror housing inspired by the surface of a golf ball, were implemented.

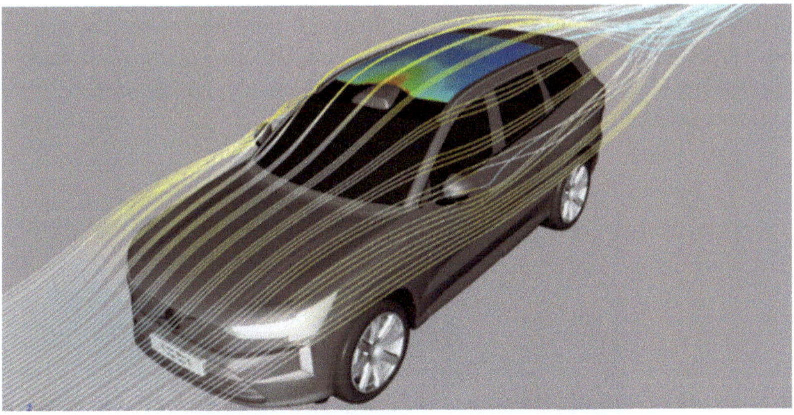

**Fig. 1.11.** Acoustic pressure map of the Glass Roof panel showing the contribution of Lidar dome wind noise.

Parallel to shaping the car exterior, different engineering concepts were developed and finalized. Effective cross-functional collaboration is crucial for providing accurate input and achieving optimal wind noise performance in the final product. The EX90 features a unique flush side glass concept that enhances its clean design language. This flush design required a new engineering concept for the sealing system.

The wind noise performance of the EX90 is among the best in its class (see Fig. 1.12), complementing the overall driving experience when cruising at motorway speeds.

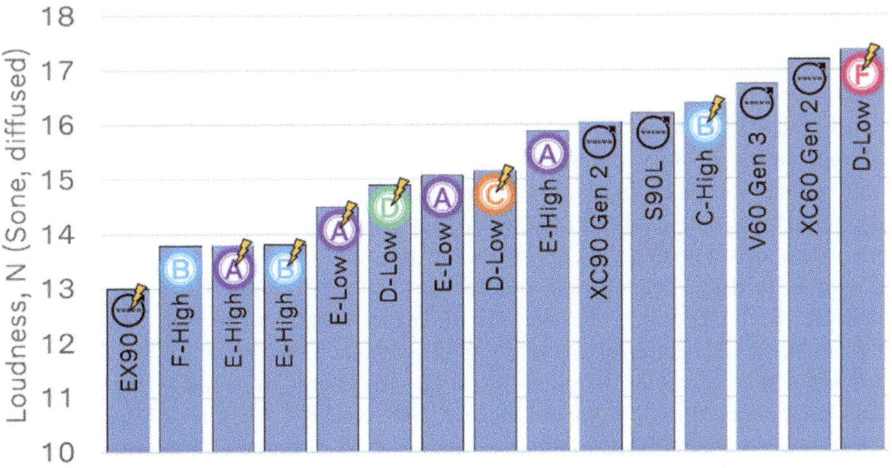


**Fig. 1.12.** Wind noise performance, premium segment. Tested in acoustic wind tunnel at Volvo Cars standard speed, driver's seat, road conditions.

### 1.2.4 Propulsion NVH

The electric drive units (EDU) in the Volvo EX90 posed many challenges regarding NVH. Both the front (Fig. 1.13) and rear EDU rely upon Volvo Cars 2nd generation of compact co-axial planetary one speed (but two-stage) transmissions which are combined with 1st generation of in-house designed permanent magnet synchronous machines (PMSM) and bolt-on inverters. The rear EDU also features a torque-vectoring dual-clutch (TVDC) unit, enabling superb vehicle dynamics properties but also range improvements.

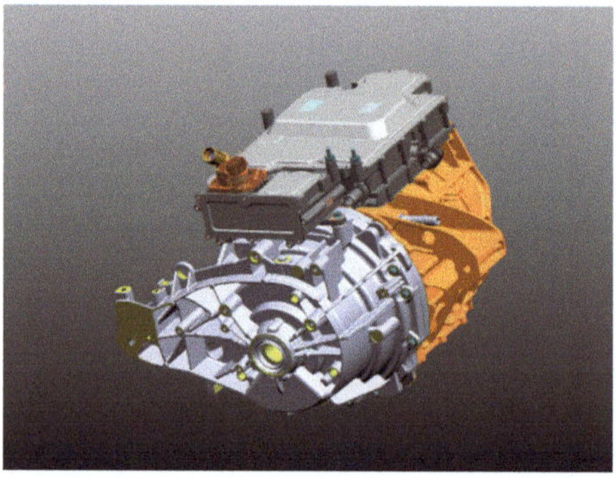

**Fig. 1.13.** The front EDU, with transmission housing, e-motor housing (in orange) and top-mounted inverter.

Even though electric powertrains have fewer sub systems and parts compared to conventional ICE powertrains, there are many aspects that need full attention regarding NVH engineering. There are high customer expectations on the interior quietness of the whining orders, particularly when they appear in the>1 kHz frequency ranges where perceived annoyance easily can be triggered [6]. Furthermore, the high peak torques of the e-machines (420/490 Nm front/rear) have several implications regarding whine orders, i.e. each gear wheel is exposed to a large torque span and consequentially must be designed to provide tolerable transmission error levels while meeting the durability requirements. Radial electromagnetic forces and current harmonics give rise to excessive noise excitation for the very high torques, potentially leading to severe e-motor and inverter/PWM noise if not mitigated properly.

Propulsion NVH engineering at Volvo Cars consists of system CAE and testing and complete vehicle CAE and testing. A key to being successful is cross-functional work, often between departments and even business units (i.e. as for system CAE and system testing). Frameworks for how to make the best use of each sub team's data were established, e.g. the transfer of virtual point transformed blocked forces [7] from the EDU test rigs to the complete vehicle CAE team for in-car structure borne noise predictions or the many loops of correlations of sound power of whining orders originating from system testing and from system CAE. Cross-functional work is often needed also between vehicle attributes. One example is mount tuning which influences not only structure borne EDU NVH but also road NVH, performance &

drivability and ride comfort. Another example is the choice of drive shaft joints which influence both shudder (another important category of propulsion NVH) and driving range. Motor efficiency and peak torque can stand in conflict with a favorable electromagnetic design regarding NVH. For the latter however, a topology optimization of the rotor geometry (Fig. 1.14) was conducted where the optimization objectives were to minimize the tooth force harmonic order 48 while not deteriorating the motor performance by more than 2% [8]. This optimization significantly reduced the radiated noise below the motor's base speed. In addition, a stator-to-rotor air gap increase of 0.3 mm significantly reduced the radiated noise in the field weakening speed range (above base speed). In total, a sound power reduction of ~ 10 dB was accomplished in a wide speed range for high torques. Figure 1.14 depicts the rotor geometry change as a result of the topology optimization.

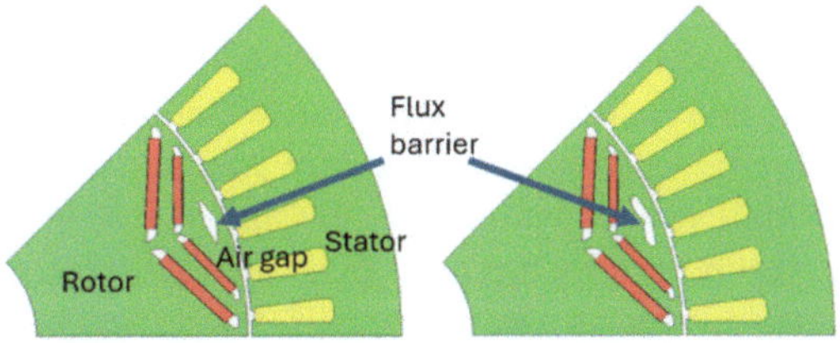


**Fig. 1.14.** Baseline (left) and NVH optimized (right) central flux barrier, reducing the 48th order electromagnetic forces.

To summarize, achieving a premium refinement of BEV propulsion NVH requires attention to many details but small changes to the EDU internal parts can prove to be very effective as noise mitigation actions. The low levels of broadband road and wind noise in the Volvo EX90 yield low tonal masking which puts higher restraints on the EDU whining orders.

## 1.3 Media Reviews

The media review of the new Volvo EX90 has been overwhelming, and the journalists praised the quiet cabin and the comfort of the ride.

From English speaking media the Volvo EX90 is consistently praised for its exceptional quietness, setting a new standard for EVs. Reviewers from Jalopnik [9] even compared its serene cabin to a Rolls-Royce. Top Gear [10] highlights the notable absence of wind, tire, and even suspension noise. Car and Driver [11] emphasize that the EX90's quietness is on another level, even for an electric vehicle and MotorTrend [12] confirms Volvo's claim that it's their quietest vehicle ever.

German media outlets have also highlighted the serene driving experience offered by the Volvo EX90. Auto Bild [13] described it as "very quiet" and noted the comfortable ride. Handelsblatt [14] states that the EX90 is primarily one thing: refined and comfortable. And praises the chassis and noise development. Car-Editors Content [15] writes: The interior is so quiet that you feel like you're in a rolling luxury lounge—perfect for long highway drives.

## 1.4 Conclusions

A passenger car is one of the most complex consumer product to develop. A successful result is dependent on the chosen technical concepts and the integration of those to a complete vehicle. The key to success is also to have a mutual understanding of the goal with the product and a cross functional collaboration from early concept phases to the finalization of the project. Contradicting attributes need to be addressed early and developed in collaboration to define concepts that enable the wanted level of attribute balancing. Once the concepts are chosen, focusing on details and leaving nothing to change is crucial.

In this paper the development of the Volvo EX90 to reach world class NVH and Ride comfort has been described. A prerequisites to start with has been to address and control the structure born road NVH from early concept phases and at the same time secure prerequisites for premium vehicle dynamics and ride comfort. Key enablers have been the body and chassis stiffness, to secure vibration isolation between the body and the chassis and secure a bushing design that fulfills durability targets and enable the right balance between road NVH, ride comfort and steering & handling. The integration of the high voltage battery into the rest of the car, really making use of its mass and stiffness, has been important for the NVH performance of the car.

To get a good balance between low and high frequency noises, high focus has been put on the development of the sound pack reducing the airborne noise and achieving a spatial balance of the interior noise. Low levels of wind noise have been reached, achieving the optimal balance between noise reduction, aerodynamic performance, and maintaining the design language. Effective cross-functional collaboration has been crucial for providing accurate input and achieving optimal wind noise performance in the final product.

Low interior tonal masking noise levels put new tough requirements on propulsion NVH, as high frequency whine noise. Achieving a premium refinement of BEV propulsion NVH required attention to many details but small changes to the EDU internal parts proved to be very effective as noise mitigation actions.

## References


1. Lindberg, E. Hörlin, N-E. and Göransson, G. (2013). Component mode synthesis using undeformed interface coupling modes to connect soft and stiff substructures. Shock and Vibration, 20(2), 157–170. https://doi.org/10.3233/SAV-2012-0734
2. T. Sakata, H. Morimura, H. Ide, Ettects of tire cavity resonance on vehicle road noise, Tire Science and Technology 18 (1990) 68–79.
3. Pasma, E. A., Klaassen, S. W. B., Passanesi, A. V. V., & Westlund, J. (2022). Advanced complex power TPA studies for road noise NVH performance on a rear axle. Presented at ISMA2022: International Conference on Noise and Vibration Engineering, Leuven, Belgium
4. Lindberg, E. Tyre-Road Noise—Experimental Component Investigation of the Structural Dynamic Behaviour of the Rim, Presented at AIA-DAGA 2013: Conference on Acoustics, Merano Italy

5. Skoglund, F. Strategies for road profile in adaptive suspension control. Master of Science Thesis in Electrical Engineering, Department of Electrical Engineering, Linköping University, 2020. LiTH-ISY-EX--20/5317--SE
6. Lennström, D.: Assessment and control of tonal components in electric vehicles. PhD thesis, Luleå University of Technology, 2015, https://doi.org/10.13140/RG.2.1.3739.4001
7. Van der Seijs, M., Harvie, J Harvie, Song, D.P.: Road Noise NVH: Embedding Suspension Test Benches in NVH Design using Virtual Points and the TPA Framework, International Modal Analysis Conference (IMAC) XXXIXAt: Orlando, Florida, 2021
8. Cederlund, J., Nategh, S., Lennström, D.: Topology Optimization of Electrical Machines for NVH Purposes in E-mobility Applications—Part 1, IECON 2021—47th Annual Conference of the IEEE Industrial Electronics Society, 2021, https://doi.org/10.1109/IECON48115.2021.9589483
9. JALOPNIK 2025. 2025 Volvo EX90 Is So Quiet And Nice You Hopefully Won't Mind The Software Bugs. https://www.jalopnik.com/2025-volvo-ex90-is-so-quiet-and-nice-you-hopefully-won-1851638593/ [2025-05-19]
10. BBC TOPGEAR 2025. Volvo EX90 review. https://www.topgear.com/car-reviews/volvo/ex90 [2025-05-19]
11. CAR AND DRIVER (2025). 2025 Volvo EX90 Is the Electric SUV You Didn't Know You Needed. https://www.caranddriver.com/reviews/a62019773/2025-volvo-ex90-drive/ [2025-05-19]
12. MOTORTREND (2025). 2025 Volvo EX90 First Drive: Modern to a Fault. https://www.motortrend.com/reviews/2025-volvo-ex90-first-drive-review [2025-05-19]
13. AUTO BILD (2025). Dieser Schwede hat die Ruhe weg. https://www.autobild.de/artikel/volvo-ex90-schwedisches-elektro-flaggschiff-26420163.html [2025-04-28]
14. Handelsblatt (2025). Nordisch nobel—So fährt sich das neue Volvo-Flaggschiff. https://www.handelsblatt.com/mobilitaet/elektromobilitaet/volvo-ex90-im-test-so-faehrt-sich-das-neue-volvo-flaggschiff/100105758.html [2025-05-16]
15. Car Editors Content (2025). Fahrbericht Volvo EX90: Luxus-Lounge mit Ladefrust. https://www.car-editors.net/artikel/detail/67026 [2025-05-16]

# The Ford Transit Custom—Electrifying Ford's Successful Commercial Vehicle

Daniel Kierspel[1(✉)], Marcus Grünewald[1], and Martin Hildebrandt[2]

[1] Ford Werke GmbH, 50769 Spessartstrasse, Germany
{dkierspe,mgrunewa}@ford.com

[2] AKAD Hochschule—staatl. anerkannt, Heilbronner Str. 86, 70191 Stuttgart, Germany

**Abstract.** Although the BEV related registration numbers in Germany currently run through some ups and downs, the global longterm rising trend of e-mobility seems to be unstoppable. This is a key driver in further electrification research, development and optimization efforts regarding efficiency, range, cost, weight, charging and comfort.

The Ford Transit Custom has been the most successful and popular light duty vehicle for many years now with a wide variety of usage patterns. While it was so far only available with a highly efficient EcoBlue Diesel powertrain, a PHEV and a BEV is going to be offered in the shape of a completely redesigned successor.

One of the challenges is not only the integration of an ICE-HEV-BEV powertrain into one platform but also satisfying a wide requirement portfolio for derivatives like Van, Double Cab-in-Van, Chassis Cab, Bus as well as the Nugget Camper Van.

Within the NVH target setting process a drive simulator has been integrated to provide the development team with a simulation of the expected realistic interior sound. Potential error states can be diagnosed even for various road profiles and traffic situations. This method allows an improved target finalization and agreement process.

Frontloaded assessments of the NVH performance are conducted by using a tailored model that is making use of data from different sources. The input data is frequently updated depending on the program progress status. While running the necessary optimization loops, other attributes like weight, safety or durability are also taken into account. Jury evaluations in the drive simulator support immediate directional decisions from the management board.

Finally, the physical prototype testing starts at a high level of maturity due to the frontloaded model-based assessments in conjunction with the drive simulator. This has been found to be extremely helpful in reducing the occurrence of "late surprises".

The paper concludes with a summary of the NVH-related aspects and is rounded off by a future outlook.

**Keywords:** Ford Transit · Electrification · NVH


A. Heintzel (Ed.): Automotive Acoustics Conference 2025, Proceedings, pp. 16–31, 2026. https://doi.org/10.1007/978-3-658-51249-1_2

## 2.1 Global and European Market Trends of LCVs

Customers increasingly demand Light Commercial Vehicles (LCVs) that offer an excellent balance of functionality, reliability, fuel efficiency, advanced safety features, cutting-edge technology and sustainability at a reasonable total cost of ownership (TCO). In response to this demand and as a show of support for growing environmental concerns and stricter emission regulations in many regions, manufacturers have expanded their product portfolios particularly by introducing more electric and hybrid vehicles. Governmental incentives are further accelerating the adoption of eco-friendly vehicles in some countries.

Furthermore, the rapid expansion of e-commerce and the increasing digitalization of supply chains implies the need for efficient last-mile delivery solutions in urban areas leading to an increased interest in LCVs, as companies seek reliable vehicles for deliveries in time and a problem-free access of zero-emission zones. Other utilization areas where LCVs seem to be indispensable are urban transportation, small business operations, public services, maintenance fleets, carriers for tools and materials or camper van applications.

LCVs are an important line of business within the Ford product family and have a leading position in the global market, Fig. 2.1.

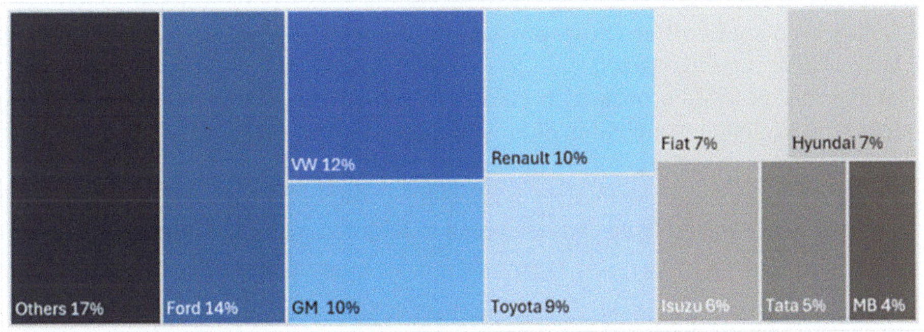


**Fig. 2.1.** Global LCV unit sales 2023 by manufacturer. Source: https://de-statista-com.gw.akad-d.de/statistik/studie/id/176017/dokument/leichte-nutzfahrzeuge-marktdaten-und-analyse/

At present, numerous companies are running test fleets to investigate the benefits of alternative propulsion techniques like electricity, hydrogen, biofuels, synfuels etc. vs the proven Diesel powertrain applications using conventional fossil fuels.

However, take rates of electrified propulsion systems are still on a relatively low level compared to conventional powertrain applications, see Fig. 2.2. One of the reasons for the low uptake in Germany might be the wait-and-see attitude of users, waiting to see if further incentives are introduced. Due to a multitude of global risks and uncertainties it is nearly impossible to project robust forecasts of electrified vehicle take rates even for the near future (Fig. 2.2).

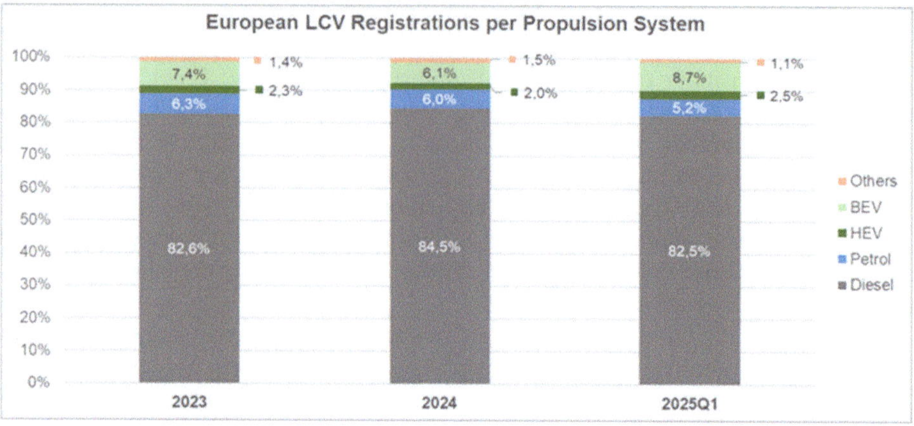


**Fig. 2.2.** EU LCV unit sales by propulsion system. Source: https://www.acea.auto/figure/vans-eu-fuel-type/ (06/05/2025)

## 2.2 The New Ford Transit Custom and Its Derivatives

The new Ford Transit Custom—International Van of the Year 2024—is expected to fully support a wide variety of usage patterns. So far it was only available with different highly efficient EcoBlue Diesel powertrains, but now a completely new PHEV and BEV derivative is being offered.

While the integration of an ICE-HEV-BEV propulsion system into one platform is already a challenge, the combination with a broad portfolio of body types shown in Table 2.1, is an even more complex mission.

Different wheelbases, individual body setups, single or twin sliding doors, and hinged rear doors or a tailgate can be chosen in addition to various body interior configurations (Fig. 2.3).

**Table 2.1.** Offered Bodystyles of the new Ford Transit Custom. Source: ford.co.uk; ford.de

| Body Type | Theme | Description |
|---|---|---|
| Cargo Panel Van | Tried. Tested. Tough | Functional load carrier |
| Double Cab-in-Van | Transport your team and your tools | Combines cabin comfort with cargo capacity, up to 6 seats |
| Kombi Van | Reliable. Robust. Versatile. Delivered | Configurable to transport tools, equipment or up to 9 people |
| Multicab | Work. Play. Sorted | Multitude of configs making it ideal for daily business or action-packed weekends |
| Tourneo Bus | Comfort. Hightech. Space | Multifunctional comfortable vehicle with up to 9 seats for private and business use |
| Nugget Camper Van | Adventure Without Limits | Home comfort with 5 seats, 4 sleeps, 33 l of fresh (hot) water |

**Fig. 2.3.** The new Ford Transit Custom, Van-of-the-Year 2024. Source: Ford Media

There is a choice of 11 powertrain variants offering the customer a tailor-made light commercial vehicle for any individual use case (Table 2.2 and Fig. 2.4):

**Table 2.2.** Offered Propulsion System Configurations. Source: ford.co.uk; ford.de

| Engine | Transmision | Drive |
|---|---|---|
| 2.0l EcoBlue (81 kW) | 6-speed manual | FWD |
| 2.0l EcoBlue (100 kW) | 6-speed manual | FWD |
| 2.0l EcoBlue (110 kW) | 6-speed manual | FWD |
| 2.0l EcoBlue (100 kW) | 8-speed auto | FWD |
| 2.0l EcoBlue (100 kW) | 8-speed auto | AWD |
| 2.0l EcoBlue (125 kW) | 8-speed auto | FWD |
| 2.0l EcoBlue (125 kW) | 8-speed auto | AWD |
| 2.5l Duratec PHEV (171 kW) | eCVT | FWD |
| E-Drive (100 kW) | 1-speed auto | RWD |
| E-Drive (160 kW) | 1-speed auto | RWD |
| E-Drive (210 kW) | 1-speed auto | RWD |

**Fig. 2.4.** The new Ford Transit Custom BEV as Cargo Van. Source: Ford Media

In total there are more than 110 different variants of the Ford Transit Custom and Tourneo Custom just based on bodystyle, wheelbase and payload, not taking trim series and other features into consideration.

In order to manage this amount of variants properly, a well-structured development process and efficient design methods are key.

## 2.3 NVH Development of the New Ford Transit Custom BEV

### 2.3.1 System Engineering Cascade

Due to the different body styles of the Ford Transit Custom combined with ICE/HEV/BEV FWD and AWD powertrains, the target setting is complex as the Sound Quality requirements need to be intelligently aligned for the whole portfolio on both, vehicle and system level.

Whereas typical SQ metrics like Loudness, Articulation Index, Sharpness, Tonality etc. are most useful within the process of vehicle benchmarking, they cannot be directly cascaded to system or component level.

Accordingly, a target cascade for tonal noises—which is one of the dominant NVH challenges—has been applied. At the same time, well-elaborated NVH related design rules and checkpoints for the propulsion system and its installation are used to guide the team through the process.

At the starting point of the target development a vehicle level spectral masking surface over the whole speed range is generated based on the current model or a prototype while coasting on different tracks with engine switched off. Adjustments referring to the future product performance can be included either through hardware modifications or based on a data matching process.

Applying the sections of the relevant orders allows to transform the 3D map into a 2D representation which results in the masking level for a certain motor/gear/wheel order. Subsequently a tonal order amplitude can be derived that stands in direct relation to the masking level and which is applied as an initial target after some modification of the signal-to-noise ratio [1, 2] (Fig. 2.5).

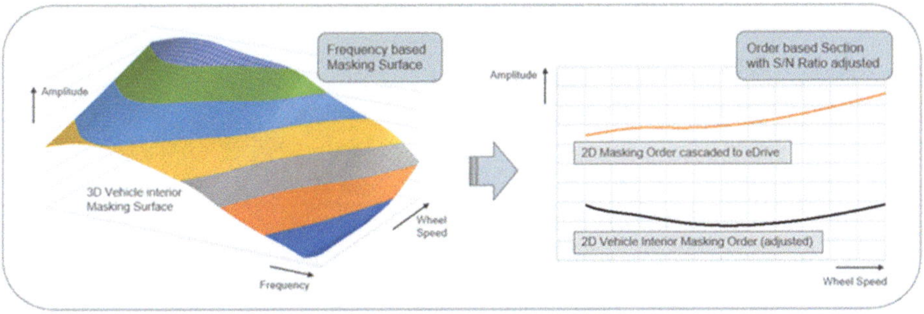


**Fig. 2.5.** Example 3D vehicle interior masking surface to derive the relevant masking orders from for vehicle and system level

In order to achieve a proper alignment between the numeric targets and their subjective impressions, the targets are further refined by making use of a drive simulator. Several loops of adjustment help to finalize the vehicle level targets in preparation to receiving approval by the senior management.

Airborne and structure-borne transfer functions, either achieved through testing or by using simulation tools, help developers to cascade the relevant order targets to system/component level, i.e. electric motor or gearbox, pumps, compressors etc.

For assessing the development status over time, the use of the drive simulator is again very beneficial.

In Fig. 2.6 shows an example of potential requirements of an e-axle.

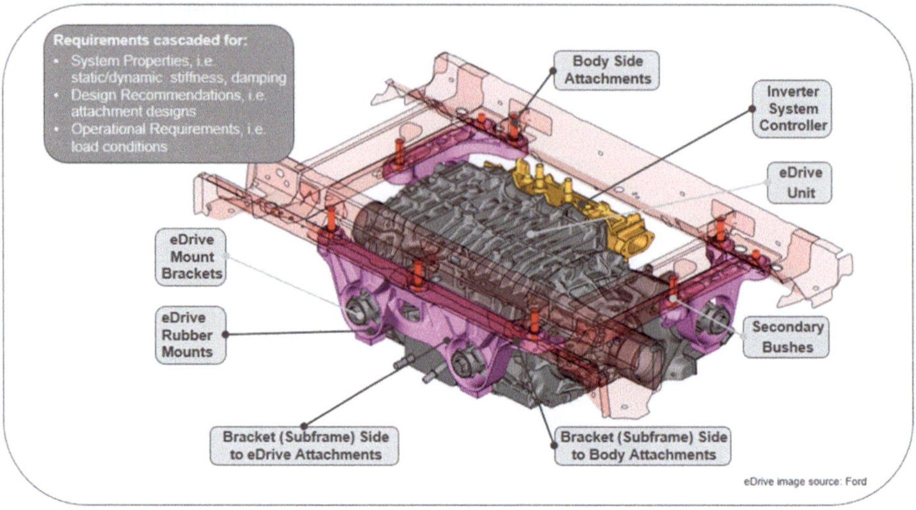


**Fig. 2.6.** Important eDrive unit related system level requirements

In addition to the target setting for the eDrive unit, components like compressors pumps, fans etc. have to be re-assessed in the BEV environment and especially during "silent" operational modes with low masking noise levels, i.e. at a red light stop or while parking. The system level target setting process also follows the approach of the calculation scheme: excitation → transfer function → response.

For launching the vehicle on the market, European legal requirements like AVAS (UN ECE-R138) and Passby Noise (UN ECE-R51), need to be homologated. Early involvement into the vehicle design and intensive development efforts are required for AVAS to identify the optimal loudspeaker positions and to compose and assess the specific warning sounds. Simulation tools are used to avoid intrusion into the vehicle cabin while fulfilling the legally required sound levels at the same time [2].

### 2.3.2 Frontloaded NVH Simulation

The automotive industry is experiencing a transition from reliance on physical measurements toward sophisticated analytical models, enabling companies to make faster decisions while reducing cost and development time. This shift has been accelerated by advances in computational power as a basis for simulating complex physical phenomena with increasing accuracy.

As BEV related high frequency acoustic phenomena present a particular challenge for this transition an enhanced TPA (Transfer Path Analysis) process was established for assessing Powertrain NVH behavior (Fig. 2.7).

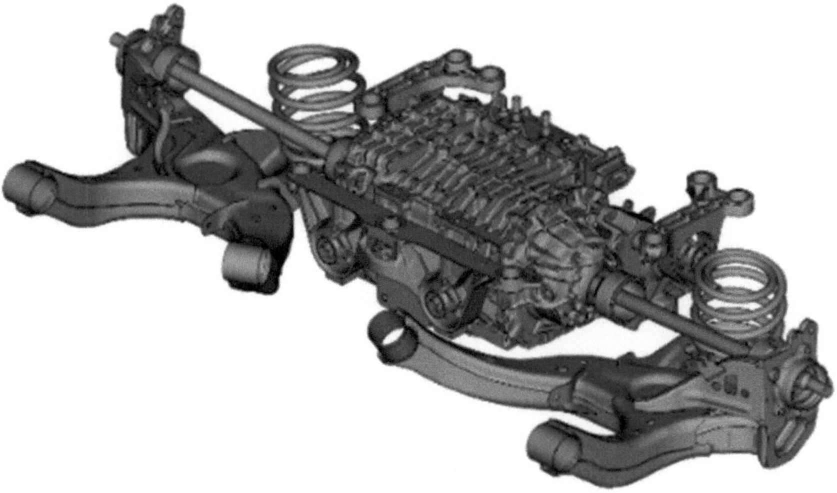

**Fig. 2.7.** Finite Element model as basis for subsequent TPA-model

In early program phases the TPA-model input data was extracted from traditional Finite Element based models, which had to be adjusted for high frequency calculations, to predict airborne and structure-borne contributions to the overall sound pressure level.

When system level measurement data (e.g. eDrive unit mount vibrations) becomes available, the TPA-model can be updated and transformed towards a hybrid model.

### 2.3.3 Assessments Using Frequent Model Updates

**Comparison of Design Concepts.** The CAE process, as described above, was successfully executed when the decision about the concept for the installation of the eDrive unit was made.

The initial concept was based for integrating the eDrive unit into the vehicle using a large frame structure that carries the eDrive unit and connects it to the rear underfloor body structure. The eDrive unit was isolated from the frame structure by a set of conventional rubber bushes while the frame structure itself was isolated from the trimmed body by a secondary layer of rubber isolators (Fig. 2.8).

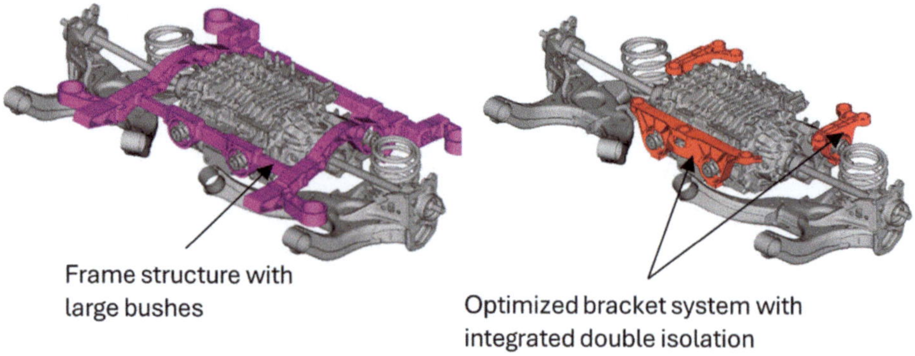


**Fig. 2.8.** Comparison of initial and weight-optimized design

To improve cost and weight the decision was taken to integrate the frame structure into the rear floor assembly of the trimmed body and connect the eDrive unit with the rear floor by a set of brackets. For this design the eDrive unit is isolated from each bracket by conventional rubber mounts while each bracket itself is isolated from the trimmed body by smaller rubber isolators to maintain a double isolated system.

The comparison of interior noise levels between both concepts in Fig. 2.9 reveals that cost and weight improvements come with broadband degradation in structure-borne noise performance. Following objective and subjective evaluations, it was concluded that the significant improvements in cost efficiency and weight reduction adequately justify the compromise in NVH characteristics.

Crucially, the acoustic experience remains within the threshold of what customers perceive as comfortable and satisfactory for their driving experience which is in line with the target curve in Fig. 2.9.

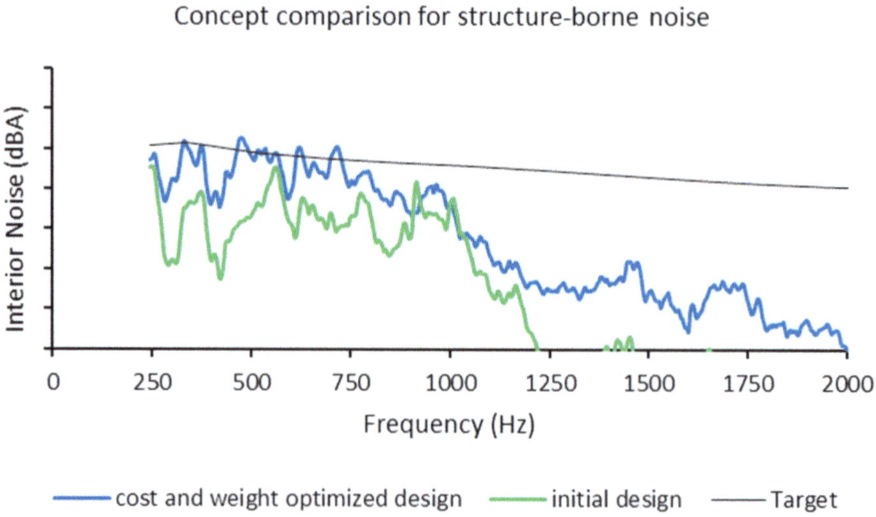


**Fig. 2.9.** Structure-borne interior noise comparison of initial design and final design

After assessing both concepts based on objective data against the targeted performance, additional NVH-simulator drive events were conducted. The required simulator models were generated from the outcomes of the TPA-models for each concept [3] (Fig. 2.10).

**Fig. 2.10.** NVH Simulator

The NVH-simulator enables comprehensive evaluation sessions with experts and management representatives, allowing for thorough subjective assessments of vehicle NVH-performance characteristics. The immersive simulation environment provides stakeholders with a consistent platform for experiencing various vehicle configurations, ultimately supporting more informed decision-making as to the best design concepts.

**Design Iterations.** In developing a cost and weight optimized design, numerous design iterations had to be assessed by a cross-functional team with members from various engineering disciplines including NVH, Durability, Manufacturing, and Safety. Results were discussed and solutions were proposed on a regular basis throughout the process, ensuring that the final system design achieved widespread acceptance and approval.

To maintain a satisfying NVH performance it was critical to deliver sufficient mount isolation. The dynamic stiffness of each bracket is a key metric to achieve this goal. Figure 2.11 illustrates how the design evolved over time while improving dynamic stiffness.

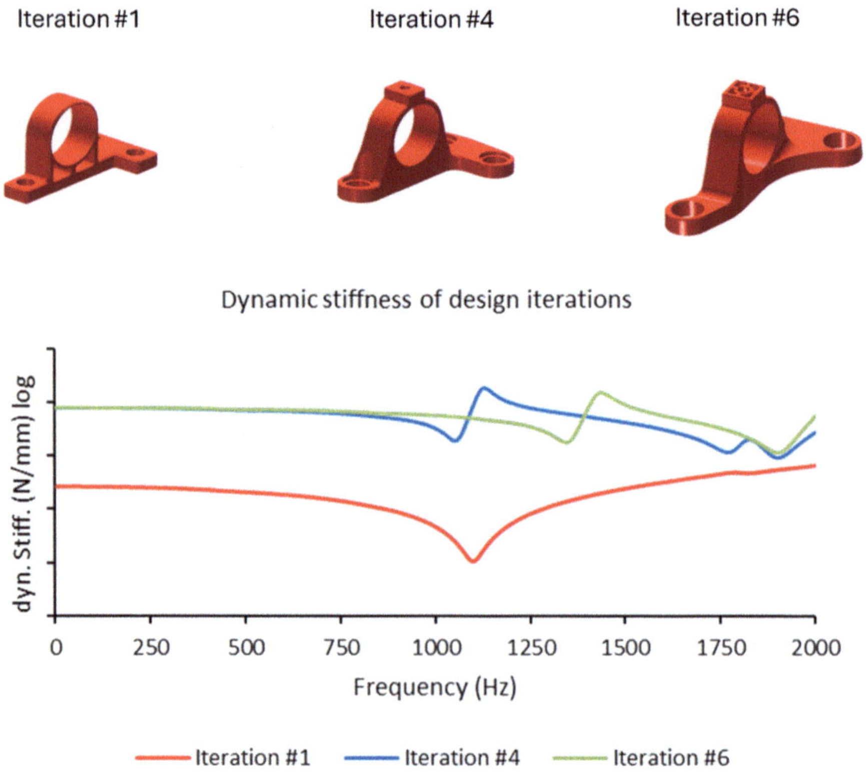


**Fig. 2.11.** Bracket optimization

To facilitate an efficient engineering process, the required assessments needed to be carried out and analyzed within short turnaround times. This was achieved by using Craig-Bampton sub-structuring techniques and automating the TPA process [4–6] (Fig. 2.12).

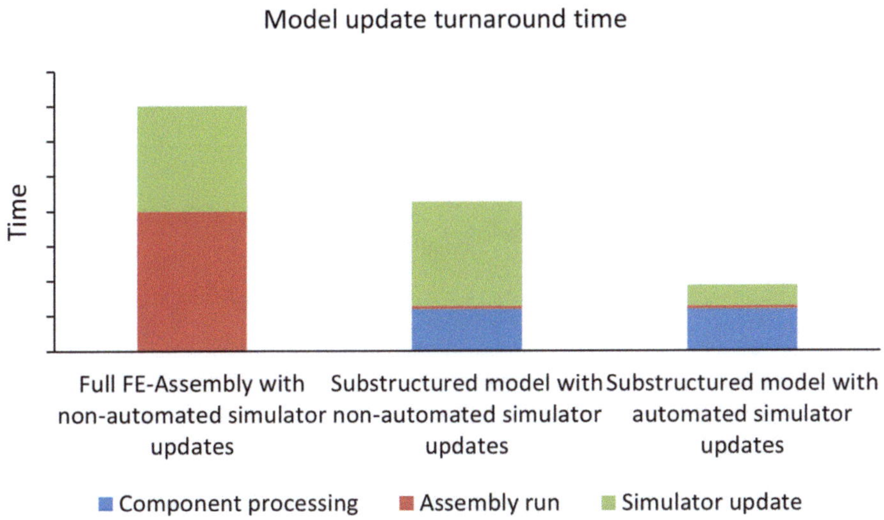


**Fig. 2.12.** Efficiency gain using sub-structuring techniques and process automation

The sub-structured model provides the required efficiency gain compared to Finite Element runs. To further improve the turnaround time, the TPA-process was partially automated. This was particularly effective for delivering simulator model updates for subjective assessments.

## 2.4 Physical Prototype Verification

### 2.4.1 Simulation Model vs Prototype—Representation and Alignment

In the demanding world of engineering and product development, predicting and analyzing NVH phenomena poses a complex challenge.

Ensuring correlation between real-world NVH measurements and their computational counterparts is essential. Sound and vibration phenomena involve complexities that can become difficult to capture in digital space. Yet the promise of accurate virtual NVH prediction is compelling: faster development cycles, reduced prototype quantities and more comprehensive design exploration.

Engineers today navigate a landscape where physical testing provides a high veracity standard but is costly and time-intensive, while virtual modeling offers flexibility and insight but demands rigorous validation. The high correlation between these approaches determines whether virtual NVH models serve as reliable design tool or merely rough approximations of acoustic and vibrational behavior.

**Interior Noise Correlation.** As displayed in Fig. 2.13, the TPA model predictions match well with physical test measurements over a broad frequency spectrum.

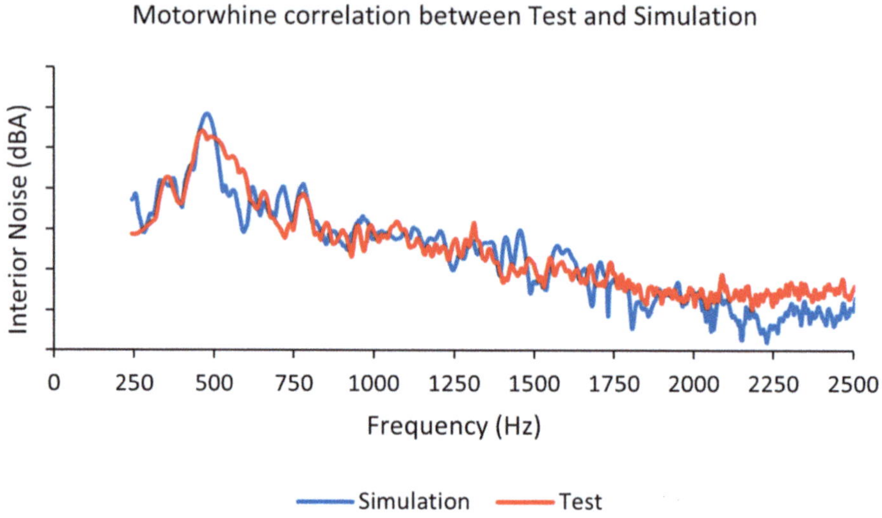


**Fig. 2.13.** Correlation between Test and Simulation during regen load

The interior noise prediction results accurately capture not only the primary resonances but also the general trend which generates confidence in the assessments and diagnostics of further simulation studies.

**System Level Correlation.** To further evaluate the accuracy of interior noise predictions, a comparison of system level data was conducted. As illustrated in Fig. 2.14, the results reveal a good correlation with the test data, further strengthening the confidence in the model's predictive capabilities and overall performance.

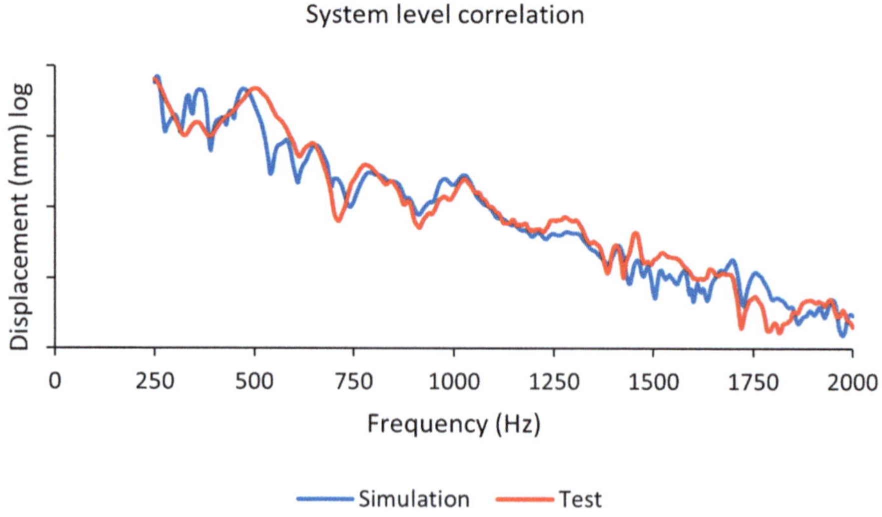


**Fig. 2.14.** System level correlation

**Correlation Summary.** Rather than serving as a rough approximation requiring extensive physical validation, the model now functions as a reliable predictor capable of supporting definitive design decisions even for high-frequency NVH phenomena. The generally good correlation is also a basis for reliable subjective assessments in the NVH-simulator.

### 2.4.2 Examples of Specific Noise Phenomena

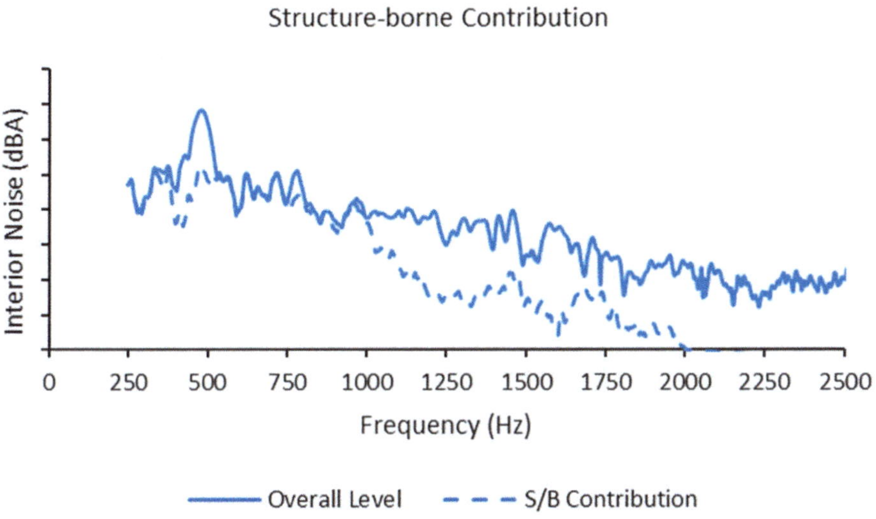


**Fig. 2.15.** Structure-borne contribution to interior noise level

The measurement data and the simulation data show a prominent amplitude peak at 500 Hz, as shown in Fig. 2.13, which was assessed as a risk for customer expectations in objective and subjective evaluations.

Figure 2.15 reveals that the structure-borne contribution is not the dominant driver for the experienced noise level when analyzing the data by utilizing the TPA-model. This leads to the conclusion that the overall sound pressure level is dominated by air-borne contribution.

When further analyzing the problem, the TPA-model and its underlying Finite Element models proved to be an excellent framework for tracking down the root cause. Once the root cause was identified a set of countermeasures were introduced which improved the interior noise level substantially (Fig. 2.16).

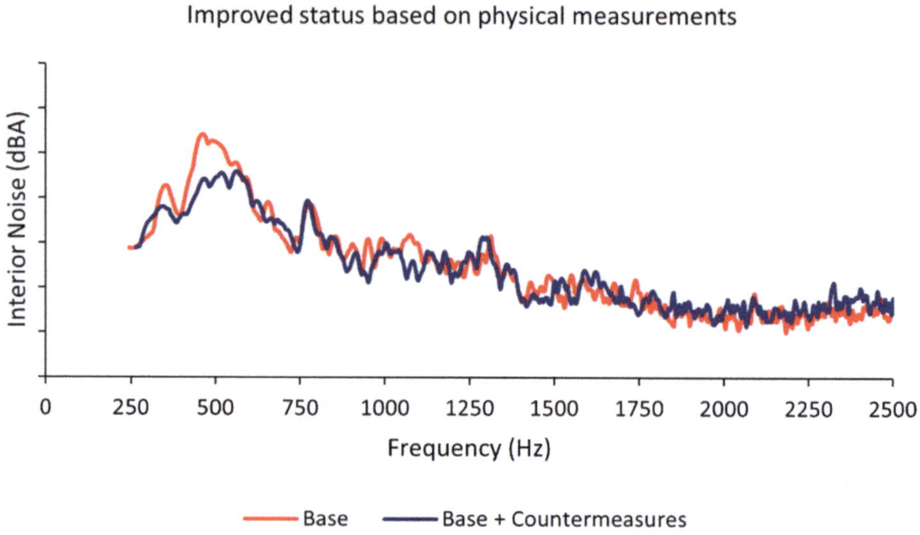


**Fig. 2.16.** Interior noise status with improvements vs. baseline

## 2.5 Conclusion and Outlook

The continued trend in the automotive development is for increased efficiency and reduction of costly physical prototypes. Hence, analytical toolsets are required to enable design development and verification as early as possible in the development process. Especially in the context of electrified vehicles this poses some significant challenges due to the wide frequency range this analytical approach needs to cover.

The high frequency CAE methodology used in the development of the new Ford EV Transit allowed the team to verify the NVH status in the course of the development phase and to support the team with a toolset for issue resolution and root cause investigations. In addition to these objectively measurable outcomes, the team was furthermore able to support subjective expert and management drive assessments by linking these analytical results to the NVH simulator.

For future applications and to further shorten development time the used model approach will have to be expanded by moving to analytical airborne and structure-borne transfer functions, which remains a challenging topic for the near future.

## References

1. Clapper, M., Blommer, M.: "Masking Perception Analysis Software (MPAS) for Tonal Level Setting in Powertrain NVH," SAE Technical Paper 2003-01-1500, 2003
2. Hildebrandt, M.; Heinrichs, R.: Electrification of a Ford Transit—Ford's Response to a cleaner London", Aachen Acoustics Colloquium, Aachen 2018

3. Heppelter, M., Schuster, C., Grünewald, M., Kierspel, D., Störig, C., Hildebrandt, M.: "NVH Development of the new Ford Puma GEN-E® utilizing the VI-GRADE Desktop Simulator", 2025 VI-GRADE Zero Prototypes Summit, May 2025, Udine, Italy
4. NASTRAN Reference Guide, https://nexus.hexagon.com/documentationcenter/de-DE/bundle/MSC_Nastran_2022.2/page/Nastran_Combined_Book/refman/refmanpreface/refmanpreface.xhtml#TOC_Preface5, last accessed 2025/05/22
5. NASTRAN Superelement Guide, https://nexus.hexagon.com/documentationcenter/de-DE/bundle/MSC_Nastran_2022.2/page/Nastran_Combined_Book/Superelements_User_s_Guide/SuperIntro/SuperIntro.xhtml#TOC_Introduction_and, last accessed 2025/05/22
6. NASTRAN DMAP Programmer's Guide, https://nexus.hexagon.com/documentationcenter/deDE/budle/MSC_Nastran_2022.2/page/Nastran_Combined_Book/dmap/dmappreface/dmappreface.xhtml#TOC_Preface4, last accessed 2025/05/22

# Noise Transmission Path Analysis and Sound Package Optimization for Electric Vehicle Using Numerical Simulation Contribution Title

Kichang Jo[1(✉)], Kichang Chae[1], Hyunwoo Kim[1], Jinung Jang[1], Theophane Courtois[2], and Thibault Lafont[2]

[1] Hyundai Motor Company, Hyundaiyoenguso-Ro 150, Hwaseong-Si, Republic of Korea
{dudreamck,kschae,nunu90,jujang21}@hyundai.com

[2] Autoneum Management AG, Schlosstalstrasse 43, 8406 Winterthur, Switzerland
{theophane.courtois,thibault.lafont}@autoneum.com

**Abstract.** Compared to conventional ICE (Internal Combustion Engine) vehicles, electric vehicles have many differences in the location and frequency characteristics of noise sources and vehicle body structure. Due to these differences, there are many differences in the noise transfer path between the ICE vehicles and the electric vehicle. Especially in the case of electric vehicles, the amount of noise inflowed through the luggage area increases, making the sound absorption performance in this area more important. However, Since the rear part of a vehicle has a complex structure and consists of various parts, it is not easy to accurately identify the path through which noise is transferred via vehicle test and efficiently find a method to reduce interior noise.

Thus, in this study, panel contribution analysis and target cascading analysis on interior noise were conducted using the Statistical Energy Analysis (SEA) analysis model of actual vehicle sound absorption performance to identify the noise transfer path through the luggage space of electric vehicles. In this process, materials for interior/exterior sound insulation and absorption components were modeled based on the Biot's parameters of materials which construct components, and the model correlation has secured using the result of components performance and vehicle contribution evaluation. In addition, based on the target cascading results and the noise inflow path analysis results, it was confirmed that the actual vehicle improvement plan could be devised and applied to secure the target level of performance without increasing the overall weight.

**Keywords:** SEA Simulation · Panel Contribution Analysis · Target Cascading Analysis · Sound Package Optimization


A. Heintzel (Ed.): Automotive Acoustics Conference 2025, Proceedings, pp. 32–45, 2026. https://doi.org/10.1007/978-3-658-51249-1_3

## 3.1 Introduction

Electric vehicles have many differences from ICE(Internal Combustion Engine) vehicles. First of all, as the power source is changed from engine to motor, the noise from power source changed from mid-low frequency to high frequency. And in the ICE vehicle, the engine was mainly located in the front of the vehicle, but in the electric vehicle, the main motor is usually located at the rear side (under the luggage room) of the vehicle, and it is common for the traditional engine bay to either have no motor or only a low-capacity motor. And as the high-weight battery has positioned at the bottom of the floor panel, the road excitation force increases and the transfer characteristics of the vibration and noise through the floor change. Like these, the electric vehicle has many differences in the structural and excitation source aspects of the vehicle from the ICE vehicles, and these differences cause significant changes in the transfer path which external noise inflow to the cabin. Especially as the amount of noise transferring through the luggage area increases in the electric vehicle, the sound insulation performance in this area becomes more important. Many studies have been conducted on the sound insulation and absorption performance of dash and floor, so high-quality acoustic analysis models and development guides have been created. But compared to the dash and floor system, there are not many studies on the sound insulation and absorption performance on luggage side, and there is not enough performance database of components. Further, since the luggage of a vehicle has a complicated structure and is composed of various components, it is difficult to accurately identify a transfer path of noise inflowed through the luggage side only by an actual vehicle evaluation, and it is not easy to identify which parts need to be improved in order to efficiently reduce indoor noise.

In this study, we developed a SEA model to identify the noise transfer path on the luggage side of electric vehicles and suggested optimized sound package of the vehicle based on the target cascading result of the SEA model. And also, we validated the suggestion through the actual vehicle prototyping test. In this progress, to improve the accuracy of the SEA model the coupling surfaces were modeled based on the material properties measurement and sound insulation and absorption performance evaluation of the actual components. Additionally, the panel contribution test on actual vehicle was performed to secure correlation of the SEA model.

## 3.2 Development of Target Cascading Analysis Model

### 3.2.1 SEA Modeling

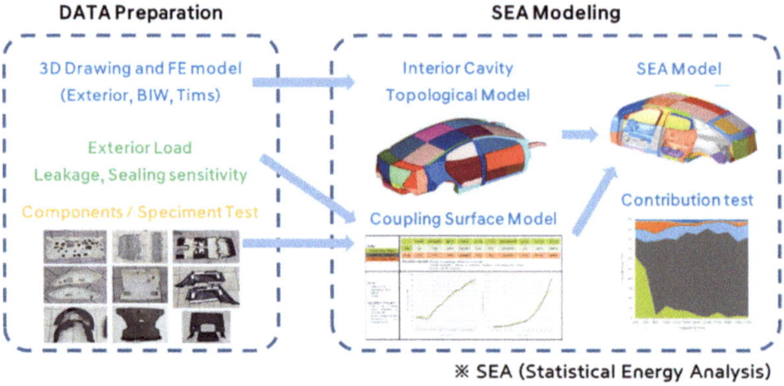


**Fig. 3.1.** SEA modeling process for Target Cascading Analysis

Figure 3.1 shows the entire SEA modeling process for target cascading analysis in this study. The SEA (Statistical Energy Analysis) model defines the sound transmission and absorption characteristics of the coupling surfaces between cavities that make up the topological model. By doing so, it enables the identification of how the transmitted noise through the panels propagates through various cavity paths into the cabin. This makes the SEA model a suitable technique for high-frequency noise Target Cascading (Fig. 3.2).

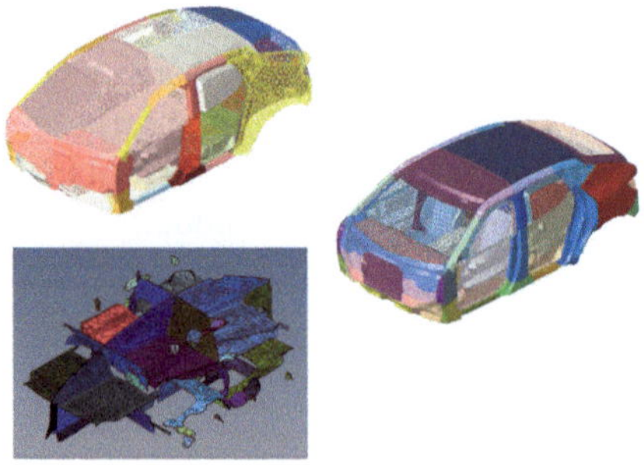

**Fig. 3.2.** Topological Model

A topological model for SEA was built using 3D drawings of the vehicle to model cavities of cabin and spaces between the trims and panels. The newly developed SEA model for this study consisted of 52 cavities and 262 coupling surfaces. And the types of coupling surfaces that make up each cavity are largely classified as virtual wall (the side where Cavity and Cavity are directly adjacent), panel, and interior trim. Virtual walls can be defined as an open surface that does not have any acoustic performance like sound insulation and absorption since it is a surface that connects two cavities. Panels have no sound absorption performance, and if there is thickness and shape information of panels, the sound insulation performance can be obtained relatively easily using analysis. However, in the case of interior trims, it is necessary to model the acoustic performance of the coupling surfaces since the performance of sound insulation and absorption varies depending on the materials, layout structure and shape of the trims. Table 3.1 shows how the coupling surfaces are modeled in this study considering whether it is possible to evaluate the acoustic performance of the components separately and whether it is possible to evaluate the physical properties.

**Table 3.1.** Acoustic Performance

| Panel group | TL Simulation | $\alpha$ Simulation |
|---|---|---|
| Dash | VSISAB 3D | VSISAB 3D |
| Floor | VSISAB 3D | VSISAB 3D |
| Wheelhouse | VSISAB 3D | VSISAB 3D |
| Roof | VSISAB 2D | VSISAB 2D |
| Pillars | Mass law | Database |
| Console | Database | Database |
| Seats | Database | Database |
| Tailgate | Mass law | Database |
| Trunk floor | Mass law | Database |
| Trunk side trim | Mass law | Measurements |
| Glazing | Database | Database |
| Doors | Database | Database |
| BIW | Mass law | Database |

Here, TL refers to the transmission loss, which represents the sound insulation performance of the surface, and $\alpha$ is the sound absorption coefficient indicating the sound absorption performance. In addition, VSISAB (VisualSISAB) is a simulation tool developed and used by Autoneum to calculate the sound insulation performance and sound absorption performance of parts using the physical properties of the material.

### 3.2.2 Coupling Surface Modeling

As shown in Table 3.1, analytical and experimental methods were both applied in this study to model the sound insulation and absorption performance of the coupling surfaces. For the components with complex shapes and large sizes, experimental evaluation of sound insulation and absorption performance requires significant man-hours and is prone to measurement errors. Therefore, an analytical approach was adapted. In addition, components that were difficult to obtain in this study were modeled using a database of material and performance of components previously available.

For analytical modeling of the sound insulation and absorption performance, thickness and weight of each layer of the component and the Biot' parameter of porous material (see Table 3.2) are required. This Biot' parameters are classified into acoustic Biot' parameters such as porosity, air-flow resistivity, torsion, viscous characteristic length, thermal characteristic length and structural Biot' parameters such as Young's modulus and Poission's ratio. In order to obtain these properties, material modeling was conducted first through the process in Fig. 3.3.

**Table 3.2.** Biot' Parameter & Measurement system

| 구분 | 항목 | 평가장비 |
|---|---|---|
| Acoustic Biot Parameter | Porosity ($\varphi$) | PORPOS |
| | ARF(Air Flow Resistivity) ($\sigma$) | AFR |
| | | Care+ |
| | Tortuosity ($\alpha$) | ELWIS-A |
| | Viscous Characteristic Length ($\Lambda$) | |
| | Thermal Characteristic Length ($\Lambda'$) | |
| | Pore Shape Factor (c) | |
| Structural Biot Parameter | Young's Modulus ($E$) | ELWIS-S |
| | Poisson's Ratio ($\nu$) | |
| | Damping Loss Factor ($\eta$) | |

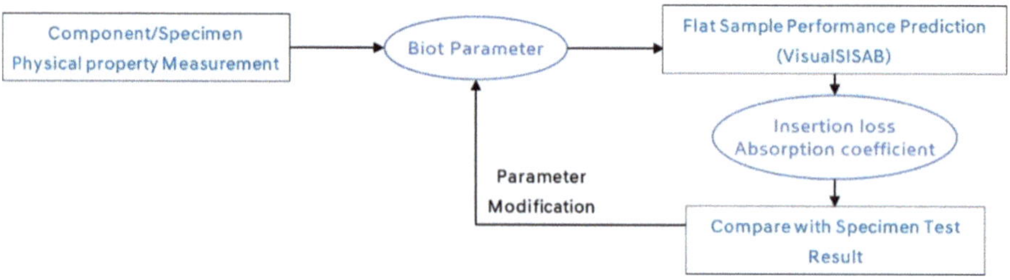


**Fig. 3.3.** Process of Material Modeling

Samples are taken from components such as dash insulation pads, floor carpets, floor pads, luggage insulators, wheel arch insulators, etc., which are the main subjects of this study, and the Biot' parameter of the material was measured using equipment such as Autoneum's Care+and the ELWIS-A/S. In parallel, a flat sample of each part was prepared to evaluate the sound insulation and absorption performance of these specimens. By utilizing the measured Biot' parameters, the sound insulation and absorption performance of the component specimen was analytically obtained via VisaulSISAB. By comparing these analytical results with the performance obtained from actual specimen tests, the material properties of the components were refined, and the accuracy of the analysis was improved. As the result, as shown in Fig. 3.4, a highly consistent analytical result with the actual evaluation was achieved.

Based on the obtained material properties and 3D drawings of the parts, the sound insulation and absorption performance of the coupling surfaces of SEA model was modeled using VisualSISAB.

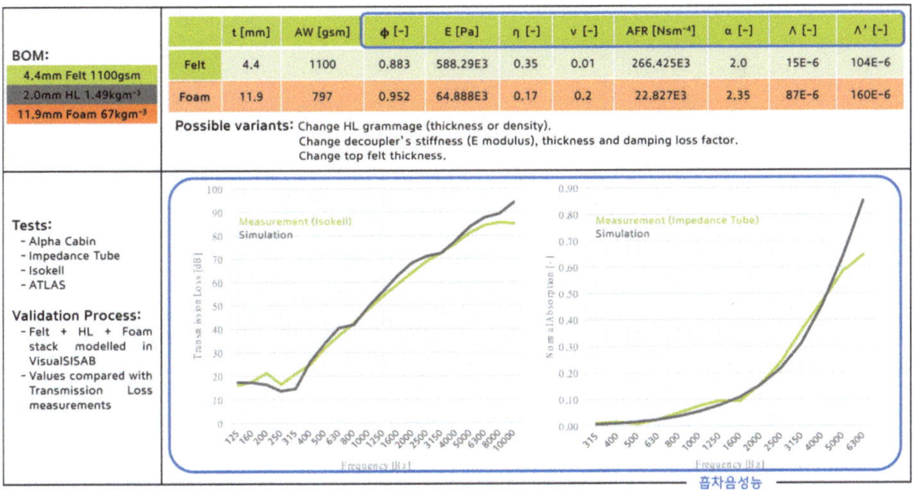

| | t [mm] | AW [gsm] | ϕ [-] | E [Pa] | η [-] | ν [-] | AFR [Nsm⁻⁴] | α [-] | Λ [-] | Λ' [-] |
|---|---|---|---|---|---|---|---|---|---|---|
| Felt | 4.4 | 1100 | 0.883 | 588.29E3 | 0.35 | 0.01 | 266.425E3 | 2.0 | 15E-6 | 104E-6 |
| Foam | 11.9 | 797 | 0.952 | 64.888E3 | 0.17 | 0.2 | 22.827E3 | 2.35 | 87E-6 | 160E-6 |



**Fig. 3.4.** Example of Material Modeling

### 3.2.3 Model Validation

Through the process of paragraphs Sects. 3.1.1. and 3.1.2, a preliminary SEA model for Target Cascading was established. However, the model is not sufficient to proceed with target casing using this model since the consistency of the model has not yet been verified. In the case of the coupling surface modeled based on mass-law or historical test database, it is necessary to check the consistency of the performance and make corrections if necessary. Therefore, an actual vehicle contribution evaluation was conducted, and the analysis results were compared with the test result to correct the sound absorption performance of the necessary coupling surfaces.

The actual vehicle contribution evaluation was conducted using the ATF (Acoustic Transfer Function) test from exterior to interior of the vehicle, and the contribution of each area was identified using the windows method. As shown in Fig. 3.5, the area of the vehicle was divided into eight sections. Under the full MAX-Package condition, the interior noise measurements were conducted with acoustic excitation applied at the tire and rear PE. And the change in interior noise when the MAX-Package of each area is removed represents the contribution of the specific area. Analytically, it is assumed that setting the exterior load of specific area to '0' is equivalent to applying the MAX-Package treatment to the area, and interior noise analysis was conducted based on this assumption.

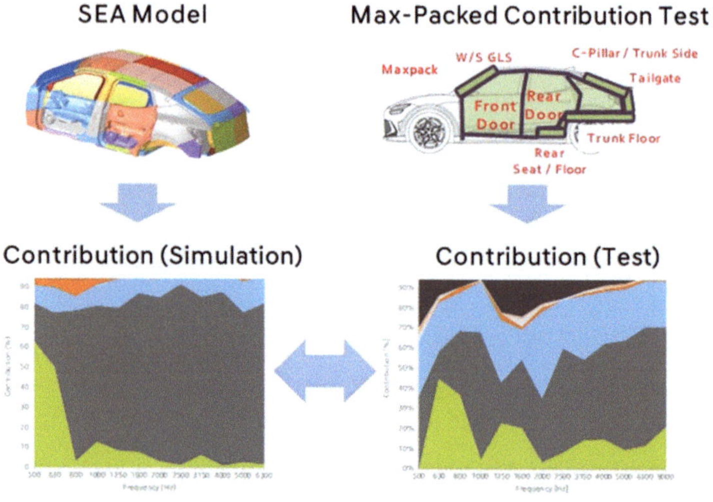


**Fig. 3.5.** Contribution Analysis Comparison between Test and Simulation

When the contribution analysis was conducted using the preliminary SEA model, there were many differences in the results with actual vehicle test as shown in Fig. 3.5. In order to find out the cause of this difference, energy flow analysis was conducted to examine how acoustic energy is transferred between each cavity. Based on the results, the sound insulation and absorption performance of the coupling surfaces were adjusted. As a result, as shown in Fig. 3.6, consistency was achieved between the actual vehicle ATF test results and the contribution analysis.

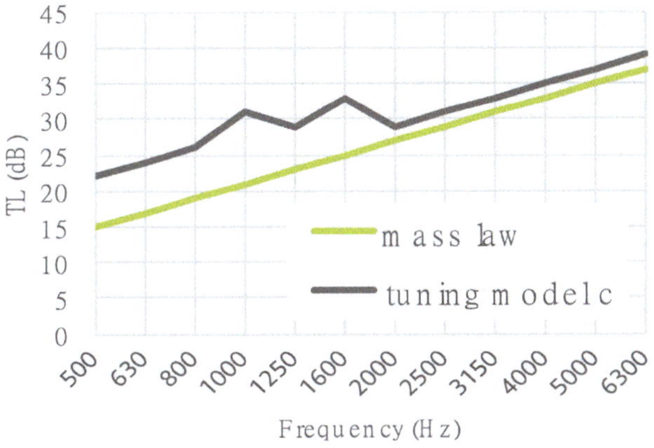


(a) Example of Boundary Surface Transmission Loss Performance Correction

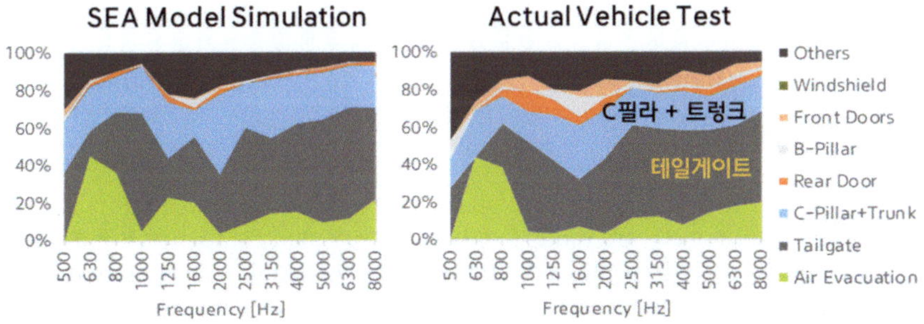


(b) Contribution Analsys

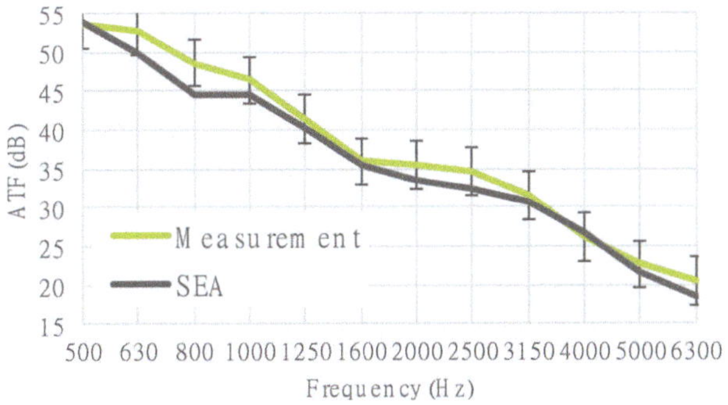


(c) Acoustic Transfer Function

**Fig. 3.6** Ensuring Consistency with the Test Results of the SEA Model (a) Example of Boundary Surface Transmission Loss Performance Correction (b) Contribution Analsys (c) Acoustic Transfer Function

## 3.3 Target Cascading Analysis

### 3.3.1 Performance Target Setting

The representative performance in this study is the acoustic transfer function (ATF) from the tires and rear PE to the ear position of the rear seat passenger, and the target performance is the ATF of the vehicle with the highest level of sound insulation performance among competitors. The difference is shown in Fig. 3.7 below.

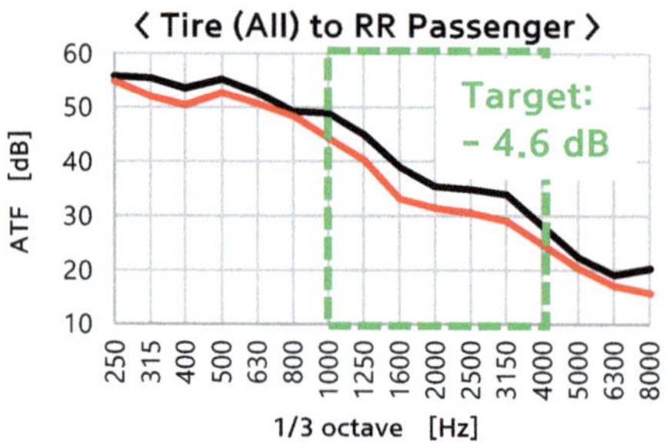


**Fig. 3.7.** ATF Target Setting

Both ATF performances are more disadvantageous than competitors in the full frequency domain. In order to achieve the target performance, it is necessary to improve tire-induced noise by about 4.6 dB and rear PE-induced noise by about 9.6 dB based on the average value of 1 to 4 kHz.

### 3.3.2 Result of TCA and Identification of Noise Transfer Path

Using the SEA model developed in Sect. 3.1, target cascading analysis was performed to achieve the target performance set in Sect. 3.2.1. And the Fig. 3.8 shows the results.

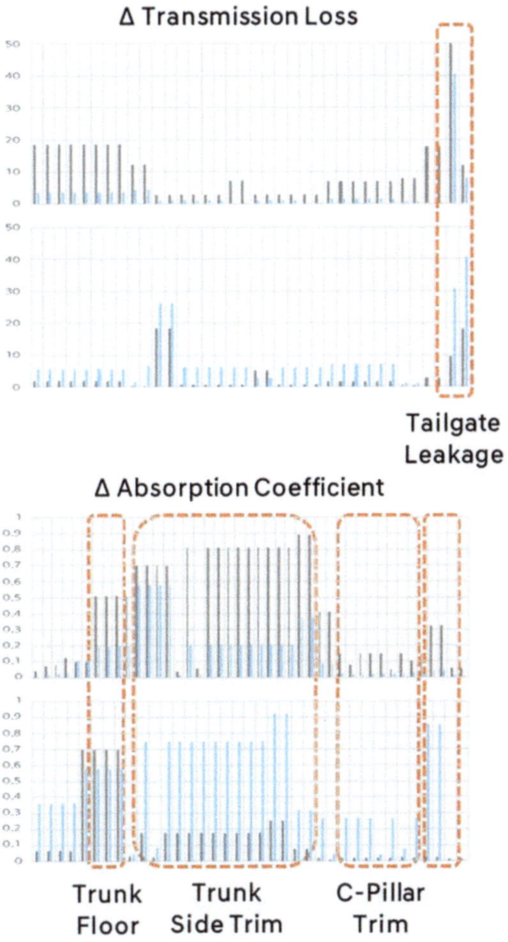


**Fig. 3.8.** Result of Target Cascading Analysis

The above result indicates which performance (sound insulation or absorption) should be improved for each area in order to achieve the target performance. Figure 3.8 shows that in the case of transmission loss, improvement of the leakage (weatherstrip or parting gap) at the tailgate is most necessary, and in the case of sound absorption ratio, it is necessary to improve the sound absorption performance of the luggage area such as the trunk floor, side trim, and C pillar.

Based on the contribution analysis results and target cascading analysis results, it is possible to estimate through which path the noise generated from tires or rear PE flows into the cabin, as illustrated in Fig. 3.9 below.

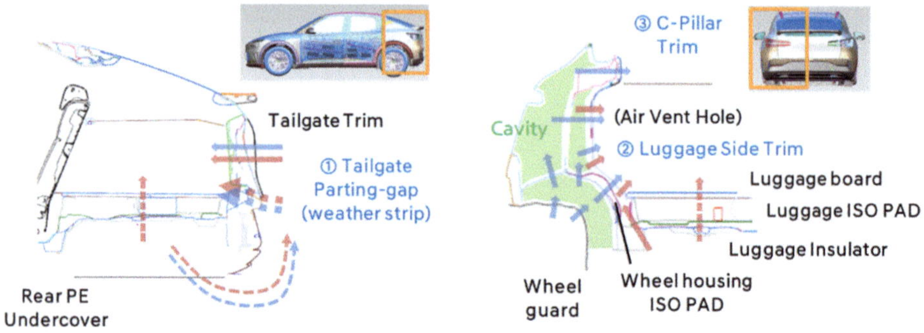


**Fig. 3.9.** Main Noise Inflow Path through Luggage

Path ① is a case in which noise generated from tires or rear PE is transmitted to the outside of the tailgate through the outside of the vehicle, and the noise is directly inflow into the interior through a leakage such as a weatherstrips and parting gap. Paths ② and ③ are cases in which tires and PE noise inflow into the cavity behind the luggage side panel is transmitted to the interior through the luggage side trim and the C pillar trim.

Another thing to note in Fig. 3.8 is that there is too much need for improvement. Such as about 40 dB improvement in transmission loss of the tailgate leakage and 0.7 improvement in sound absorption ratio of the trunk side trim are virtually impossible. This means that in order to achieve the target performance, not only the performance of the interior but also the performance of the exterior parts need to be improved at the same time.

## 3.4 Sound Package Optimization

### 3.4.1 Concept for Sound Package Optimization

Based on the results of Sect. 3.2.2, the actual vehicle performance improvement evaluation was conducted. The concept of this performance improvement plan is classified shown in Table 3.3 below.

In the case of the interior trim, in order to increase the sound insulation performance of the luggage side trim, which is judged to be the main path, it is necessary to increase the coverage ratio of the trim and increase the thickness and weight. But it is difficult to apply the improvement plan. And due to the complicated luggage structure, it is impossible to block all noise flowing into the room. Therefore, it was judged that it is efficient to absorb noise flowing into the room or into the cavity between the trim and the panel by increasing the sound absorption performance rather than improving the sound insolation performance of the trim.

**Table 3.3.** Concept for Performance Improvement

| Classification | Improvement plan |
|---|---|
| Interior trims | Enhancement of interior sound absorption performance to reduce noise transmitted into the cabin or cavity areas. |
| Exterior trims | Mitigation of external acoustic load by applying sound-absorptive exterior trim components |
| ETC. | Development of felt-based acoustic components for light weight and enhanced sustainability |

As mentioned in Sect. 3.2.2, it is difficult to achieve the target performance only by improving the performance of the interior trim. And, most of the noise inflowed through moving parts such as tailgates and doors is usually directly inflow into the interior through weather strips or leakage, which is difficult to change since various conflicting performances are related. Therefore, it is difficult to improve the acoustic exterior loads. A method of reducing the external load applied to the vehicle body using sound absorbing exterior trims.

Lastly, since the importance of dash area and floor insulation performance is reduced compared to the ICE vehicles, and there is not need to improve the dash and floor in the target cascading result, felt materials are applied to these areas to improve sustainability and light weight.

### 3.4.2 Prototyping

**Fig. 3.10** Components Prototyping

In accordance with the improvement concept identified in Sect. 3.3.1, the parts shown in Fig. 3.10 were prototyped.

A total of 10 improved parts were prototyped. 3 for light weight and sustainability improvements, 5 for enhanced sound absorption performance of cavities, and 2 for external load mitigation. The specifications of the improved parts were determined based on the sensitivity test, ensuring that they meet the results of the analytical Target Cascading analysis.

Figure 3.11 below illustrates the change in total trim weight following the application of the prototype test.

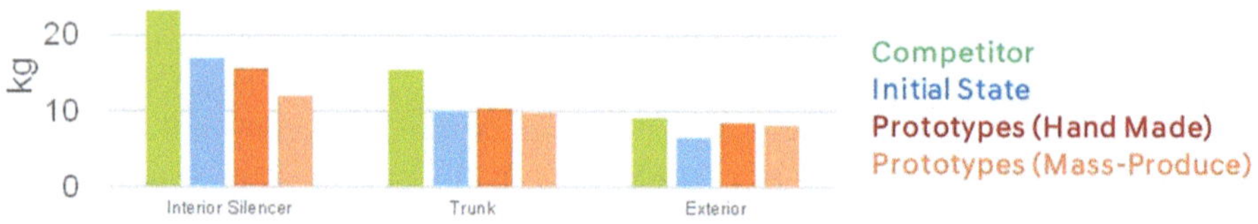


**Fig. 3.11** Weight Change according to the application of the Prototyping Components

On the left side of the graph is the weight of interior tirms in the cabin like the dash insulation pad and floor carpet, and in the middle, "Trunk" is the total weight of the trims on the luggage side of the vehicle, and "Exterior" is the weight of the exterior trims such as the undercovers. As can be seen from the graph, the original components of test vehile already have weight advantage over the compeitor even in the initial state, and the application of the prototyped components allowed for the futher reduction in the weight of trims. In the addition, since th prototypes were made by hand, the weigh of prototypes includes the weigh of auxiliary materials such as adhesives, so if the improved components are mass-produced, the weight reduction effect will be more certain. Although the weight of the exterior trims slightly increased due to the addition of sound absorbing materials, the total trim weight was still reduced compared to the initial state.

### 3.4.3 Vehicle Verification Evaluation

The results of ATF evaluation by applying the above improvements to the actual vehicle are shown in Fig. 3.12 below.

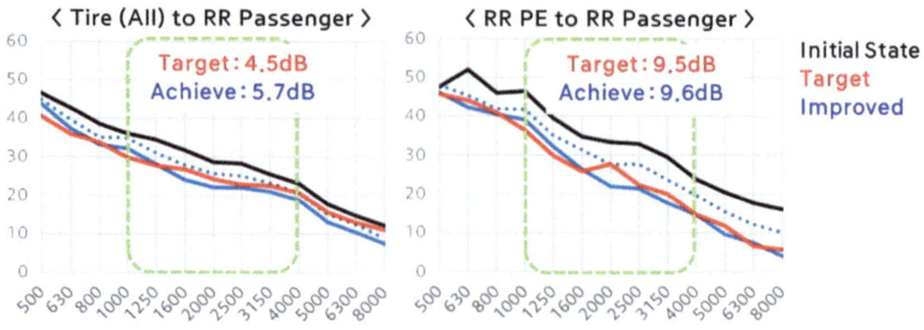


**Fig. 3.12** ATF Test Result of the Performance Improvement

Here, the blue line is the result of the ATF evaluation when the improved product is applied, and the dotted line is the graph when only the indoor trim improvement plan is applied, and the solid line is the graph when the exterior trim is applied additionally. As shown in Fig. 3.12, it was confirmed that the noise transfer function exceeded the previously targeted tire noise transfer function of 4.5 dB and the noise transfer function of 9.5 dB for the rear PE by applying the improved product.

## 3.5 Conclusion

In this study, the physical properties of interior and exterior trims applied to electric vehicles were measured and used to develop a SEA model for contribution analysis and target cascading analysis. And based on the analysis result, we identified the primary noise transmission paths into the cabin. For our electric vehicle, it was confirmed that a significant portion of interior noise is transmitted through the luggage area, including the tailgate and luggage side trims.

Based on this finding, the following three optimization strategies for improving full-vehicle acoustic performance were established:

1) Reduction of external noise load through improved sound absorption performance of exterior trims.
2) Reduction of transmitted noise energy through improved sound absorption inside the luggage compartment
3) Weight reduction and increase sustainability by applying felt materials to components with low acoustic contribution

Using these strategies, improvements were identified, and prototypes were developed for a total of 10 parts, followed by full-vehicle testing. As a result, without increasing overall weight, the tire-induced noise transfer function was improved by 5.7 dB and the rear PE-induced noise transfer function by 9.6 dB, achieving best-in-class acoustic performance for a production vehicle.

## References

1. Claudio Bertolini and Theophane Courtois, An SEA-based procedure for the Optimal Definition of the Balance between Absorption and Insulation of Lightweight Sound Package Parts, SAE International 2013
2. Thibault Lafont, Application of Statistical Energy Analysis on a car: from the vehicle modeling to parts targeting, INTER-NOISE 2016
3. Federico Di Marco, A Critical Review of Some Panel Contribution Analysis Method Used in Interior Vehicle Acoustics, ISNVH 2024

# SEA Hybrid P/Q Correlated Modelization Process for Vehicle Transient "On-Track" Noise Assessment

Jean-Baptiste Prunet[1](✉), Johann Löwen[2], Nicolas Aubert[1], Cristina Campagna[3], Salvatore Carducci[4], Pascal Iborra[1], and Patricia Strasser[5]

[1] Centre Technique de Belchamp, Stellantis France, Mail Box SX.BP.30, F-25218 Montbéliard, France
{jeanbaptiste.prunet,pascal.iborra}@stellantis.com, nicolas.driot@hexagon.com

[2] TCCE Bahnhofsplatz IPC S1-4, Stellantis, D-65423 Rüsselsheim, Germany
johann.loewen@stellantis.com

[3] Corso Giovanni Agnelli, Stellantis, 200, I - 10135 Torino TO, Mirafiori, Italy
cristina.campagna@stellantis.com

[4] Ia Ex-Aeroporto, Stellantis, ISn-Building n4, Pomigliano d'Arco, NA 80038, Italy
salvatore.carducci@stellantis.com

[5] Stellantis, 212 Bd Pelletier, 78955 Carrières-sous-Poissy, France
patricia.strasser@mpsa.com

**Abstract.** For OEMs, the urgency of designing even lighter but even more acoustically performing vehicles remains a major target, considering constraints such as reaching carbon neutrality target, and the rise of electrical car motion (with drawback of time-consuming battery reloading process).

Traditionally the SEA tool is delt in a global manner so that the projected vehicle receives pre dimensioned draft sound package. However, for extended applications, operating on a validated model requires a correlation with measurements. A straightforward way to complete the correlation is to compare the p/Q transfer functions, which should not be limited to the sources to passengers' analysis. The risk is that the model contents would not represent enough faithfully the real physics of the vehicle.

In fact, since the performance of the insulation of the vehicle is highly dependent on the smaller components' contributions, SEA appears as unable to lead to noise prediction inside the cockpit, because SEA deals with assembled large volumes not considering the details of the vehicle. Therefore, an operational model was achieved including detailed components performances, and realistic physical parameters.

Because the question of the noise prediction at passengers' ears remains in the scope of the designer and in the purpose of making a computational model of the car, we also operated an on-track analysis including on-track


A. Heintzel (Ed.): Automotive Acoustics Conference 2025, Proceedings, pp. 46–57, 2026. https://doi.org/10.1007/978-3-658-51249-1_4

noise measurements at noise sources. Sources are first conditioned and then "Pressure imposed" on the SEA model through steady speeds and transient speed sequences.

**Keywords:** SEA modelization · Air borne transfers · Model correlation · Transient speed modelization · Noise predictivity

## 4.1 Scope of the Study

The study consists in the building and exploitation of an SEA model dedicated to insulation performance assessment of the full car and evaluate its performance in comparison with valid experimental data.

The first step is the model building with its specific approach trying to reach a complete realistic description of the vehicle. The same physical vehicle is submitted to experimental tests for purpose of correlation of the model. The tool that is used for correlation is the airborne transfer function in p/Q format. Once the model correlated, the same vehicle is measured on-track at passengers' ears but also at sources (tires and e-motor).

The purpose is to introduce these noise sources in the SEA synthetic model. These measurements are first operated at steady speeds and then at transient speeds on different tracks. The sources measurements are obviously impacted with the noise of the air flow around the vehicle. This aerodynamic noise at sources (tire & motor) must not be applied on the SEA model since we plan to use the contribution of the aerodynamic noise that is measured at passengers' ears in a semi-anechoic wind tunnel as a component of the noise synthesis (since aero-acoustic noise computations are not enough reliable for our needs).

Once the aerodynamic flow noise at external microphones is retrieved from the full noise when driving, the true road noise sources and the true e-motor noise are available. They can be injected in the SEA model in a "Pressure imposed" mode.

Either for steady speeds or for the transient run ups or run downs, the contributions are computed at passengers' ears individually and summed up with aerodynamic noise data obtained from wind tunnel. The result is compared to the real measurement of the vehicle. The computed contribution is expected to be at lower level than the real vehicle on track. Hence the difference between the computed synthesis and the real noise of the car is expected to be representing the structure borne noise contribution however melted with other background noises.

This process can be summarized in the following graphics at Figs. 4.1 and 4.2.

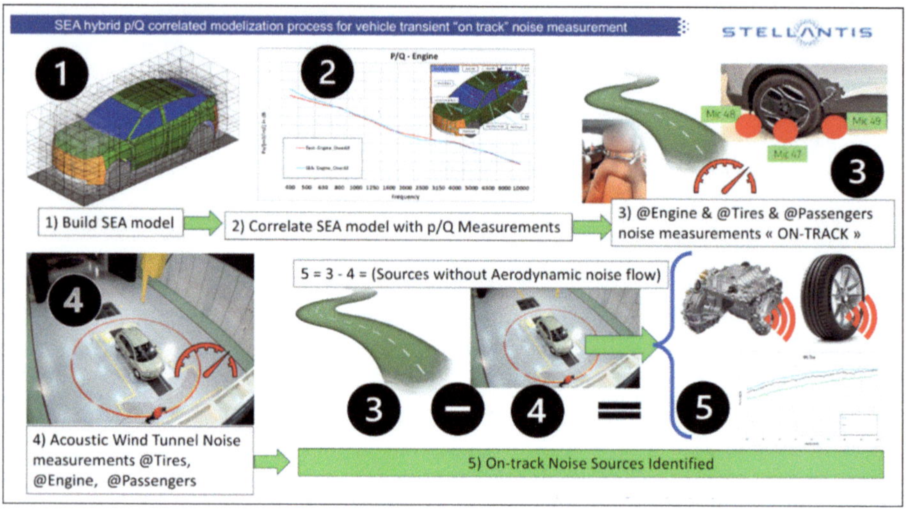


**Fig. 4.1.** Steps no 1 & 2 of SEA model correlation process, followed by on-track noises sources and passengers's pressure measurements (no 3). A wind tunnel is also used to estimate aerodynamic contributions (no 4) at the same sensors' positions. Step 5 is the re-computation of the noise sources without aerodynamic airflow contribution.

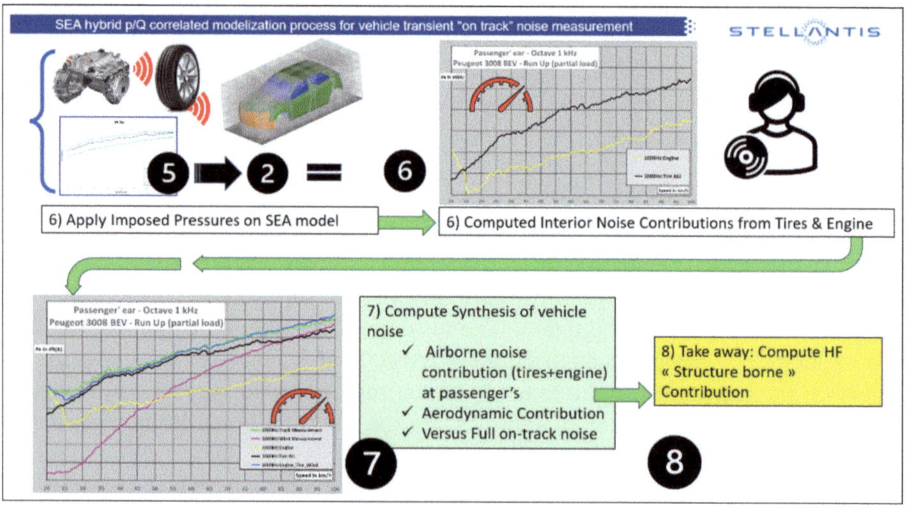


**Fig. 4.2.** Step no 6 consists in applying tires and engine sources to the SEA model and get independant contributions. Step no 7 deals with the summation of all known contributions (Tires, Engine, Aeroacoutic noises). Step no 8 is an attempt to estimate the structure borne noises contributions.

## 4.2 SEA Model Description

Selected Bandwidth is from 400 Hz to 10 kHz. The Spectral format is 3rd of octave.

The space in which the modelized vehicle is placed is semi-anechoic (free field boundary condition and floor reflecting).

All panels (steel sheets, glasses, Polypropylene walls) are described using either tabulated data but also especially for glasses measured data coming either from suppliers or from Stellantis.

About the insulation layers, the format is Insertion Loss data, originated from combination of measured data (material properties), 3D design of the part, and computational tools developed internally at Stellantis.

About the absorptive silencers, they follow the same scope as the insulation layers above, but absorption properties are evaluated as equivalent absorption area. Random incidence format is selected for this purpose.

All other components are either computed or measured when no model is available. The format is Insertion Loss. This is the case for the bulb joints, sealing lips, compressed foams (doors, hood, instrument panel). The passthroughs IL are measured in as special tool (Insulation cabin). Rarely a component needs to be estimated. Also, all sealing plugs, stapples, adhesive tapes, what we call repetitive parts are known by IL measurements. The baffles have acoustic properties that are measured thanks to an internal standard or can be estimated with its wall property (TL).

All the components that are listed above are assembled in a sub-structuring approach so that macro elements can bet manipulated in the model, with respect of the expected assumptions that makes the validity of the SEA approach.

For the low frequency domain, the lack of adaptable data for the model, led us to introduce some structural damping effect and some modal statistical behavior of a few panels (Coupling factor) using some of the tabulated data from VAONE.

The creation of the model was elaborated thanks to available 3D FEM NVH meshes. The cells that were created for SEA model, follows faithfully the 3D geometry of the car, not too much simplified. Finally, the model appears as a highly detailed model which is generally not the case of usual SEA models that acoustic engineers manipulate as shown in Figs. 4.3 and 4.4.

The outcome of the model reaches around 1000 cavities, 3000 plates, 5000 junctions and nodes, 100 TL & 70 Salpha spectra.

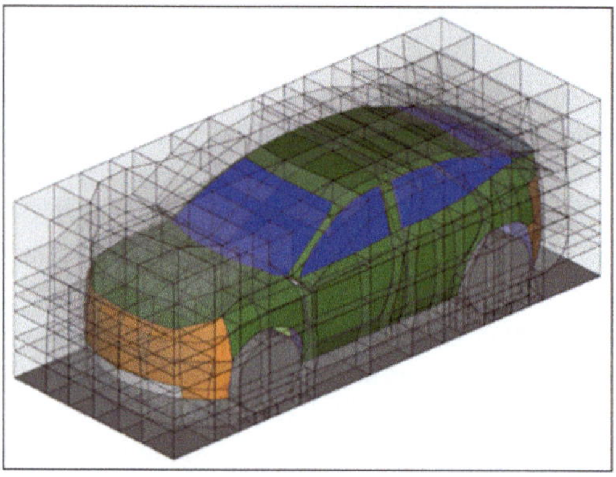

**Fig. 4.3.** Outer picture of the modelized vehicle through SEA cellular mesh.

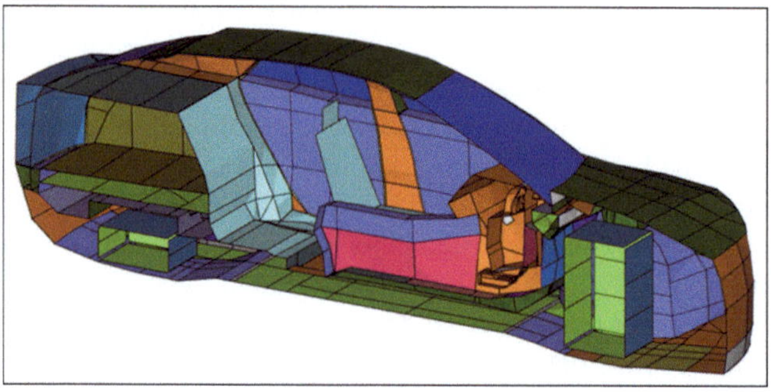

**Fig. 4.4.** Inner picture of the modelized vehicle through SEA cellular mesh.

## 4.3 Acoustic Air Borne Transfer Functions Measurement Problem

The SEA tool allows to compute the p/Q transfer functions between two cells very easily. At this step, we meet a tricky problem: The pressure that is computed inside each cell is a unique spectrum. But the reality of the physical volume is much different from SEA. Due to the boundaries and modal effects, the pressure will vary significantly inside the cell depending on the position of the microphone. Micro-localization of the energy flow also modifies the result. Another parameter is that when changing the position of the microphone in the physical volume corresponding to the SEA cell in the model, one gets different results, also due to the fact the cell sizes are much wider than the wavelengths above 4 kHz (Semi-wavelength around 3,6 cm). This means that changing the microphone position inside each SEA cell will lead to a changing p/Q transfer value. Which position of the sensor is the representative one?

For this reason, it was decided to elaborate a specific testing procedure that is applied to transfer function measurements especially for SEA correlation activity. The cavities on which a correlation is needed, will be scanned with the microphone so that the variability of the transfer function inside the volume can be estimated. An average of the spread positions is computed.

The other aspect of the transfer function correlation problem is that correlation between sources and ears are not a sufficient descriptor of the quality of the model. So, to be sure that the cockpit is correctly designed, transfer functions are measured not only at sources but also in many places all around the vehicle and inside the volume.

Some illustration of the microphones selected positions for the SEA correlation study in Fig. 4.5a and b.

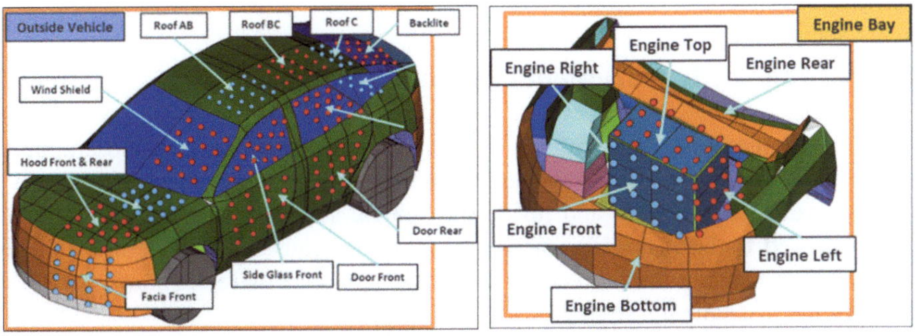


**Fig. 4.5.** Selected sensors' positions around the vehicle (a) and inside engine bay (b) dedicated to transfer function measurements.

## 4.4 Results of the P/Q Correlation Study

We present some of the spectra resulting from the correlation study. In the engine bay and around tires best correlation appears as results and give insurance on the model quality. However, at distance from the passengers, the level of the correlation remains of very high quality. The question of getting a correct respect of the shape of the measurement curve is a good criterion. Figures 4.6 and 4.7 show typical correlation curves that the model allows to reach. They are representative of the correlation level reached by other curves.

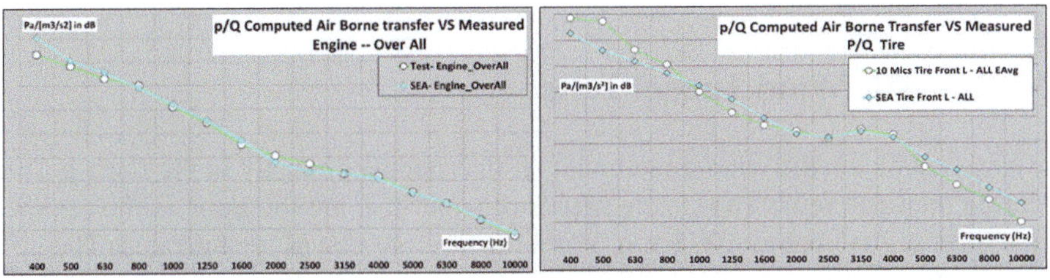


**Fig. 4.6.** p/Q transfer function computed vs measured at engine (a) and front left tire (b).

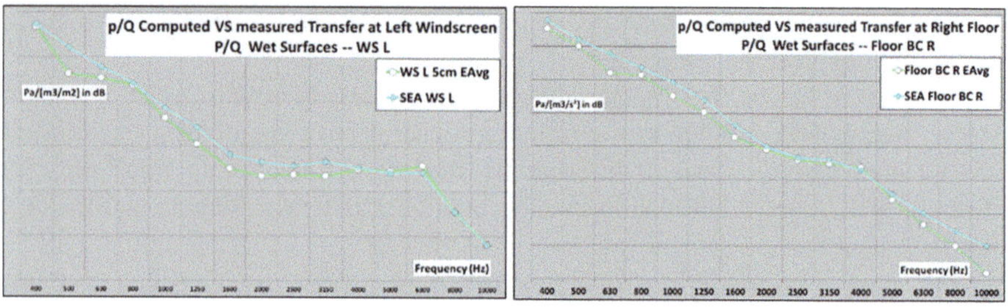


**Fig. 4.7.** p/Q transfer function computed vs measured at windscreen outer (a) and middle outer floor (b).

## 4.5 On Track Measurements Process

The car is measured on track with set of microphones around the tire and around the engine and at passengers' ears. All the measurements are practiced synchronously at all points around the vehicle and inside the vehicle so that coherence of the data is much improved between measurement points: Better causality can be acquired that way between noises sources and noise inside the cockpit. Figure 4.8 show some illustrations of the instrument's settings.

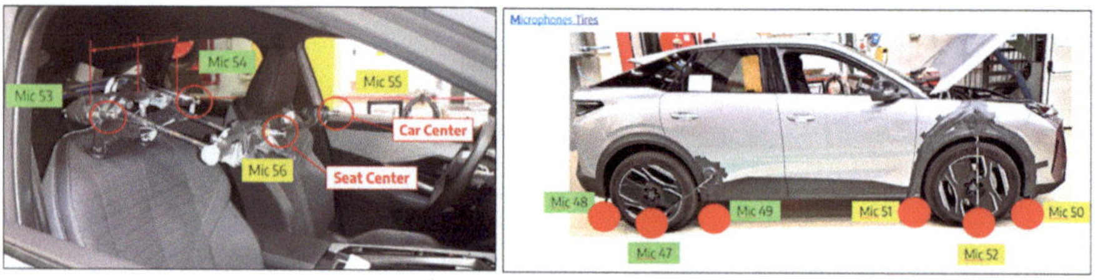


**Fig. 4.8.** Sensors positions inside cockpit **a** and around tires **b**

Measurements were taken with 4 steady speeds below 100 km/h, and run ups and run downs, partial load, full load, run down no brake or with energy recuperation mode. The floor noise (background noise) of instrumentation has also been measured to verify that the level of noise when running remains high enough to be considered as valid data.

## 4.6 Acoustic Wind Tunnel Tests

The car is set inside the wind tunnel with all microphones exactly in the same position as during the track tests. The steady speeds of wind flow are the same as the car speed on the track always below 100 km/h. Wind run ups and downs are operated. The influence of external microphones at sources was assessed by retrieving them and repeat the measurements. It appeared that the microphones influence is weak regarding inside cockpit effects which was simplifying the study a lot, as shown on two graphs below as examples (Fig. 4.9).

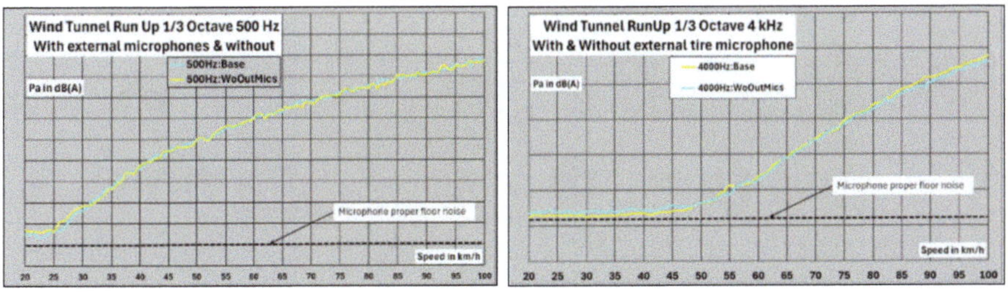


**Fig. 4.9.** Wind tunnel noise at passenger's ear with external microphones in place and removed.

## 4.7 Recomputing Noise Sources so That They Can Be Applied to the SEA Model

The noise sources pressure levels, that were measured on track, cannot be applied directly on the air borne model including the noise of the air flow around the car. The reason is that we chose to operate the synthesis of noise at passengers' ears using the inside cockpit microphones directly for the air flow noise contribution as measured in wind tunnel. And because the aerodynamic noise loads are not quantized around the car, so that we cannot expect to use them as pressure inputs to the SEA model. Not retrieving the air flow noise at sources would introduce a bias on the result, like applying twice the air flow noise load (at least partially). Therefore, the air flow aerodynamic noise must be retrieved at sources from the rest of the noise measured on track as shown in Fig. 4.10.

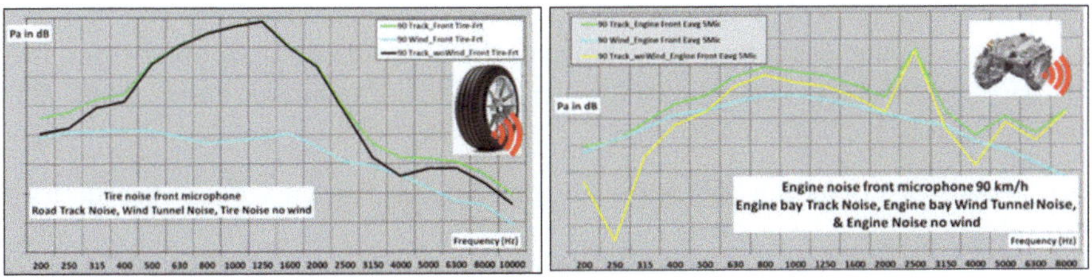


**Fig. 4.10.** Measured noise at sources are re computed retrieving the aerodynamic flow noise at the external microphones.

## 4.8 Applying Sources Pressure Levels to the SEA Model

The pressure of sources can be applied to the airborne SEA model. The result is interesting in terms of vehicle physical transmission property: The process allows to estimate the relative impact of the various sources inside the cockpit.

Figure 4.11 shows a computation of the airborne sources' contributions to the ears.

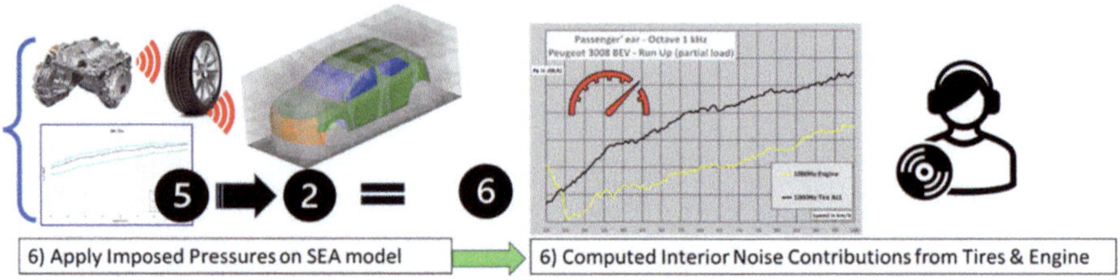


**Fig. 4.11.** Measured noise sources are applied in "Pressure Imposed" mode on SEA model so that internal contributions are predicted inside the cockpit.

## 4.9 Synthesis of All Known Contributions vs Real Noise on Track

The following graphics show the individual contribution of tires, e-motor, and aerodynamic air flow noise in wind tunnel at ears compared to real noise on track. A remarkable satisfying fitting of the computed synthesis with the real running vehicle is stated.

However, some remarks appear (please refer to graphics below)

- At higher speed, the contribution of the wind tunnel appears as slightly higher that the vehicle running on track (at partial load). The effect can be attributed to discrepancies between the wind tunnel acoustic field and the real track acoustic field, and the SEA model boundary at floor.
- Engine, tires, airflow noise contributions are not the topic of the article, but this method allows us to analyze the contributions with respect of the vehicle design.
- At high frequencies 4–8 kHz, with such level of insulated vehicle like the tested car, it appears that the proper floor noise of the microphone is a limit for our analysis. However, the level being very low, more performing measurement might not be our first priority.

All these considerations can help acoustic engineer defining an acoustic profile for the future vehicle (Figs. 4.12 and 4.13).

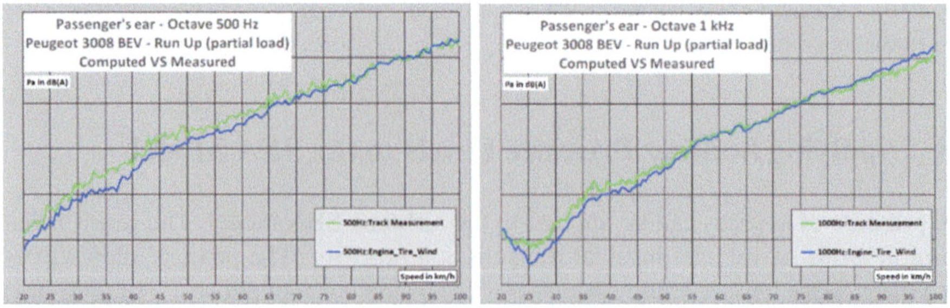


**Fig. 4.12.** Computed contributions (Tires and Engine and Aerodynamic airflow noise) VS measured on-track shown on Fig. 4.12 (Octave 500 Hz) and (1 kHz).

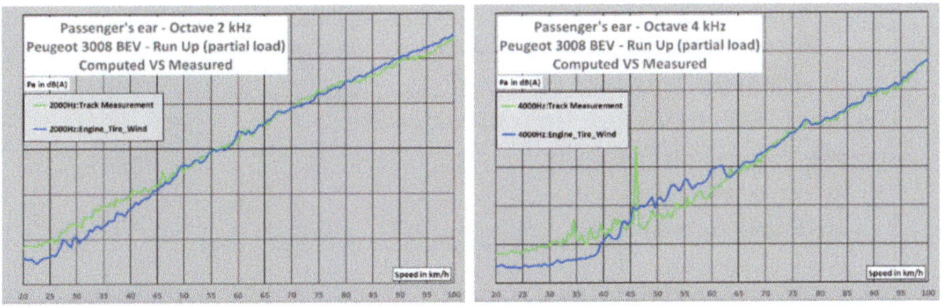


**Fig. 4.13.** Computed contributions (Tires and Engine and Aerodynamic airflow noise) VS measured on-track shown on Fig. 4.12 (Octave 2 kHz) and Fig. 4.12 9 (4 kHz).

## 4.10 Take Away the Structure Borne Contributions?

Once were gathered all partial results from this study, one could expect that the information about the structure borne contribution inside the car, could be reached. The structure borne contribution could theoretically be easy to compute by retrieving all known contributions from the on-track measurement. Off course, an attempt was done to compute this contribution.

However, when the computed synthesis overpasses the noise of the real car (even slightly), the result cannot be computed. The problem is that the wind tunnel data cannot be equalized with a real on track measurement because of the tire noise is always active during track tests. We tried to "force" the level of the computed contribution to be set slightly lower than the on-track noise. Off course, in those case the structure borne contribution is very low, and we know that the structure borne contribution will depend on how the forcing below track level was operated.

Other problem is also generated by other unexpected functional or external noises such as some gravel noises due to impacts, even if the track is cleaned up before test session.

Till now, the method that will help us separate wind noise around the car from other sources is not ready. But the question is a key to the synthesis problem. Such a computing case is shown on Fig. 4.14 showing that above 35 km/h the results are not trustworthy. A gravel impact appears at 46 km/h and is understood as structure borne or other contribution.

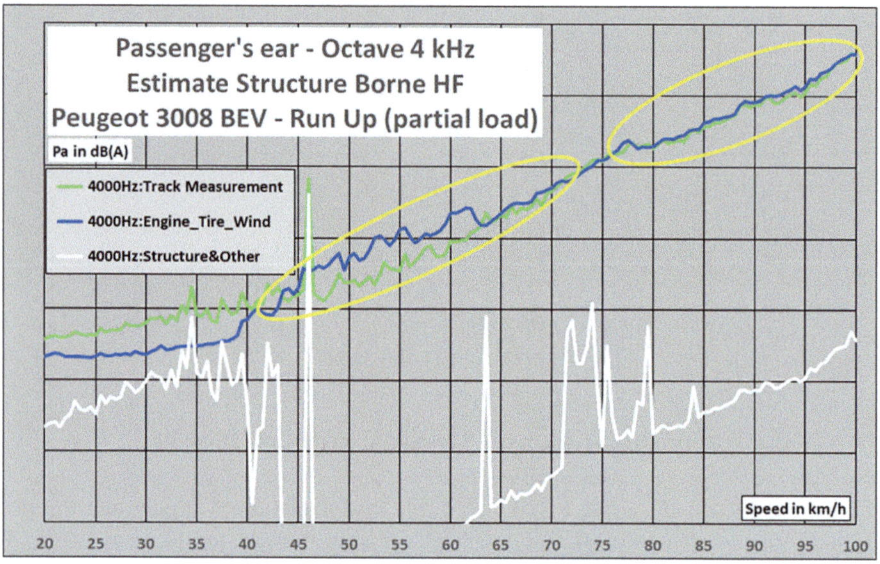


**Fig. 4.14.** Attempt to estimate global structure borne contribution shows limits above 45 km/h (and below), due to disturbing complementary contributions such as microphone floor noise impact, wind tunnel contribution over-estimated, and other noises. Gravel impact appears on road at 34 km/h and 46 km/h are understood as contribution.

## 4.11 Conclusions

SEA models whenever they are carefully built up can deliver very realistic results that can be used to compute a credible estimation of the real noise inside the car, assuming the acoustic sources are well known.

The use of the p/Q transfer function is a very performing tool to complete first step correlation, but it should be operated not only at sources, but in the whole volume of the SEA model for best credibility. For this purpose, the real modelized vehicle is needed for test. And since the SEA model delivers only one dynamic pressure spectrum inside each cell of the model, the measurement of the p/Q transfer function must be adapted to describe correctly the result inside the cell. This process allows establish full confidence on the model as a proper description of the real car.

Applying the methodology using the real on-track conditions is a major step that allows confirm the confidence established on p/Q correlation and allows to predict with a good precision the noise any place in the vehicle volume.

## 4.12 Thanks


- Many thanks to Pascal Iborra and Nicolas Aubert for their significant contributions with the operated measurements on vehicle for this study.
- Many thanks to Patricia Strasser who contributed to the creation of the SEA model building procedure.

- Many thanks to Salvatore Carducci who lead the worldwide team in defining what the p/Q transfer function process should be in general and for SEA correlation application, and who wrote the measurement procedure.
- Thank you to Cristina Campagna for her help in the whole process and extending it to the real life of projects.

# Recent Advances in Efficient Fully Trimmed Body NVH Analysis

Katrin Hoge[1(✉)], Petr Svec[2], Gaetan Jacquemin[3], Julien Verhaegen[3], and Benoit Van den Nieuwenhof[3]

[1] Audi AG, 85045 Ingolstadt, Germany
Katrin.Hoge@audi.de
[2] Mladá Boleslav II, Skoda Auto a.s., tř. Václava Klementa 869, 293 01 Mladá Boleslav, Czech Republic
Petr.Svec@skoda-auto.cz
[3] Hexagon AB, 9 Rue Emile Francqui, 1345 Mont-Saint-Guibert, Belgium
{gaetan.jacquemin,julien.verhaegen, benoit.vandennieuwenhof}@hexagon.com

**Abstract.** Due to electrification as well as more stringent vibration and noise regulations, NVH numerical models show constantly growing sizes and need to incorporate more detailed noise control and damping treatments. In this context, commercial frameworks such as Nastran and Actran, have brought significant innovative improvements to their trim components modelling capacities, both in terms of physical modelling, easiness and automation of model setup and computing performance. Industrial-sized fully trimmed car body models now deliver results in much less than a nighttime.

This paper reviews several recent key features that allow to reach this unmatched efficiency. First, a methodology that properly sequences the modal extraction process for the undamped in-vacuo car body and rigid-wall cavity model and reduced trim impedance matrix computation (RIM) is presented. Care is taken to avoid unnecessary modal data recovery to speed-up the process without compromise on numerical predictions. Additional compression methods are exposed that further accelerate the solving process by reducing the I/O operations. Second, a new air-gap coupling condition on trim interfaces is presented that allows to reduce the size of the finite-element (FE) trim models. Finally, robust 3D to 1D automated trim model conversion makes it easy to extend the frequency range of analysis of an existing FE setup.

After the brief presentation of the theoretical aspects of each of those improvements and the illustration of their individual benefits, an industrial-sized fully trimmed car body model at Audi is finally presented that leverages on this state-of-the-art modelling process. The review focuses on both the overall accuracy and the computational performance.

**Keywords:** Fully trimmed body NVH · Reduced impedance matrix · Trim modelling


A. Heintzel (Ed.): Automotive Acoustics Conference 2025, Proceedings, pp. 58–72, 2026. https://doi.org/10.1007/978-3-658-51249-1_5

## 5.1 Introduction

In the domain of noise, vibration, and harshness (NVH) modeling, acoustic trim components are essential for dampening and insulating lightweight passenger cabin structures. Therefore, being able to represent them through finite element models (FEM) is crucial for many industries, including automotive. The FEM allows the description of complex geometries and the complex behavior of poro-elastic materials.

The increasing demand for handling large models and extended frequency ranges intensifies the need for enhanced efficiency and accuracy. As a result, the reduced impedance matrix (RIM) method emerged as a solution, capturing the complexity of trim components, modeled in physical degrees of freedom (DOFs), while reducing the size of the assembled problem solution [1, 2]. The method quickly became a reference solution for such problems and has been successfully applied to various industrial applications [3–6]. However, technological limitations arose as modern models expanded in size and higher frequency ranges were analyzed.

The method relies on computing frequency-dependent RIMs for each trimmed interface, these matrices might be stored, either in RAM or on disk. However, contemporary models with modal components represented by up to 30,000 modes posed challenges, since each RIM is fully dense, and square with size corresponding to the total number of modes. It can lead to trim databases exceeding 1TB and performance challenges as it is costly to handle such large matrices and their I/O-related operations.

In response to these challenges, a method to further compress these RIMs has been recently presented [7] and implemented in commercial packages solving fully trimmed car-body models, namely MSC Nastran PEM and Actran for Trimmed Body. Because it has its foundation on a robust filtering of the modal information before projection of the RIM in modal space, the approach implicitly assumes the availability of the in-vacuo structural modes and rigid-wall cavity modes before RIM computation. Given that the efficiency of such modal extractions is primarily governed by the size of the modal data recovery space, an appropriate sequencing of the operations allows to reach an optimal performance. This sequencing is presented in this paper.

Also, the flexibility to connect any 3D trim finite element model to the car-body structure and the passenger compartment is of prime importance. In particular, trim components are usually made of different materials, ranging from visco-elastic damping layers, internal air cavities to poroelastic foams or fibrous layers. Their connection to the cavity may involve communicating pores or closed pores and may be characterized by glued or sliding connection to the structural parts. The paper presents and discusses a new boundary/contact condition that facilitates the trim component installation: an airgap connection with the structural part, which is equivalent to an infinitely thin air layer between the trim component and the structural part.

Finally, worth mentioning is the possibility to leverage on an existing fully trimmed model to extend the frequency range of analysis validity or to perform dedicated component-level analyses. In this context, the paper demonstrates how an existing trim model can be converted from a 3D finite element model to a 1D transfer admittance model, with spatially variable thickness and material stacking. Such a model allows for high-frequency transmission loss characterization, integration into an SEA model as a component characterized by its transmission loss properties. Also, it is possible to capitalize on an existing 3D trim model to perform component-level optimization and/or solve local issues in an efficient manner.

In this paper, the RIM is first introduced, and the developed compression methodology is explained. The modal-based workflow for performance improvement is then detailed. The new air-gap boundary conditions for trim component installation in the assembled model is then presented. The effectiveness of the proposed approaches is finally assessed utilizing a realistic industrial car model. The results of a noise transfer simulation up to 800 Hz for the new Audi Q6 e-tron are presented. The simulation model encompasses the vehicle's structure, cavity, and main trim parts. The study's results are obtained using commercial codes Actran [8] and MSC Nastran [9], integrated into the comprehensive trimmed solution MSC Nastran PEM. The discussion covers results, performance, accuracy evaluations as well as the limitations of the method.

## 5.2 Theoretical Aspects

### 5.2.1 Reduced Impedance Matrix

The RIM methodology offers a systematic approach for detailed modelling of complex trimmed interfaces within designated regions of interest, adhering to specific design criteria [1–6]. Initially, impedance matrices are computed to accurately represent the frequency-dependent characteristics of the defined trims. These impedance matrices are then condensed onto the physical DOFs associated with each coupling surface. Subsequently, the reduced impedance matrices can be further minimized by projecting them onto the global modes of vibrations of the components to which the trim is attached.

To better understand this concept, consider a simple plate-trim-cavity system. In this scenario, the trim component is connected to a plate and a fluid component by different coupling surfaces, as shown in Fig. 5.1.

The equation of motion of the trim can be described, as:

$$Z_T(\omega) \cdot d_T = h_T \tag{5.1}$$

where $Z_T$ is the frequency-dependent acoustic impedance of the trim, $d_T$ the physical DOFs vector and $h_T$ the excitation vector.

The complete discretized vibroacoustic system can be represented by the structural inner displacement, $(u_T)_i$, the fluid inner acoustic pressure $(p_T)_i$ and respectively the structural displacement and fluid pressure coupling DOFs $(u_T)_{\Gamma_s}$ and $(p_T)_{\Gamma_f}$.

**Fig. 5.1.** Representation of the plate-trim-cavity system, highlighting the coupling surfaces of the trim with the structure and fluid components in red lines.

Therefore, the problem can be expressed as a function of inner and coupling DOFs, such as:

$$(u_T, p_T)^T = \left((u_T)_i, (p_T)_i, (u_T)_{\Gamma_s}, (p_T)_{\Gamma_f}\right)^T = ((d_T)_i, (d_T)_\Gamma)^T \tag{5.2}$$

With the definition of inner and boundary DOFs, the frequency-dependent impedance of the system can be partitioned as:

$$Z_T(\omega) = \begin{bmatrix} (Z_T)_{i\times i}(\omega) & (Z_T)_{i\times\Gamma}(\omega) \\ (Z_T)_{\Gamma\times i}(\omega) & (Z_T)_{\Gamma\times\Gamma}(\omega) \end{bmatrix} \tag{5.3}$$

where $(Z_T)_{i\times i}$ corresponds to the impedance of the inner DOFs, $(Z_T)_{i\times\Gamma}$ and $(Z_T)_{\Gamma\times i}$ are corresponding to the cross-impedance of the inner and coupling DOFs and $(Z_T)_{\Gamma\times\Gamma}$ the impedance of the DOFs at the interface.

To compute the reduced impedance matrix, the Schur complement [1, 2] of $Z_T/(Z_T)_{i\times i}$ is computed which defines:

$$(Z_T^{red})_{\Gamma\times\Gamma} = (Z_T)_{\Gamma\times\Gamma} - (Z_T)_{\Gamma\times i} \cdot (Z_T)_{i\times i}^{-1} \cdot (Z_T)_{i\times\Gamma} \tag{5.4}$$

In this configuration the matrices are condensed to the physical coupling DOFs of the interface $\Gamma$. If the main structural and fluid components are defined in physical coordinates, then the reduced impedance can be directly stored.

However, if the component global modes of vibration are available, i.e. both the main structural and fluid parts are represented in modal coordinates, the modal shapes localized at the coupling surfaces $\phi_{n\times\Gamma}$ can be used to further reduce the matrices in the space of global modes, such that:

$$(\tilde{\boldsymbol{Z}}_T^{red})_{n\times n} = \boldsymbol{\phi}_{n\times\Gamma} \cdot (\boldsymbol{Z}_T^{red})_{\Gamma\times\Gamma} \cdot \boldsymbol{\phi}_{\Gamma\times n}^T \tag{5.5}$$

Note that the evaluation of this RIM in modal coordinates requires the knowledge of the modal shapes on the interface only.

For preparing the assembled solution process, Actran generates a trim database containing the frequency-dependent RIMs for each trimmed interface. This simplifies the model by eliminating the need to discretize the trim components in the main vibroacoustic analysis. It also allows modifications in the components to which they are connected, as long as the coupling surfaces and modal basis remain unchanged. Consequently, it functions as a reusable database that may be integrated into an optimization process.

To consider the trim information at the solution stage, the physical RIMs structural and acoustic contributions are added to the main system accordingly:

$$\begin{bmatrix} Z_s(\omega) + Z_{T_s}^{red}(\omega) & C_{sf} + Z_{T_{sf}}^{red}(\omega) \\ C_{fs} + Z_{T_{fs}}^{red}(\omega) & Z_f(\omega) + Z_{T_f}^{red}(\omega) \end{bmatrix} \cdot \begin{pmatrix} u \\ p \end{pmatrix} = \begin{pmatrix} F_s \\ F_f \end{pmatrix} \tag{5.6}$$

where:

- $Z_s$ and $Z_f$ represent the frequency-dependent structural and fluid impedance of the car structure and acoustic cavity, respectively.
- $C_{sf}$ and $C_{fs}$ denote the fluid-structure coupling terms.
- $Z_{T_{sf}}^{red}$ and $Z_{T_{fs}}^{red}$ the fluid-structure reduced impedance.
- $u$ and $p$ denote the structural displacements and fluid pressures, respectively.
- $F_s$ and $F_f$ represent the external forces applied to the structure and acoustic cavity, respectively.
- $Z_{T_s}^{red}$ refers to the added structural impedance of the trim and $Z_{T_f}^{red}$ denotes the added acoustic impedance of the trim component.

Typically, a modal projection takes place, in which case physical DOFs are turned into modal coordinates and the modal RIMs in Eq. 5.5 are involved.

Since the RIMs may not computed for each solution frequency, for computational performance reasons, they are linearly interpolated on the intermediate frequencies $\omega$ in the range $[\omega_1, \omega_2]$ of the analysis, before the solution step:

$$\boldsymbol{Z_T^{red}(\omega) = Z_T^{red}(\omega_1) + \frac{\omega - \omega_1}{\omega_2 - \omega_1}\left(Z_T^{red}(\omega_2) - Z_T^{red}(\omega_1)\right)} \tag{5.7}$$

### 5.2.2 Compressed Reduced Impedance Matrix (CRIM)

The reduction basis produced for structural or acoustic components adheres to a systematic process [7]. Initially, modal amplitudes are localized on the target coupling surface, yielding a matrix $\Phi_{n\times\Gamma}$, where n represents the number of modes and $\Gamma$ the coupling DOFs. Before projection in the modal space (Eq. 5.5), the matrix undergoes an SVD decomposition, such that:

$$\phi_{n\times\Gamma} = U_{n\times n} \cdot S_{n\times\Gamma} \cdot V_{\Gamma\times\Gamma}^T \tag{5.8}$$

where U are the left singular vectors, S the singular values, and V the right singular vectors.

The selection of the relevant singular vectors is carried out by considering the vectors above a certain threshold ($\sigma$). This threshold can be interpreted as the contributing vectors associated with singular values exceeding a certain percentage relative to the maximum singular value:

$$S \geq (1 - \sigma) \times S_{max} \tag{5.9}$$

where $S_{max}$ is the maximum singular value, $\sigma$ the compression tolerance and S the singular value.

### 5.2.3 Modal-Based Solution Workflow

Because it has its foundation on a robust filtering of the modal information before projection of the RIM in the modal space, the modal-based approach and its compressed RIM variant implicitly assume the availability of the in-vacuo structural modes and rigid-wall cavity modes before RIM computation.

On the other hand, the performance of the modal extraction process, which nowadays is unmatched thanks to multi-level techniques such as ACMS, strongly depends on both the problem physical size, the number of requested eigenmodes and the size of the modal data recovery space. An efficient way to accelerate the process, both in terms of CPU time and of required disk storage, is thus to finely identify the regions where modes should be computed.

Finally, when intensive computing resources are available, the question of the optimal usage of them must be addressed. Some parts of the computation may benefit from distributed memory parallelism (DMP), from shared memory parallelism (SMP) or concurrent disk access, while other may simply not leverage on parallelism.

For all the reasons above, the solution process has been completely revamped into the following workflow:

- Pre-processing of the user input, structural FE model build up, acoustic fluid FE model build up, build up of the FE models for each of the trim components.
- Through mesh mapping processes, identification of the list of grids for which modes should be computed in the structural model and the acoustic model.
- Modal extraction of the in-vacuo structural modes and rigid-wall acoustic modes and data recovery of the modal amplitude exclusively at the previously identified grids. This step is significantly accelerated by SMP and moderate DMP.
- Generation of the RIM database for each trim component. Compression using the CRIM technique. Storage on disk. This step is significantly accelerated by SMP, DMP and concurrent disk storage access.
- Through a restart technique, computation of the modal coordinates and synthesis of the physical response at user-selected data recovery grids. This step is significantly accelerated by SMP and DMP.

### 5.2.4 Trim Boundary Conditions

The flexibility to connect any 3D trim finite element model to the car-body structure and the passenger compartment is of prime importance. Trim components are usually made of different materials, ranging from visco-elastic damping layers, internal air cavities, membranes, thin shells to poroelastic foams of fibrous layers. Their connection to the cavity may involve communicating pores or closed pores and may be characterized by glued or sliding connection to the structural parts. Usually, the user attributes to the coupling surfaces of the trim component a connection type that describes the expected nature of the coupling, e.g. SGLUED, SSLIDING, SOPEN or SIMPERVIOUS in MSC Nastran PEM. This paper presents a new contact/boundary condition that facilitate the trim component installation.

An air-gap connection with the structural part (illustrated in Fig. 5.2), which is equivalent to an infinitely thin air layer between the trim component and the structural part can be defined for cases where a poroelastic foam needs to be connected to a structure, alternatively to glued or sliding connections, which only deal with the skeleton of the foam. In that case, additional DOFs need to be automatically created to model this coupling in the RIM computation process:

- normal displacement $u_n$ DOFs are created and serve as reduction DOFs;
- a first fluid-structure coupling contribution $u$-$p$ is added between the surface $p$ DOFs and the surface $u$ DOFs;
- a second one is added between the $p$ DOFs and the new reduction $u_n$ DOFs.

It should be noted that this connection emulates a locally reacting air layer between the foam and the structure to which it is connected. Whenever 3D effects are expected to happen in this thin air layer, a detailed 3D modelling of it, including spatially varying thickness, should be alternatively selected.

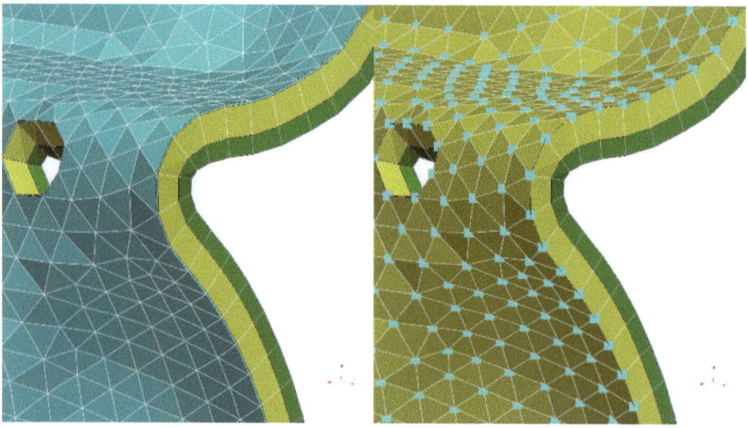

**Fig. 5.2.** Equivalent modelling approaches: thin air layer represented with 3D elements (left) or with the air-gap boundary condition (right).

### 5.2.5 1D Trim Modelling

The effort to generate the 3D FE meshes of trim components, even with the currently available very efficient meshing tools, is not negligible. The question then arises to leverage as much as possible on their availability. One typical usage is to borrow them from fully trimmed models to component-level analyses where optimization work, detailed analysis and/or higher frequency analysis are targeted.

For faster analysis, or when 3D effects are not dominant anymore, e.g. at higher frequencies, a 1D modelling can alternatively be selected. The question then is to convert the 3D mesh to a 1D model with spatially varying thickness and material stackings.

A unique, robust and efficient tool has recently been developed in Actran to perform this operation. It relies on the following steps:

- Based on the coupling surfaces with the structure car-body and the acoustic cavity, identify the local thickness of the trim component (elementwise). Appropriate fallback and threshold strategies are needed to save robustness when dealing with complex trim models such as firewalls, headliners and floor trim components.
- Build, in the normal direction of the structure coupling surface, elementwise, edges with appropriate length, and identify the intersections of those edges with the 3D trim component mesh.
- From the intersections, identify the material stacking and corresponding thicknesses. Here again, fallback rules and only a few min/max user parameters are needed to keep the process robust and efficient.

## 5.3 Numerical Application

### 5.3.1 Industrial Model Description

The finite element model of the Audi Q6 e-tron is composed of three main parts: the structural car body, the cabin fluid, and the main trim parts. The latter includes firewall insulation, floor lining, trunk lining, trunk loading panel, trunk side trims, and seats. The structural model of the Audi Q6 e-tron is shown in Fig. 5.3 on the left and the incorporated trim parts on the right side.

The structural and fluid model comprises over 5,800,000 nodes and 7,100,000 elements, with the degrees of freedom (DOFs) of the main model totaling over 30,000,000. For the added trim parts the number of coupled degrees of freedom adds up to over 125,000.

In the following sections, the performance of a typical noise transfer simulation is used for comparison. In this analysis 126 subcases are evaluated, covering a frequency range of 10–800 Hz with a 1 Hz step. The reduced impedance matrices are evaluated at 42 pre-defined frequencies within this range. In the ACMS module of the analysis, a total of 12.010 eigenmodes (including residual vectors) for the structure and 2188 eigenmodes for the main fluid are extracted.

The system is solved with MSC Nastran version 2024.1. This and later versions of Nastran notably integrate MUMPS solver within the ACMS module to allow perfect repeatability of the modal extraction sequence. Parameter UPFACT = 3.0 is used.

All simulations are executed on a cluster with 32 cores, 2 TB RAM and 9.9 TB disk space.

### 5.3.2 Trimmed-body Results

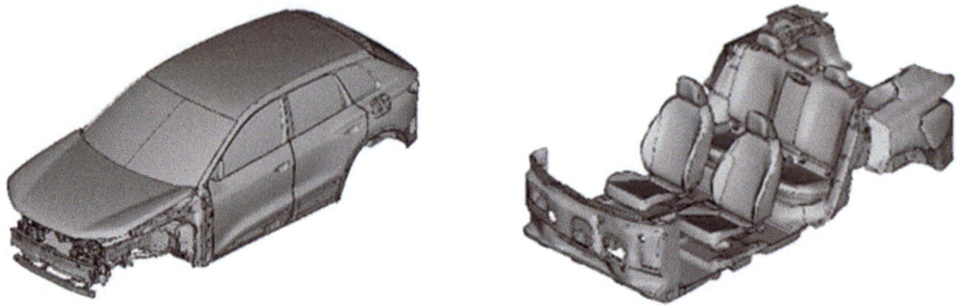

**Fig. 5.3.** Finite element model of the Audi Q6 e-tron, with body structure and cabin fluid on the left, main trim parts on the right

The model described in the above section is run with acoustic treatments effect considered by the RIM method described in Sect. 5.2.1. At first, CRIM and modal-based solution workflow, respectively described in Sects. 5.2.2 and 5.2.3 are not activated. This constitutes the configuration later referred to as the "original" or "baseline" configuration.

The same model is now run with the new modal-based workflow described in Sect. 5.2.3. As shown in Fig. 5.4, noise levels are unchanged compared to the original workflow.

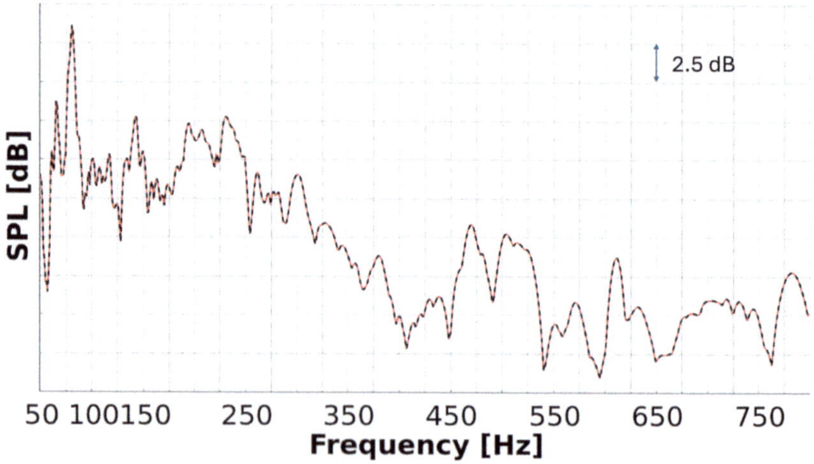


**Fig. 5.4.** Comparison of SPL results between the original (solid red curve) and modal-based workflow (dotted black curve).

The corresponding calculation times are listed in Table 5.1. While the simulation results remain in perfect concordance between the old and new workflows, the computation time is significantly reduced with the implementation of the new modal-based workflow. As evidenced by Table 5.1, the modal-based workflow offers substantial performance enhancements, reducing the computation time by approximately 25%, from an initial 8 hours to 6 hours and 6 minutes.

### 5.3.3 Compressed RIM

The model is now run with RIM compression, as described in Sect. 5.2.2. The impact of compression tolerance $\sigma$ with values of 0.9, 0.95 and 0.99 is evaluated and compared to the reference results without compression, which corresponds to a compression tolerance of 1. All configurations display good agreement in terms of narrow-band SPL, as shown in Fig. 5.5. As expected, noise levels converge to the reference results as the compression rate increases. In the presented case, the maximum SPL difference between baseline and iterations with compression rates of 0.9 and 0.95 amounts to around 2 dB. For 0.99 compression rate, the SPL difference is negligible.

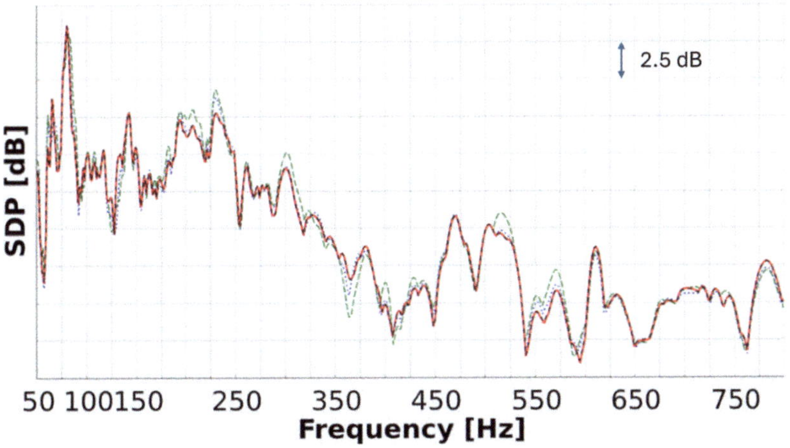


**Fig. 5.5.** Impact of CRIM on SPL results with different compression rates: no compression (solid black curve), 0.9 compression rate (dashed green curve), 0.95 (dotted blue curve) and 0.99 (dashed red curve).

As shown in Table 5.1, using CRIM leads to additional performance improvement. Specifically, RIM compression further reduces the computation time obtained with the configuration described in 3.2, yielding an additional 30-minute reduction. Overall, this leads to a final computation time of 5 hours 15 minutes compared to the original 8 hours, which is a 34% reduction in time.

Besides these additional performance benefits, the CRIM method notably reduces disk storage requirements. Table 5.2 details the disk storage usage for the RIM in the baseline configuration, the modal workflow without compression, and compression rates of 0.99, 0.95, and 0.9, respectively. Significant disk space savings are achieved as the disk storage requirements for compressed RIMs are negligible compared to those without compression, as listed in Table 5.2. Without compromising accuracy, employing a 0.99 compression reduces the disk usage for RIM storage from 650.1 to 1.2 GB, providing a 99.8% improvement over the baseline configuration.

**Table 5.1.** Computation times for all configurations.

| Configuration | Computation time (hours) | Improvement from Baseline configuration |
|---|---|---|
| Baseline | 8.0 | — |
| Modal Workflow | 6.1 | 24% |
| Modal Workflow + 0.90 Compression | 5.4[1] | 33% |
| Modal Workflow + 0.95 Compression | 5.4[1] | 33% |
| Modal Workflow + 0.99 Compression | 5.3[1] | 34% |

**Table 5.2.** Disk space used for RIM storage.

| Configuration | Disk space [GB] | Improvement from Baseline configuration |
|---|---|---|
| Baseline | 650.1 | — |
| Modal Workflow | 383.0 | 41.1% |
| Modal Workflow + 0.90 Compression | 0.6 | 99.9% |
| Modal Workflow + 0.95 Compression | 0.8 | 99.9% |
| Modal Workflow + 0.99 Compression | 1.2 | 99.8% |

[1] The computation time is expected to slightly increase with the compression rate, but this is not observed here. This is attributed to computation time variability between runs.

### 5.3.4 Air-gap Modelling

As described in Sect. 5.2.4, the air-gap boundary condition can be used to account for a thin layer between the structure panel and the acoustic treatment. From a modelling point of view, this can save engineering time as the air layer does not need to be meshed. A comparison between both approaches applied to all trims—explicit meshing of the air layer and air layer boundary condition – is shown in Fig. 5.6. Results show good agreement despite some local discrepancies.

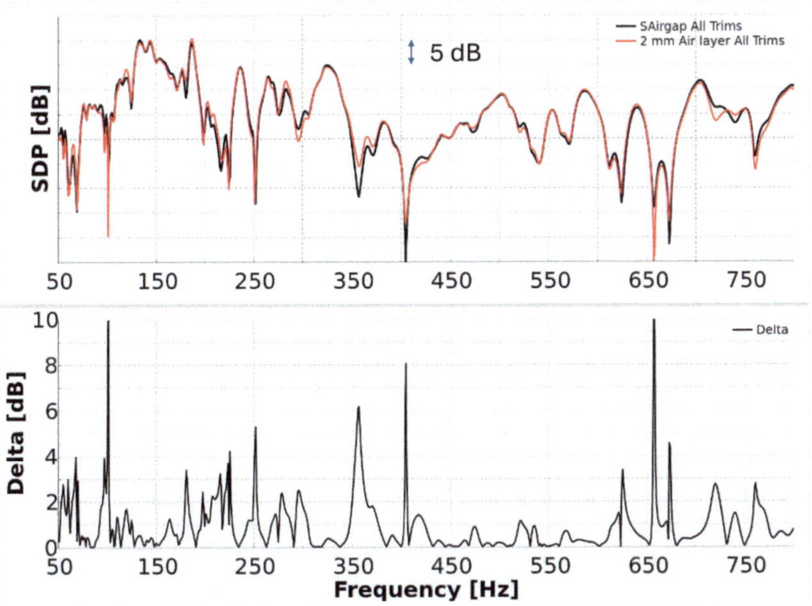


**Fig. 5.6.** Comparison between explicit meshing of the air layer and air-gap coupling condition applied to all trims

As a 2 mm thick air layer was used, on can expect volumetric effects to be the cause of these differences. Iterations on the firewall trim are shown in Fig. 5.7 and display the convergence between both approaches as the thickness of the meshed layer is decreased from 2 to 1 and 0.5 mm.

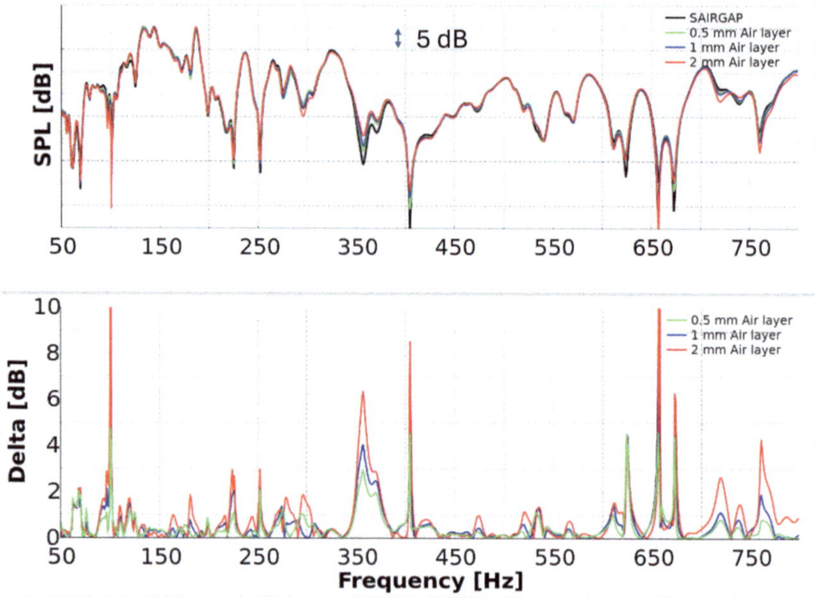


**Fig. 5.7.** Impact of meshed air layer thickness of the firewall trim on SPL results. Comparison with air-gap coupling condition.

### 5.3.5 3D to 1D Trim Conversion

As explained in section 2.5, acoustic treatments can be modelled with a 1D approach that allows both to improve performance and save engineering time in the model preparation phase compared to the classic 3D approach. Here, an innovative tool is presented that allows to convert a 3D trim to its 1D equivalent. From the pre-existing layers directly meshed with volume elements, multiple materials stackings are automatically detected. Each stacking corresponds to a different combination of layers, and to each layer, a spatially dependent thickness field is attached. The newly constructed 1D trim can be straightforwardly substituted in the main analysis to change the modelling approach.

This conversion is performed on the floor trim and displayed in Fig. 5.8. As displayed on the original 3D model, the floor trim includes a layer of foam with highly variable thickness (in yellow) and two thinner layers on top of it (in orange and purple). Inner cavities are included as well (in light blue). The algorithm properly detected the areas covered by different stackings and captures the original vertical layers arrangement. As displayed in the converted model:

- The largest surface in green represents the area covered by the stacking with the foam and the two thin layers.
- The dark blue surface is the stacking including the fluid from the inner cavities, the foam and the two thin layers.
- The smaller magenta areas represent the stacking with the fluid and the two thin layers.

**Fig. 5.8.** Trim conversion from the original 3D model (left) to its 1D equivalent (right). Colors represent the different materials on the 3D model exploded view and the different stackings on the 1D model.

For all layers, the algorithm computes the spatially dependent thickness. This is displayed in Fig. 5.9 for the foam layer included in the green stacking represented in Fig. 5.8.

Going further, one can consider leveraging this newly constructed 1D representation of the floor trim and substitute it in their trimmed body model to increase performance and reach higher frequencies. In the framework of Actran software, this can not only be done in the classic Actran for Trimmed Body module, but also the Virtual SEA module, a FEM-based SEA modelling approach tailored to handle these 1D trims.

**Fig. 5.9.** Foam layer thickness field in the green stacking surface

## 5.4 Conclusions

In response to a global trend for NVH numerical models to show constantly growing sizes and to incorporate more detailed damping treatments, significant improvements have been brought over the last two decades to FE-based modelling techniques. In matter of fully trimmed car body modelling, the RIM technique is now a well established approach.

This paper has reviewed and demonstrated several recent innovations that allow this approach to reach unmatched efficiency, both in terms of accuracy and computational solution time. On the industrial-sized trimmed car-body model of Audi, the computational time drops by far under the nighttime as it reaches 5 to 6 hours.

First, the paper has shown that a proper sequencing of the modal extraction process for the undamped in-vacuo car body and rigid-wall cavity model, coupled to an adequate selection of the modal data recovery regions, allows for an optimal evaluation of the RIM representing the trim components. Second, an additional compression method, the CRIM, has been exposed that further accelerates the solving process by dramatically dropping the disk storage and therefore reducing the I/O-related operations. Result accuracy has been shown to be lightly impacted by the CRIM strategy.

Third, a new boundary condition (air-gap connection to the structure) on trim interfaces has been presented. This allows to greatly simplify the trim model build up and to reduce the size of the finite-element (FE) trim models, as a physical representation of a thin air layer is no longer needed. However, numerical analysis has shown that it should be used with care, as the air cavity that the boundary condition replaces by an infinitely thin air layer assumption may induce 3D effects that significantly impact numerical results.

Finally, a robust 3D to 1D automated trim model conversion has been presented that makes it easy to extend the frequency range of analysis of an existing FE setup. Further work is planned to investigate the usage of this tool in higher frequency range models, such as Virtual SEA setups, as well as in optimization component-level projects.

## References

1. Coyette J.P., Lielens G., Van den Nieuwenhof B., Bertolini, C. et al.: From Body in White to Trimmed Body Models in the Low Frequency Range: A New Modeling Approach. In: SAE Paper 2007-01-2340 (2007).
2. Van den Nieuwenhof B., Lielens G., and Coyette J.P.: An Updated Modal Approach for Dealing with Vibroacoustic Damped Models under Random Excitation. In: Noise-Con 2008 Proceedings, Dearborn, MI, 2008.
3. d'Udekem D., Acher F., Liu W., Shu K.T. et al. : Efficient Analysis of a Fully-Trimmed Car Body using Modal Approaches. In : Noise-Con 2008 Proceedings, Dearborn, MI, 2008.
4. Van den Nieuwenhof B., Lielens G., Coyette J.P., Acher F. et al.: Efficient Analysis of Large Trimmed Configurations Using Modal Approaches. In: ISMA, Leuven, BE, 2008.
5. d'Udekem D., Saitoh M., Van den Nieuwenhof B., and Yamamoto T.: Numerical Prediction of the Exhaust Noise Transmission to the Interior of a Trimmed Vehicle by Using the Finite/Infinite Element Method. In: SAE Paper 2011-01-1710 (2011)
6. Zhou Z., Jacqmot J., and Vo-Thi G.: Evaluation of Trim Absorption to Exterior Dynamic and Acoustic Excitations Using Hybrid Physical-Modal Approach. In: *SAE Int. J. Passeng. Cars—Mech. Syst.* 7, no. 3 (2014): 1205–1211
7. Paiva A., Verhaegen J., Lielens G. and Van den Nieuwenhof B.: Frequency Response Analysis of Fully Trimmed Models Using Compressed Reduced Impedance Matrix Methodology. In; SAE Paper 2024-01-2947 (2024).
8. Hexagon, Actran – Reference Manual, 2025.
9. Hexagon, MSC Nastran – Quick Reference Guide, 2025.

# Coherence and Statistics-Based Optimal Sensor Set Expansion for Active Noise Control Applied to HVAC System in a Vehicle

Daniela Bellinetto[1,2](✉), Alexander Reinbach[2], Thomas Sailer[2], Marc von Elling[3], Sven Höber[3], and Christian Bohn[1]

[1] TU Clausthal, Clausthal-Zellerfeld, Germany
daniela.bellinetto@bmw.de, cboh@tu-clausthal.de
[2] BMW AG, München, Germany
{Alexander.AR.Reinbach,Thomas.Sailer}@bmw.de
[3] recalm GmbH, Hamburg, Germany
{marc.vonelling,sven.hoeber}@recalm.com

**Abstract.** Active Noise Control (ANC) is an increasingly prominent technology applied across diverse domains, from vehicle cabin acoustics to building environments. A critical challenge in deploying effective ANC systems lies in the placement of input sensors, which capture the so-called reference noise. Given the wide range of applications in which ANC is now employed, the necessity for a flexible and effective method for sensor selection, applicable across different domains, is required.

This study introduces a versatile method for the optimal selection of microphones used as input to an ANC algorithm targeting the noise of a Heating, Ventilation, and Air Conditioning (HVAC) system in a vehicle. The proposed method identifies an optimal arrangement of sensors that demonstrate high multiple coherence with the target microphones, expanding the optimal sensors set on the base of statistical indicators, to deliver the best performance over the various HVAC operating conditions. Additionally, the technique allows for the prioritization of specific frequency ranges where ANC performance is to be enhanced. The effectiveness and adaptability of this optimization strategy are experimentally validated by applying the ANC system to HVAC noise, focusing on either the driver's or passenger's ear microphones, or both. The results are compared with those obtained using a conventional sensor selection strategy, based solely on multiple coherence, which is normally applied to identify the best sensor set for a single operating condition.

**Keywords:** Active noise control · HVAC noise · Optimal sensors selection


A. Heintzel (Ed.): Automotive Acoustics Conference 2025, Proceedings, pp. 73–90, 2026. https://doi.org/10.1007/978-3-658-51249-1_6

## 6.1 Introduction and Motivation

Active noise control (ANC) is a rapidly advancing technology within the automotive sector, primarily utilized to mitigate noise generated from the interaction between the road and the tires, as well as engine noise [1, 2]. Its adaptability and ease of implementation render it a highly advantageous solution compared to traditional material insulation methods, as it can diminish noise levels at the listener's position, thereby enhancing cabin comfort without necessitating the addition of extra passive insulation, aside from the sensors required to monitor reference noise. Given the significant potential of this technology, its application is being broadened to various other domains, and numerous strategies aimed at enhancing its effectiveness have been proposed in recent years. These include the development of novel control algorithms [3], the integration of speakers in headrests [4], the incorporation of head-tracking technology [5], and the optimization of sensors placement [6, 7]. The latter issue is recognized as particularly impactful, as it poses a theoretical constraint on the overall performance of ANC.

The issue of sensor positioning is significant across various domains, including structural health monitoring (SHM) [8], modal analysis [9, 10], ANC, and several others [11, 12]. In each of these areas, the fundamental challenge lies in identifying the optimal sensors configuration, both for what concerns their number and positions. Therefore, the process to achieve this involves several steps, including selecting the appropriate type of sensor, identifying potential locations, and formulating an optimization function along with its input parameters [8], all while considering time constraints and budget limitations. The first two aspects of this process are highly case-dependent. Generally, a solid understanding of the physics of the system is essential for determining the most suitable type of sensors and their potential locations. In the context of SHM, sensors may include accelerometers, strain gauges, Bragg's gratings, or temperature and velocity sensors, typically positioned near the structure nodes [8]. In ANC, where noise emanates from a vibrating source, the sensors are usually either accelerometers or microphones, placed at locations where noise propagates through the structure. For instance, in road noise control (RNC), the sensors used to monitor the vibrations produced by the road-tire interaction and transmitted into the cabin, are typically accelerometers mounted on the front and rear axles of the vehicle [1], where vibrations from the road are mechanically conveyed into the cabin. In the specific case examined in this paper, the noise generated by the HVAC system is primarily of aeroacoustics nature, resulting from the interaction of airflow with the system's ducts and vents [13]. The sensors selected for this application are microphones, and the candidate positions for these sensors are distributed around the dashboard, where the airflow is directed towards the passengers.

The second crucial step in the sensor selection process involves the choice of the optimization strategy and its input parameters [8]. The majority of the optimization strategies have been formulated in the field of SHM, where the dimensions and the complexity of the structures makes of the optimal sensors positioning (OSP) a particularly challenging task, due to a significantly high number of initial candidate sensors. Consequently, many of the methodologies employed for OSP, such as the modal assurance criterion (MAC) [14], the information entropy (IE) [15], the Fisher information matrix (FIM) [16] and many others [17–19], originate from this domain. Some of these strategies have also been adapted for applications in ANC [6, 7], where typically smaller systems are analyzed.

In the context of selecting the most suitable optimization strategy, also the cost function plays a fundamental role. In ANC, the aim is to reduce as much as possible the noise that remains at the ears location, therefore a logical approach would be to utilize directly an expression of the residual noise to build the cost function that enables the selection of the best sensors [7]. However, this approach is only feasible when the control strategy is predetermined, which is not universally applicable in the automotive sector, where several stakeholders may participate in the vehicle design process. Another parameter that could be used to obtain the OSP is the multiple coherence function (MCOH) [20] between the input sensors signal and the output sensors noise, recorded inside the cabin. Nevertheless, relying solely on MCOH as a parameter for OSP, would entail the calculation of the MCOH for all the possible combinations without repetition of all the candidate input sensors in the desired number of positions in the optimal set. The computational resources required for this operation are substantial and increase exponentially with the number of candidate sensors. Consequently, more efficient methodologies have been developed. Building on the effective independence method (EFI) [21] and the construction of the FIM, a novel method is introduced in [6]. This method expands the sensors set starting from the one with highest MCOH and then selects one by one the sensors which add the most information, according to the FIM matrix. This approach has been further refined in [7], where the overall noise reduction (OANR), produced by a causally constrained Wiener filter, is utilized to construct the FIM, potentially offering better performances compared to MCOH-based methods, which are derived from unconstrained frequency domain analyses [20]. However, as previously noted, certain prerequisites must be met for the applicability of the FIM-OANR approach: both the control strategy and the secondary path transfer function from the speaker to the listener's position must be known in advance. Since these conditions cannot be always fulfilled, a new strategy for the expansion of the optimal sensor set, starting from the MCOH, is developed.

The need to develop a different strategy also comes from the several operating conditions which exist for systems such as HVAC systems. For specific operating conditions, in which different vents are open on the dashboard, or different settings of the HVAC are selected, the optimal sensor configuration can vary significantly. Consequently, the optimization of the sensor set must be conducted with direct consideration of multiple measurements obtained from several operating conditions. To address this challenge, statistical parameters are introduced alongside the MCOH, in such a way to produce a coherence-statistical value for each candidate sensor in the selection process. To account for possible differences in the sound spectrum generated by the HVAC in different operating conditions, this new MCOH-statistical approach also enables the selection of specific frequency ranges where the MCOH indicator should be maximized. An additional benefit of this methodology is the calculation of an improvement parameter, which quantifies the extent to which the inclusion of new sensors enhances the MCOH indicator. Thus, rather than predefining the number of sensors, the method can terminate the expansion of the sensor set when the contribution of the newly added sensor to the overall MCOH falls below a predetermined threshold.

The remainder of this paper is structured as follows. Section 6.2 provides an explanation of the conventional MCOH method as well as of the newly developed MCOH-statistical method. In Sect. 6.3, a case study demonstrating the application of these methods to HVAC systems is presented, with a comparison of the simulation results obtained with the two approaches. Sect. 6.4 showcases the validation results from an application within a vehicle. Finally, Sect. 6.5 outlines the conclusions drawn from the study and discusses further development.

## 6.2 Methodology

### 6.2.1 Traditional MCOH-Method

This method directly selects sensors based on the highest coherence between their signals and the signals at the listener's position. In scenarios where the control strategy is a feedforward controller, specifically of the FxLMS type [20], the achievable noise reduction (ANR) can be calculated directly from the coherence as [22]

$$ANR = 10log_{10}\left|1 - \gamma_{xy}^2\right|, \tag{6.1}$$

where $\gamma_{xy}^2$ represent the ordinary coherence between the input signal $x$ and the output signal $y$. This formulation derives from the calculations of the optimal weights of the controller for a Wiener filter not causally constrained, developed in the frequency domain [22].

In cases in which the control strategy being used for ANC is unknown, however, coherence becomes the primary indicator, which can tell how much of the output signal is caused by the input signals [23]. Therefore, it can predict whether the selected sensors are adequate for providing sufficient information about the noise captured at the listener's position.

In scenarios where the system comprises only one input sensor and one output sensor, the formulation of ordinary coherence for a single-input single-output (SISO) system,

$$\gamma_{yx}^2(f) = \frac{\left|S_{xy}(f)\right|^2}{S_{xx}(f)S_{yy}(f)} \tag{6.2}$$

is applied, where $S_{xy}$ is the cross-power spectral density (CPSD) between signals $x$ and $y$, and $S_{xx}$ and $S_{yy}$ are the auto-power spectral densities (PSDs) of the input and the output signals respectively. Having, as in this paper, a MIMO system with input sensors that measure signals correlated one with each other, the ordinary coherence between each input and target sensor would represent the portion of the output signal that comes from that one sensor and the other sensors correlated with it [23]. For this reason, the ordinary coherence formulation should be extended to the MIMO case, so that all the contributions of the input set of sensors $X$, can be considered simultaneously. The MCOH can be calculated as [24]

$$\gamma^2_{y_j:X}(f) = \frac{S^{\dagger}_{Xy_j}(f)S^{-1}_{XX}(f)S_{Xy_j}(f)}{S_{yy}(f)} =$$

$$\begin{bmatrix} S^*_{1y_j}(f) \dots S^*_{ky_j}(f) \end{bmatrix} \begin{bmatrix} S_{11} & \cdots & S_{1k} \\ \vdots & \ddots & \vdots \\ S_{k1} & \cdots & S_{kk} \end{bmatrix}^{-1} \begin{bmatrix} S_{1y_j}(f) \\ \vdots \\ S_{ky_j}(f) \end{bmatrix} \frac{1}{S_{y_jy_j}(f)} \tag{6.3}$$

where $\gamma^2_{y_j:X}$ is the MCOH between the output signal $y_j$ and the entire set of the $n$ signals in $X$, $S_{Xy_j}$ is the CPSD between the whole set of input $X$ and the output $y_j$ and $S_{XX}$ is the matrix of the PSDs of the $k$ input signals in the set $X$. It is important to note that these calculations are frequency-dependent, meaning the resulting vectors contain one value for each frequency point between $f_{min}$ and $f_{max}$. Moreover, since the system considered is MIMO, $\gamma^2_{y_j:X}(f)$ obtained from Eq. (6.3) is a value valid only for the $j$-th sensor of the $m$ output sensors present in the system. This means that to obtain a final MCOH measure which is valid for the whole set of $m$ output sensors, an average over the MCOH obtained for each $j$-th output sensor is required [6]

$$\gamma^2_{Y:X}(f) = \frac{\sum_{j=1}^{m} \gamma^2_{y_j:X}}{m}, \tag{6.4}$$

For a standard MCOH method, which relies solely on the overall multiple coherence, the objective is to identify the optimal combination of sensor locations that results in the maximum overall coherence among all possible positions. The number of potential combinations within this optimal set is determined by the desired number of input sensors and the available candidate positions, adhering to the equation for combinations without repetition

$$C_{n,k} = \binom{n}{k} = \frac{n!}{k!(n-k)!}, \tag{6.5}$$

where $k$ is the number of sensors desired in the optimal set, and $n$ is the number of candidate positions.

For instance, in a scenario like the one discussed in Sect. 6.3, where 16 positions are available, and four or six sensors need to be selected, the available combinations without repetitions are 1,820, for four sensors and 8,008 for six sensors. Considering the computational burden of repeating the calculation in Eq. (6.1) for all the available combinations, over the entire frequency range from $f_{min}$ to $f_{max}$, a new strategy is developed to make the selection more efficient, particularly in the presence of several operating conditions.

### 6.2.2 Proposed MCOH-Statistical Method

The method presented in this section is designed to identify the optimal set of sensors when a finite number of known operating conditions exist for the system under analysis. In constructing the final indicator for sensor selection, the method assigns different weights to specific frequency ranges where improved performances of ANC are required. A comprehensive diagram outlining each step of this method is illustrated in Fig. 6.1.

The first step of this methodology consists of the choice of the operating conditions to be considered for the selection. The total number of operating conditions is equal to $L$, and for each operating condition a number $p$ of measurements is performed.

In the second step, the selection of the frequency ranges in which the performance of ANC is to be maximized is carried out. This can be done looking at the $m$-th output spectra. The selection of the ranges is not considered as part of the methodology and is out of the scope of this paper. Therefore, it is sufficient to state that the frequency ranges in which the MCOH indicator needs to be enhanced are designated as selected frequency ranges (SFR), and an unlimited amount of them can be specified: $SFR_q = \left[f_{rq,\mathrm{i}}, \ldots, f_{rq,\mathrm{e}}\right]$, where $f_{rq,\mathrm{i}}$ and $f_{rq,\mathrm{e}}$ represent respectively the initial and final frequencies of the selected range $q$. Once the ranges are selected, the corresponding weights $\psi_q$ must also be determined. Subsequently, based on these weights and the frequency ranges, a weight vector $\Psi$ is constructed over the frequencies, such that the values within this vector represent the weight that the MCOH should assume at every point in frequency. For instance, if two frequency ranges $SFR_1 = [f_{r1,\mathrm{i}}, \ldots, f_{r1,\mathrm{e}}]$ with weight $\psi_1$ and $SFR_2 = [f_{r2,\mathrm{i}}, \ldots, f_{r2,\mathrm{e}}]$ with weight $\psi_2$, are selected, then the weights in vector $\Psi$ will be equal to $\psi_1$ in the positions of $\Psi$ corresponding to the frequency points of $SFR_1$, equal to $\psi_2$ in the positions of $\Psi$ corresponding to the frequency points of $SFR_2$, and equal to one otherwise. Therefore $\Psi$ will look like

$$\Psi = [1, \ldots, 1, \psi_1, \ldots, \psi_1, 1, 1, \ldots, 1, \psi_2, \ldots, \psi_2, 1, \ldots, 1]. \tag{6.6}$$

In the third step, the first sensor is added to the optimal set. A weighted coherence vector is produced by taking the scalar product of the weight vector $\Psi$ with the SISO ordinary coherence between each available input and the outputs, computed as in Eqs. (6.2) and (6.4). This operation is performed for every input sensor available, and for every operating condition. To retrieve a final quantitative indicator of the utility of each candidate position, the values in each weighted coherence vector are summed up. Consequently, for every $k$-th measurement considered of the $N = p \times L$, only one representative coherence-statistical value $\vartheta_{k,g}$ is obtained for each candidate sensor, over the whole frequency range from $f_{\min}$ to $f_{\max}$ as

$$\vartheta_{k,g} = \sum_{i=f_{\min}}^{f_{\max}} \psi_i \gamma_{xy_i}^2 \quad \text{with}\, k = 1, \ldots, N. \tag{6.7}$$

In the fourth step, for each $g$-th sensor, an average of these $\vartheta_{k,g}$-values is calculated over the $N = p \times L$ measurements, to obtain $\overline{\vartheta}_g$ average values, one for each candidate sensor. To ensure that excessive weight is not assigned to a measurement that produces an indicator significantly different from the others, the $\vartheta_{k,g}$ values that, for each sensor, exceed the standard deviation of the sample more than three times, are discarded. The results are $\overline{\vartheta}_g$ average values,

$$\overline{\vartheta}_g = \frac{\sum_{k=1}^{N} \vartheta_{k,g}}{N}, \tag{6.8}$$

where $g$ indicates the sensor to which this indicator corresponds.

The fifth step consists of the calculation of the reliability of the mean value $\overline{\vartheta}_g$ for each candidate sensor in the sample of $N$ measurements. The reliability is verified using the standard error of the mean (SEM) and the coefficient of variation (CV). This operation produces a reliability-parameter $\alpha_g$ between 0 and 1, which is multiplied with the respective $\overline{\vartheta}_g$, to produce the final value $\Theta_g$ used for the selection of the sensor,

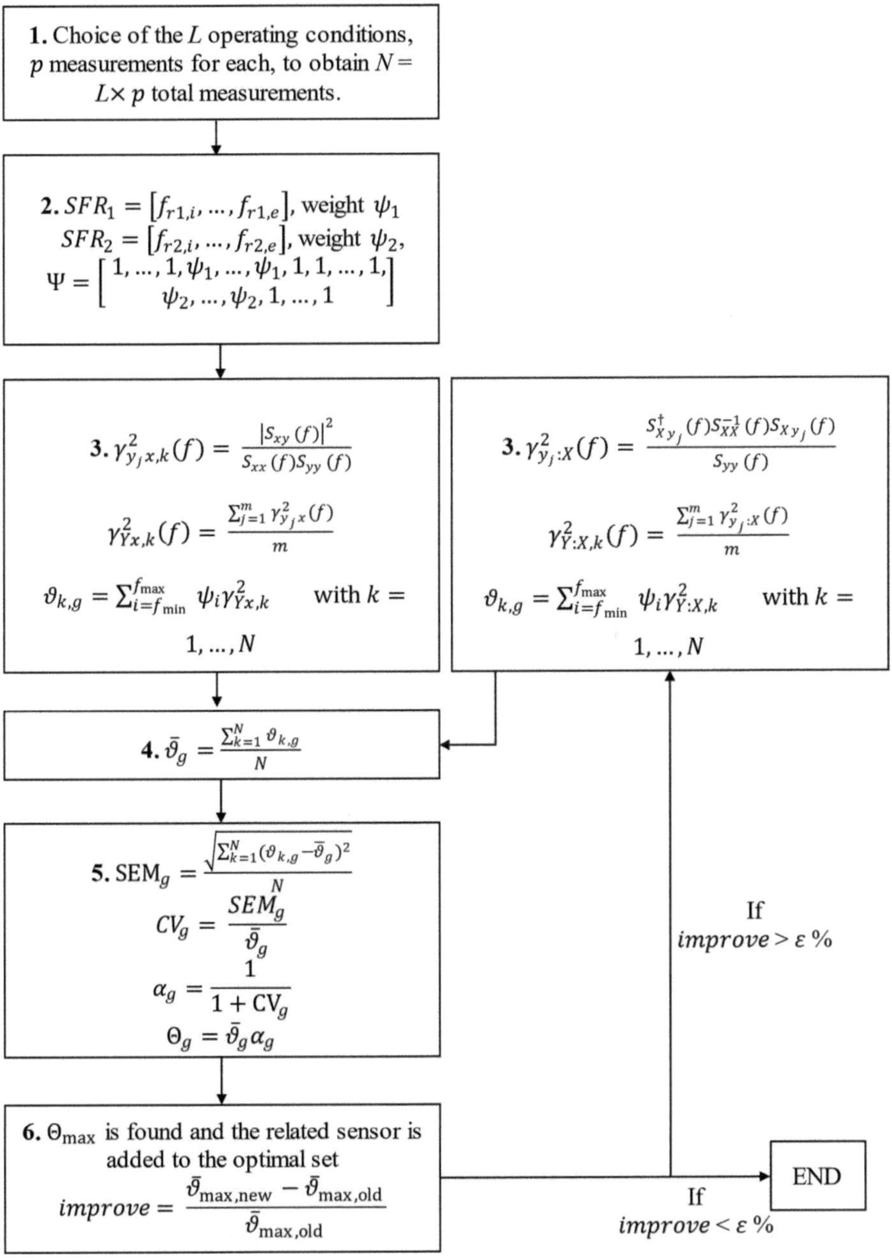


**Fig. 6.1.** Flowchart of the main calculations required to add new sensors to the optimal set.

$$\mathrm{SEM}_g = \frac{\sqrt{\sum_{k=1}^{N} \left(\vartheta_{k,g} - \overline{\vartheta}_g\right)^2}}{N}, \tag{6.9}$$

$$\mathrm{CV}_g = \frac{SEM_g}{\overline{\vartheta}_g}, \tag{6.10}$$

$$\alpha_g = \frac{1}{1 + \mathrm{CV}_g}, \tag{6.11}$$

$$\Theta_g = \overline{\vartheta}_g \alpha_g. \tag{6.12}$$

Finally, the $\Theta_{\max}$ is found among the calculated $\Theta_g$. The sensor associated with this $\Theta_{\max}$ is included in the optimal set and discarded from the initial candidate set. In this step, it is also possible to assess the overall improvement caused by the newly added sensor, as in Eq. (6.13), and terminate the algorithm if this improvement falls below a predetermined threshold. In Fig. 6.1 this threshold is represented by $\varepsilon\%$ and compared with

$$improve = \frac{\overline{\vartheta}_{\max,\,\mathrm{new}} - \overline{\vartheta}_{\max,\mathrm{old}}}{\overline{\vartheta}_{\max,\mathrm{old}}}. \tag{6.13}$$

The steps three to six are repeated every time a new sensor needs to be added to the optimal set, with the only difference that in step three, the coherence calculation is executed using the MCOH formulation given by Eqs. (6.3) and (6.4). This calculation includes all the sensors that have been previously selected, along with a new candidate at a time, to get a new $\Theta_{\max}$ and add a new sensor to the optimal set. In Fig. 6.1 is shown how step three is modified from the second iteration on.

## 6.3 Case Study

### 6.3.1 HVAC System and Operating Conditions

This section provides a description of the HVAC system and outlines the potential operating conditions that will later be utilized to assess the effectiveness of the MCOH-statistical method. In the case being examined, the HVAC unit features four vents directed towards the driver and the passenger, three vents pointing at the feet, and one large vent oriented towards the windshield, as depicted in Fig. 6.2. All measurements are taken in a battery electric vehicle (BEV), ensuring that the HVAC noise is the dominant sound, free from engine noise interference. The measurements are taken in a semi-anechoic chamber, to limit any other possible disturbances.

The sources of noise within such a system are multiple [25], and relevant in different frequency ranges. The fan is the main contributor to the HVAC noise, closely followed by the aerodynamic interaction of the air with the vents and the ducts [13]. To determine optimal locations for input sensors for control purposes, an initial approach could involve positioning them as near as possible to the noise source. For instance, placing sensors directly within the ducts may be considered. With this solution, however, it could be difficult to avoid a high level of noise due to the turbulence of the flow. Alternatively, as implemented in this paper, sensors can be placed around the dashboard, where they are closer to the output microphone at the ears position, and it is still possible to satisfy causality constraints. In total, 16 microphones are distributed in the car, a quantity selected to maximize coverage of the dashboard surface. Figure 6.2 shows the arrangement of the candidate microphones.

**Fig. 6.2.** Arrangement of the 16 candidate microphones.

Regarding the operating conditions of the HVAC system, several factors can influence the noise generated. First, the operating ducts and vents vary based on the user-selected HVAC setup. Second, even when the same setup is selected, the flow through the ducts, and therefore the noise produced, may vary because of a different temperature in the cabin and outside the vehicle. This inherent variability necessitates a comprehensive database of measurements to ensure that all relevant conditions are considered in the placement of the sensors. In this specific study, eight distinct conditions have been evaluated, with four measurements conducted for each condition, resulting in a total database of 32 measurements available for each sensor. A detailed description of the setup utilized is provided in Table 6.1, where each condition is characterized by the settings for air conditioning (AC), recirculation (REC), desired temperature (DT), cabin temperature (CT), and active vents, which are categorized into active vent feet (AVf), active vent body (AVb), and active vent windshield (AVws). The spectra of the output sensor signals corresponding to these eight operating conditions are shown in Fig. 6.3. This graph reveals that the frequency range between 110 Hz and 250 Hz frequently exhibits elevated noise levels compared to the rest of the spectrum. Additionally, an examination of the bandwidth between 320 Hz and 440 Hz indicates that the sound pressure level (SPL) presents a steep peak in this range, in most of the operating conditions. Consequently, these two frequency ranges are identified as priorities for the simulations.

**Table 6.1** Operating conditions of the HVAC considered to build the database.

| Operating condition | AC | REC | DT | CT | AVf | AVb | AVws |
|---|---|---|---|---|---|---|---|
| A | V | V | 16° | 20° | open | open | open |
| B | V | V | 28° | 20° | open | open | open |
| C | V | V | 16° | 20° | open | closed | closed |
| D | V | V | 16° | 20° | closed | open | open |
| E | V | V | 28° | 2° | open | open | open |
| F | | V | 16° | 20° | open | closed | closed |
| G | V | | 28° | 20° | open | closed | open |
| H | V | | 16° | 20° | open | open | open |

### 6.3.2 Simulation

**Simulation—Setup**
In this paragraph, the application of the methodology outlined in Sect. 6.2 to the HVAC case study is explained. To achieve this, all six steps described in Fig. 6.1 are systematically executed, utilizing the specific data derived from the HVAC description provided in Sect. 6.3.1. This means, according to Fig. 6.1, that the number of operating conditions is $L = 8$, the number of measurements per case is $p = 4$, and this results in a total number of measurements $N = 32$. The candidate initial set comprises $n = 16$ microphones, while the optimal set is constrained to a maximum of $k = 6$ microphones.

A minimum improvement threshold of $\varepsilon = 3\%$ is also established, to evaluate the efficacy of adding further microphones to the optimal set. As mentioned in Sect. 6.3.1, two frequency ranges are selected in which the performance of ANC should be enhanced: $SFR_1 = [110, \ldots, 250]$ Hz with weight $\psi_1 = 3$, and $SFR_2 = [320, \ldots, 440]$ Hz with weight $\psi_2 = 2$. For the measurement, a sampling frequency of 12 kHz is utilized.

**Simulation—Results**
The MCOH-statistical method is then executed iteratively to find the best sensor configuration for the several operating conditions outlined in Table 6.1. This optimization process is conducted multiple times, focusing on three different configurations of the output microphones: at first only the driver microphones are used, then only the passenger microphones are selected, and finally, both sets are utilized. It is indeed expected that optimizing the input sensor configuration for a single seat in the front row will result in a different selection of microphones compared to the optimization conducted for both the driver and passenger simultaneously.

Table 6.2 presents a comparison of the input microphones selected by the new methodology alongside those chosen by the traditional MCOH method, only for conditions A, C, D and G to enhance clarity. In the case of the new MCOH-statistical method, the selected sensors remain consistent across all operating conditions, with changes occurring only when the output sensors are modified. Conversely, the classical MCOH method yields a distinct optimal set for each output configuration and operating condition.

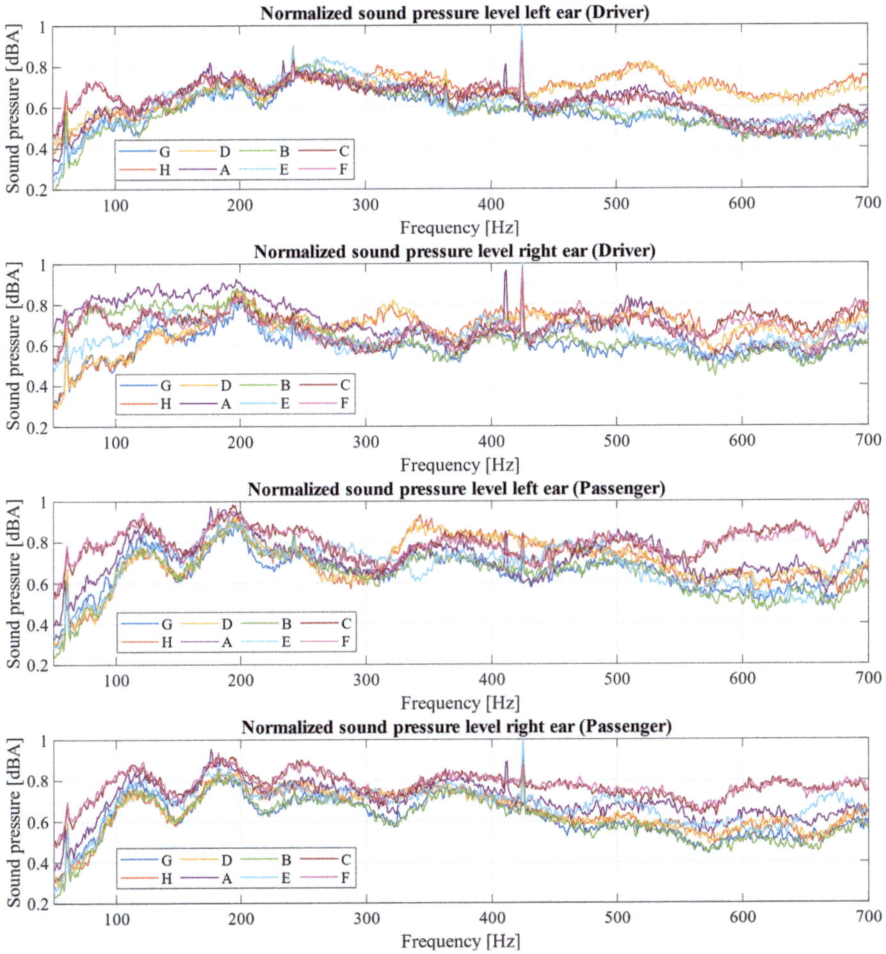


**Fig. 6.3.** Output spectra of the four microphones placed at the left and right ears of the driver and the passenger.

The final comparison focuses on the ANR achievable under conditions A, D, G as illustrated in Fig. 6.4. Specifically, Fig. 6.4 (a) compares the ANR with the driver ears being the output, with the HVAC set on condition A. Figure 6.4 (b) utilizes the passenger ears as output, and the HVAC set on operating condition D, while Fig. 6.4 (c) includes both passenger and driver microphones as outputs, and the selected operating condition for the HVAC is G. Each of these graphs features five lines, where the bold black line represents the ANR achieved with the sensors selected by the new MCOH-statistical method. The remaining, colored, lines depict the ANR for the optimal set determined by the classical MCOH method, based on measurements from either the A, C, D or G operating conditions. The analysis is confined to these four conditions also due to the significant computational demands of the classical MCOH method,

**Table 6.2** Comparison of the sensors selected by the classical MCOH method and by the new MCOH-statistical method.

| Operating condition | Classical MCOH method | | | Proposed MCOH-statistical method | | |
|---|---|---|---|---|---|---|
| | Driver | Passenger | Both | Driver | Passenger | Both |
| A | 1, 3, 4, 5, 9, 13 | 2, 4, 6, 10, 14, 16 | 2, 7, 8, 13, 14, 16 | 1, 3, 5, 7, 9, 10 | 2, 6, 8, 10, 14, 16 | 1, 2, 8, 10, 14, 16 |
| C | 1, 3, 5, 7, 9, 15 | 2, 4, 6, 7, 14, 16 | 1, 4, 13, 14, 15, 16 | | | |
| D | 6, 7, 8, 10, 14, 16 | 2, 7, 8, 10, 13, 14 | 2, 7, 8, 10, 14, 16 | | | |
| G | 1, 6, 7, 8, 9, 13 | 2, 6, 7, 8, 14, 16 | 1, 2, 6, 8, 14, 16 | | | |

associated with processing results for six sensors, three different output microphone positions, and four distinct operating conditions, as outlined in Table 6.2.

The behavior expected from these graphs is that the ANR obtained with the sensors selected by the MCOH-statistical method consistently ranks as the second best, surpassed only by the configuration derived from the classical MCOH method when employing measurements acquired under the corresponding operating condition. Specifically, in Fig. 6.4 (a), the red line, representing the sensor configuration selected based on condition A, is expected to exhibit superior performance compared to the MCOH-statistical subset, which nonetheless outperforms sensor selections derived from measurements in all other operating conditions. A similar conclusion is valid also for Fig. 6.4 (b), where the best subset is found from measurements taken in condition D and Fig. 6.4 (c), where the sensors selected from the classical MCOH method from measurements taken in condition G will yield the best performance. The graphical data, obtained from simulations, generally confirms the expected pattern across most tested conditions, with notable deviations occurring only within certain frequency bands. A more comprehensive synthesis of the average ANR across the selected frequency range is provided in Table 6.3, which confirms that the observed results largely align with the predicted behavior.

Additional considerations can be made about the frequency ranges in which the performance needs to be optimized, which in this application are $SFR_1 = [110, \ldots, 250]$ Hz with weight $\psi_1 = 3$, and $SFR_2 = [320, \ldots, 440]$ Hz with weight $\psi_2 = 2$, as stated in Sect. 6.3.1. In this case, it possible in principle, for the proposed MCOH-statistical method, to outclass also the performance of the sensors set found by the classical MCOH method in that specific condition. A demonstration of this is shown in Table 6.4, where, for the same conditions analyzed before (A—Driver, D—Passenger, G—Driver & Passenger), the ANR obtained by the new MCOH-statistical method and by the classical MCOH method are compared in the two frequency ranges $SFR_1$ and $SFR_2$.

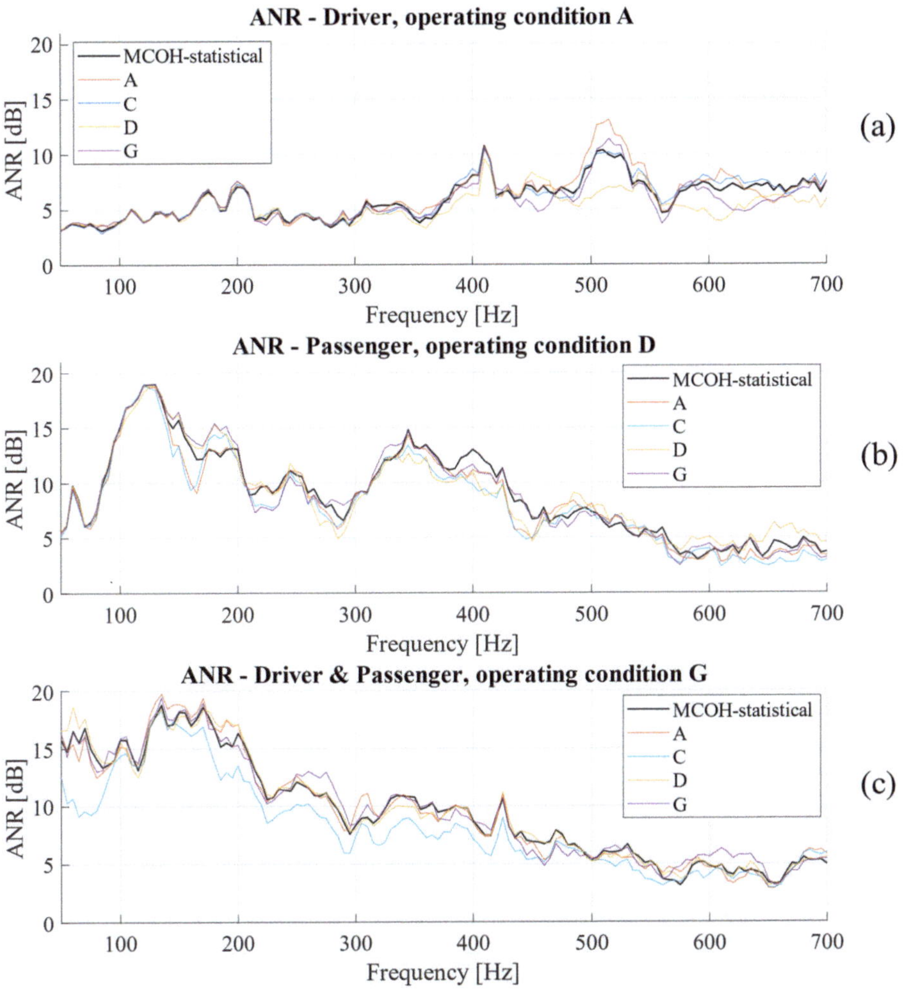


**Fig. 6.4.** ANR obtained for three different operating conditions A, D, G, over the entire frequency range, utilizing as input set the one coming from the MCOH-statistical method, and the ones found in operating conditions A, C, D, G from the classical MCOH method.

## 6.4 ANC Application in the Vehicle

The MCOH-statistical strategy is then employed to find the optimal set of sensors to be utilized as input to an ANC system running in real-time in the car, with the aim to suppress the noise coming from the HVAC.

**Table 6.3** Average ANR obtained with different optimal input sets over the whole frequency ranges selected.

| Operating condition utilized for validation | Operating condition for sensors selection | Average ANR [dB] | Operating condition for sensors selection | Average ANR [dB] |
|---|---|---|---|---|
| | Classical MCOH method | | Proposed MCOH-statistical method | |
| A—Driver | A | 5.76 | All | 5.61 |
| | C | 5.65 | | |
| | D | 5.11 | | |
| | G | 5.39 | | |
| D—Passenger | A | 7.00 | All | 7.41 |
| | C | 6.63 | | |
| | D | 7.56 | | |
| | G | 7.22 | | |
| G—Driver & Passenger | A | 7.65 | All | 7.74 |
| | C | 6.78 | | |
| | D | 7.72 | | |
| | G | 7.88 | | |

**Table 6.4** Average ANR obtained with different optimal input sets over the two frequency ranges where performances need to be optimized.

| Frequency range considered | Operating condition utilized for validation | Operating condition for sensors selection | Average ANR [dB] | Operating condition for sensors selection | Average ANR [dB] |
|---|---|---|---|---|---|
| | | Classical MCOH method | | Proposed MCOH-statistical method | |
| $SFR_1$ | A—Driver | A | 4.91 | All | 4.93 |
| $SFR_1$ | D—Passenger | D | 12.26 | All | 12.26 |
| $SFR_1$ | G—Driver &Passenger | G | 14.23 | All | 14.34 |
| $SFR_2$ | A—Driver | A | 5.94 | All | 5.70 |
| $SFR_2$ | D—Passenger | D | 10.85 | All | 12.37 |
| $SFR_2$ | G—Driver &Passenger | G | 9.53 | All | 9.66 |

Due to technical limitations, a maximum of eight input microphones can be recorded simultaneously, with the output signals exclusively acquired from microphones located at the driver's position. Additionally, the car is moved to a different environment with respect to where the measurements were initially taken for simulations, consequently, a new measurement campaign is conducted to determine an optimal subset of four microphones from the available eight inputs which can be recorded concurrently. Following the notation introduced in Fig. 6.1, the experimental conditions are characterized by $L=4$ operating conditions, $p=2$ number of measurements per case, therefore $N=8$ measurements overall are available. The candidate set of eight microphones from which the optimization begins in this case, is selected starting from the initial sixteen through the MCOH-statistical approach applied to prior data, targeting an optimal configuration of eight microphones out of sixteen. This candidate set comprises microphones numbered 1, 2, 3, 5, 7, 9, 10 and 13. Table 6.5 presents the final sensor sets of four, derived from both the MCOH-statistical and classical MCOH methods, limited to operating conditions A, C, and G due to practical constraints related to time and resource availability.

**Table 6.5** Comparison of the sensors selected by the classical MCOH method and by the new MCOH-statistical method starting from a subset of eight candidate positions.

| Operating condition | Classical MCOH method | Proposed MCOH-statistical method |
|---|---|---|
| A—Driver | 1, 5, 10, 13 | 1, 3, 7, 10 |
| C—Driver | 1, 3, 9, 13 | |
| G—Driver | 1, 3, 9, 10 | |

Using these four selected sensor configurations, ANC is implemented to suppress HVAC noise within the vehicle cabin. The system employs a headrest speaker as anti-noise source.

The noise attenuation performance, averaged over the entire frequency range, is provided in Table 6.6 for operating condition G, showing the results at the left and the right driver's ears. Once again, in most of the cases, the sensors set resulting from the MCOH-statistical method, is the one yielding the second-best performance, surpassed only by the microphones selected in the same operating condition G by the conventional MCOH method.

## 6.5 Conclusions

In this study, a flexible and robust methodology for sensors set expansion in ANC applications is presented. The proposed method is based on an iterative procedure which adds a sensor to the optimal set starting from a MCOH-statistical indicator. This approach is designed to offer an efficient solution for the selection of an optimal combination of input sensors, particularly in scenarios in which a defined number of known operating conditions exist for the noise-emitting system, and the control algorithm utilized for ANC is not predetermined. Another notable advantage of this methodology with respect to a traditional MCOH selection strategy is the reduction of the computational burden. The effectiveness of the proposed strategy is verified through simulations and real-time application on an HVAC system. HVACs systems in vehicles are a suitable application for this methodology, since they can operate in a broad range of operating conditions, tunable by the user.

**Table 6.6** Performance of ANC obtained at the left and right ear in the car, using the sensors set found with the MCOH-statistical strategy or the classical MCOH method.

| Operating condition utilized for validation | Operating condition for sensors selection | Average ANC left ear [dB] | Average ANC right ear [dB] | Operating condition for sensors selection | Average ANC left ear [dB] | Average ANC right ear [dB] |
|---|---|---|---|---|---|---|
| | Classical MCOH method | | | Proposed MCOH-statistical method | | |
| G—Driver | A | 0.49 | 0.99 | All | 0.61 | |
| | C | 0.82 | 0.91 | | | 1.14 |
| | G | 0.91 | 1.18 | | | |

Simulation results demonstrate that, for the target system considered in this study, the proposed methodology can yield an average improvement of up to 1 dB in ANC performance over a conventional MCOH-based sensors selection strategy, across the whole frequency range considered. Notably, enhanced performance is observed within frequency bands where higher noise attenuation is specifically required, with additional ANC gains reaching up to 1.5 dB. These improvements are evident even when compared to sensor selections optimized directly for the same operating conditions used in the validation phase.

These results are partly reproduced also in the real-time application where, because of technical limitations, only a partial validation of the strategy is performed, with four sensors selected out of eight. This makes the advantage of the proposed methodology less evident, since the sensors selected are more similar one to each other with respect to the combinations of six sensors selected from an initial candidate set of sixteen. A potential improvement of this work could involve a more in-depth investigation into the real-time application, with a validation campaign extended to a wider range of operating conditions, to assess the robustness of the proposed strategy.

In general, the proposed MCOH-statistical method presents a viable alternative to traditional MCOH-based approaches, offering significant reductions in both computational complexity and elaboration time, as well as flexibility for selecting sensors in systems where multiple known operating conditions exist. A prospective enhancement to this framework involves replacing the MCOH criterion with an estimator of noise reduction tailored to the causally constrained filter structure used in the ANC algorithm employed. Such an extension could further align the sensor selection process with actual ANC performance objectives.

## References

1. Sutton, T.J., Elliott, S.J., McDonald, A.M., Saunders, T.J.: Active control of road noise inside vehicles. Noise Control Engineering Journal, 42 (4), (1994).
2. 2. Lee, S.K., Lee, S., Back, J., Shin, T.: A new method for active cancellation of engine order noise in a passenger car. Applied Sciences 8 (8), 1394 (2018).
3. Redel-Macías, M.D., Cubero-Atienza, A.J., Salas-Morera, L., Arauzo-Azofra, A., García-Hernandez, L.: Review of algorithms for active noise control. In: 13th International conference on project engineering, Bajadoz, 1553, (2009).

4. Chang, C.Y., Chuang, C.T., Kuo, S.M., Lin, C.H.: Multi-functional active noise control system on headrest of airplane seat. Mechanical Systems and Signal Processing, 167 (A), (2022).
5. 4. Elliot S.J., Jung, W., Cheer, J.: Head tracking extends local active control of broadband sound to higher frequencies. Scientific Reports, 8, 5403 (2018).
6. Park, Y.S., Cho, M.H., Oh, C.S., Kang, Y.J.: Coherence-based sensor set expansion for optimal sensor placement in active road noise control. Mechanical Systems and Signal Processing, 169, (2022).
7. Cheng, C., Liu, Z., Li, X., Lu, C., Chen, W.: An optimal sensor layout method based on noise reduction estimation for active road noise control. Mechanical Systems and Signal Processing, 220, (2024).
8. 7. Wang, Y., Chen, Y., Yao, Y., Ou, J.:Advancements in Optimal Sensor Placement for Enhanced Structural Health Monitoring: Current Insights and Future Prospects. Buildings, 13, 3129, (2023).
9. Heylen, W., Lammens, S., Sas, P. Modal Analysis Theory and Testing; Katholieke Universiteit Leuven: Leuven, Belgium, 200 (1997).
10. Larson, C.B., Zimmerman, D.C., Marek, E.L.: A comparison of modal test planning techniques: Excitation and sensor placement using the NASA 8-bay truss. In Proceedings of the 12th International Modal Analysis Conference, Honolulu, 2251 (205), (1994).
11. 10. Hintermüller, M., Rautenberg, C.N., Mohammadi, M., Kanitsar, M.: Optimal Sensor Placement: A Robust Approach. SIAM Journal on Control and Optimization 55, 3609, (2017).
12. Meyer zu Westerhausen, S., Kyriazis, A., Hühne, C., Lachmayer, R.: Design methodology for optimal sensor placement for cure monitoring and load detection of sensor-integrated, gentelligent composite parts. Proceedings of the Design Society, 4, 673, (2024).
13. 12. Rašuo, B., Dinulović, M., Trninić, M., Milošević, N., Ćurčić, N.: A study of aerodynamic noise in air duct systems with turning vanes. FME Transactions, 49 (2), 308, (2021).
14. 13. Lee, E.T., Eun, H.C.: Optimal Sensor Placement in Reduced-Order Models Using Modal Constraint Conditions. Sensors, 22(2), 589, (2022).
15. Ye, S.Q., and Y.Q. Ni.: Information Entropy Based Algorithm of Sensor Placement Optimization for Structural Damage Detection. Smart Structures and Systems 10(4_5), 443, (2012).
16. Lee, E.T., Eun, H.C.: Optimal sensor placements using modified Fisher information matrix and effective information algorithm. International Journal of Distributed Sensor Networks, 17 (6), (2021).
17. 16. Kim, J.T., Ryu, Y.S., Cho, H.M., Stubbs, N.: Damage identification in beam-type structures: Frequency-based method vs modeshape-based method. Engineering Structures 25 (1), 57, (2003).
18. 17. Gui, C., Lei, J., Duan, Z., Zhang, X.: Optimal sensor placement by using MSSP/DM-MAC method. Chinese Journal of Applied Mechanics, 36 (6),1499, (2019).
19. 18. Lin, T.Y., Tao, J., Huang, H.H.: A Multiobjective Perspective to Optimal Sensor Placement by Using a Decomposition-Based Evolutionary Algorithm in Structural Health Monitoring. Applied Sciences, 10, 7710, (2020).
20. 19. Nelson, P.A., Sutton, T.J., Elliott, S.J.: Performance limits for the active control of random sound fields from multiple primary sources. Proceedings of the Institute of Acoustics, 12 (1), 677, (1990).
21. 20. Kammer, D.C.: Sensor set expansion for modal vibration testing. Mechanical Systems and Signal Processing 19 (4), 700 (2005).
22. Sutton, T.J., Elliott, S.J., McDonald, A.M., Saunders, T.J.: Active control of road noise inside vehicles. Noise control engineering, 42 (4), (1994).

23. Howes, N. H., Allen, M. S., Farina, D., Charles, S. K.: A generalized coherence framework for quantifying input contributions in multi-input systems with correlated or uncorrelated inputs. Biomedical Signal Processing and Control, 104, (2025).
24. Rabiner, L.R., Gold, B. Theory and Application of Digital Signal Processing. Englewood Cliffs, NJ, Prentice-Hall (1975).
25. Silva, Y.A.R., He, J., Song, J., Kniaz, O.: Application of local active control for HVAC system noise in an electric vehicle. In: Proceedings of the 49th International congress and exposition on noise control engineering (INTERNOISE 2020).

# Comparison of Different Noise and Vibration Reduction Measures on a Cover for Power Electronics

Sebastian Rieß[1(✉)], Nikolai Kleinfeller[1], Armin Weber[2], José Cóndor López[3], Heiko Atzrodt[1], and Karsten Finger[3]

[1] Fraunhofer Institute for Structural Durability and System Reliability LBF, Darmstadt, Germany
{sebastian.riess,nikolai.kleinfeller, heiko.atzrodt}@lbf.fraunhofer.de

[2] Institute for Automotive Engineering (Ika), RWTH Aachen University, Aachen, Germany
armin.weber@rwth-aachen.de

[3] Mercedes-Benz AG, Sindelfingen, Germany
{jose_gabriel.condor_lopez, karsten.finger}@mercedes-benz.com

**Abstract.** Electric vehicles feature specific noise, vibration and harshness (NVH) challenges due to the lack of the acoustic masking of a combustion engine. The emitted airborne sound of the electric powertrain is one of the main challenges the industry faces right now. In particular, the cover of the power electronics compartment is often acoustically suspicious due to its thin walled, broad shape and its close installation to the electric engine, which is the main excitation source. For the NVH design of such covers, standard measures like alubutyl and bitumen reach their limit, motivating the application of novel techniques for noise and vibration reduction such as vibroacoustic metamaterials (VAMM) and acoustic black holes (ABH). The presented research quantifies the potentials of such novel methods by comparing them to common state-of-the-art measures. A reference cover is equipped with two different add-on VAMM and modified with a structurally integrated VAMM and an ABH. The structural dynamic behavior of the cover variants is evaluated experimentally in the lab and compared to the ones of a bare cover and a cover equipped with an alubutyl layer. Within the so-called stop band of the VAMM variants a large vibration reduction of more than 20 dB in comparison to a cover with alubutyl is achieved. The cover with integrated ABH leads to extensive damping while being lighter as the cover with an alubutyl layer. For a vehicle test under driving conditions, an add-on VAMM is selected and compared to a state-of-the-art cover regarding the cover's vibrations and sound pressure level in the inside of the passenger compartment at the driver's ears. Here, a reduction of tonal components of up to 11 dB is achieved for the sound pressure level within the passenger compartment.


A. Heintzel (Ed.): Automotive Acoustics Conference 2025, Proceedings, pp. 91–108, 2026. https://doi.org/10.1007/978-3-658-51249-1_7

**Keywords:** NVH · Cover for power-electronic · Vibroacoustic metamaterial · Acoustic black hole

## 7.1 Introduction

The transition in the automotive industry from internal combustion engines (ICs) to electric engines causes major challenges in terms of noise, vibration, and harshness (NVH). The excitation spectrum of an electric engine differs considerably in frequency and amplitude from the one of an IC. In electric vehicles (EVs), for example, there is no masking effect of noises arising from auxiliary units, which are then more audible to the passengers. Due to the wide range of the rotational speed of an electric engine, sounds especially in the higher frequency range become more audible and can disturb the passengers as howling noises.

Another significant change in the automotive industry is the increased emphasis on lightweight materials and structures. This is necessary to reduce $CO_2$ emissions while driving and during the production. While this trend is advantageous for environmental protection, it concurrently poses challenges regarding driving noise. Lightweight structures are more prone to vibrations and therefore are more likely to emit disturbing noise. In the past these problems are often tackled by heavy layers or heavy damping materials, which, today, are not desired anymore, since they counteract the lightweight efforts. Additionally, they lead to other disadvantages in the recycling process of vehicles, since the additional NVH-materials need to be removed from metallic parts of the body or the drive train.

These challenges lead to the demand in the industry for novel approaches for noise and vibration reduction in vehicles. The approaches need to be more performant, lightweight and ideally composed of just one material. Vibroacoustic metamaterials (VAMMs) and acoustic black holes (ABH) are novel technologies that hold considerable promise in addressing the stated problems. Since both approaches are not widely used in the industry yet, this article aims at demonstrating their potential based on the example of a cover for power electronics in an electric drivetrain. Different approaches of VAMMs and an ABH are compared to a state-of-the-art measure by means of an alubutyl damping layer. At the end, the results of a vehicle test with a cover equipped with an add-on VAMM are presented. The work extends the authors' prior works [1–3] by the comparison of different covers with add-on VAMM to a cover with structurally integrated VAMM and a cover with integrated ABH.

### 7.1.1 Vibroacoustic Metamaterials

Vibroacoustic metamaterials are a subgroup of metamaterials, which are rationally designed artificial materials that feature effective properties which cannot be found in nature [4]. Metamaterials exist in different physical domains, such as optics, electromagnetism, mechanics, and acoustics [4]. The concept of metamaterials roots back to a theoretical work by Veselage in 1968 [5] on metamaterials with negative electric permittivity and magnetic permeability. The special properties found by Veselage, later were confirmed experimentally by Pendry et al. in the late nineties [6, 7].

VAMM with local resonances is a promising approach for vibration reduction within the field of metamaterials. They feature a negative effective mass in a designable frequency range. Within this frequency range, referred to as stop band, no free propagation of structure-borne sound waves is possible. Consequently, a VAMM therefore acts as a stop band filter for the propagation of structure-borne sound waves. A VAMM is composed of a host structure onto which mechanical resonators are attached. A schematical representation can be seen in Fig. 7.1a. The behavior of such a VAMM is shown in Fig. 7.1c for a 1 mm thick steel plate in terms of the accelerance frequency response function (FRF) with and without attached resonators (Fig. 7.1b). The VAMM-plate is excited on the left side with a force $F$ (•), and the out of plane vibrations are measured in the red (•) points on the right side. The plate with VAMM exhibits a frequency range (approx. 300 Hz–500 Hz) with a large reduction in vibration amplitudes, the so-called stop band, as indicated in grey (■).

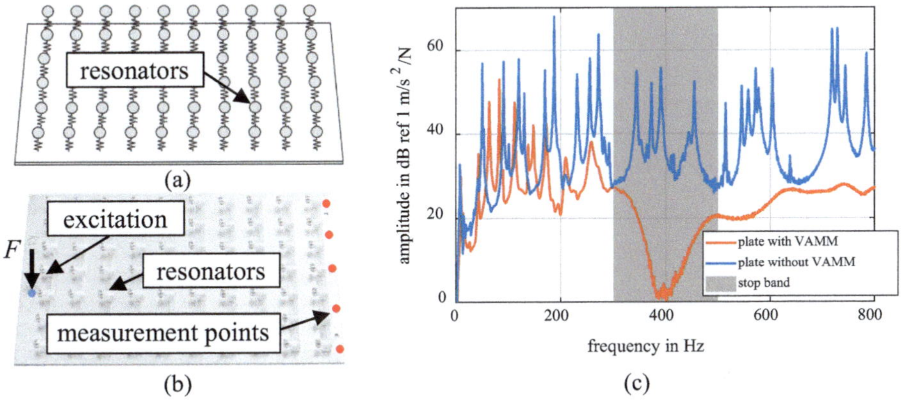


**Fig. 7.1.** Concept of a VAMM with plate-like host structure (a). Steel plate with VAMM (b). Comparison of FRFs with and without VAMM (c). *Source* Compare [8], ©Fraunhofer LBF, Raapke.

The potential of VAMMs to reduce vibrations and structure-borne sound waves on automotive structures has been demonstrated in different scientific works. The application of VAMMs to body parts is shown for doors in [9–11], for the roof in [12, 13], for a wheelhouse and strut dome in [14, 15] and for a firewall in [16, 17]. The application of VAMMs to rubber pars is demonstrated for tires [18] and rubber bushings [19]. In the works [9, 20–22] by the authors of this study, different configurations of a steel cover for power electronics of an EV with integrated VAMMs and common measures are compared. It is demonstrated that with VAMMs, larger vibration reductions can be achieved than with common measures without enlarging the component weight. A structurally integrated VAMM leads to a reduction of 15 dB [20] and 27 dB [21] in comparison to a cover with a bitumen or an alubutyl layer, respectively.

While VAMMs prove to be well suited for vibration reduction in lightweight structures exceeding common measures, the production of VAMMs is seen as a large obstacle. Due to this some of the above-mentioned works highlighted the aspect of production. Production of VAMMs is demonstrated by thermoforming [12], punching and bending [9–11], waterjet cutting [9, 20–22] and injection molding [23]. In the meantime, the Mercedes-Benz Group has proved the large-scale production of VAMMs by adding a VAMM to a fire wall of a production vehicle [17].

### 7.1.2 Acoustic Black Holes

While VAMMs lead to a reflection of waves and thereby act as a stop band filter for vibrations and structure-borne sound waves, ABHs [25–30] represent an efficient way to damp vibrations. The theory of ABHs originates from a work by MIRONOV in 1988 on retarding structures [26]. Here, he investigates a beam with a wedge of decreasing thickness $h$ following a power-law function [26]. The idea underlying an ABH is to use a phenomenon occurring in black holes in space to reduce the vibrations in structures. Black holes are characterized by a so-called event horizon [28]. This means that photons, for example, can enter the black hole but cannot escape from it once having passed the event horizon [24]. Bending waves in an elastic medium behave in a similar way like photons, so it is possible to transfer the principle of black holes of capturing photons to capture vibrations in mechanical structures [25–30].

The propagation speed $c_\mathrm{B}$ of a bending wave

$$c_\mathrm{B}(x) = \left[\frac{E \cdot \omega^2}{12\rho \cdot (1 - \nu^2)}\right]^{1/4} \cdot \sqrt{h(x)} \tag{7.1}$$

depends, among other parameters like YOUNG's modulus $E$, the POISSON's ratio $\upsilon$ and the mass density $\rho$, on the thickness $h$. If the thickness $h(x)$ is asymptotically reduced to "zero", the wave is trapped, and no reflection of the wave occurs at the boundaries [25, 26]. The profile for the reduction of the thickness $h(x)$ can follow a power law. Along this profile, the propagation speed of a bending wave is gradually reduced in a way that ultimately leads to infinite times of propagation or respectively a trapping of the bending wave. The effect depends on the wavelength $\lambda_\mathrm{B}$ and is only effective above a cut-on-frequency $f_{\text{cut-on}}$ [27]

$$f_{\text{cut-on}} = \frac{h}{2\pi \cdot L_{\mathrm{ABH}}^2} \cdot \left[\frac{E \cdot (40 - 24\upsilon)}{12\rho \cdot \left(1 - \upsilon^2\right)}\right]^{1/2}. \tag{7.2}$$

Here, $h$ is the thickness, $L_{\mathrm{ABH}}$ a theoretical length of the ABH and the material properties are defined by the YOUNG's modulus $E$, the mass density $\rho$, and the POISSON's ratio $\upsilon$. Since in a real-world application a non-zero thickness h is required, damping material is used in the region of largely reduced wave velocity, leading to a very efficient local absorption of the bending wave and suppression of reflection. Consequently, the wave is prevented from running out of the ABH again and can be dissipated in a narrow region of a structure with only a small amount of damping material, which typically would be spread over the whole structure [25–30]. Figure 7.2a and 2b show an example of a beam with such a thickness variation following a power law with a damping layer attached to the tip of the wall-thickness reduction. In the back of Fig. 7.2b a beam without this wall-thickness reduction is shown for comparison. Both beams are excited on the left side with a shaker under free-free boundary conditions. The resulting vibrations were measured in 18 points in the portion of the beam without the thickness reduction.

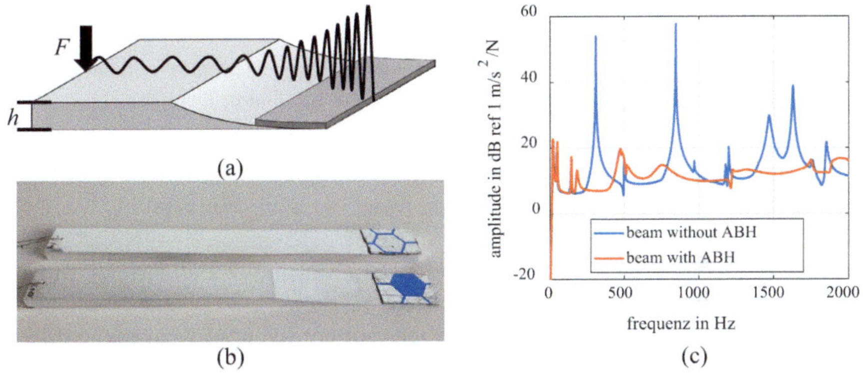


**Fig. 7.2.** Theory of an ABH for a beam-like structure (a). Hardware model of a beam structure with (front) an ABH and without (back) an ABH (b). Comparison of FRFs of beam structures with and without ABH (c).

Figure 7.2c shows the resulting FRFs in terms of accelerance of the two beams. The ABH-beam with a reduction of the wall-thickness leads to a decrease in the averaged vibration amplitudes of 30 dB, proofing the damping effect of the ABH. Even though, the concept of ABH gained lots of interest in engineering there is only a limited amount of scientific works on ABH in the context of automotive application like for an oil pan [28], a motor cover [29] or a car undercarriage [30] is investigated.

## 7.2 Cover Under Investigation

The cover under investigation in this study is depicted in Fig. 7.3. The cover seals a power electronic compartment which is mounted on the electric motor of an EV (Fig. 7.3a). The cover is made from casted aluminum and is reinforced by seams on its back (not displayed in Fig. 7.3). In between the stiffening seams, the cover has a thickness of 3 mm. Measurements at a test vehicle have shown, that the cover is acoustically conspicuous in the frequency range of 1000 Hz to 1500 Hz.

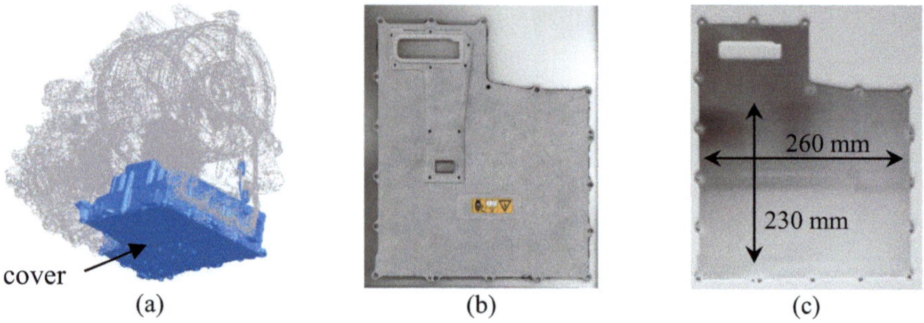


**Fig. 7.3.** Cover under investigation attached to the power electronics compartment of an electric motor (a). Original cover (b) and simplified cover (c). *Source* Compare [1]

As only a limited amount of reference covers made from casted aluminum are available, simplified reference covers made from 3 mm aluminum sheets were used in this study. A comparison of the structural dynamic behavior of these covers with free-free boundary conditions shows a good correlation in the overall structural dynamic behavior [1]. The resonance frequencies of the reference cover made from aluminum sheets are lower by approx. 130 Hz in the relevant conspicuous frequency range compared to the casted aluminum cover. This appears due to the lack of stiffening seams of the simplified cover in contrast to the casted aluminum cover.

## 7.3 Cover with Add-On VAMM

In previous works by the authors [1–3] two cover configurations with an add-on VAMM have been developed and investigated. One VAMM was obtained from sheet metal and the other from a thermoformed plastic film. In contrast to a structural integrated VAMM design, that is presented in section 4 of this study, the add-on VAMM are not integrated into the cover structure but are applied to the cover as an additional component in a similar way as an alubutyl layer. Both add-on concepts are shortly described in the following sections. For further information on the design, the reader is referred to [1].

### 7.3.1 Add-On VAMM from Sheet Metal (VAMM-SM)

The cover with applied VAMM obtained from sheet (VAMM-SM) is shown in Fig. 7.4. The resonators are cut into a steel sheet by a waterjet cutting machine. To give the resonators freedom to oscillate, the steel sheet with integrated resonators is supported by a plastic frame in the regions where no resonator is located. The frame is attached to the resonator sheet and the cover by a double-sided adhesive tape (TESAFIX 4965).

The resonance frequency of the resonators can be adjusted via the area of the mass element and the width of its connection to the frame. The resonators are tuned by numerical simulations to have a stop band in the frequency range of 1230 Hz to 1420 Hz [1]. The obtained add-on VAMM adds a weight of 498 g to the cover.

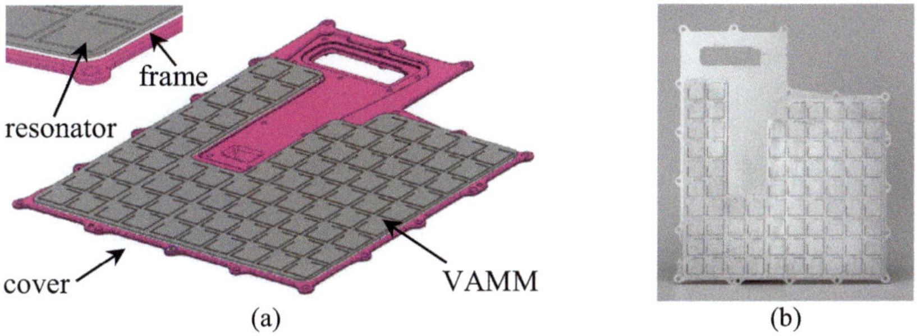


**Fig. 7.4.** Close-up view of the cover with attached add-on VAMM made from sheet metal (a). Cover with attached add-on VAMM (b). *Source* Compare [1], ©Fraunhofer LBF, Raapke.

### 7.3.2 Add-On VAMM from Plastic Film (VAMM-PF)

The second add-on VAMM is made from a thermoformed plastic film (VAMM-PF, Fig. 7.5). This design is elaborated to obtain a lighter add-on VAMM compared to the aforementioned VAMM-SM. The plastic film is designed to form an array of cavities, that can be used with inserted masses as membrane resonators. The steel masses are glued with a double-sided adhesive tape (TESAFIX 4965) into the cavities. The volume of the mass elements and the surface of the adhesive tape can be used to tune the resonators. The add-on VAMM made from plastic film weights 242 g.

**Fig. 7.5.** Resonators and VAMM made from plastic film (a). Cover with attached add-on VAMM made from plastic film (b). *Source* Compare [1], ©Fraunhofer LBF, Raapke.

## 7.4 Design of Cover with Structurally Integrated VAMM

Both above described VAMMs add weight to the cover, which is not desired. Therefor a structurally integrated VAMM is developed, where the resonators are cut into the cover. A similar concept already was presented by the authors in [9, 20–22] on a 2 mm steel cover. The resonators are cut into the cover via a water jet cutting machine. The resulting slits are sealed in one configuration with silicone and in another configuration with a plastic film. Besides the sealing functionality, damping is added to the VAMM leading to an amplitude reduction in the frequency range above the stop band.

For the cover under investigation in this study, a similar approach is evaluated to compare it to the add-on VAMM variants. The chosen unit cell in its first eigenfrequency is shown in Fig. 7.6a. For the comparison with the add-on VAMM, the resonator is tuned to a frequency of 1250 Hz. The stop band is predicted by calculating the dispersion relation by numerical simulations. For this purpose, the Bloch's theorem [31] is applied to the unit cell. Within the dispersion curve, a stop band can be identified as a frequency range in which no free wave propagation is possible. Figure 7.6b shows the calculated dispersion curve of the unit cell, indicating a stop band for transversal waves in the frequency range of 1250 Hz to 1930 Hz. The curves intersecting the stop band are attributed to shear (continuous line) and longitudinal waves (dotted line), which are in-plane waves. These waves can be neglected, since they are not the predominant source of the sound radiation of the cover. For tuning the resonator, the parameters (width and length) of the mass and stiffness element have been varied.

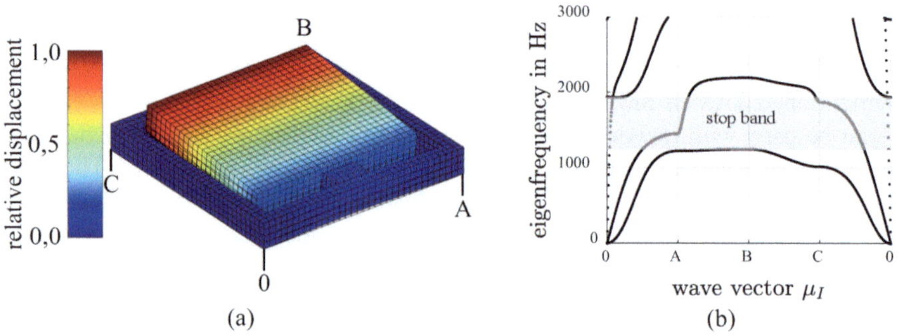


**Fig. 7.6.** Resonator of the structurally integrated VAMM in its first eigenmode (a) and corresponding dispersion curve (b) with stop band indicated in grey.

The resulting cover with integrated resonators is shown in Fig. 7.7. The slits are cut into the cover by a water jet cutting machine. To seal the cover and increase the damping, the slits are filled with silicone (PCI Silcoferm® S). As found in [8] for a similar resonator and stop band frequency range, the silicon used does not increase the stiffness of the resonator considerably. For this reason, it was neglected in the numerical design. The obtained cover weights 541 g and therefor is 10% lighter than the reference cover.

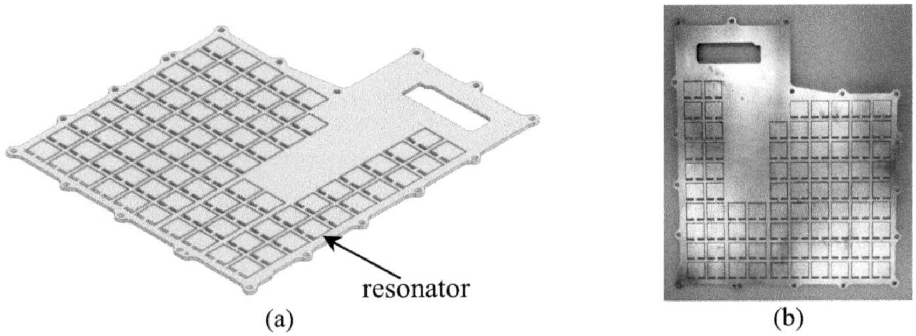


**Fig. 7.7.** Cover with structurally integrated VAMM (a) and hardware demonstrator (b).

## 7.5 Design of Cover with ABH

As described in the introduction, ABH are a promising approach for broad band damping and can be structurally integrated into components. To quantify the vibration reduction achievable with an ABH, this approach is applied to the reference cover.

First, the cut-on frequency $f_{\text{cut-on}}$ of an ABH is determined using Eq. (7.2). The cut-on frequency is the lowest frequency above which an ABH leads to a damping effect [27]. Here, thickness $h = 3$ mm, a YOUNG's modulus $E = 70$ GPa, mass density $\rho = 2700\ \text{kg/m}^3$ and a POISSON's ratio $\nu = 0{,}34$ are chosen for the aluminum cover. Figure 7.8 shows the cut-on frequency $f_{\text{cut-on}}$ over the theoretical length of the ABH $L_{\text{ABH}}$. The cut-on frequency $f_{\text{cut-on}}$ decreases with increasing length $L_{\text{ABH}}$. Since the cover is acoustically conspicuous in the frequency range in between 1000 Hz and 1500 Hz, the aim is to achieve a cut-on frequency $f_{\text{cut-on}}$ below 1000 Hz. Consequently, a theoretical length of the ABH $L_{\text{ABH}} = 110{,}8$ mm is selected, which gives a cut-on frequency $f_{\text{cut-on}} = 343$ Hz.

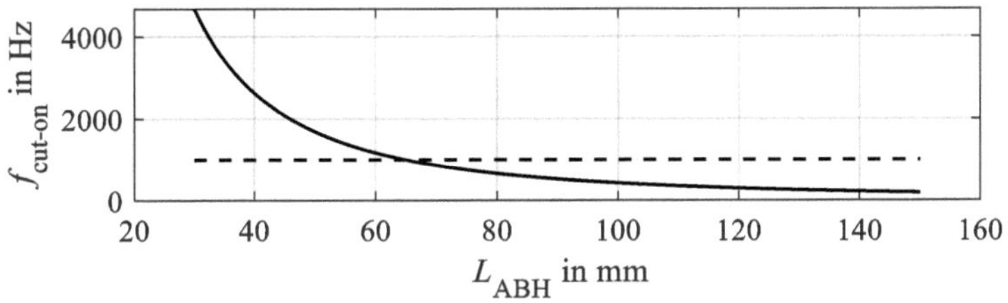


**Fig. 7.8.** Dependence of the cut-on frequency $f_{\text{cut-on}}$ on the length of the ABH $L_{\text{ABH}}$

Next, the theoretical thickness reduction $h_x$ following the power law Eq. [27]

$$h(x) = h \cdot \left[\frac{L_{\text{ABH}} - x}{L_{\text{ABH}}}\right]^4 \tag{7.3}$$

is evaluated and plotted in Fig. 7.9 (dotted line). Since such a profile is practically impossible to manufacture, it must be cut off beforehand. Thus, the minimum thickness $h_{\text{min}}$is set to 0,5 mm, which is a compromise between the structural durability and the effectiveness of the ABH. Consequently, the theoretical ABH design is cut to achieve the real ABH design which is plotted in Fig. 7.9 (continuous line).

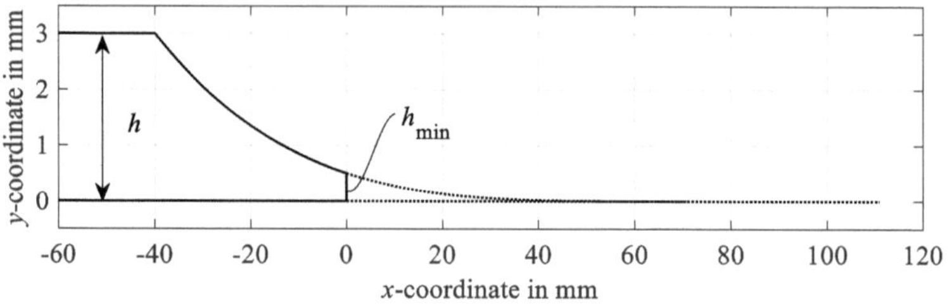


**Fig. 7.9.** Comparison of the ideal theoretical ABH design (dotted line) and the ABH design chosen for the cover (continuous line).

The cover has the outer dimensions of 260 mm × 230 mm. This means that a total of four ABHs with the same design can be placed on the cover. Figure 7.10 shows the cover with the four integrated ABHs as a CAD model and the hardware demonstrator used for the experimental investigation. Please note that the design of the ABH cover is approached in a practical manner. Enhanced results are anticipated with the application of optimization methods.

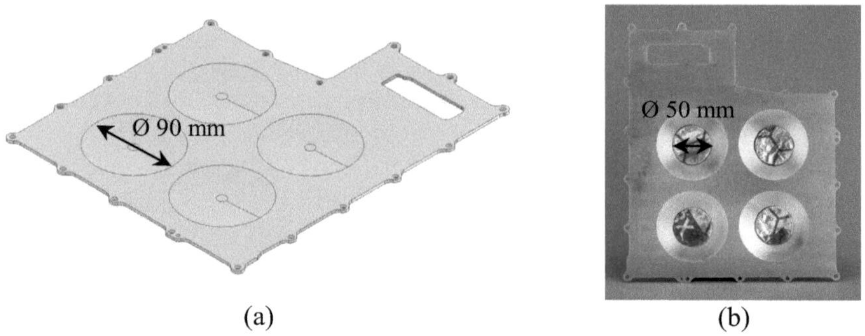


**Fig. 7.10.** Cover with four integrated ABH as CAD model (a) and hardware demonstrator (b). ©Fraunhofer LBF, Raapke.

## 7.6 Experimental Investigation of Structural Dynamic Behavior

The above-described covers are experimentally investigated and compared regarding their structural dynamic behavior and weight with the reference cover and two covers with an alubutyl-layer (Reckhorn ABX 2 mm) with a thickness $h_{\mathrm{Damp}}$ of 2 mm. An overview of the different cover variants together with their mass and relative mass differences to the reference cover and the alubutyl covers are given in Table 7.1. The first cover with alubutyl (Cover + Alubutyl 1) represents the state-of-the-art configuration. On the second cover with alubutyl (Cover + Alubutyl 2), the alubutyl is added using four circular patches with a diameter of 50 mm which is the same amount of alubutyl added to the Cover + ABH.

**Table 7.1.** Overview of the weights of the seven different covers under investigation.

| Reference cover | Cover+ Alubutyl 1 | Cover+ Alubutyl 2 | Cover+ VAMM-SM | Cover+VAMM-PF | Cover+ VAMM-SI | Cover+ABH |
|---|---|---|---|---|---|---|
|  | 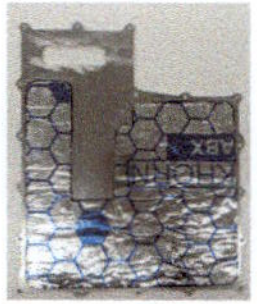 | 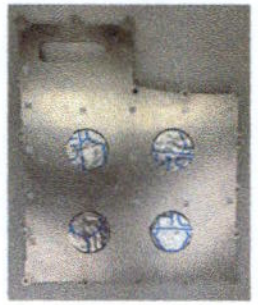 |  |  | 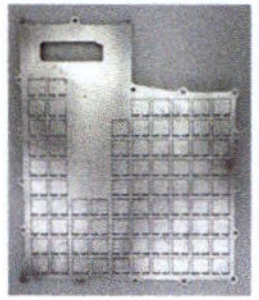 | 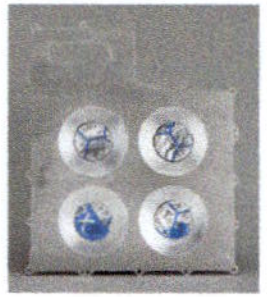 |
| 601 g | 778 g | 632 g | 1083 g | 843 g | 541 g | 539 g |
| 0% baseline | +29% | +5% | +80% | +40% | −10% | −10% |
| −23% | 0% baseline | −19% | +39% | +8% | −30% | −31% |
| −5% | +23% | 0% baseline | +71% | +33% | −14% | −15% |

Adding alubutyl to the reference cover increases the weight by 29 % compared to the reference cover. The add-on VAMM variants add 80 % (VAMM-SM) and 40 % (VAMM-PF) compared to the reference cover. The cover with integrated VAMM and the ABH-cover both weight 10 % less than the reference cover and 30 % to 31 % less than the reference cover with alubutyl as used in the vehicle. Compared to the cover with circular alubutyl patches, the ABH-cover is 15 % lighter.

### 7.6.1 Experimental Setup

The covers are experimentally investigated regarding their structural dynamic behavior according to an experimental setup described in [1](see Fig. 7.11). The covers are softly suspended by foam on their edges, ensuring a free-free boundary condition. For exciting the covers, an automatic impulse hammer is used. The covers are excited in the blue (•) point as indicated in Fig. 7.11. The excitation force is measured by means of a piezoelectric force-transducer. The resulting out-of-plane vibrations are evaluated in the red (•) points on the right side of the cover using a laser DOPPLER vibrometer. In the evaluation of the FRFs the measured signals in the five measurement points are averaged.

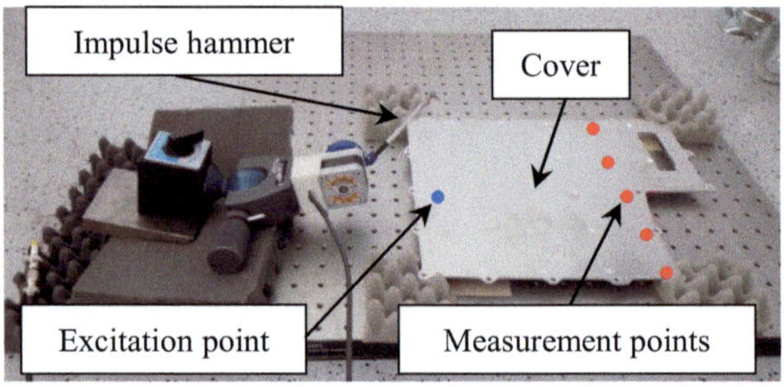


**Fig. 7.11.** Experimental setup for the FRF determination between the excitation point (•) and the measurement points (•) of the different covers. Impulse hammer used for impact excitation. The scanning laser DOPPLER vibrometer is not shown in the picture. Source: compare [1].

### 7.6.2 Experimental Results

The measurement results of the different VAMM variants in comparison to the reference cover and the cover with alubutyl 1, are shown in Fig. 7.12. Adding alubutyl to the reference cover leads to a damped structural behavior, as expected. The amplitudes in the resonances are reduced by approx. 23 dB in the frequency range in between 1000 Hz and 1500 Hz.

The cover with VAMM-SM shows a distinct stop band in the frequency range of approx. 1050 Hz to 1800 Hz with an associated amplitude reduction of approx. 69 dB compared to the reference cover and 46 dB compared to the Cover+Alubutyl 1. Below the stop band frequency range, a reduction in amplitude is visible as well, which is due to the added mass of the add-on VAMM as well as constrained layer damping attributed to the adhesive tape and plastic frame. Above the stop band, the behavior of the cover with VAMM-SM is comparable to the Cover+Alubutyl 1.

The second add-on VAMM variant (VAMM-PF) shows a stop band in the frequency range of approx. 700 Hz to 1560 Hz. The stop band is wider, but associated with a lower amplitude reduction than the VAMM-SM. Both effects occur due to the larger damping of the plastic foil in comparison the sheet metal of the VAMM-SM. The reductions in amplitude of the VAMM-PF variant compared to the reference cover and Cover+Alubutyl 1 are 50 dB and 26 dB, respectively within the frequency range of interest. Since the VAMM-PF is lighter than the VAMM-SM, the amplitudes in the frequency range below the stop band are larger. In the frequency range above the stop band the behavior, again, is comparable to the one of the Covers+Alubutyl.

The widest stop band is reached with the VAMM-SI, which reaches from approx. 1140 Hz to 2400 Hz and is associated with a maximum amplitude reduction of 58 dB compared to the reference cover and 37 dB compared to the Cover+Alubutyl 1. This behavior can be explained by the fact that extracting the resonators from the cover structure, the remained cover structure becomes lighter. The behavior of the VAMM-SI in the frequency range below the stop band is due to the same effect. Here, the amplitudes are greater compared to the other VAMM variants despite the added damping. Above the stop band frequency range, the amplitudes of the VAMM-SI are also larger than the amplitudes measured on the Cover+Alubutyl 1 by approx. 8 dB.

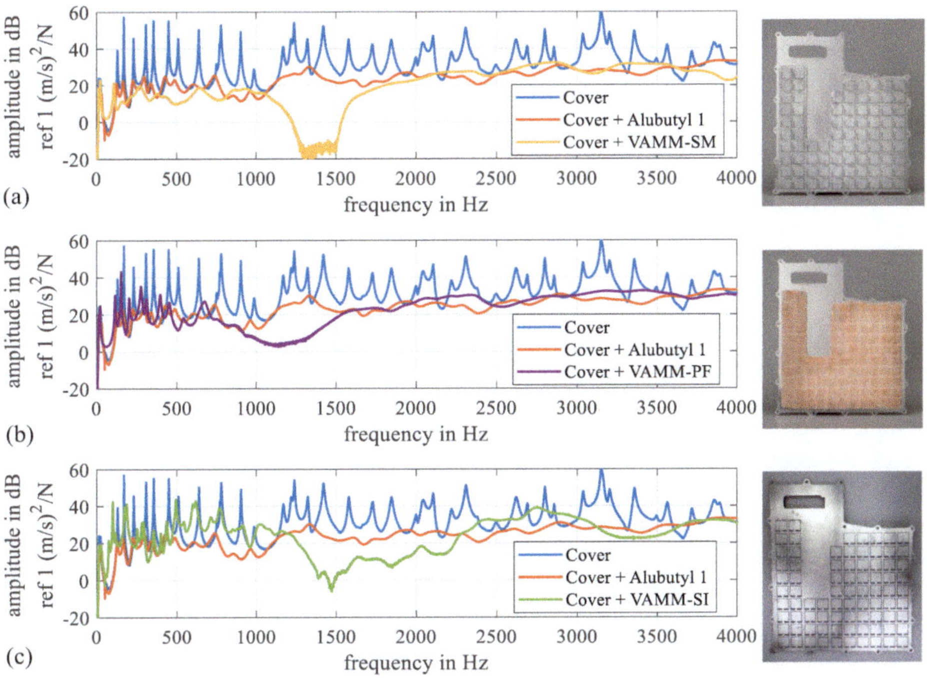


**Fig. 7.12.** Measurement results in terms of accelerance of the different VAMM variants in comparison to the reference cover and the cover with alubutyl. (a) VAMM-SM, (b) VAMM-PF, (c) VAMM-SI, ©Fraunhofer LBF, Raapke.

In the same way as described above, the results of the Cover + ABH are depicted in Fig. 7.13. The Cover + ABH shows a distinctly damped behavior above 1500 Hz. Above this frequency, the amplitude reduction in comparison to the reference cover is similar to that of the Cover + Alubutyl 1, even though the mass of applied alubutyl is much lower (81 % less) for the ABH-cover. Due to the complexity of the cover in comparison to a simple infinite plate as assumed by the Eq. (7.2), the cut-on frequency seems to be higher than initially calculated.

The effect of the ABH becomes clear when comparing the Cover+ABH to the Cover+Alubutyl 2 (same amount of alubutyl as the ABH-cover). The ABH leads to a larger amplitude reduction (approx. 20 dB) above approx. 1500 Hz compared to the reference cover, than the Cover+Alubutyl 2 compared to the reference cover (approx. 5–10 dB).

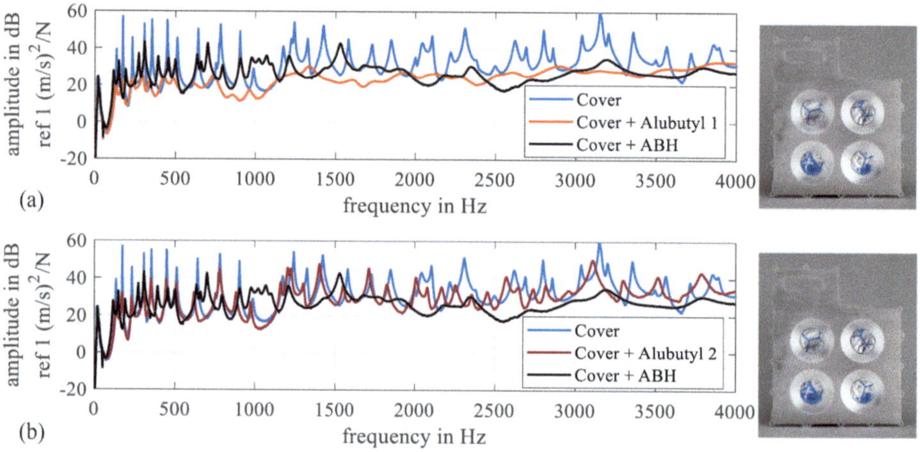


**Fig. 7.13.** Measurement results in terms of accelerance of the ABH cover in comparison to the reference cover and the cover with alubutyl 1 (a) and the cover with alubutyl 2 (b).

## 7.7 Experimental Investigation in Test Vehicle

Vehicle tests have been performed to validate the function of the abovementioned novel approaches for noise and vibration reduction on automotive structures. Since the VAMM-SM shows the largest effect in the structural dynamic measurements, this configuration is chosen for the vehicle test. The results of this investigation have been presented in depth in the publications [1–3] by the authors. In the following, the setup and results are demonstrated briefly, to give the reader an overview.

Figure 7.14a shows the add-on VAMM-SM applied to the cover of power electronics in the test vehicle. To measure the vibrations on the cover, an acceleration sensor is placed on the cover (not shown in Fig. 7.14). Additionally, the sound pressure within the passenger compartment is measured with microphones at the driver's and co-driver's ears, as seen in Fig. 7.14b–e.

The measurement results for a half-throttle acceleration from 0 km/h to 80 km/h are shown in Fig. 7.15 as acceleration and sound pressure level (SPL) spectrum for the cover with and without applied VAMM-SM. A critical order is visible in the acceleration frequency spectrum of the cover without VAMM-SM (Fig. 7.15a), intersecting with resonances of the cover at approx. 38 km/h. The same order is well visible in the spectrum of the SPL inside the vehicle. The acceleration and SPL amplitudes of the order are reduced by approx. 20 dB and 5–6 dB, respectively, within the stop band frequency range when applying the VAMM-SM to the cover.

The measurement is repeated at a constant speed of 38 km/h, to evaluate the performance at the critical driving speed of 38 km/h. Here, a reduction of tonal components at the driver's right ear at a magnitude of 11 dB are achieved [1].

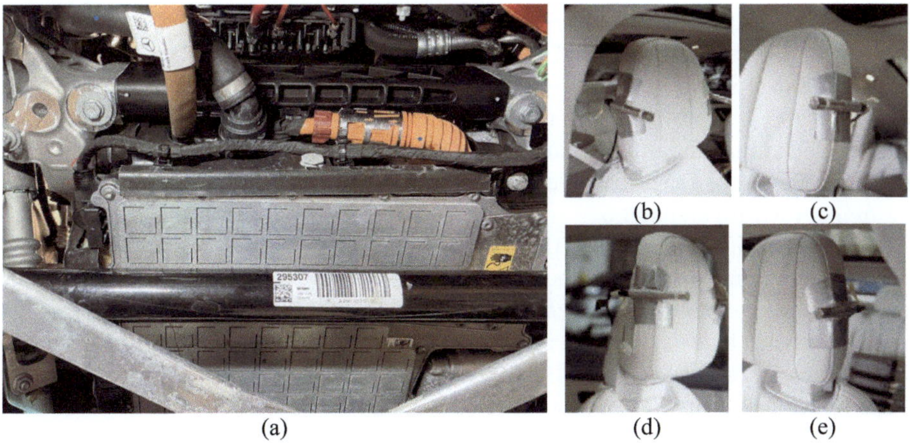


**Fig. 7.14.** Test vehicle with installed VAMM-SM (a). Microphones mounted at the headrests of the driver and co-driver: Driver's seat right ear (b), driver's seat left ear (c), co-driver's seat left ear (d), co-driver's seat right ear (e). *Source* Compare [1–3].

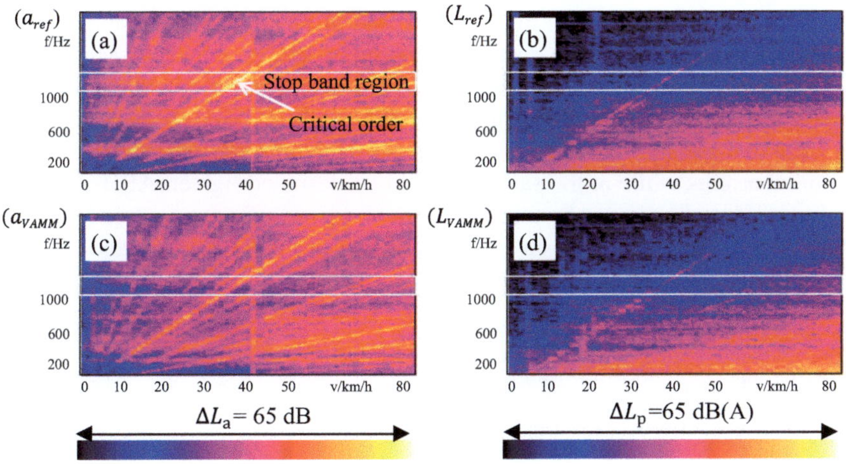


**Fig. 7.15.** Frequency spectrum of the measured acceleration on the cover without VAMM-SM (a) and with VAMM-SM (c). Sound power level (SPL) in the passenger compartment of the test vehicle without VAMM-SM (b) and with VAMM-SM (d). *Source* Compare [1–3].

## 7.8 Conclusion

In this study, a cover for power electronics in an EV is treated with different common, VAMM and one ABH approaches for vibration and noise control.

Two add-on VAMMs are elaborated, one made from sheet metal, and one made from a plastic film, both adding weight to the reference cover. Additionally, a cover with a structurally integrated VAMM and an ABH are designed, both weighting less than the reference cover. These covers are compared regarding their component weight and structural dynamic behavior with a bare reference cover and two reference covers equipped with alubutyl. While the covers with add-on VAMMs weight 39% (VAMM-SM) and 8% (VAMM-PF) more than a cover with a common alubutyl layer, the structurally integrated VAMM (VAMM-SI) and the ABH-cover weight 30% and 31% less, respectively.

Within the desired frequency range of 1000 Hz to 1500 Hz, the VAMM variants achieve a maximum amplitude reduction of 26 dB to 46 dB, compared to the common alubutyl layer. Above their stop band frequency range, the behavior is comparable to that of a cover with alubutyl.

The cover with integrated ABH features a cut-on frequency of approx. 1500 Hz and has a similar amplitude reduction compared to the bare reference cover, as the cover with common alubutyl layer. Below the cut-on frequency the amplitude reduction against the bare reference cover is lower than that achieved by the common alubutyl layer. Comparing the ABH cover with a reference cover, having the same amount of alubutyl applied, as the ABH cover, the ABH cover outperforms this configuration by 10–15 dB in amplitude reduction above approx. 1130 Hz. Below that frequency, the behavior is comparable.

For validation in a test vehicle, the VAMM-SM is chosen, because it achieved the best amplitude reduction in the relevant frequency range of 1000 Hz to 1500 Hz. With this add-on VAMM applied to the cover in the test vehicle, the SPL in the passenger compartment is reduced within one critical order by 5–6 dB during a run up from 0 km/h to 80 km/h. Tonalities at the driving speed of 38 km/h, where the critical order intersects with resonances of the cover, are reduced by 11 dB with the VAMM approach.

The study shows that VAMM and ABH are promising measures for noise and vibration reduction on automotive structures that can be tailored to specific problems. They offer the possibility for achieving larger vibration and noise reductions than common measures while at the same time can be designed lighter. ABH can contribute to saving damping layers by utilizing them in a more efficient manner.

Further research will focus on the optimization of the ABH cover and the consideration of its production by die casting.

## References

1. Rieß, S., Weber, A., Hülsebrock, M., Cóndor López, J., Kleinfeller, N., Uszynski, O., Atzrodt, H.: Vibration and noise reduction on a cover for power electronics for electric vehicles with vibroacoustic metamaterials. In: 53rd International Congress & Exposition on Noise Control Engineering INTER-NOISE 2024. Nantes, France, (2024).

2. Weber, A., Rieß, S., Cóndor López, J., Uszynski, O., Atzrodt, H., Hülsebrock, M., Hansen, T., Eckstein, L.: Erhöhung des NVH-Komforts von Elektrofahrzeugen durch Einsatz von vibroakustischen Metamaterialien. In: DAGA 2024 - 50. Jahrestagung für Akustik, pp. 111–114. Hannover, Germany (2024)
3. Weber, A., Rieß, S., Cóndor López, J., Atzrodt, H., Hülsebrock, Böttcher, T., Eckstein, L.: Application concepts and potentials of vibroacoustic metamaterials for NVH comfort. In: Aachen Acoustics Colloquium 2023. Aachen, Germany (2023).
4. Kadic, M., Milton, G. W., van Hecke, M., Wegener, M.: 3D metamaterials. Nature Reviews Physics 1, 198–210 (2019).
5. Veselago, V.: The electrodynamics of substances with simultaneously negative values of ε and μ. Soviet Physics Uspekhi 10(4), 509–514 (1968).
6. Pendry, J. B., Holden, A.J., Stewart, W. J., Youngs, I.: Extremely Low Frequency Plasmons in Metallic Mesostructures. Physical Review Letters 76(25), 4773–4776 (1996).
7. Pendry, J., Holden, A., Robbins, D., Stewart, W.: Magnetism from Conductors, and Enhanced Non-linear Phenomena. Microwave Theory and Techniques IEEE Transactions 47(11), 2075–2084 (1999).
8. Rieß, S.: Schwingungs- und Lärmreduktion an Kreissägeblättern mit strukturintegrierten vibroakustischen Metamaterialien. Dissertation, TU Darmstadt, Darmstadt, Germany (2024).
9. Rieß, S., Droste M., Atzrodt, H., Appel, A., Cóndor López, J., Finger, K., Erraji, A., Troge, J.: Vibroacoustic metamaterials for noise and vibration reduction in the automotive industry. In: Aachen Acoustics Colloquium 2022, pp 85–94. Aachen, Germany, (2022).
10. Rieß, S., Droste, M., Manushyna, D., Melzer, S., Druwe, T., Georgi, T., Atzrodt, H: Vibroacoustic Metamaterials for enhanced acoustic Behavior of Vehicle Doors. In: 2021 Fifteenth International Congress on Artificial Materials for Novel Wave Phenomena (Metamaterials), pp. X-374–X-376. New York City, NY, USA (2021).
11. Droste M., Manushyna, D., Rieß, S., Atzrodt, H., Druwe, T., Melzer, S., Troge, J., Struß, A., Lühring, A., Clausen, J.: Application of vibroacoustic metamaterials in a vehicle door. In: DAGA 2022 - 48. Jahrestagung für Akustik, pp. 233–235. Stuttgart, Germany (2022).
12. de Melo Filho, N. G. R., Claeys, C., Deckers, E., Desmet, W.: Realisation of a thermoformed vibroacoustic metamaterial for increased STL in acoustic resonance driven environments. Applied Acoustics 156, 78–82 (2019).
13. Wan, Q., Shao, R.: A hybrid phononic crystal for roof application. Journal of the Acoustic Society of America 142(5), 2988–2995 (2017).
14. KU Leuven Mecha(tro)nic System Dynamics (LMSD), Lightweight vibro-acoustic metamaterial patches for structure borne noise reduction in a vehicle, https://www.youtube.com/watch?v=MdHVCnvj9x0&feature=emb_title, last accessed 2024/03/24.
15. Sangiuliano, L., Reff, B., Palandri, J., Wolf-Monheim, F., Pluymers, B., Deckers, E., Desmet, W., Claeys, C.: Low frequency tyre noise mitigation in a vehicle using metal 3D printed resonant metamaterials. Mechanical Systems and Signal Processing 179, 109335 (2022).
16. Chang, K.-J., Jung, J., Kim, H.-G., Choi, D. R., Wang, S.: An Application of Acoustic Metamaterial for Reducing Noise Transfer through Car Body Panels. SAE Technical Paper 2018–01–1566 (2018).
17. Zobec, B., Weiss, J., Bäuerle, T., Johler, M.: Meta Design — die weltweit erste industrielle Anwendung von Meta Material im Automobil. Metamaterialien in der Akustik - 17. DEGA-Symposium, Nürnberg, Germany (2024).
18. Sangiuliano, L., Boukadia, R., Deckers, E., Desmet, W., Claeys, C.: Reduction of Structure-Borne Tyre/Road Noise through Rubber Resonant Metamaterials in Tyres. SAE International Journal of Advances and Current Practices in Mobility 5(2), 909–920 (2023).

19. Cóndor López, J. G., Hansen, T., Weber, A., Kleinfeller, N., Finger, K.: Gezielte Modikation des Übertragungsverhaltens von Elastomerlagern durch den Einsatz von vibro-akustischen Metamaterialien. In: DAGA 2024 - 50. Jahrestagung für Akustik, pp. 115–118. Hannover, Germany (2024).
20. Rieß, S., Droste, M., Erraji, A., Cóndor López, J., Finger, K., Atzrodt, H.: Noise and vibration reduction with vibroacoustic metamaterials on a cover for power electronics of an electric powertrain. In Rottengruber, H., Luft, T. (eds.) Aggregate- und Antriebsakustik: 12. Magdeburger Symposium. Otto von Guericke University Library, Magdeburg, Germany (2023).
21. Rieß, S., Droste, M., Atzrodt, H., Cóndor López, J., Finger, K., Erraji, A., Alasadi, E.: Vibroacoustic Metamaterials for Noise and Vibration Reduction on the Cover of Power Electronics of an electric Vehicle. In: 2023 Seventeenth International Congress on Artificial Materials for Novel Wave Phenomena (Metamaterials), pp. X-294–X-296. Chania, Greece (2023).
22. Droste, M., Rieß, S., Cóndor López, J. G., Erraji, A.: Vibroacoustic Metamaterials for Reduced Sound Radiation of an Electric Powertrain. MTZ Worldwide 84(4), 44–47 (2023).
23. Steijvers, K., Claeys, C., Van Belle, L., Deckers, E.: Improving Locally Resonant Metamaterial Performance Predictions by Incorporating Injection Moulding Manufacturing Process Simulations. Recent Trends in Wave Mechanics and Vibrations 125, 672-680 (2022).
24. Schwarzschild, K.: Über das Gravitationsfeld eines Massenpunktes nach der Einsteinschen Theorie. Sitzungsberichte der Königlich-Preussischen Akademie der Wissenschaften, pp. 189–196, Berlin (1916).
25. Pelat, A., Gautier, F., Conlon, S. C., Semperlotti, F.: The acoustic black hole: A review of theory and applications. Journal of Sound and Vibration 476, 115316 (2022).
26. Mironov, M. A.: Propagation of a flexural wave in a plate whose thickness decreases smoothly to zero in a finite interval. Soviet Physics Acoustics-USSR 34(3), 318–319 (1988).
27. Aklouche, O., Pelat, A., Maugeais, S., Gautier, F.: Scattering of flexural waves by a pit of quadratic profile inserted in an infinite thin plate. Journal of Sound and Vibration 375, 38–52 (2016).
28. Hoffmann, S., Rothe, S., Langer, S. C.: Acoustic black Holes – Modelling, Shaping, Placement and Application. In: Eberhard, P. (eds) Calm, Smooth and Smart. Lecture Notes in Applied and Computational Mechanics 102. Springer, Cham (2024).
29. Zhang, Y., Tong, M., Huang, K., Yang, F., Rui, X.: Experimental Investigation into the Acoustic Black Hole Vibrating Absorbing Effect and Its Automotive Applications. In: Rui, X., Liu, C. (eds.) Proceedings of the 2nd International Conference on Mechanical System Dynamics, ICMSD 2023. Lecture Notes in Mechanical Engineering, pp 1077–1086. Springer, Singapore (2024).
30. Prill, O., Roos, C., Busch, R.: Finite element Simulations of Acoustic Black Holes as Lightweight damping treatments for automotive body panels with application to full vehicle interior wind noise predictions. In: INTER-NOISE and NOISE-CON Congress and Conference Proceedings, InterNoise16, pp. 5210–5221. Hamburg, Germany (2016).
31. Bloch, F.: Über die Quantenmechanik der Elektronen in Kristallgittern. Zeitschrift für Physik 52, 555–600 (1929).

# Optimizing e-axle Noise and Vibration Through Virtual Simulation and Experimental Validation

Iacopo Stoppato(✉), Angelo Montanino, and Giuseppe Credo

Dumarey Automotive Italia S.p.A, Torino, Italy
{iacopo.stoppato,angelo.montanino,
giuseppe.credo}@dumarey.com

**Abstract.** This paper presents the development and validation of a flexible multibody simulation model for an E-axle system. Validation was conducted through an ex-tensive experimental campaign, which involved detailed vibration and acoustic measurements under various operational conditions to ensure accurate representation of the system's key dynamic and acoustic behaviors. Sensitivity analyses were carried out to improve correlation between the simulation model and experimental results.

The validated model was then used to evaluate design modifications of the E-axle covers, with particular emphasis on enhancing acoustic performance. Virtual simulations analyzed multiple configurations and optimized critical design parameters, achieving significant improvements in noise and vibration characteristics. The study demonstrates the efficiency of integrating virtual simulation with experimental validation, significantly reducing the reliance on physical testing while enhancing vibro-acoustic performance.

In conclusion, this research highlights the pivotal role of virtual simulation technologies in the development of modern electrified propulsion systems. Incorporating experimental data into the simulation process facilitated faster iterations and more informed design decisions, showcasing the potential to reduce development cycles and effectively achieve high NVH performance standards.

## 8.1 Introduction

The rapid transition toward electric mobility has significantly transformed automotive propulsion system design and engineering. Central to this transition are electric axles (E-axles), which have become prevalent due to their highly integrated, compact design combining electric motors, power electronics, and transmission components. While these systems greatly enhance packaging efficiency and energy performance, they introduce new Noise, Vibration, and Harshness (NVH) challenges, necessitating targeted development methodologies.


A. Heintzel (Ed.): Automotive Acoustics Conference 2025, Proceedings, pp. 109–124, 2026. https://doi.org/10.1007/978-3-658-51249-1_8

The absence of internal combustion engines, traditionally responsible for masking drivetrain noises, has made tonal noise components from gear meshing, electromagnetic forces, and structural resonances more noticeable in battery electric vehicles (BEVs) [1, 2]. This change creates a new NVH landscape where previously negligible excitation sources now significantly influence in-cabin acoustic experiences, particularly at low to medium vehicle speeds.

To meet customer expectations for comfort and refinement, a simulation-driven approach to NVH development is increasingly critical. Recent studies emphasize the importance of integrating multibody dynamics (MBD), finite element analysis (FEA), and acoustic simulations, enabling early-stage evaluations and reducing the need for physical prototypes [3, 4]. For example, research by Mehrgou et al. [5] highlighted the efficacy of advanced CAE tools in simulating noise radiation from high-speed E-axles, enabling early validation of subsystem NVH targets.

Further emphasizing the efficacy of simulation methods, Marano et al. [6] demonstrated the benefits of flexible multibody modeling, particularly the Craig-Bampton model reduction technique, to investigate how spiral bevel gear mounting distances influence radiated noise and dynamic accelerations. Their findings underline the importance of manufacturing precision and alignment in shaping the acoustic characteristics of E-axle housing.

Structural optimization also significantly contributes to enhancing NVH performance. Klarin et al. [4] presented a robust process combining root cause analysis, modal correlation, and topology optimization to effectively reduce radiated sound power from electric drivetrains. By addressing key eigenfrequencies and reducing structural amplifications through targeted design changes, substantial noise reductions were achieved.

At the system integration level, Wellmann et al. [7] emphasized the necessity of addressing NVH characteristics specific to Electric Drive Units (EDU) notably tonal noises originating from electric machines and gearboxes during vehicle development. They advocated for cascading NVH targets across subsystems and implementing strategies such as vibration isolation and optimized mounting configurations to control noise propagation and ensure refined vehicle-level NVH performance.

Building upon these established methodologies, this paper proposes a virtual validation framework for E-axles, integrating flexible multibody simulations with experimental NVH testing. The validation and correlation efforts specifically focus on the gear mesh of both the first and second stages, demonstrating how a validated digital twin can accelerate development, enhance efficiency, and achieve substantial improvements in NVH performance.

## 8.2 Numerical Model

### 8.2.1 System Modeling

The virtual model of the E-axle system was developed using Romax, a specialized multibody simulation software distributed by Hexagon [8]. To accurately capture the system's dynamic behavior, all structurally significant components were integrated into the model, each detailed according to its specific contribution to overall system dynamics. The comprehensive set of components included in the virtual model (illustrated in Fig. 8.1) were:

- Electric motor, Inverter
- Shafts gears and bearing
- Mounts
- Housing and cover
- Accessories (oil pump, filter, cooler)

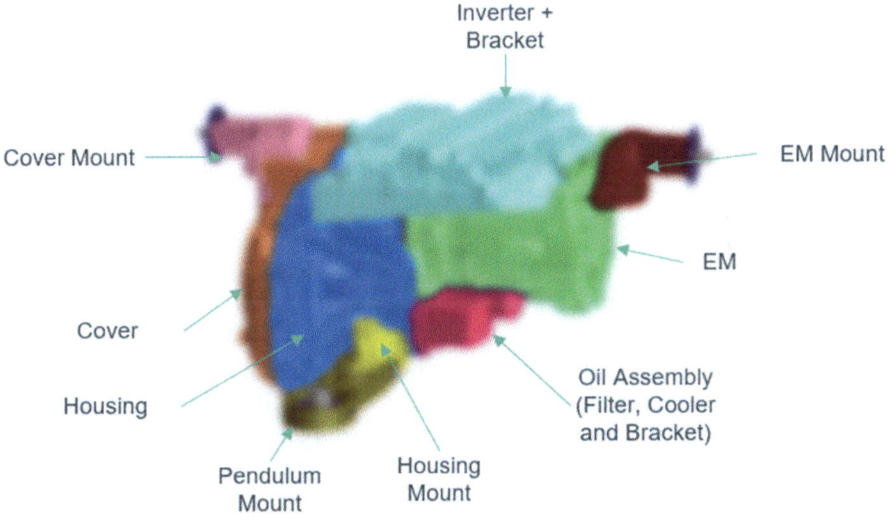


**Fig. 8.1.** E-axle system's virtual model

These components were modeled using detailed three-dimensional finite element (FE) representations. Subsequently, these FE models underwent a dynamic condensation process, effectively retaining essential stiffness and mass properties, which significantly enhanced computational efficiency during subsequent dynamic simulations [9]. Bearing elements received particular attention, modeled with high precision to account for nonlinear stiffness characteristics that vary based on load and operational conditions. This detailed bearing modeling enabled accurate predictions of complex phenomena such as misalignments, contact patterns, and dynamic interactions, which play crucial roles in influencing the system's vibrational behavior. Components without structural stiffness significance were accounted for by employing mass balancing techniques, thereby ensuring the accurate inertial characterization of the entire system.

### 8.2.2 Vibroacoustic Analysis: Conditions and Outputs

The vibroacoustic analysis initiated with a modal analysis aimed at identifying the system's natural frequencies and corresponding mode shapes, providing insights into its inherent dynamic properties. Following this, a detailed dynamic analysis was conducted, primarily focusing on Gearbox Transmission Error (GBTE) as the principal excitation source. Unlike Transmission Error (TE), which relates to individual gear pairs and does not consider influences from other gear meshes, GBTE incorporates highly coupled interactions among gear meshes within the complete gear train or epicyclic arrangements. This coupling is essential to accurately represent realistic gear dynamics, particularly in complex systems where the torque and load sharing among gear meshes vary dynamically with rotation (Table 8.1).

**Table 8.1.** Differences between TE and GBTE

| Single Gear Pair TE | GBTE—Gear Train and Epicyclic Gears |
|---|---|
| No influence from other gear meshes | Highly coupled interactions between meshes |
| Constant for a given load case | Time-variable due to changing load sharing between planets as the carrier rotates |
| Misalignment and tilt excitation assumed constant | Mesh misalignment and lead correction considered, contributing to excitation |

GBTE calculations integrated detailed assessments of gear micro-geometry, variations in mesh stiffness, torque distributions, and mesh misalignments. Although Romax offers an advanced Loaded Tooth Contact Analysis (LTCA) method that accurately considers local bending stiffness coupling and the Poisson effect, it is only applicable for calculating individual gear pair TE. Due to software limitations, this advanced LTCA method could not be directly applied to GBTE. In any case, the use of GBTE remained indispensable for comprehensive dynamic analysis since it effectively captures the critical coupled interactions among gear meshes necessary for realistic vibration predictions.

Realistic boundary conditions were simulated through soft-mounted constraints, closely mimicking actual installation scenarios. Diverse operational conditions, including varying torque levels and rotational directions representative of realistic duty cycles, were thoroughly evaluated. The analysis provided extensive vibrational metrics such as displacements, velocities, accelerations, and bearing forces. Additionally, the acoustic performance was quantified by calculating Equivalent Radiated Power (ERP), measuring the noise emissions under different operating conditions. The ERP calculation utilized a spatially uniform radiation efficiency assumption to derive the Pseudo Sound Power, offering a quantitative indicator of the system's vibroacoustic behavior.

The ERP was calculated using the following relationship:

$$w = \int_S \boldsymbol{I} \cdot \boldsymbol{n} da = \int_S I_n da = \Delta a \sum_{j=1}^{N} I_{n_j} = \eta_S^2 \rho_0 c A \frac{\sum_{j=1}^{N} V_{MS_j}^2}{N} = \eta_S^2 \rho_0 c A \overline{V}^2{}_{MS} \quad (8.1)$$

where:

- $w$ = Sound Power
- $I_n$ = Normal Acoustic Intensity
- $A$ = Surface Area
- $\eta_S$ = Radiation efficiency

## 8.3 Experimental Test

The primary objective of the experimental campaign was to thoroughly characterize the NVH behavior of the fully assembled E-axle system under representative operational conditions, encompassing both stationary and transient scenarios. The testing campaign was methodically divided into distinct vibration and acoustic assessments, each employing specific instrumentation and performance criteria derived from internal guidelines and customer specifications. Instrumentation adhered strictly to ISO 17025 [10], ensuring equipment management, periodic calibration, and performance verification, thus guaranteeing accurate and repeatable measurement outcomes. Acoustic measurements were conducted within a semi-anechoic chamber environment compliant with ISO 3744 [11] standards.

### 8.3.1 Test Environment and Measurement Setup

The Device Under Test (DUT) was a fully assembled gearbox tested on a dynamometer bench (illustrated in Fig. 8.2), carefully designed to replicate real-world boundary conditions, installation constraints, and project-specific oil temperature and pressure requirements.

**Fig. 8.2.** DUT in semi-anechoic test bench

Acoustic measurements were collected using a precisely configured array of nine microphones arranged in a parallelepiped pattern (see Fig. 8.3), placed at a standardized distance of 50 cm from the DUT, ensuring consistency and reliability for engineering-grade acoustic evaluations.

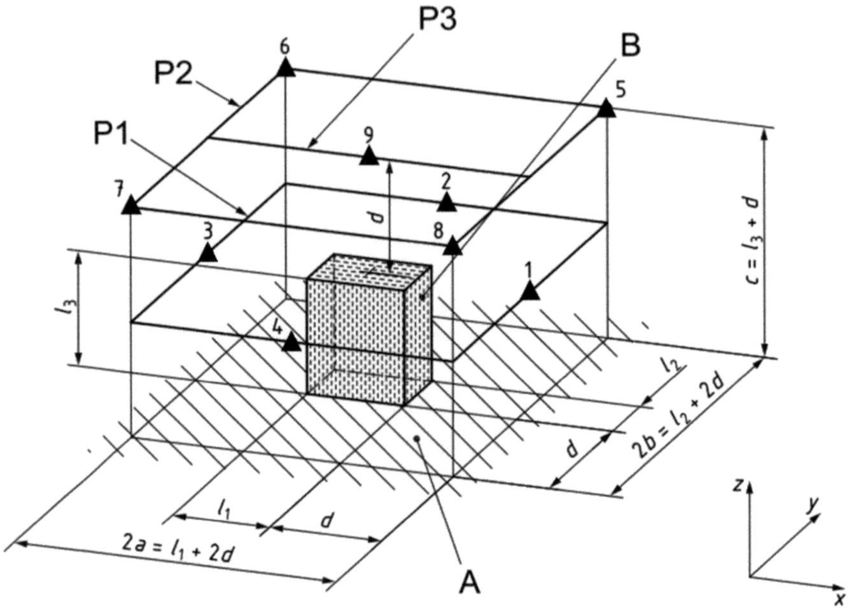


**Fig. 8.3.** Nine-microphone layout

Key:

- ▲ microphone positions
- A reflecting plane
- B reference box
- 2a measurement surface length
- 2b measurement surface width
- c measurement surface height
- d measurement distance
- $l_1$ reference box length
- $l_2$ reference box width
- $l_3$ reference box height

According to ISO 3744, two correction factors were applied to the acquired acoustic measurement data to guarantee accuracy. The first factor, known as background noise correction $K_1$, adjusts the mean of the time-averaged sound pressure levels to compensate for ambient background noise. The second factor, environmental correction $K_2$, was employed to account for reflected and absorbed sound within the test environment. The environmental correction factor is calculated as follows:

$$K_2 = 10 \cdot log[1 + 4(S/A)]dB \tag{8.2}$$

where:

$$A = 0.16(V/T_r) \tag{8.3}$$

- V is the volume of the chamber [$m^3$]
- $T_r$ is the reverberation time of the camber [s]
- V is the area of the measurement surface [$m^2$]

An additional near-field microphone was positioned close to the inverter to specifically capture tonal acoustic contributions resulting from inverter switching. Vibration measurements employed triaxial accelerometers placed at critical points, including cantilevered components, bearing housing, gearbox housing, and mounting brackets.

The testing campaign consisted of speed sweeps ranging from 0 rpm to the maximum rotational speed allowed by the software calibration at that stage of the project. Tests were performed in both motoring (positive torque) and regenerative (negative torque) modes, under strictly controlled thermal conditions. During each test, vibration (accelerometers), acoustic signals (microphones), torque and rotational speed (encoders/tachometers), as well as oil pressure and temperature data were acquired simultaneously.

### 8.3.2 Data Analysis and Order Tracking

The acquired experimental data underwent rigorous post-processing tailored to meet specific customer requirements, encompassing defined frequency ranges, resolution, windowing, and A-weighting methodologies. Sound power levels ($L_w$) were meticulously calculated following ISO 3744 standards, using the relationship:

$$L_W = \overline{L_p} + 10log\frac{S}{S_0}dB(A) \tag{8.4}$$

$\overline{L_p}$ is the surface time-averaged sound pressure level calculated by correcting the mean time-averaged sound pressure level for background noise ($K_1$) and for the influence of the test environment ($K_2$) using equation:

$$\overline{L_p} = \overline{L}'_{p(ST)} - K_1 - K_2 \tag{8.5}$$

where $\overline{L}'_{p(ST)}$ is the mean A-weighted time-averaged sound pressure level from the array of microphone positions over the measurement surface, with the noise source under test in operation, in decibels.

Whine order analyses were specifically performed to extract gear meshing orders, magnetic orders, and unbalance orders, each defined based on the transmission architecture and electromagnetic design specifications. Acoustic results were comprehensively presented in terms of sound power levels [W], expressed in dB(A), systematically plotted against input shaft rotational speeds. This analytical approach facilitated the identification of critical harmonics, speed intervals, and torque conditions. Waterfall diagrams (Fig. 8.4) effectively represented the total Sound Power from the conducted speed sweep tests, explicitly highlighting the most significant NVH orders.

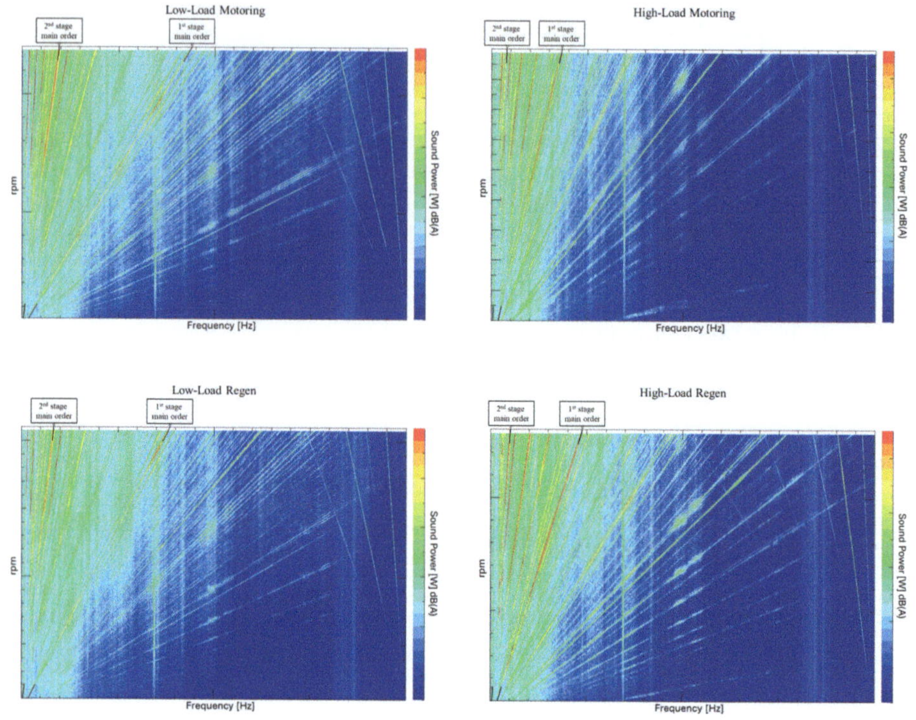


**Fig. 8.4.** E-Axle Sound Power Waterfall plots

## 8.4 Correlation

### 8.4.1 Comparison Between Simulation and Experimental Results

This section presents an in-depth comparison between the simulation results and experimental measurements conducted on the E-axle system. The correlation assessment covered two distinct operational scenarios: motoring and regenerative braking (regen), with focus on gear meshing excitations, specifically the 1st Stage Main Order and the 2nd Stage Main Order of the gearbox. Both stages were thoroughly evaluated under low-load and high-load torque conditions, aiming to capture the system's comprehensive behavior accurately.

**Motoring Condition—1st Stage Main Order**

Under low-load conditions, the simulation exhibited good correlation with experimental data. This correlation can be attributed primarily to the lower energy input into the system at reduced torque levels, which results in minimal component deformation and reduced gear misalignment. At lower RPMs, experimental data are notably influenced by background noise present in the test environment, an effect inherently absent in the simulation. In contrast, the correlation under high-load conditions deteriorates significantly. Higher torque loads introduce substantial component deformation and misalignment, presenting phenomena challenging to accurately replicate with current modeling approaches (Fig. 8.5).

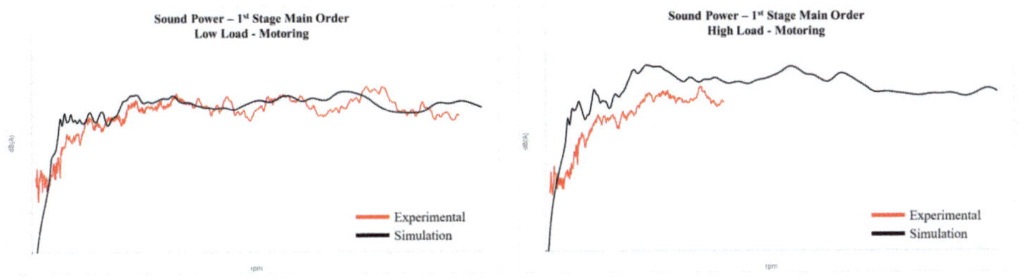


**Fig. 8.5.** 1st Stage order Sound Power Motoring Plots

### Motoring Condition—2nd Stage Main Order

For the 2nd Stage Main Order under motoring conditions, correlation remained consistently good across both low-load and high-load scenarios. This correlation is primarily due to the lower frequency range and simpler vibrational modes inherent in this stage, facilitating more accurate simulation predictions (Fig. 8.6).

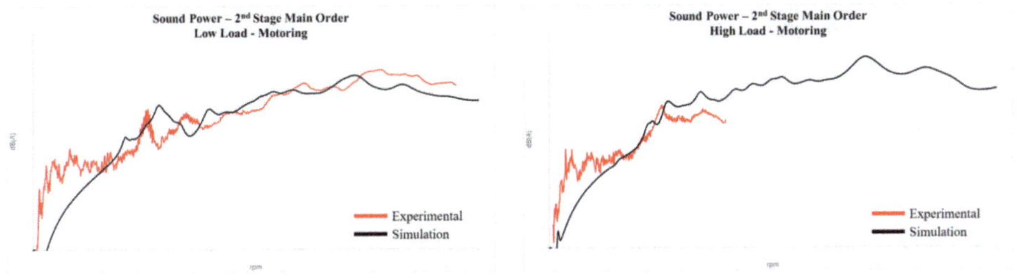


**Fig. 8.6.** 2nd Stage order Sound Power Motoring Plots

### Regen Condition—1st Stage Main Order

Similar to motoring conditions, the correlation in low-load regenerative braking scenarios remained good, benefiting from the lower energy states leading to minimal deformation and misalignment. However, under high-load conditions, notable discrepancies emerged, emphasizing the need for further model refinement to accurately represent the complex behaviors associated with elevated loads (Fig. 8.7).

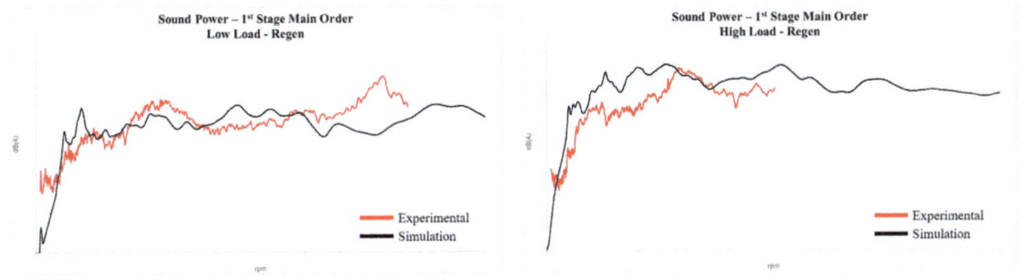


**Fig. 8.7.** 1st Stage order Sound Power Regen Plots

**Regen Condition—2nd Stage Main Order**

Correlation for the 2nd Stage Main Order during regenerative braking conditions remained consistently good across both low-load and high-load scenarios. The lower frequency dynamics in this stage simplified vibrational behaviors, enabling effective simulation predictions even at higher loads Fig. 8.8.

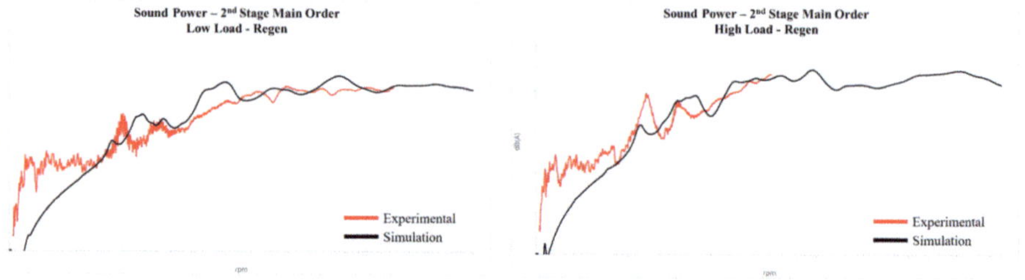


**Fig. 8.8.** 2nd Stage order Sound Power Regen Plots

Overall, these analyses underline the simulation model's effectiveness at lower torque loads, particularly for second-stage gear mesh orders. They also identify critical areas for enhancement, notably in modeling nonlinear stiffness, deformation, and dynamic misalignments at higher operational loads and RPM ranges, particularly affecting the first-stage correlation accuracy.

### 8.4.2 Linear Transmission Error vs. Tilt Transmission Error

In Romax, Transmission Error (TE) and Gearbox Transmission Error (GBTE), are always separated into Linear Transmission Error (LTE) and Tilt Transmission Error (TTE). LTE denotes displacement along the gear tooth line of action, influenced mainly by mesh stiffness variations and manufacturing imperfections. Conversely, TTE represents rotation across the gear teeth face, arising from shifting tooth force centroids creating moments perpendicular to the line of action. TTE significantly impacts gear mesh behavior, particularly under high torque conditions characterized by pronounced tooth and shaft deformation.

Romax's comprehensive GBTE analysis integrates variations in misalignment, tooth stiffness, and torque distribution within the entire gear train. Although the advanced Loaded Tooth Contact Analysis (LTCA) methodology is available for TE calculations, it is not available for GBTE analysis due to software limitations. The advanced LTCA method significantly enhances accuracy by considering bending stiffness coupling and the Poisson effect, capturing phenomena such as anticlastic curvature and load redistribution between adjacent teeth. In contrast, the basic LTCA method uses an uncoupled, simplified approach that does not fully account for these intricate coupling effects.

However, for dynamic analyses where all stiffnesses, deformations, and gear interactions must be simultaneously considered, GBTE remains necessary despite the absence of the advanced LTCA option. To address this limitation and improve model correlation at higher loads, GBTE values were adjusted based on insights from the advanced LTCA method. A sensitivity analysis was conducted focusing on damping, LTE, and TTE parameters. Increasing damping reduced peak amplitudes, mitigating model overestimation but risking the omission of critical dynamic details. Initially, modifying only the LTE component by the percentage indicated by the advanced LTCA method provided minimal improvements. Conversely, adjusting the TTE component significantly enhanced simulation accuracy, notably improving correlation for the 1st Stage Main Order under high-load motoring conditions (Fig. 8.9). This underscores the critical role and sensitivity of accurately modeling TTE within GBTE simulations.

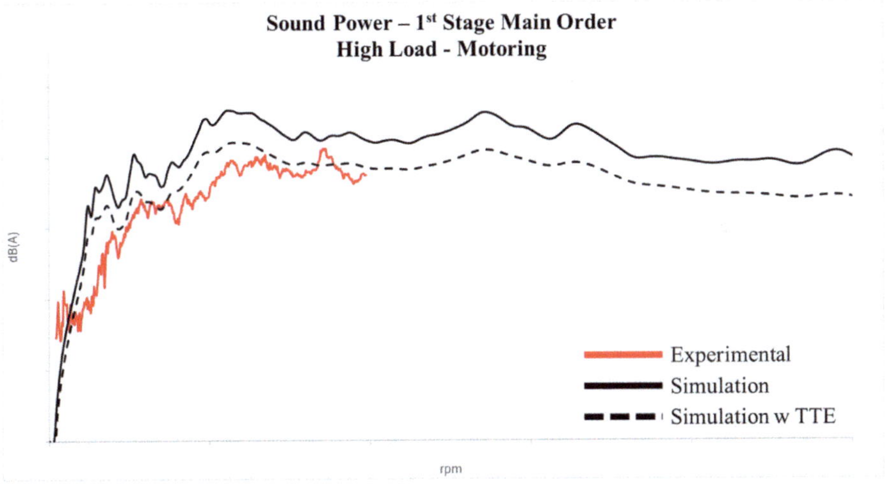


**Fig. 8.9.** Impact of Tilt Transmission Error (TTE) on model correlation

## 8.5 Structural Optimization for Enhanced Dynamic Performance

### 8.5.1 Dynamic Structural Optimization

Structural rigidity plays a critical role in reducing energy transmission from primary excitation sources such as gear meshing vibrations. Enhancing structural stiffness decreases vibration amplitudes transmitted through mechanical components, thereby improving overall Noise, Vibration, and Harshness (NVH) characteristics. A more rigid structure effectively isolates and attenuates the propagation of undesirable vibrational energy. Additionally, increased structural stiffness beneficially influences a structure's natural frequencies and corresponding modal shapes. Optimized stiffness typically shifts natural frequencies away from operational frequencies, effectively avoiding resonance conditions. Resonance occurring when operational excitation frequencies align with structural natural frequencies, significantly amplifies vibrations, potentially leading to structural fatigue and acoustic discomfort. Therefore, managing modal characteristics is essential to ensure both structural integrity and acoustic comfort.

### 8.5.2 Structural Topology Optimization Approach

Topology optimization is a numerical technique designed to identify optimal material distribution within a predefined spatial region, aiming to achieve maximum structural efficiency. The method iteratively redistributes material, systematically removing unnecessary mass while preserving or enhancing structural performance within specified constraints and objectives.

In this study, topology optimization was carried out using Altair Optistruct, renowned software extensively utilized for structural optimization and NVH analysis. Optistruct integrates finite element methods with advanced optimization algorithms, efficiently producing lightweight yet structurally robust solutions [12].

The optimization process involved explicitly defining two critical regions: the Design Space, where the optimization algorithm can modify or remove material, and the Non-Design Space, representing regions with fixed geometry due to functional, manufacturing, or packaging constraints.

The accurate identification and definition of these regions significantly impacted optimization outcomes. Well-defined design spaces provided the optimization algorithm with adequate freedom to propose innovative solutions, while explicitly designated non-design spaces ensured practical feasibility and compliance with external constraints, including assembly, manufacturing processes, and maintenance requirements [13].

### 8.5.3 Case Study

The presented case study focused on optimizing the gearbox housing, a crucial automotive component directly influenced by gear meshing vibrations, making it an ideal candidate for structural optimization. The definition of design and non-design spaces (illustrated in Fig. 8.10) was carefully conducted in collaboration with the design engineering team.

Considerations included packaging constraints, assembly processes, manufacturing limitations, and service accessibility. Non-design spaces predominantly included bolt connections, bearing interfaces, and critical sealing surfaces, whereas remaining regions were classified as design spaces.

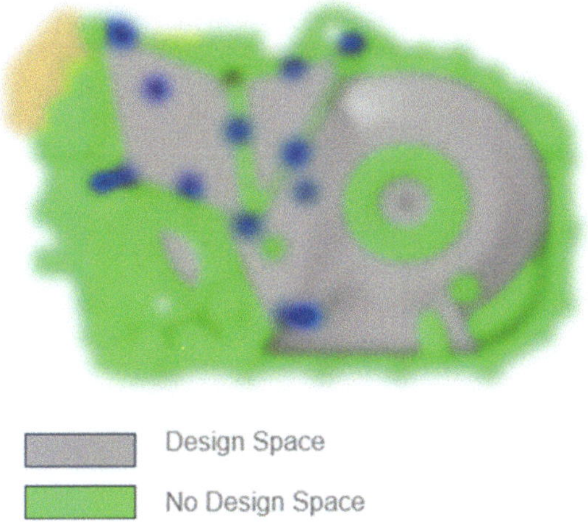


**Fig. 8.10.** Design end No design space selection

The optimization objective was set to maximize structural stiffness (minimizing compliance at bearing and mount interfaces), with a strict mass constraint to maintain lightweight design requirements. Additionally, modal frequency constraints were imposed to ensure operational frequencies avoided resonance conditions. Objective and constraint selections were guided by operational performance criteria, practical production requirements, and best practices from prior optimization studies. This combination ensured a balanced approach, emphasizing structural rigidity, dynamic robustness, and practical manufacturability.

Topology optimization yielded several promising solutions, notably enhancing structural stiffness and improving modal behaviors.

A detailed comparative analysis of candidate designs evaluated stiffness-to-weight ratios, natural frequency distributions, manufacturing feasibility, and assembly considerations. The selected final design represented the optimal balance between enhanced dynamic performance, reduced mass, and practical manufacturability.

The optimized designs showed significant improvements over the initial baseline, effectively shifting critical modal frequencies away from operational excitation ranges and leading to a reduction in Pseudo Sound Power levels (see Fig. 8.11).

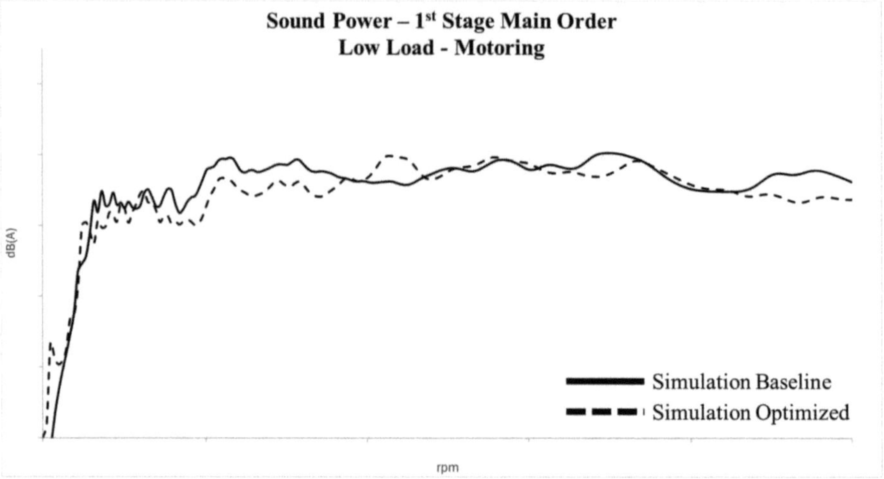


**Fig. 8.11.** Optimized Design vs Baseline

# 8.6 Further Developments and Conclusions

## 8.6.1 Further Developments

The correlation analyses identified critical areas for further improvement, particularly under high-load operational conditions. Subsequent development activities will focus specifically on detailed sensitivity analyses aimed at refining Tilt Transmission Error (TTE) modeling, given its substantial influence on correlation accuracy as demonstrated by previous analyses. Future software releases are anticipated to directly integrate the advanced Loaded Tooth Contact Analysis (LTCA) methodology into GBTE calculations, allowing more accurate predictions of gear mesh interactions by capturing local bending stiffness coupling, Poisson's effect and adjacent tooth interactions.

Sensitivity analyses will further investigate the impact of bearing preload adjustments, as defined within the simulation environment, to evaluate their influence on nonlinear bearing stiffness characteristics, gear misalignment, and overall system NVH responses. By systematically adjusting preload parameters within realistic operational bounds, these analyses aim to precisely capture these influential factors.

Moreover, extending beyond gear mesh excitations, subsequent activities will incorporate additional excitation sources from electric motors and inverters into the NVH simulation framework. These analyses have exclusively addressed gear meshing excitations. The inclusion of these broader excitation phenomena will provide a more comprehensive understanding of NVH performance, significantly enhancing the predictive reliability and accuracy of the developed simulation methodologies under realistic operational scenarios.

### 8.6.2 Conclusions

This study presented a structured methodology integrating advanced multibody simulation techniques, structural topology optimization, and comprehensive experimental validation, effectively addressing Noise, Vibration, and Harshness (NVH) challenges specific to E-axle systems. The virtual model developed demonstrated good correlation with experimental results under lower load conditions and lower-frequency gear mesh orders, affirming its predictive capability and reliability for NVH evaluations.

At higher torque conditions, discrepancies emerged primarily due to increased structural deformation, which affected dynamic misalignments and complicated the representation of higher-frequency vibration modes. Sensitivity analyses, based exclusively on gear meshing excitations, underscored the critical importance of accurately modeling Tilt Transmission Error (TTE), identifying it as a key factor in improving correlation accuracy under demanding operational conditions.

The rigorous correlation activity significantly validated the numerical model, enabling its confident application across subsequent development phases. This approach facilitated considerable reductions in development costs and timelines by minimizing reliance on physical prototypes. Structural topology optimization notably enhanced structural stiffness and effectively shifted critical natural frequencies away from operational excitation ranges, mitigating resonance risks.

Overall, this integrated virtual-experimental strategy established a robust and efficient framework for NVH-driven design optimization, highlighting once again the fundamental role of advanced simulation and virtualization techniques in achieving consistent, high-quality vibroacoustic performance within modern electrified automotive propulsion systems.

## References

1. Ulz, A., Graf, B., Priestner, C., Mehrgou, M., Criteria for the NVH Development of Electric Drive Units, ATZ Worldwide 07–08/2021.
2. Chengping, Z., Guo, F., Senhai, L., Qingshuang, C., Ping, X., Deng, X., Xianzhong, Y., Analysis and Optimization of Whine Noise of Light Bus E-axle Sys-tem, SAE Technical Paper 2023–01–1102.
3. Holehouse, R., Shahaj, A., Michon, M., and James, B., Integrated Approach to Electro-Mechanical System NVH Analysis, SAE Technical Paper 2018–01–1499, https://doi.org/10.4271/2018-01-1499.
4. Klarin, B., Resch, T., Grozdanovic, I., Pevec, D., Root Cause Analysis and Structural Optimization of E-Drive Transmission, SAE Technical Paper 2020–01–1578, https://doi.org/10.4271/2020-01-1578.
5. Mehrgou, M., Pohn, J., Graf, B., Priestner, C. et al., Advanced CAE Methods for NVH Development of High-Speed Elec-tric Axle, SAE Technical Paper 2020–01–1501, https://doi.org/10.4271/2020-01-1501.
6. Marano, D., Giese, T., Ebrahimi, S., NVH Analysis of an Axle Drive with Bevel Gearset, AGMA Technical Paper 21FTM12, 2021.
7. Wellmann, T., Tousignant, T., Govindswamy, K., Tomazic, D. et al., NVH Aspects of Electric Drive Unit Development and Vehicle Integra-tion, SAE Technical Paper 2019–01–1454, https://doi.org/10.4271/2019-01-1454.

8. Romax Spectrum 2023.1 User's Guide
9. Jain, T., Desai, A., & Kurna, S. (2024). Methodology Development for E-Axle NVH Performance Evaluation through Virtual Simulation (SAE Technical Paper 2024-26-0120). SAE International. https://doi.org/10.4271/2024-26-0120
10. International Organization for Standardization & International Electrotechnical Commission. (2017). General requirements for the competence of testing and calibration laboratories (ISO/IEC Standard No. 17025:2017). https://www.iso.org/standard/66912.html
11. International Organization for Standardization. (2010). Acoustics—Determination of sound power levels and sound energy levels of noise sources using sound pressure—Engineering methods for an essentially free field over a reflecting plane (ISO Standard No. 3744:2010). https://www.iso.org/standard/52055.html
12. Altair Optistruct User Manual, Altair Engineering Inc.
13. Bendsoe, M. P., & Sigmund, O. (2004). Topology Optimization: Theory, Methods, and Applications. Springer.

# NVH Optimal Dedicated Hybrid IC Engine

Bernhard Graf(✉), Werner Biermayer, Iacopo Zallio, and Josef Gojo

AVL List GmbH, Graz, Austria
{bernhard.graf,werner.biermayer,
iacopo.zallio,josef.gojo}@avl.com

**Abstract.** The optimization of internal combustion (IC) engines for hybrid vehicles is critical to achieving superior noise, vibration, and harshness (NVH) characteristics while maintaining high efficiency and low emissions. Usually, the target focus of these Dedicated Hybrid Engines (DHE) is on fuel efficiency and cost with NVH taking a back seat.

This study focuses on the conceptual development and target-setting for an NVH-optimized dedicated hybrid engine. The process begins with identifying the unique requirements of hybrid vehicle applications, such as frequent start-stop cycles, operation under varying loads and integration with electric drivetrains. Key considerations as number of cylinders, cylinder displacement, balancing requirements, combustion concept and combustion noise reduction, torque fluctuations and isolation strategies, are addressed focusing on the NVH performance of the engine. By setting specific NVH performance targets while considering the engine's hybrid system demands, this research outlines a pathway for achieving quieter and smoother hybrid powertrains. The findings contribute to enhancing passenger comfort and driving experience, underscoring the importance of NVH optimization in future hybrid vehicle technologies.

**Keywords:** NVH · Dedicated Hybrid Engine · Range Extender

## 9.1 Introduction

For passenger cars the actual BEV sales have been significantly lower than expected in most regions. In Europe the BEV sales even dropped versus 2023 [1]. Changing circumstances have a dramatic impact on expected customer's technology preferences as can be seen when comparing technology predictions from 10/2024 with those one year earlier, Fig. 9.1. As shown in the graph, the sales volumes in the three large markets Europe, China and North America have been corrected to lower volumes respective to BEV shares. Thus, significant sales volumes—corresponding to a cumulative volume of about 100 million units (Europe: 29 Mio, China: 57 Mio, N. America: 17 Mio)—are shifted from BEV towards ICE based powertrains from 2025 to 2035.

Accordingly, the focus of powertrain development is shifting from battery electric to hybrid and range extended vehicles especially in the Asian market. This results in


A. Heintzel (Ed.): Automotive Acoustics Conference 2025, Proceedings, pp. 125–140, 2026. https://doi.org/10.1007/978-3-658-51249-1_9

development of internal combustion engines (ICE) dedicated to hybrid applications, the so-called Dedicated Hybrid Engines (DHE).

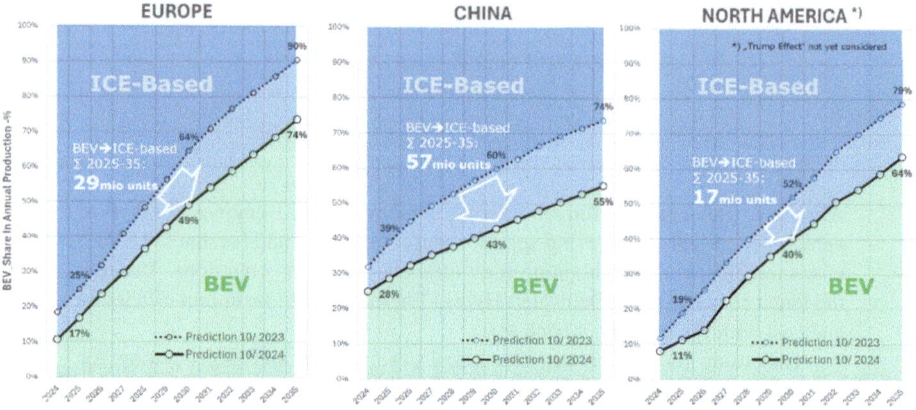


**Fig. 9.1.** Change of expected Technology Split, BEV share from 10/2023 to 10/2024

The expected change of technology split in the outlook requires a reimagination of propulsion concepts, supporting the transition from conventional powertrains towards pure BEVs. The high demand for ICE based powertrains and the variety of applications are enablers for the development of new DHE concepts and the optimization towards NVH, efficiency, packaging, cost and operating strategy.

The primary goal regarding Noise, Vibration, and Harshness (NVH) development for DHE applications is achieving stealth operation of the power unit (ICE & generator), making it imperceptible both inside and outside the vehicle. The ICE in the power unit is the main source of NVH, radiating airborne noise and introducing vibration (structure-borne noise) into the vehicle body. A careful consideration and selection of the engine type including an optimization of the ICE's NVH behavior is crucial for low perception levels during operation. As attributes like cost, weight, packaging and efficiency often take priority over NVH in DHE development, NVH is compromised too much in many applications leading to an inacceptable NVH behavior in the vehicle. To increase awareness for considering NVH as a main target already in the concept phase, this paper presents the most important features for a design concept of an NVH optimal dedicated hybrid engine application.

## 9.2 Specification/Definition of Dedicated Hybrid Engines

The concept of conventional IC engines (without hybridization) was mainly based on maximum performance and efficiency (fuel consumption), corresponding to displacement, number of cylinders and aspiration type fitting to the vehicle application. NVH was typically not part of the first three top priorities at least for the concept decision. Other aspects directing the selection of the ICE in traditional applications were also the branding of the OEM's, marketing drivers like sound or trends like downsizing.

For dedicated hybrid engines these requirements have been changed in the way that the ICE is a pure power source in DHE applications. The operating profile of the DHE is usually "disconnected" from the driver request as the hybrid control unit is managing the power flow in the system, automatically switching the ICE on and off (start and stop) on demand e.g. based on the SOC level of the battery. This "random operation" is the reason for the importance of the lowest possible NVH perception for DHE during operation, such as frequent start-stop cycles, operation under varying loads, integration with electric and non-correlated IC operation with driver requests. Ideally, noise and vibration of IC engine should not be detectable in the vehicle interior and external noise radiation from IC should be low [2].

The further one moves from traditional ICE towards "purely electric" drive (i.e., BEV) in the degree of hybridization, the more dedicated the ICE becomes. The most extreme version is the range extender engine (REX), which is solely driving a generator to produce electric power for the electric drive unit (EDU) propelling the vehicle, Fig. 9.2.

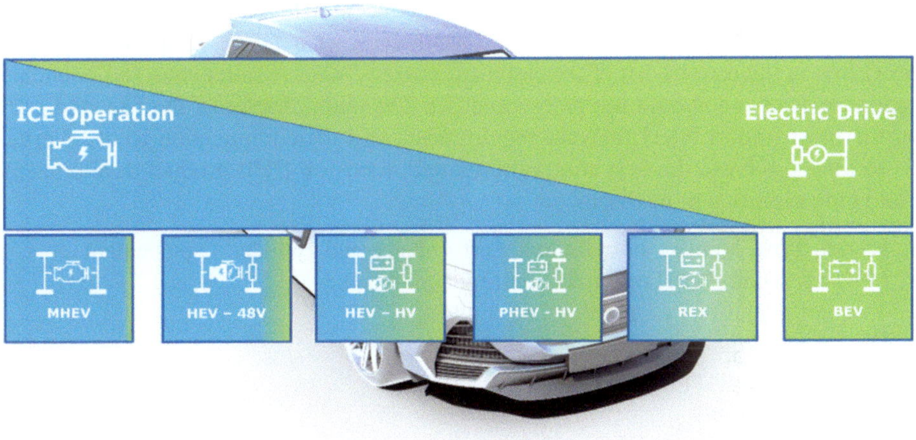


**Fig. 9.2.** Overview of hybrid powertrain concepts

To meet these requirements, NVH should be strongly considered already in the concept phase and therefore also in the decision for the base attributes of an IC engine like number of cylinders, aspiration (NA or TC) and combustion system (Atkinson or Miller). The concept for DHE described below considers the balance or trade-off between NVH, fuel consumption, cost, weight, packaging but focusing on NVH behavior.

Another important fact is the NVH integration concept of the DHE. The key requirement is ensuring low perceptibility, especially when concessions are made for ICE optimization. First hybrid units were developed for conventional ICE or hybrid vehicle architectures, utilizing existing engine compartment space for better NVH management. However, new DHE and especially REX units are integrated into BEV platforms, posing challenges for NVH integration due to limited space and different design requirements resulting from new vehicle architectures.

For cost-sensitive applications, integrating the REX and EDU into a single package can be suboptimal for NVH. Introducing a decoupling layer between the components can mitigate vibration transfer. In luxury vehicles, separating the REX and EDU units into different compartments allows for further NVH improvement. Measures on intake and exhaust systems, including tuned resonators and decoupler integration, are essential for minimizing airborne noise. Ultimately, the strategy to operate the REX unit for the ideal NVH experience is influenced by decoupling quality, encapsulation effectiveness and intake/exhaust noise reduction, in conjunction with vehicle operational conditions.

## 9.3 Basic Requirements of Dedicated Hybrid Engines Regarding NVH

Involving NVH considerations already in the concept selection and definition phase are crucial for the successful product development of a DHE. Even when putting NVH first in the priority list of development criteria, other requirements and boundary conditions can and should not be ignored [3]. The main chapter of the paper focuses on the most import NVH relevant requirements and their implication to other criteria to finally achieve the best compromises with focus on DHE product development. Following aspects will be discussed:

- Engine Type, Number of cylinders, Performance and cylinder displacement
- Load vs. speed (NVH optimal operating condition)
- Torque fluctuations vs. number of cylinders
- Design requirements
- Combustion concept vs. NVH
- Engine integration

### 9.3.1 Engine Type, Number of Cylinders, Performance and Cylinder Displacement

To select the right engine type, amongst other aspects the main excitation mechanism coming from the firing order and from the orders of unbalanced mass forces/moments needs to be considered. From NVH point of view the firing order should be high and the mass forces and moments of the engine should be fully balanced. Therefore, the ideal IC engine configuration would be an inline 6-cylinder, which is inherently balanced and has relatively small reciprocating masses due to the higher number of cylinders (see Fig. 9.3).

| Engine Type | Firing Order | Orders of Unbalanced Mass Forces / Moments | | Lowest Order | Highest Order |
|---|---|---|---|---|---|
| | | Mass Force | Mass Moment | | |
| 1-Cylinder | 0.5 | 1, 2 | | 0.5 | 2 |
| 2-Cylinder in-line (odd firing) | 0.5 | 2 | 1 | 0.5 | 2 |
| 2-Cylinder in-line (even firing) | 1.0 | 1, 2 | | 1.0 | 2 |
| 3-Cylinder in-line | 1.5 | | 1, 2 | 1 | 2 |
| 4-Cylinder in-line | 2.0 | 2 | | 2 | 2 |
| 5-Cylinder in-line | 2.5 | | 1, 2 | 1 | 2.5 |
| 6-Cylinder in-line | 3.0 | | | 3 | 3 |
| V6-Cylinder | 3.0 | | 1, 2 | 1 | 3 |
| V8-Cylinder | 4.0 | | 1 | 1 | 4 |
| 8-Cylinder in-line | 4.0 | | | 4 | 4 |

**Fig. 9.3.** Engine types, number of cylinders and excitation orders

Assuming 0.4l displacement per cylinder as a baseline, Fig. 9.4 illustrates the respective suggested engine types for different power requirements. The lower end in power requirement of each engine type can be covered with naturally aspirated (NA) engines with Atkinson combustion, whereas the mid to higher power versions of each engine type will be covered by turbocharged (TC) engines with a Miller combustion system.

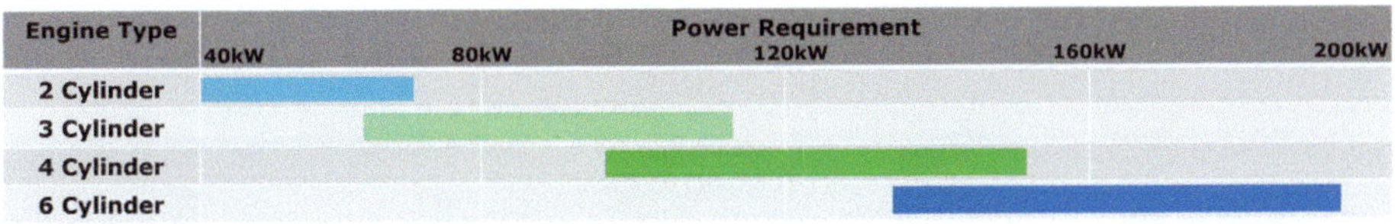


**Fig. 9.4.** Number of cylinders vs. power requirement (0.4l cylinder displacement and specific power of 50–80 kW/l)

Choosing a smaller displacement per cylinder is possible but will lead to reduced fuel efficiency in part load operation (Fig. 9.5). Choosing a larger cylinder displacement would have a small benefit in terms of fuel consumption, but going too big has a negative impact on NVH since this would result in the same performance with lower number of cylinders.

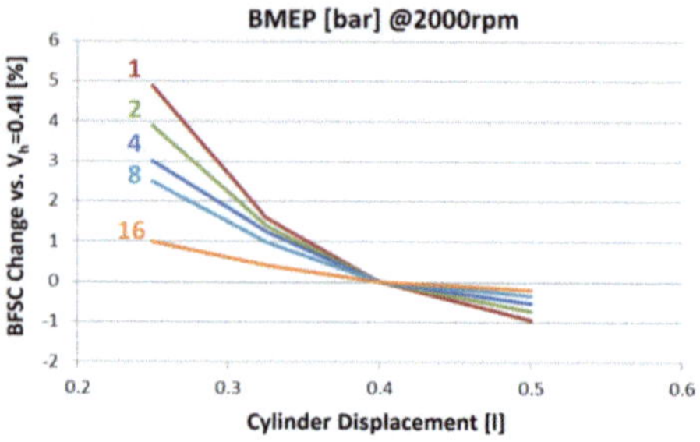


**Fig. 9.5.** BFSC change vs. cylinder displacement in dependence of BMEP

As the power demand for dedicated hybrid engines in general is moderate in relation to the power which could be delivered by a usual inline 6 engine, using an inline 6-cylinder configuration would then result in a rather small displacement per cylinder leading to increased fuel consumption especially in medium/part load operation (see Fig. 9.5). Therefore, it is apparent, that an inline 6-cylinder is not always an ideal selection for majority of a DHE with optimal NVH performance, even when considering relevant engine attributes and their effects on NVH for different cylinder configurations (Fig. 9.6).

Since a 2-cylinder configuration features high speed irregularity and requires a mass balancing unit and big volumes for intake and exhaust silencers to meet acceptable NVH levels, it is also not preferred for the application as an NVH optimal DHE. The benefit of its smaller packaging size cannot compensate for these attributes' detrimental effects on NVH.

| Engine Attributes | 2 Cylinder | 3 Cylinder | 4 Cylinder | 6 Cylinder |
|---|---|---|---|---|
| Speed Irregularity | high | medium-high | medium | low |
| Mass Balancing Unit | required | required | not required | not required |
| Packaging Engine | small | small-medium | medium | big |
| Packaging Intake /Exhaust Silencer | big | medium-big | medium | small |

**Fig. 9.6.** Engine attributes and NVH effects vs. number of cylinders (0.4l cylinder displacement and specific power of 50–80 kW/l)

Consequently, and as can be seen in the market, DHE applications are typically inline 4-cylinder (I4) or inline 3-cylinder (I3) engines with a cylinder displacement of approx. 0.4 L per cylinder. For NVH an I4 engine is strongly preferred compared to an I3 engine. To avoid critical vibration behavior in an I3 engine, a mass balancing unit is necessary. The addition of balancer shafts to the I3 engine, however, negate its advantages in friction and weight compared to an I4 engine. Even though a balanced I3 retains its smaller length versus the I4, this packaging benefit cannot be realized in an overall consideration, because larger volumes for silencing the intake and exhaust orifice noise need to be considered.

As we have seen in Fig. 9.5 smaller displacement per cylinder can lead to increased fuel consumption. With an identical total engine displacement an I4 features a smaller displacement per cylinder as an I3 and therefore a potential for less efficiency.

As displacement is closely related to the engine power, a decision on the number of cylinders must be made considering the tradeoff between efficiency (fuel consumption) and NVH.

If efficiency (fuel consumption) is the priority, the cylinder displacement should not be below 0.4l, which means if a performance of e.g. 80kW is needed, an I3 engine with a total displacement of 1.2l should be chosen. Figure 9.5 also shows, that at high loads (i.e. high BMEP) the fuel consumption disadvantage is minimal.

If NVH is the priority, an I4 with 0.25l cylinder displacement can be selected with less than 1% fuel consumption disadvantage at high loads (Fig. 9.5). As discussed in the next chapter, higher loads are not as critical for NVH (when compared to higher speeds to achieve identical power). Additionally, a DHE engine is not operated in its entire BMEP and speed map. Not only for NVH it is much more beneficial to operate the engine at or close to its optimum BMEP/speed operating point, which is usually at high load and medium speed.

If the power demand is lower than can be achieved with a 0.25l cylinder displacement in an I4, a naturally aspirated I4 should be considered before switching to an I3 engine to retain the I4's NVH benefits while saving weight and cost by eliminating the charging system.

### 9.3.2 Load vs. Speed (NVH Optimal Operating Condition)

In conjunction with the selection of the right engine size, it is important to consider the fact that DHEs are designed to be operated in a smaller speed/load map compared to traditional ICEs. DHE engines do not need to be operated e.g. in idle or at max performance points but in the so called "sweet spot" operating range for best fuel consumption. From NVH perspective the DHE should be operated at the lowest possible speed within this sweet spot area as generally, the noise increase vs. engine speed is higher than the noise increase vs. load (see Fig. 9.7). Accordingly, for an identical power demand low engine speed and high load is preferable for good NVH performance.

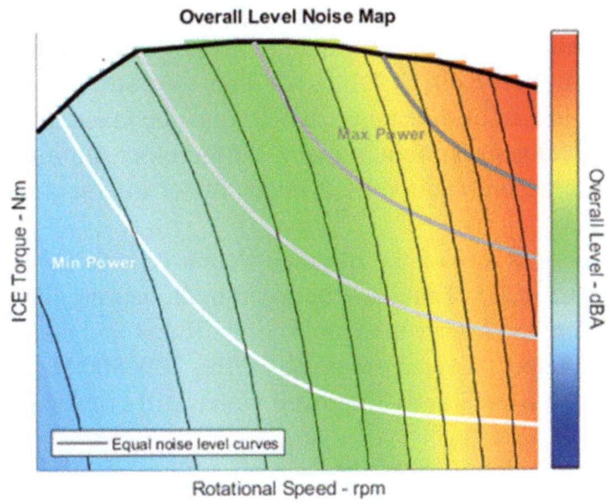


**Fig. 9.7.** Overall level noise map

### 9.3.3 Torque Fluctuations vs. Number of Cylinders

Another important aspect in selecting the right engine type and especially the right number of cylinders are torque fluctuations. The disadvantage of an I3 engine compared to an I4 is the higher torque fluctuation (or speed irregularity). Achieving the same mean torque in an I3 requires a considerably higher peak torque, which also occurs less frequently per crankshaft rotation (Fig. 9.8). This results in higher forces on the engine mounts, that hence need to be stiffer to support this torque. Higher stiffness of the engine mounts, however, reduces their vibration isolation properties. Not only must the engine mounts of an I3 be stiffer compared to an I4, the layout of the powertrain mounting is also more difficult, because separation of powertrain modes is critical.

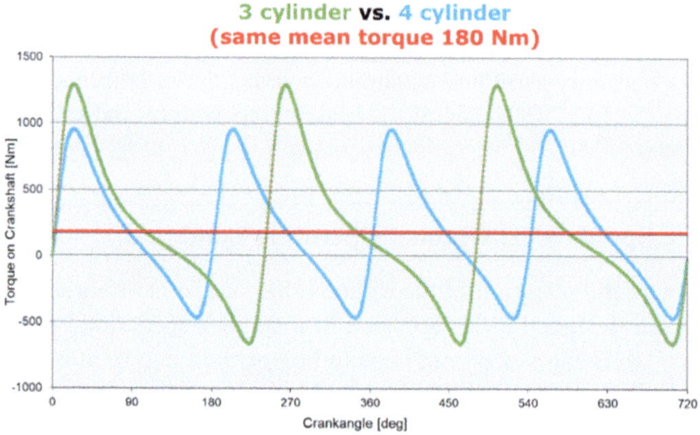


Fig. 9.8. Torque fluctuation 3 cylinder vs. 4 cylinder

### 9.3.4 Design Requirements

Designing the cranktrain of DHE for optimal NVH performance results in a relatively long conrod and stroke/bore ratio of 1,1 to 1,2. This long stroke design leads to a relatively big packaging height of the engine, which must be taken into consideration and balanced against the long stroke requirement. Furthermore, limiting piston slap noise can be achieved with a combination of cylinder offset in combination with a piston pin offset.

Crankshaft stiffness is one of the most important criteria for NVH. The weakest point of the crankshaft is the connection between main bearing journal and crank pin. The higher the overlap, the higher the crankshaft stiffness and consequently the better the NVH (Fig. 9.9). Especially engine noise in the 200–800 Hz range, usually perceived as rough, can be significantly improved with a stiff crankshaft design.

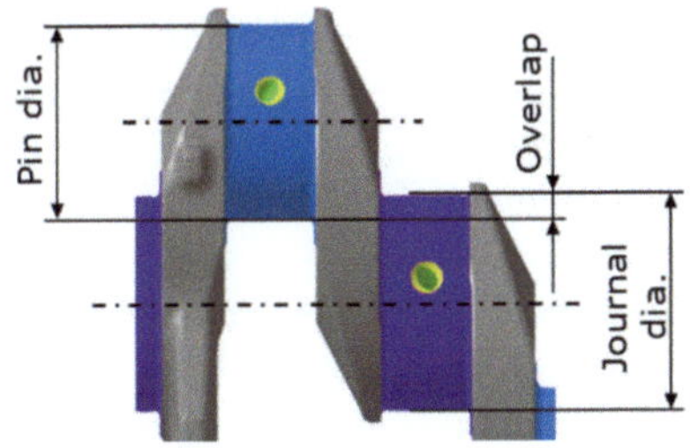


**Fig. 9.9.** Main bearing & crank pin overlap

An increase in main bearing and crank pin diameter also increases friction, but this influence is rather small on the overall fuel consumption. Figure 9.10 shows a friction walk converting a conventional engine into a DHE. The main steps for reducing friction are resulting from the removal of the FEAD (Front End Alternator Drive) belt system, a changed oil type (lower viscosity) and the electrification of components like e.g. the coolant pump. Removing the FEAD reduces noise radiation tremendously, since the crankshaft pulley of the belt drive acts as a kind of loudspeaker excited by the crankshaft vibrations. Additional advantage of eliminating the FEAD is the reduced friction; not only by reducing crankshaft load, but also because the sealing ring in the front cover can be omitted.

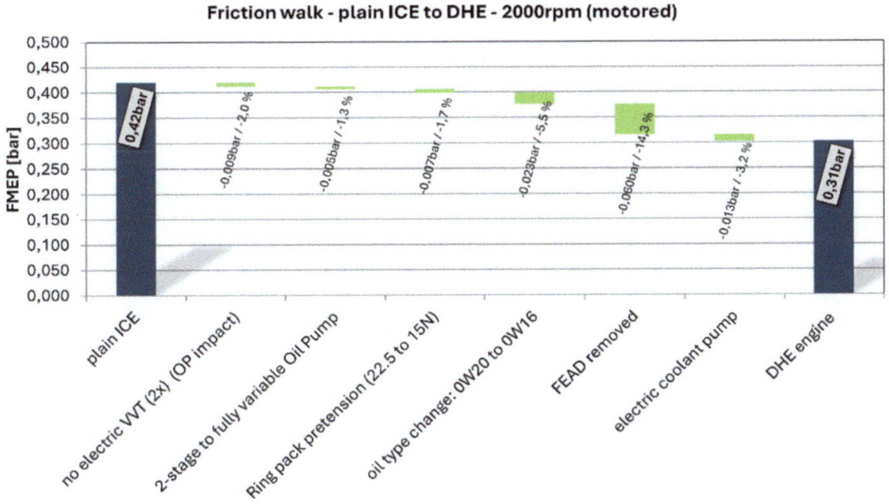


**Fig. 9.10.** Friction walk conventional ICE to DHE

The high voltage electrical system in a DHE application facilitates switching the accessories like A/C compressor, vacuum pump, coolant pump, etc. to electric drive independent of the crankshaft.

A further important feature are the counterweights. Here it is also beneficial to use a symmetric number of counterweights, so e.g. using 8 counterweights instead of 4 in an inline 4 cylinder or 6 instead of 4 counterweights in an inline 3 cylinder. Using a symmetric number of counterweights is beneficial, because the edge loading on the individual main bearings could be reduced [4].

Engine block design and especially the bottom end design containing the structure of the crankshaft bearings is important for NVH. Best solution for NVH is a bedplate design, which provides best stiffness with a disadvantage of added weight. A good alternative is a deep skirt design. If the crankshaft is quite stiff, the stiffness disadvantage for the deep skirt design compared to the bedplate is small with a clear benefit in powertrain weight.

The preferred solution for the oil pan is a 2-layer sandwiched sheet steel oil pan to reduce airborne noise radiation. Alternatively, a stiff aluminum oil pan can be used, but it will need local acoustic shielding of the oil pan surface area.

Minimum requirement for NVH regarding the timing drive design is a silent chain to avoid whine noise from chain drive. A belt-in-oil timing drive is preferable from NVH point of view, but increased service efforts to ensure durability often do not allow for this. The same principles should be applied to the oil pump drive, which for safety reasons is still engine driven and not yet electrified.

Another important NVH relevant parameter especially for controlling the mechanical excitation are the clearances. To minimize force impact into the engine structure, the clearances of main bearing, the conrod bearing as well as the piston to liner clearance need to be as small as possible. A minimum backlash in all engine gear drives is also mandatory for lowest possible mechanical noise.

### 9.3.5 Combustion Concept vs. NVH

The main influencing factor on noise caused by the combustion process is the pressure rise rate. Pressure rise vs. engine noise is always a compromise depending on the priority between NVH and efficiency (fuel consumption). On the one hand high pressure rise is beneficial for low fuel consumption, on the other hand high pressure rise causes high noise radiation and vibration levels on the engine structure. For maximum pressure rise values below 3.5 bar/°CA combustion noise typically has low influence on total engine noise (Fig. 9.11). Maximum pressure rise values above 5 bar/°CA can lead to high combustion noise contribution to the overall engine noise (especially rough or “growling” noise). How critical maximum pressure rise values are affecting the NVH depends a lot on the engine structure (stiffness behavior of crankshaft, engine block, etc.). A general rule is higher engine stiffness allows for higher pressure rise rates and still reasonable NVH.

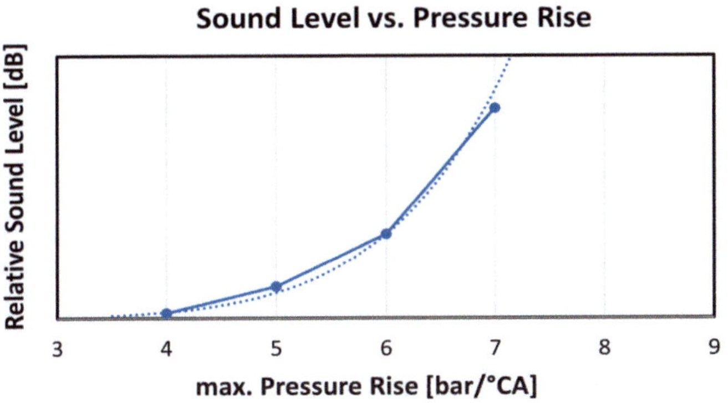


**Fig. 9.11.** Sound level vs. pressure rise

Regarding the combustion cycle, the Miller process is preferred in turbocharged engines due to its better thermodynamics especially at high and low loads, which leads to higher specific power. In a naturally aspirated engine, the Atkinson process is preferred, since the Miller valve lift curve with short duration/low lift would affect the air flow too much. Reconciling the trade-off between low fuel consumption and slow pressure rise is feasible with exhaust gas recirculation (EGR). EGR systems, however, add cost and complexity in form of higher efforts for calibration.

### 9.3.6 Engine Integration

The most vital part for an optimal NVH integration of the power unit into the vehicle is the powertrain mounting. The layout and design of the powertrain mounting depends on the chosen (hybrid) powertrain concept and packaging possibilities.

If the powertrain has no direct connection from the ICE to the wheels—i.e. a range extender serial hybrid—mounting the ICE and generator power unit separately from the propulsion EDU is the preferred solution for NVH. In this case the engine mounts can be quite soft, since no static torque needs to be transferred, and therefore have a high isolation effect. To further reduce vibration transfer via the mounts a double layer decoupling system is recommended, where the ICE and generator power unit is mounted to a subframe with decoupling elements and the subframe is again decoupled from the vehicle body.

In case of a parallel hybrid application, with a mechanical direct drive connection to the wheels, the powertrain installation is more like conventional ICE powered vehicles with a typical split of using powertrain mounts to carry the load caused by the weight of the powertrain in combination with a torque arm supporting the positive or negative torque load caused by the wheels. Compared to serial hybrid or REX application also in this case a double layer decoupling is the preferred solution for NVH.

The main factor in mitigating the vibration transfer from the power unit to the vehicle body are the mounting brackets. They need to be well-integrated into the power unit structure (e.g. the engine block) and should be extremely stiff to reduce vibration transfer. Also, the vehicle body needs to be stiff enough at the connection points of the powertrain mounts to avoid high vibration and noise transfer into the body.

Secondly, intake and exhaust systems are another considerable source for noise in a DHE vehicle. For turbocharged engines, possibly flow noise dampers (high frequency absorbers) should be applied and in the intake system resonators (Helmholtz or pipe resonators) should be used to reduce noise radiation from the intake orifice.

The damping volume for the exhaust system should be rather high to avoid noise radiation from the exhaust orifice. Necessary heat shields should consist of a double layer concept and mounted with isolation (e.g. metal wire mesh washers) from the engine structure.

Additionally, secondary measures to mitigate noise radiation should be considered. Especially, if the noise of the IC engine should not be detectable for the vehicle passengers at all. Secondary measures consist of local shielding or part encapsulation of critical parts like fuel rail, fuel pump, oil pan or other local areas of high noise radiation using molded foam or double layer concepts with absorption/mass damper effect as well as total encapsulation of the power unit.

## 9.4 DHE Operating Strategy for Optimal NVH

Optimization of the engine operating strategy of dedicated hybrid engines depends on the type of application and differs based on hybrid architecture. In principle it can be split into two main base systems, the serial hybrids (where the REX represents a typical application) and parallel hybrids. In general, the NVH target for operating the DHE is linked to the interior noise level apparent in pure electric drive mode [5]. This level is limiting the operation of the DHE independent of the hybridization concept (see Fig. 9.12).

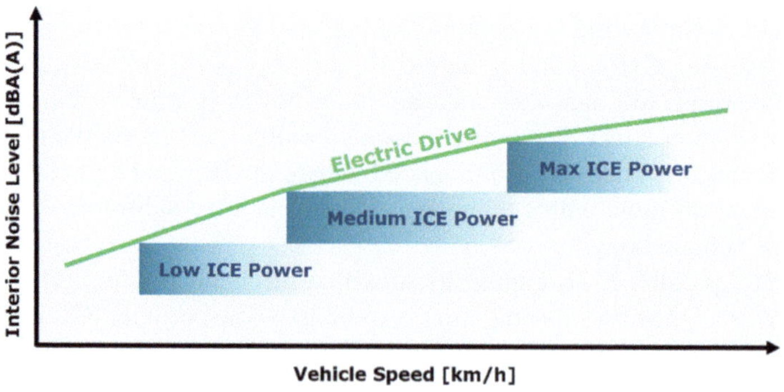


**Fig. 9.12.** Interior noise limit for DHE ICE operation

### 9.4.1 Serial Hybrid—Range Extender

A key characteristic of Serial hybrid or range extender is that all propulsion power comes from an electric motor with a power electronic inverter, which takes its power from a direct current link. The power to the DC link, besides coming from the battery, can also come directly from a generator, which is driven by a combustion engine.

The engine speed can be selected freely, irrespective of the vehicle speed. In REX applications the most common operating strategy is determined by the sequence of charge depletion (CD) and charge sustaining (CS) mode, where the IC engine is activated and deactivated as function of the battery state of charge.

The DHE is preferably operated at the maximum efficiency power point, turning the engine off when the SOC has reached a defined upper limit (Fig. 9.13). To ensure an optimal NVH performance, another control variable must be introduced, the vehicle speed. The IC engine is activated at high vehicle speeds taking advantage of the masking noise level caused by the vehicle, making the engine noise less prominent inside the passenger compartment.

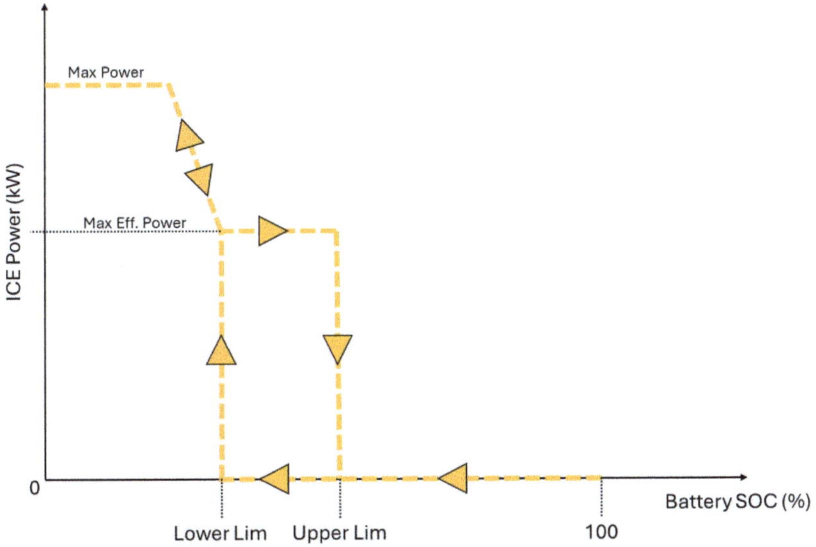


**Fig. 9.13.** ICE torque demand during REX operating strategy

When the IC engine is used to recharge the battery, it is preferable to reach the target power by increasing the torque rather than the engine speed (noise level increase vs. speed is much higher than vs. load). The start/stop NVH behavior of the IC engine should not be recognizable, because it could happen independent from user action. Load and speed must be varied in slow progression to ensure a smooth transition in the noise and vibration level.

For operating conditions like stand still and at low vehicle speed, the best strategy is to turn off the IC engine and leverage the electric propulsion to propel the vehicle.

During braking, the best strategy both for energy consumption and NVH is again to switch off the IC engine.

As the engine operates at a fixed rpm point or in a limited speed range, its components must be designed in a way so their resonances are outside the frequency range of firing order and combustion energy excitation.

### 9.4.2 Parallel Hybrids

The word parallel refers to the fact that the torque from the electric machine is added to the torque from the combustion engine and the total torque propels the vehicle. For this reason, the engine and possibly the electric motor are connected to a transmission that typically allows gear shifting.

Parallel hybrids allow, depending on the architecture, the removal of the starter motor and the start of the engine with the electric machine. A fast speed ramp is strongly recommended to avoid the shaking during pass through of the powertrain mounting resonances. For low loads and low vehicle speed it is recommended to keep the engine off and run in full electric mode.

When the external load requirement is higher than the electric machine maximum capability and the IC engine is consequently turned on, it must be operated at the minimum possible speed, selecting the highest possible transmission gear. The enabler for this strategy is the battery that must be used to reduce the engine power such that the operating point on the next higher gear is one of the optimal NVH operating points (Fig. 9.14).

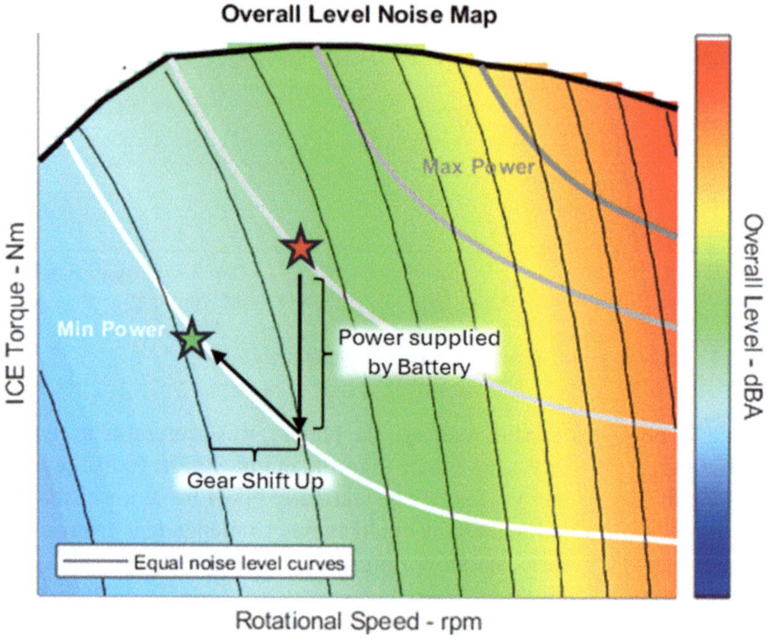


**Fig. 9.14.** ICE optimal NVH operating point in parallel hybrid

When the need to recharge the battery using the IC engine occurs, it is preferred to do this operation increasing the engine power at high vehicle speed, avoiding a relevant increase in the perceived noise. Achieving the higher power demand by torque increase is again preferred over engine speed.

At vehicle stand still, as for range extenders, the best solution for NVH is to switch off the IC engine.

During braking, it is optimal both from efficiency and NVH perspective to not operate the IC engine and to open the clutch with the result of cancelling both the combustion and the mechanical noise radiation from the IC engine.

## 9.5 Conclusion

As the global automotive market experiences a slower-than-anticipated transition to Battery Electric Vehicles (BEVs), the focus is shifting toward hybrid and range-extended powertrains. This shift has led to the need for the development of Dedicated Hybrid Engines (DHEs)—internal combustion engines designed solely for the application in hybrid systems.

Unlike traditional engines, DHEs operate under unique conditions: frequent start-stop cycles, variable loads, and asynchronous operation with vehicle speed. These characteristics make NVH critical design priority. The goal is to achieve near-invisible engine operation, both acoustically and vibrationally, to preserve the quiet driving experience expected from electrified vehicles.

**Key Technical Considerations**

- Engine Configuration: Inline 4-cylinder engines with ≤ 0.4 L displacement per cylinder are preferred for their balance between NVH, efficiency, and packaging. Inline 3-cylinder engines require additional balancing measures and often compromise NVH.
- Mechanical Design: Long-stroke engines with high crankshaft stiffness, minimized clearances, and optimized counterweight configurations reduce mechanical noise and vibration. Silent chains and belt-in-oil systems are recommended for timing drives.
- Combustion Strategy: The Miller cycle (for turbocharged engines) and Atkinson cycle (for naturally aspirated engines) are favored. Pressure rise rates should be kept below 3.5 bar/°CA to minimize combustion noise, with EGR systems used to balance efficiency and NVH.
- Integration and Mounting: Decoupled mounting systems, especially in serial hybrid (REX) configurations, are essential. Double-layer decoupling and subframe isolation help prevent vibration transfer. Acoustic treatments such as resonators, encapsulation, and shielding further reduce airborne noise.
- Operating Strategy: In serial hybrids, the engine should operate at high vehicle speeds and optimal efficiency points. In parallel hybrids, the engine should be engaged only when necessary, favoring low-speed, high-load operation to minimize noise.

The development of DHEs requires a comprehensive, NVH-focused approach that begins at the concept phase. By carefully selecting engine architecture, optimizing mechanical and combustion systems, and integrating the engine thoughtfully into the vehicle, manufacturers can deliver hybrid vehicles that meet modern expectations for comfort and refinement. NVH optimization is not a secondary concern—it is a defining factor in the success of DHEs as a transitional technology toward full electrification.

## References

1. Mark Lines (2025) Vehicle Sales Database
2. Tschöke H, Gutzmer P, Pfund T (2019) Elektrifizierung des Antriebsstrangs: Grundlagen—vom Mikro-Hybrid zum vollelektrischen Antrieb. Springer Vieweg, Berlin
3. Dr. P. Kapus, Dipl.-Ing. L. Leonhartsberger, Dipl.-Ing. (FH) F. Hoelbling, A. Eckart MSc, Dr. A. Huss (2023) Holistic View on a Dedicated Hybrid Drivetrain—Thermodynamics, Thermal Management and Controls. In: Proceedings of the 44th International Vienna Motor Symposium: 26–28 April 2023. ÖVK, Vienna, Austria
4. Yusof MY, Nik Abdul Aziz F, Zuhdi MF, et al (2009) 4 Versus 8 Counterweights for an I4 Gasoline Engine Crankshaft—Measurements of Vibration and Bearing Wear. SAE Int J Engines 2:1777–1785. https://doi.org/10.4271/2009-01-1938
5. Langjahr P, Resch M, Graf B (2020) HEV Evaluation in Simulation Phase Based on Predicted Sound Behavior. pp 2020-01–1511

# Full NVH Predictions from Day One: Enhancing Hybrid NVH Models with Benchmark Database Substructures

Christoph Steffens[1(✉)], Michael Häussler[2], Christian Speuser[3], and Rene Göbbels[3]

[1] FEV Europe GmbH, Aachen, Germany
steffens@fev.com
[2] FEV Vehicle GmbH, Aachen, Germany
haeussler@fev.com
[3] FEV Consulting, Aachen, Germany
{speuser,goebbels}@fev.com

**Abstract.** Achieving vehicle-level NVH targets at the driver's ear requires robust predictions of electric drive-unit (EDU) sound long before physical prototypes exist. Acoustic target cascading has traditionally relied on legacy-vehicle experience to derive component targets. Full NVH evaluations at the driver's ear had to wait until the first vehicle prototypes. Today's compressed development cycles benefit from modular NVH substructure models that can be snapped together like building blocks within a substructuring framework. Experimental models of benchmark vehicles are used in early development phases as placeholders where numerical models are missing or lack the required accuracy. In the resulting hybrid NVH model, experimental substructures, blocked-force data and airborne-source descriptions from benchmark vehicles are coupled with CAE models. The proposed workflow delivers realistic NVH predictions to the driver's ear from day one of the development program.

Two complementary use cases illustrate the value of the approach:

1. Supplier / EDU-developer perspective—They virtually integrate the EDU model (testbench data or simulations) into benchmark vehicles, overlay the measured road- and wind-noise masking-spectra, and can therefore judge whether expensive design refinements will deliver an audible benefit at the driver's ear and potentially avoid over-engineering.

2. Vehicle-integrator perspective—Full Vehicle NVH teams can create a digital NVH twin of a future vehicle immediately, using benchmark substructures as placeholders. As development progresses, these placeholders are seamlessly replaced by updated CAE models or early hardware data, maintaining a continuous, decision-ready NVH assessment thread.

This paper outlines the methodology, required data quality and validation strategy, and addresses key challenges such as the frequency- and preload-dependent behavior of rubber bushings. Finally, it shows how the workflow


A. Heintzel (Ed.): Automotive Acoustics Conference 2025, Proceedings, pp. 141–153, 2026. https://doi.org/10.1007/978-3-658-51249-1_10

is seamlessly embedded in our existing vehicle-benchmark program, allowing customers to tap into an ever-growing library of ready-to-use substructures.

**Keywords:** NVH · Digital Twin · Substructuring · Blocked Force

## 10.1 Introduction

The shift from combustion to electrified powertrains has removed the dominant masking noise once provided by the internal combustion engine and exposed a variety of tonal, higher-frequency NVH phenomena originating in the electric drive-unit and inverter. At the same time, product cycles have contracted to the point where the first full-vehicle prototypes appear much closer to SOP. To keep decisions "NVH-ready" throughout this compressed timeline, OEMs and suppliers are increasingly turning to hybrid, component-based modelling that couples measurement-based substructures with CAE data. Authors affiliated with BMW [1, 2] and Volvo [3] used Frequency-Based Substructuring (FBS) and blocked forces to validate the predictions of EDU sound at the driver-ear before a rolling prototype exists.

The methods formerly under research have been formalized with ISO 21955 [4] (transfer of active component forces from testbench to final receiver) and ISO 20270 [5] (in-situ blocked forces, currently under revision), underscoring the fact that modular, hybrid NVH modelling is rapidly moving from research into routine industrial practice. Hardware- and software providers in the NVH domain mirror this trend: Siemens Simcenter, VIBES Technology and HEAD Acoustics among others, now offer standard workflows for component-based TPA and Substructuring. The off-the-shelf workflows let OEM NVH teams and EDU suppliers exchange modular substructures as a common communication layer for NVH.

Additionally, new battery-electric platforms often differ so radically from legacy architectures that extrapolating NVH targets from earlier platforms is no longer reliable or results in too conservative over-engineering. A more robust starting point is to draw on benchmark vehicle derived substructures—vehicle body models, mount FRFs and blocked-force sources—from vehicles whose layout and mass distribution already resemble the planned concept more closely. Because each substructure is stored with a standard six degree of freedom (DoF) interface at coupling points, engineers can combine them like Lego bricks and run first acoustic checks on day one, without waiting for a vehicle prototype. The value of such a benchmark structure database is illustrated with two concrete use cases in the following:

**Use case 1—EDU supplier / developer**
EDU developers are typically asked to meet strict vibration limits while receiving little or no information about the customer's trimmed vehicle body transfer function. By virtually mounting their EDU model into a bushing isolation and vehicle body

from a benchmark database, they can quantify the return on investment of tighter gear tolerances, premium bearings or harmonic-current-injection (HCI) strategies before committing to costly changes in their product. Road- and wind-noise scatterbands supplied with each benchmark vehicle reveal whether further noise reduction would even be audible to the driver. Additional benefits include:

- Objective comparison with competitor EDUs in the same vehicle segment.
- Fact-based discussion of specification-sheet limits—distinguishing realistic targets from overly conservative values that simply shift all risks related to the vehicle transfer function to the supplier.
- Option to work with a generalized transfer function—an average of several receivers—to avoid optimizing towards particularities of a single vehicle.

**Use case 2—Vehicle integrator / OEM NVH team**

Full-vehicle NVH groups have to define the isolation concept—and thus excitation levels, mount stiffness, body stiffness and path sensitivities—already at concept freeze. Using benchmark placeholders, they can build a digital NVH twin on day one and compare options such as single- or double-isolation. With this prediction, they can additionally evaluate the use of active sound design to mask or shape power-train noise at an early stage. Whenever new CAE results or early-hardware data become available, the model is updated and the targets are re-evaluated, allowing the team to converge on the most sensible and cost-effective solution. This continuous thread keeps troubleshooting and root-cause analysis ahead of the first prototype cars, so the cheapest effective countermeasure can be chosen early.

Assembling the full vehicle transfer function with FBS also yields insight that physical tests cannot. Transfer paths that show a poor signal-to-noise ratio in the complete vehicle can be reconstructed by coupling accurately characterized component substructures. Figure 10.1 illustrates this principle for an electric climate-compressor: the predicted FRF from compressor to carrier (dashed black line) closely follows the validation measurement (solid blue line), while the coherence plot shows that, above about 300 Hz, even the first isolation stage cannot be measured with sufficient signal to noise rate in the full vehicle [6]. The example in Fig. 10.1 further demonstrates the flexibility of the approach: measured data, an analytical rigid-body model, and an FEM model are combined to predict the full-vehicle transfer function. In short, architectural decisions can be made—and, if required, revised—long before a physical prototype exists.

The remainder of this paper first details the methodological basis for this hybrid substructuring workflow—including the required data sets and quality-assurance checks—then addresses practical challenges such as frequency and preload-dependent mount stiffness and their mitigation, and finally shows how the approach is embedded in FEV's ongoing vehicle-benchmark program.

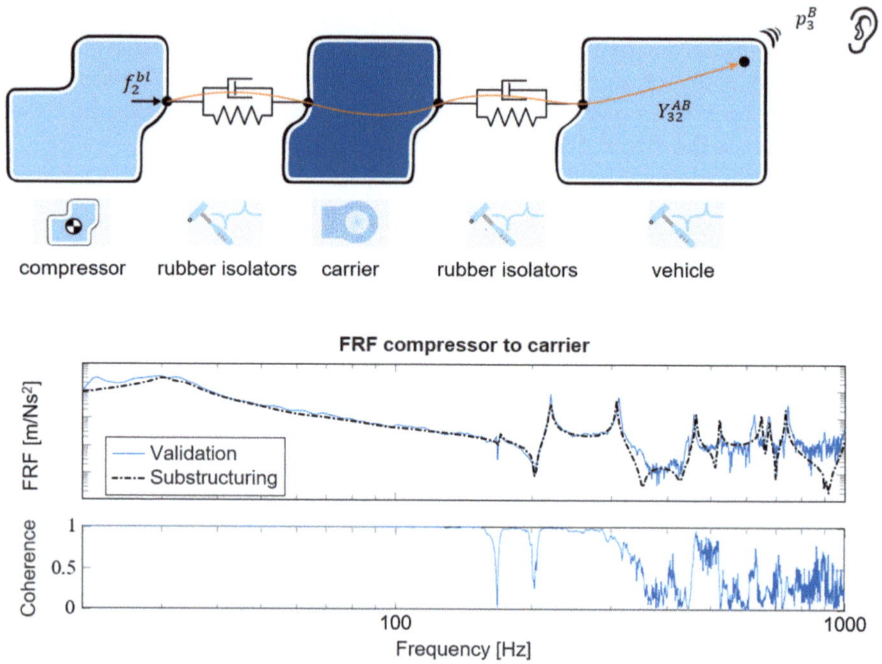


**Fig. 10.1.** Top: schematic of the two-stage compressor isolation used for a substructuring study [6]. The compressor is idealized as a rigid body, the rubber mounts are characterized by 12-DoF experimental models, the intermediate carrier is represented by a finite-element model, and the receiver vehicle body is derived from a mule vehicle. The substructure models are coupled via frequency-based substructuring to predict the transfer function. Bottom: Validation of simulation results with measurements on the prototype [6].

## 10.2 Technical Basis and Data Requirements

### 10.2.1 Frequency-Based-Substructuring (FBS) in a Nutshell

Vibration models generally describe the relationship between dynamic forces and the resulting motion. They may stem from simulation—multi-body or finite-element (FE) analyses—or from experimental frequency-response-function (FRF) measurements. Simulation is ideal for components still in design because geometry or material changes can be evaluated quickly; measurement is preferable for existing parts with complex transmission paths, as it avoids laborious modelling and validation. FBS couples such heterogeneous models in the frequency domain [7]. At every interface between two components A and B, two boundary conditions apply:

- Compatibility Translational and rotational motions are identical on both sides of the interface between component A and B.
- Equilibrium Interface forces are equal in magnitude and opposite in direction.

Expressed in matrix form, these conditions are enforced on the FRF matrices of the substructures in the assembly [7].

**Virtual Point Transformation (VPT)**
A practical challenge when creating experimental FRF models is to capture all six degrees of freedom (DoF)—three translational, three rotational—at each coupling point. VPT provides a systematic solution [8]. Multiple accelerometers and force inputs are distributed around the mounting point; the measured data is then geometrically projected onto a single virtual point that represents the interface motion in six-DoF space. Including rotational DoFs is essential for accurate predictions at higher frequencies [9].

**Quality Checks on experimental substructure models**
Because the VPT requires sensor and impact locations on the local geometry, it enables a multi-layer checking scheme during the FRF measurement campaign to ensure that the final test-based model is both consistent and physically plausible [10]:

1. *Operational-deflection overview:* With full geometry available, animated vibration shapes of the measured FRF are reviewed at the test bench to spot obvious mounting or instrumentation errors at a glance.
2. Signal integrity: All raw channels are screened for sensor overloads, short or open circuits, double hammer hits, poor signal-to-noise ratio and too low coherence. Suspect impacts are repeated, or a hammer with different mass or tip is selected.
3. *Virtual-point consistency:* The six-DoF motion reconstructed at each virtual point (VP) is back-projected onto the individual sensor channels. Large differences to the raw measurement flag mis-located sensors, high noise or a violation of the local rigid-body assumption underlying the VPT. This check also gives a clear value for the maximum frequency up to which the model is accurate [8].
4. *Virtual-point physicality:* Basic physical rules such as drive-point passivity are checked directly on the VP FRFs to verify that the transformed data respect energy conservation constraints.
5. *Reciprocity between virtual points:* For subsystems with more than one VP, the VP to VP transfer functions are compared. Good agreement indicates linear, noise-free data; significant deviations signal modelling or measurement errors that must be resolved before the substructure is released for coupling.

### 10.2.2 Rubber-Bushing Models

Rubber mounts are the decisive elements that decouple the EDU from the trimmed body and therefore deserve the same modelling care as the powertrain itself. Two test-based approaches are used in the creation of the substructure database. The EDU mount under test (denoted as I) is fixed between two rigid fixtures (denoted as A and B). The fixtures are used for vibration measurement with accelerometers and excitation with impact hammer, see Fig. 10.2. The resulting FRF matrices are transformed to VP's with six DoF on either end, so the resulting admittance matrix Y^AIB has a size of 12x12.

**Inverse substructuring (IS)**

The dynamic stiffness of the rubber mounts can then be obtained, as a quick approach, via inverse substructuring. This is done via time-integration of the acceleration signals to displacements and subsequent inversion of Y^AIB. Thereby the dynamic stiffness matrix Z^AIB of the assembly AIB (the two fixtures and the mount) is obtained:

$$Z^{AIB} = \left(Y^{AIB}\right)^{-1}$$

The off-diagonal terms of Z^AIB reveal the transfer stiffness of the rubber bushing itself, without the influence of the rigid blocks used for excitation and measurement. This can easily be illustrated when considering the dynamic stiffness matrix of a simple spring k fixed between two masses m:

$$\begin{bmatrix} -m\omega^2 + k & -k \\ -k & -m\omega^2 + k \end{bmatrix}$$

The stiffness of the spring can be read on the off-diagonal as -k. For more information, please refer to [11]. Figure 10.2 compares three mounts stiffnesses obtained via inverse substructuring. The mounts have nominally different static stiffnesses, simply denoted as soft, medium and hard in the figure. The stiffness curve highlights the pronounced stiffness rise due to internal resonances in the rubber continuum starting at 500 Hz that must be contained in the mount model when coupling the substructures into the full vehicle model, especially when dealing with high frequency EDU noise.

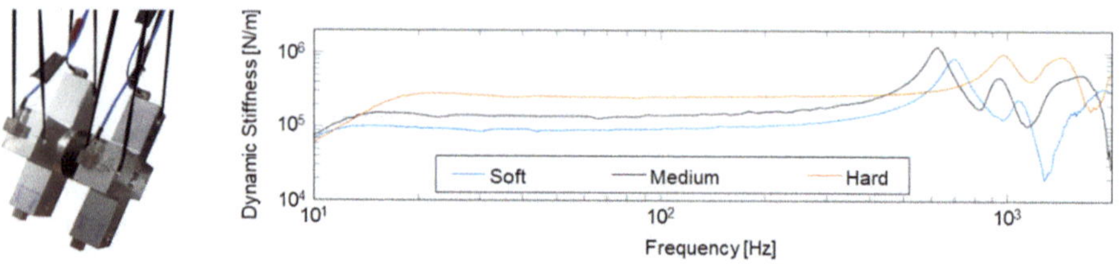


**Fig. 10.2.** Experimental 12-DoF characterization of rubber mounts. (left) test set-up with the mount clamped between two locally rigid adapters and instrumented with tri-axial accelerometers. (right) Dynamic stiffness for three mount variants. All mounts exhibit large dynamic stiffening for frequencies higher than 500Hz due to internal resonance effects—features that must be preserved when the mounts are coupled into the full vehicle model for EDU sound.

**Frequency-based substructuring (FBS): decoupling of fixtures**

When using a rubber bushing model to create a hybrid model of the full vehicle, it is more accurate to model the fixtures A and B and mathematically remove them from the measured matrix Z^AIB. This can be done by simply modeling the fixtures as rigid bodies and decouple them with Frequency based substructuring, see [11]. The resulting 12-DoF matrix of the pure rubber element preserves cross-coupling-terms and therefore delivers a more reliable basis for substructure coupling. The fixtures can be designed to behave rigid up to 2–3 kHz, and therefore the rubber bushing models are valid in the same frequency range.

**Non-linear effects**

The above-mentioned models already contain the frequency dependent stiffness behavior. However, static preload, temperature and vibration amplitude can additionally modify the rubber stiffness. These effects are not represented in the linear models described above. Yet fleet data indicate that more than 95% of real-world driving occurs at low EDU torque, where preload effects do not yet change the bushing stiffness. A linear rubber bushing model is thus adequate for most NVH predictions. If higher fidelity is required, two extensions are available:

- A high-frequency rubber testbench which allows to apply a given preload to the rubber mount as in [12]. The pre-load dependent stiffnesses shown in [12] validate that preload effects only change the bushing stiffness at higher preloads.
- An in-vehicle transmissibility approach that derives the operational stiffness from vibration ratios measured before and after the mount and local dynamic stiffness measurements on the vehicle body. An example is the established FEV VINS method [13].

### 10.2.3 Excitation Source Descriptions: Structure- & Air-Borne

**Structure-borne source description**

For the structure-borne noise excitation of the EDU, methods from component-based Transfer Path Analysis (TPA) should be used [14], as they provide an independent characteristic of the excitation that can be reused in hybrid NVH models.

- **Preferred metric—blocked forces:** 6-DoF blocked forces are identified at each mounting point of the EDU in accordance with ISO 20270 / ISO 21955. This representation can be transferred to any hybrid NVH model of vehicles under development. When the EDU is mounted on a test bench, it is typically possible to obtain all DoF at the mounting points.
- **In-vehicle alternative—pseudo forces:** If the EDU is in the benchmark car and some mounting points are inaccessible, we determine a set of pseudo forces (on accessible points on the EDU housing) from indicator responses close to the mounting points [14]. Although a little less transparent, this still retains source independence and allows transferring these forces to other assemblies for early vehicle predictions before a prototype is ready.

**Air-borne source descriptions**

- **Vehicle environment**—limited space around the EDU favors operational TPA or the FEV VINS [13] transmissibility approach. A sparse microphone array around the unit delivers operational responses which are combined with cabin transmissibility functions for prediction of the airborne noise share.
- **Test-bench environment**—with the housing fully exposed on a semi-anechoic test-bench, the sound power can be obtained as a independent description of the

airborne source strength. A more rigorous approach for transferring airborne sound descriptions into a hybrid vehicle model is to identify a set of equivalent monopoles. At least 6 monopoles, one per side of the EDU, is recommended [15, 16]. These monopoles, together with blocked forces, give a unified hybrid source model that can be coupled to any benchmark vehicle.

**Data-quality checks**
Before a source model is admitted to the benchmark database it passes three layered checks:

1. **On-board validation**—blocked-forces and monopole strengths are re-applied to the original measurement assembly. The predicted responses must stay within a limited residual to the raw measurement.
2. **XDOF Algorithm**—Evaluates the Linear Independence of Identified Forces or Monopoles and Suppresses Spurious Extra DoFs [17].
3. 3.**Interface Completeness Criterion (ICC)**—confirms that the chosen set of structural and acoustic DoFs fully describes the interface of the source with a receiver. A rising residual indicates missing paths and triggers re-measurement [9] of those paths.

Together, these steps ensure that both structure-borne and air-borne excitations are "decision-ready" for continuous NVH target setting and virtual vehicle assembly.

### 10.2.4 Masking Layer Auralization

After the substructures have been coupled, yielding the structure-borne and airborne transfer paths, applying the corresponding source quantities produces a spectrum of the pure EDU contribution at the driver's ear. To judge whether that contribution will be perceived, the prediction is overlaid with a speed-dependent scatterband of rolling- and wind-noise levels that is supplied together with each benchmark receiver. The grey envelope in Fig. 10.3 shows the spread measured for C-segment battery-electric vehicles; the three colored curves mark threshold lines for when a tonal component is merely audible, clearly apparent or fully dominant. By comparing the predicted tonal orders against these limits, engineers can set or revise EDU targets and decide whether further design effort is justified.

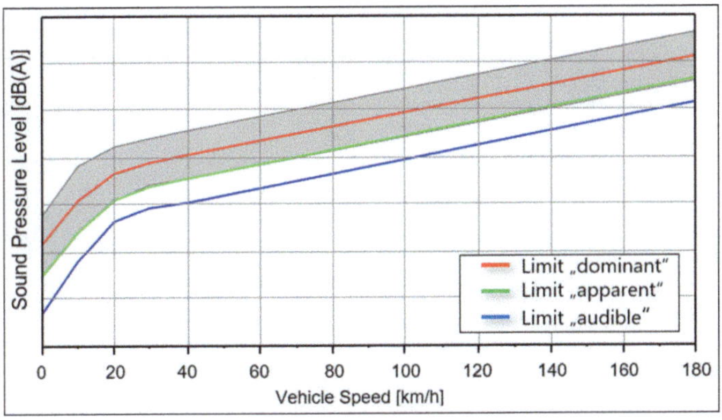


**Fig. 10.3.** FEV Speed-dependent scatterband of rolling and wind noise for C-segment BEVs (grey envelope) with derived tonal-noise thresholds: audible (blue), apparent (green) and dominant (red).

For a perceptual cross-check, the time-domain signals—EDU, background noise and, if desired, additional sound-design content—are reproduced in an NVH simulator. Listening sessions confirm whether the chosen isolation concept satisfies subjective expectations or, if necessary, help explore countermeasures such as active sound design before committing to hardware changes.

## 10.3 Practical Implementation Within the FEV NVH Benchmark Program

FEV has more than 30 years of experience in deep technical benchmarking. With four benchmark centers worldwide, more than twelve technical centers and a wide range of test facilities, FEV performs comprehensive benchmarking analysis programs. FEV offers a unique combination of testing capabilities, design assessment and cost evaluation. This enables us to analyze technical products from multiple dimensions and provide a complete picture of why they sound so well, are so efficient or can be delivered at such low cost. FEV Global Benchmarking is continuously monitoring the global market regarding latest technologies and vehicles and runs multiple benchmarking programs per year. Figure 10.4 shows some of the latest vehicles in the global benchmarking program.

**Fig. 10.4.** Latest vehicle benchmarking program from the FEV's global benchmark

The hybrid NVH workflow described here slots into that infrastructure: the vehicles are also available for substructuring purposes, next to cost, design and efficiency benchmarking, which is giving NVH engineers a broader context for their acoustic findings.

### 10.3.1 NVH Benchmark Models: Data -Generation Workflow

The first step towards the hybrid NVH models, is an optical 3-D scan of the instrumented components and mounting points. The scanned geometries of the components are aligned with respect to each other and the vehicle coordinate system, and six-degree-of-freedom virtual points are defined at each EDU mounting point. See Fig. 10.5 for an example. This geometry backbone guarantees that all subsequent FRF data are kinematically compatible for substructuring, and the above mentioned quality checks can be performed.

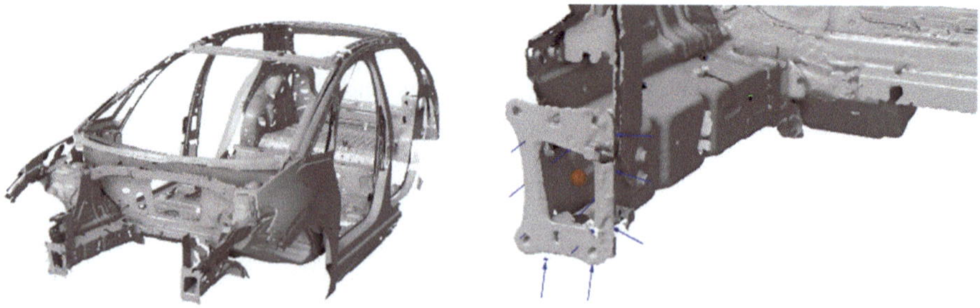

**Fig. 10.5.** (left) Optical 3-D scan of an ID.3 body-in-white (right) virtual point on the left crash bar. Blue arrows denote impact positions which are transformed to a 6 DoF virtual point (orange ball) compatible for substructuring.

**EDU characterisation**

***In-vehicle.*** A full-vehicle chassis-dyno run is used to measure operational vibratory and acoustic responses while the EDU delivers defined load to the dyno. Depending on access, structure-borne excitation is captured either as blocked forces directly at the mounting points or, where space is limited, as pseudo forces on the EDU housing. Experience has shown that transfer paths via drive-shafts and high-voltage cables can be significant, which is why they are included in the measurements and the resulting model. The Air-borne contribution is determined with an operational TPA using the VINS transmissibility method. Torque at the rollers, battery voltage (OBD-CAN) and current input to the inverter (current clamp) are logged so that the load state of the EDU can accurately be identified.

***EDU passive Substructure.*** When the EDU is removed, a free-suspended FRF campaign provides the pure substructure of the EDU which is used for substructuring of the transfer path.

***Test-bench operation.*** Control tables, linking operating conditions, like speed and torque demand, to control parameters, like current, voltage and phase angles, can be reconstructed from recordings during the vehicle testing. During the test-bench runs the control table settings can be applied to a universal inverter. This ensures that the EDU runs under an electromagnetic excitation on the rig that closely resembles the in-vehicle operation, resulting in precise blocked-force, acoustic monopole and sound power measurements.

***Design and cost teardown.*** For customers interested in deeper insight, the global benchmark program can supply a full tear-down and design benchmark of the EDU. Components such as stator, rotor, resolver, magnets and transmission are weighted, measured and costed; the findings often explain NVH performances in the NVH benchmark model. Figure 10.6 illustrates an EDU teardown for design benchmarking.

**Vehicle receiver model**

After the EDU has been disassembled from the vehicle, a hammer and accelerometer grid around each virtual point yields the coupling point and mount-to-cabin FRFs. The data are transformed to 6-DoF virtual points and checked for passivity, reciprocity and interface completeness. Thereby they are ready to use for hybrid NVH models.

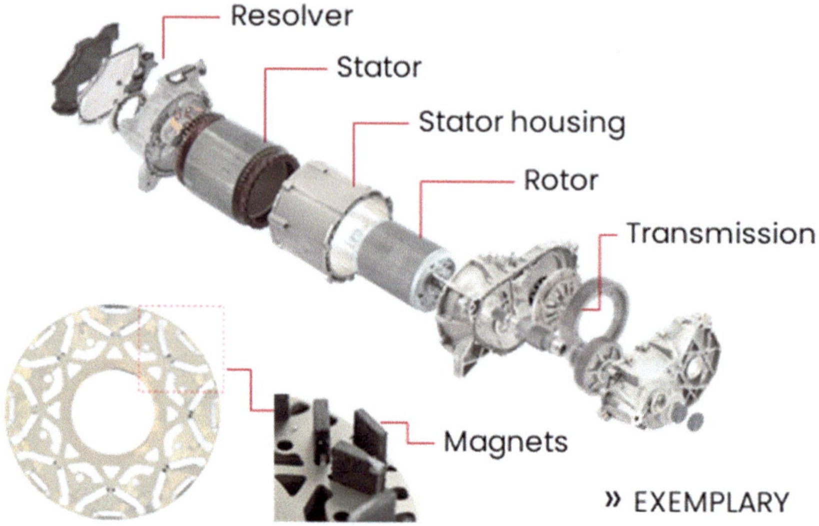


**Fig. 10.6.** Tear-Down and Design Benchmark of EDUs in the Global Benchmark Programm

**Rubber isolators**
Each mount is tested separately using the 12-DoF method. A preload sweep can be added with the in-vehicle VINS method if operational data indicates that non-linear effects are relevant.

## 10.4 Summary

FEV augments the NVH substructuring workflow by closing an important gap in today's compressed BEV development cycles. By combining test-based benchmark receiver models, rubber-mount models and blocked-force / monopole source descriptions with customers' CAE data, reliable driver-ear predictions are available early in the development program. OEM NVH teams can fix mount concepts and path sensitivities before hardware exists and keep their digital twin current as new data arrive; EDU suppliers can verify whether extra design effort truly moves the needle—or simply benchmark their unit against rival products under realistic vehicle conditions.

The required models are generated within FEV's global vehicle-benchmark program, which dismantles multiple new BEVs and PHEVs each year. Because the vehicles are already on site for design- and cost-teardowns, adding NVH substructure measurements creates a clear business case: one set of hardware serves several engineering communities, while customers avoid the expense of sourcing complete cars themselves. With an ever-growing library of ready-to-couple substructures—and optional deep-dive EDU tear-downs—the approach offers a scalable, data-driven route to faster, less risky NVH development for both integrators and component suppliers.

## References


1. P. Wagner, F. Bianciardi, P. Corbeels and A. Hülsmann, "High frequency source characterization of an e-motor using component-based TPA," in SIA NVH Comfort , Le Mans, 2021.
2. P. Wagner, A. P. Hülsmann and M. van der Seijs, "Application of dynamic substructuring in NVH design of electric drivetrains," in ISMA Conference on Noise and Vibration Engineering, Leuven, 2020.
3. D. Lennström, F. Wullens, M. Olsson and A. Nykänen, "Validation of the blocked force method for various boundary conditions for automotive source characterization," Applied Acoustics, vol. 4–5, no. 102, pp. 108-119, 2016.
4. Technical Comittee, ISO/TC 43/SC 1, ISO 21955:2021 "Acoustics—Experimental method for transposition of dynamic forces generated by an active component from a test bench to a receiving structure", International Organization for Standardization (ISO), 2021–08.
5. Technical Comittee, ISO/TC 43/SC 1, ISO 20270:2019: "Acoustics—Characterization of sources of structure-borne sound and vibration—Indirect measurement of blocked forces", International Organization for Standardization (ISO), 2019.
6. M. Haeussler, Modular Sound and Vibration Engineering by Substructuring, PhD Thesis: Technical University of Munich (TUM), 2020.
7. D. de Klerk, D. Rixen and S. Voormeeren, "General Framework for Dynamic Substructuring: History, Review, and Classification of Techniques," AIAA Journal, vol. 46(5), pp. 1169-1181, 2008.
8. M. van der Seijs, D. van den Bosch, D. J. Rixen and D. de Klerk, "An improved methodology for the virtual point transformation of measured frequency response functions in dynamic substructuring," in 4th International Conference on Computational Methods in Structural Dynamics and Earthquake Engineering, Kos Island, Greece, 2013.
9. M. Haeussler, T. Mueller, E. A. Pasma, J. Freund, O. Westphal and T. Voehringer, "Component TPA: benefit of including rotational degrees of freedom and over-determination," in ISMA International Conference on Noise and Vibration Engineering, Leuven, 2020.
10. M. van der Kooij, S. Klaassen and A. Hülsmann, "Using Dynamic Substructuring and Component TPA to Shape the NVH Experience of a Full-Electric Vehicle," in 12th International Styrian Noise, Vibration & Harshness Congress, Graz, Austria, 2022.
11. M. Haeussler, S. Klaassen and D. Rixen, "Experimental twelve degree of freedom rubber isolator models for use in substructuring assemblies," Journal of Sound and Vibration, vol. 474, 2020.
12. M. Gröne, S. Schwarzendahl, J. Troge and E. Hensel, "Transfer Stiffness Characterization of Elastomer Mounts at High Frequencies," ATZ worldwide, vol. 127, no. 1, pp. 40-43, 2025.
13. G. Eisele, K. Wolff, J. Hoppermanns and P. Genender, "15 Years of Transfer Path Analysis VINS in the Vehicle NVH Development—Selected Results," in 8th International Styrian Noise, Vibration & Harshness Congress, Graz, 2014.
14. M. van der Seijs, D. de Klerk and D. J. Rixen, "General framework for transfer path analysis: History, theory and classification of techniques," Mechanical Systems and Signal Processing, vol. 68, pp. 217-244, 2016.
15. T. Müller, M. Haeussler, S. Sedlmair and D. J. Rixen, "Airborne transfer path analysis for an e-compressor," in ISMA—International Conference on Noise and Vibration Engineering, Leuven, 2020.
16. D. Lennström, R. Johnsson, A. Ågren and A. Nykänen, "The Influence of the Acoustic Transfer Functions on the Estimated Interior Noise from an Electric Rear Axle Drive," SAE International Journal of Passenger Cars—Mechanical Systems, vol. 7, 2014.

# Analysis of Noise Transmission Routes for Advanced Development of Door Systems

Yuko Nakagawa[1], Kazumi Inoue[2], and Paolo Bertolo[2](✉)

[1] Toyota Boshoku Corporation, Kariya-Shi, Aichi, Japan
yuko.nakagawa@toyota-boshoku.com

[2] Autoneum Management AG, ATN Auto Acoustics, Oguchi-Cho, Aichi, Japan
kazumi.inoue@atn-aa.com, paolo.bertolo@autoneum.com

**Abstract.** The ongoing transition from internal combustion engines (ICE) to electric motors has deep implications on the perceived balance between the different noise sources in a car. The absence of the noise-masking effects of the ICE is increasing the perceived contribution of both wind and tire noise. As doors are essential components against wind and tire noise, their acoustic performance is becoming more and more relevant for meeting the continuous demand for better and quieter vehicles. In the development of high-frequency soundproofing materials, a combination of experimental measurements and SEA simulation models are very often used for the identification of the airborne noise transmission paths and for the definition of the appropriate counter measures. However, in conventional full-vehicle level simulation models, the structure of the doors and the corresponding acoustic parameters are usually very simplified, making these models unsuitable for soundproofing design of door systems. Therefore, the need for more elaborated models, able to offer a deeper understanding of the noise transmission routes through the vehicle doors. In this paper, at first, we describe an experimental investigation on the airborne noise transmission paths from the front tires through the front door panels and eventually into the front passengers' space. Based on these results, we then introduce an improved simulation model developed for the purpose of supporting the definition of the correct NV counter measures at the very early design stage of the door systems.

**Keywords:** Vehicle Door · NVH · SEA · Simplified Model

## 11.1 Introduction

### 11.1.1 Recent Changes in the Automotive Industry and New Challenges in Quietness

In recent years, the idea of achieving both environment conservation and sustainable development has been widely accepted worldwide. In the automotive industry, the development of zero-emission vehicles has been actively promoted, and there is


A. Heintzel (Ed.): Automotive Acoustics Conference 2025, Proceedings, pp. 154–175, 2026. https://doi.org/10.1007/978-3-658-51249-1_11

a trend of shifting from ICE (Internal Combustion Engine) vehicles to EVs (Electric Vehicles). Although the momentum of EV sales is slowing down, the share of EVs is expected to continue to grow in the long term. [1, 2] (See Fig. 11.1).

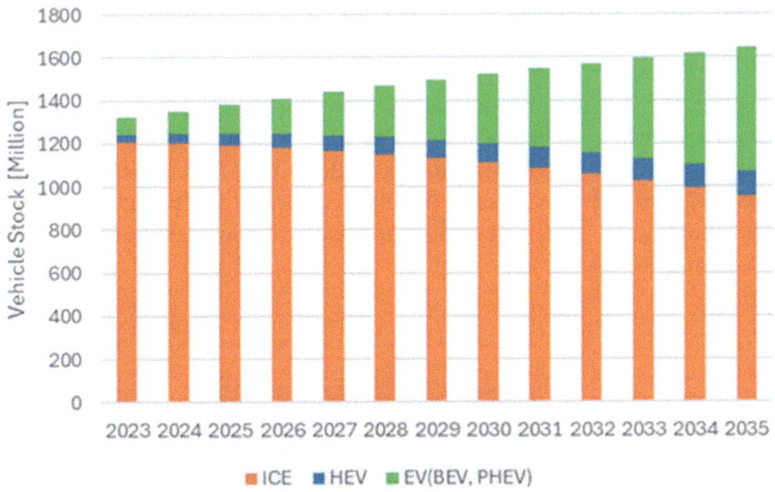


**Fig. 11.1.** Passenger light-duty vehicle stock by powertrain and scenario, 2023–2035 (IEA Global EV Outlook 2024 [2])

In this transition to EVs, new issues have emerged regarding quietness inside the vehicle cabin. With the shift to BEVs (Battery Electric Vehicles), the powertrain will change from an internal combustion engine to an electric motor, significantly reducing noise levels. However, tire and wind noise, which were previously masked by the engine noise, will become more noticeable to passengers. (See Fig. 11.2).

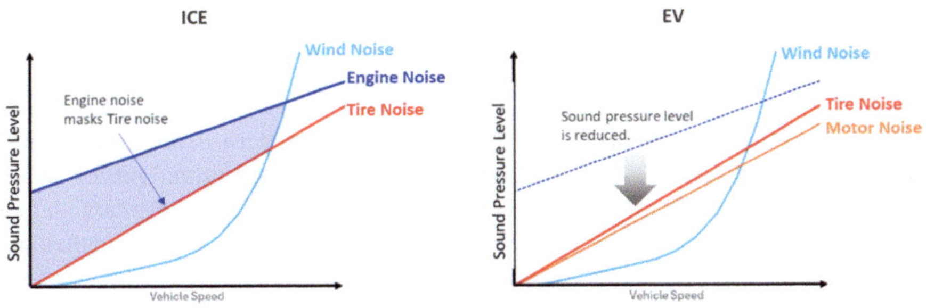


**Fig. 11.2.** Comparison of Sound Source Contribution due to Drive System [3]

Furthermore, although the efforts of automakers have improved the quietness of vehicle interiors, the shift to EVs is raising user expectations, and it is expected that automakers will set higher acoustic performance targets for EVs. In order to achieve these high targets, it will no longer be enough to focus only on the dash panel and floor, which were previously the focus of NVH prioritization, and further measures will be required for other areas as well. Figure 11.3 shows a typical example of the acoustic power contributions at the driver's ear position for airborne noise from the front tires sound source. Traditionally, doors have attracted attention for their acoustic performance in the low and mid-frequency range for solid-borne noise, and research and countermeasures have been conducted on door panels vibration targeting this frequency range. However, as shown in the example in Fig. 11.3, doors can also make a large contribution to airborne noise in the high-frequency range, and it was found that in order to meet higher target values, it is necessary to consider countermeasures also for airborne noise for doors.

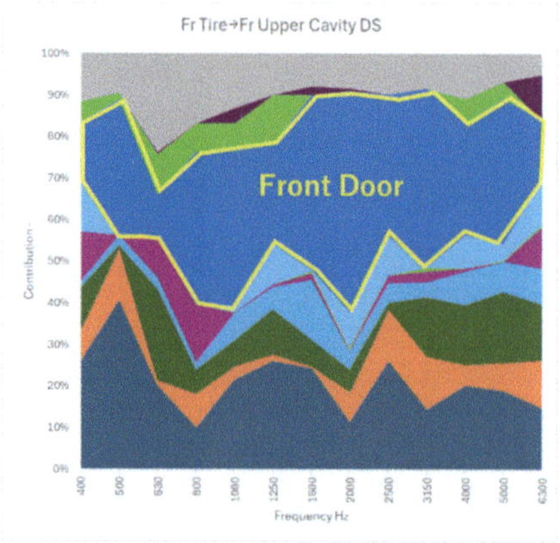


**Fig. 11.3.** Panel contribution of acoustic power at front passenger's ear position from front tires sound source (test result, airborne noise)

### 11.1.2 Issues in Developing Doors to Improve Quietness

The adoption of the so-called "V-shaped development" pattern is accelerating in the vehicle development process. By setting goals and allocating optimal performance in the upstream stages of the development process, rework in the downstream stages of the process is prevented. In the upstream stages of the process, simple mathematical models that can easily represent the phenomenon to be evaluated are sometimes used. Such simulations have the advantage that the phenomenon and the underlying physical mechanisms are easy to understand, and the essence of the problem can be identified. In addition, such analysis tools are relatively simple to operate and can be used by designers as well. If such tools could be applied to the consideration of the acoustic performance of doors, it would support smooth door development.

In particular, the statistical energy analysis method (SEA) is often used to consider measures against high-frequency noise in the vehicle cabin [4–6].

A vehicle SEA model is created to consider the placement of soundproofing materials inside the vehicle cabin. Conventional SEA vehicle models focus on the dash and floor areas and doors, which are not expected to be the subject of detailed consideration in conventional SEA vehicle models, have their structure and acoustic characteristics simplified to improve model efficiency. As consequence, with such models, it is not possible to understand the acoustic power flow inside the door and to analyze in detail the acoustic performance of the door.

To solve this problem, we are working on creating a simple acoustic performance evaluation tool for a door that allows designers to consider the acoustic performance of doors at the upstream development stage. In this paper, we introduce an example in which the sound source is the front tire sound source and the door to be considered is the front one.

This paper is structured as follows. In Sect. 11.2, as a preliminary investigation for creating the tool, we conduct sensitivity tests on each part of the door and analyze the acoustic power flow through the door, which is of particular interest. Section 11.3 describes the modeling of the door, the performance evaluation tool that was developed and the results obtained with this tool. In addition, to verify the validity of the tool, we calculate the effectiveness of measures when the sound absorption characteristics of the space inside the door are changed and compare the results obtained from the model with test results. Section 11.4 describes a method for applying the acoustic performance of the door as evaluated with the tool to a simplified door of a conventional SEA model of a vehicle.

## 11.2 Acoustic Sensitivity Analysis of a Vehicle Door

### 11.2.1 Test Conditions

First, we must understand the acoustic power flow to be modeled. An exposure test is conducted to investigate the acoustic power flow emitted from the front tire sound source, passing through the door and reaching the interior of the vehicle (at the driver's ear position). In this study, we report on the case of a mid-size SUV. The door to be evaluated is the front door on the driver's side, and the sound source is installed on the front tire on the driver's side. For the speaker, we use a volume-velocity calibrated acoustic source that can generate white noise from 200 Hz to 10 kHz (See Fig. 11.4). Parts other than the door to be evaluated are sound insulated with felt and rubber sheets, with care taken to ensure that the contribution of each part of the door can be fully confirmed (See Fig. 11.5).

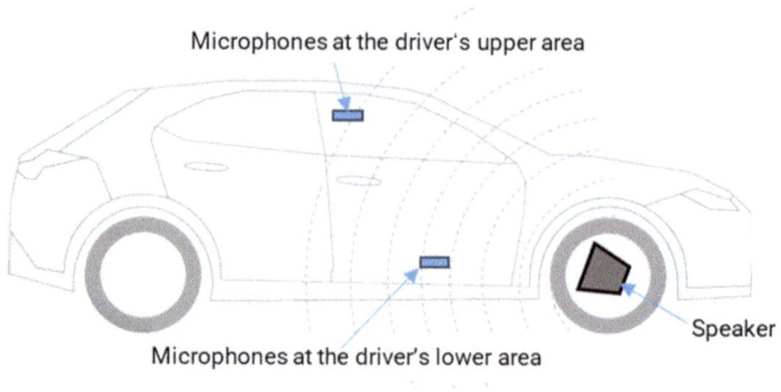


**Fig. 11.4.** The speaker and microphones at driver's area

**Fig. 11.5.** Test conditions inside the vehicle

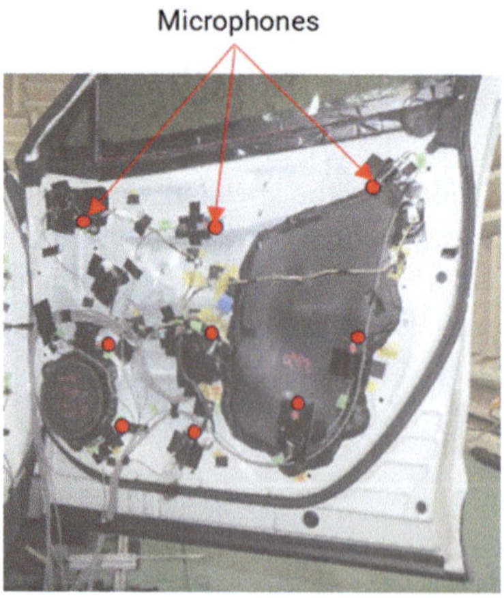


**Fig. 11.6.** Microphones in the door inner space

The space inside the door is divided into two spaces: the outer space surrounded by the outer door panel and inner door panel, and the inner space surrounded by the inner door panel and the door trim. Nine microphones are installed in each of these spaces (See Fig. 11.6). In addition, two microphones are installed in the passenger compartment, one at the driver's ear position and one in the feet area, to measure the sound levels in the upper and in the lower part of the driver's space.

### 11.2.2 Sensitivity Test Parts

The areas through which noise penetrates are classified into the following four areas. These areas are called the door window area, the door outer panel area, the door inner panel area, and the door trim area (See Fig. 11.7). In the test, each area is further divided into characteristic areas, and the sensitivity of each characteristic area is assessed (See Fig. 11.9). Each area is evaluated based on its contribution to the acoustic spaces which are in contact with it. The spaces to be investigated are called the *door outer space* surrounded by the *door outer panel* area and the *door inner panel* area, the *door inner space* surrounded by the *door inner panel* area and the *door trim* area, the *upper* part of the *driver* space, and the *lower* part of the *driver* space, and a schematic diagram of these is shown in Fig. 11.8.

1. Door window area:
   Assuming a route from outside the vehicle that passes through the door window area and reaches the passenger's ear position, we investigated the contribution of sound power to the upper part of the driver space.
2. Door outer panel area:
   Assuming a route from outside the vehicle that passes through the door outer panel area and reaches the door outer space, we investigated the contribution of sound power to the door outer space.
3. Door inner panel area:
   Assuming a route from the door outer space that passes through the door inner panel and reaches the door inner space, we investigated the contribution of sound power to the door inner space.
4. Door trim area:
   Assuming a route from the door inner space that passes through the door trim area and reaches the space below the driver, we investigated the contribution of sound power to the lower part of the driver space.

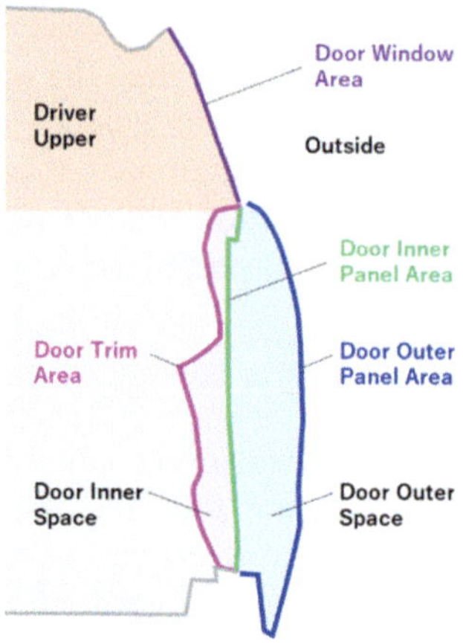


**Fig. 11.7.** Door areas, spaces inside a door

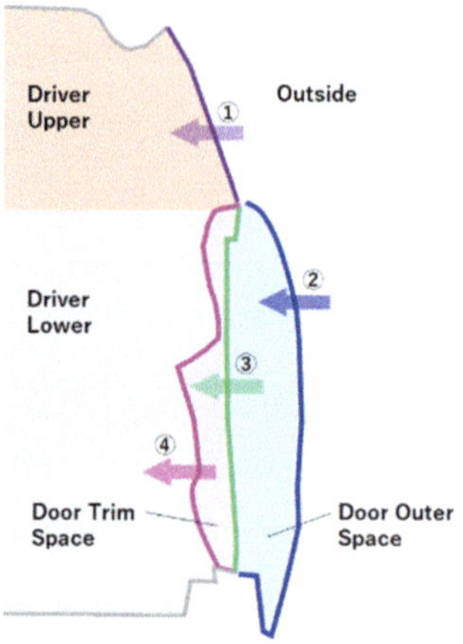


**Fig. 11.8.** Noise flow through the door areas

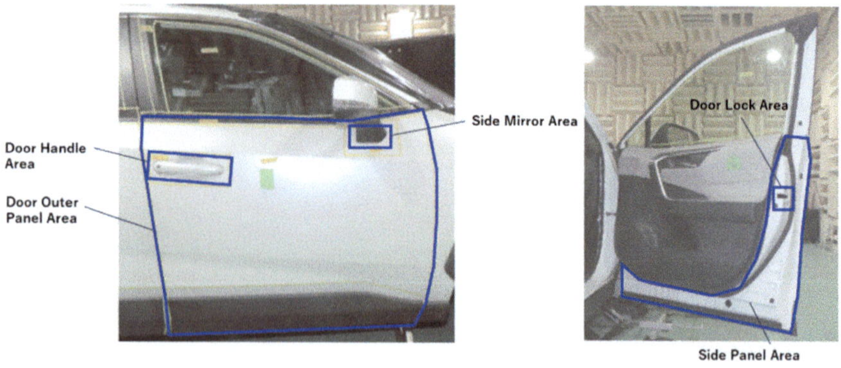


**Fig. 11.9.** Examples of part division for door sensitivity test (Door outer panel area)

The way door seals were handled in these investigations deserves a specific explanation. In the case here analyzed, door seals are installed on the edge of the door opening and on the edge of the door trim. The cross section of the front door is shown diagrammatically for the structure of the door seal in Fig. 11.10. There are mainly two possible routes for noise that pass-through door seals (See Fig. 11.11). One is the noise route that passes through both seals and enters directly inside the vehicle, as shown in the left figure of Fig. 11.11. With regard to this noise route, the intrinsic seal performance is what matters, and the improvement of this performance is in fact often considered as a possible countermeasure to improve the quietness of the passenger compartment. However, this performance does not relate directly to that of the vehicle door as such. The second route is the one that passes through the seal at the door opening, penetrates the door side panel area, and enters the door outer space, as shown in the right figure of Fig. 11.11. Unlike the previous one, this is a route that should be taken into consideration when considering the acoustic performance of the door, since it involves the door structure.

Therefore, in order to separate these paths, in this study, sound insulation material was applied to the door trim side seal to block the noise that passes through the seal and enters the vehicle interior. This allows us to evaluate only the sound that, being transmitted through the door opening seal, penetrates the door and only then enters the vehicle interior.

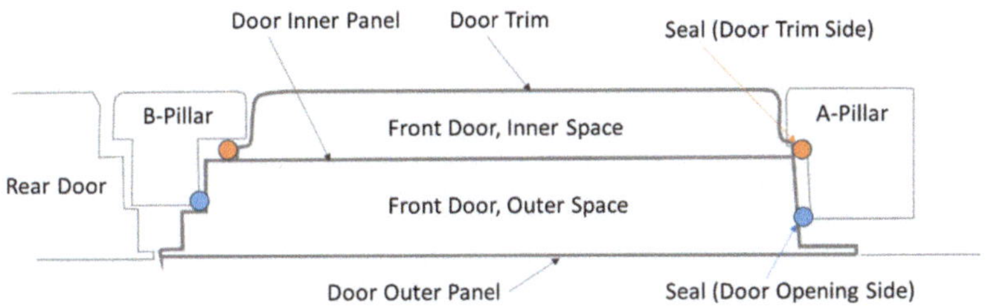


**Fig. 11.10.** Diagram of the horizontal section of the front door

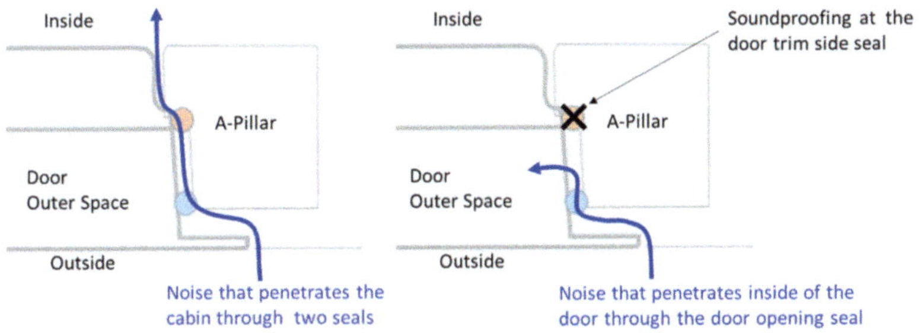


**Fig. 11.11.** Noise route image through the door seals of the front door

### 11.2.3 Test Results

As previously mentioned, each area under investigation radiates noise into a specific acoustic space it is in contact with (e.g., the door outer panel radiates noise into the door outer space).

For each area under investigation, the degree to which noise radiated by that area contributes to the noise in the corresponding acoustic space was investigated.

This was done by measuring the degree to which the sound pressure level increases when the sound insulation is removed from the area under investigation, starting from a state in which the entire area under investigation is covered by sound insulation (exposure technique).

Figure 11.12 shows an example of the measurement results for the outer door panel area. The graph shows the sound pressure level in the acoustic space on the outer side of the door. *MAX* is the result when the entire door is soundproofed. *Original* is the result when no sound insulation is applied on the door. *Outer Panel* is the result when the sound insulation applied on the door outer panel is removed from the *MAX* state. In both cases, no sound insulation is applied on the inner door panel area and door trim area. Figure 11.13 shows the state of sound insulation.

Based on the difference between the SPL in an acoustic space when a certain area under investigation is covered by sound insulation and when the same area is not covered by sound insulation (e.g., the difference between the green and the blue curves in Fig. 11.12) it is possible to obtain a ranking between the different transmission paths that contribute the most to the previously mentioned 4 areas.

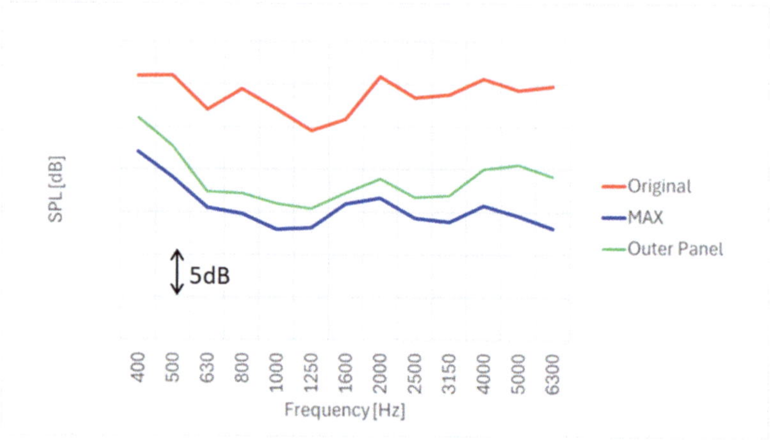


**Fig. 11.12.** Comparison of SPLs in door outer space in the sensitivity test of the outer door panel

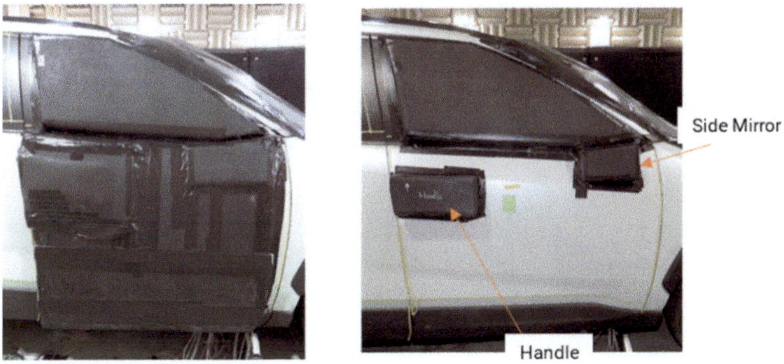


**Fig. 11.13.** Door outer panel inspection test status (Left: *MAX* status, Right: Door outer panel exposure status)

The transmission paths that contributed most were as follows:

1. Door window area:

   Door glass, glass run, door outer belt molding (See Fig. 11.14).

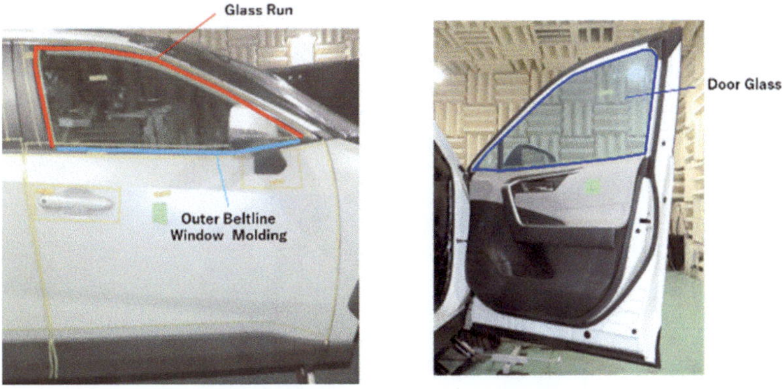


**Fig. 11.14.** Large contribution parts in the door window area

2. Door outer panel area:

   Door outer panel, door outer side panel, drain hole (See Fig. 11.15).

   (Side mirror area, door handle area, and door locks were small contribution).

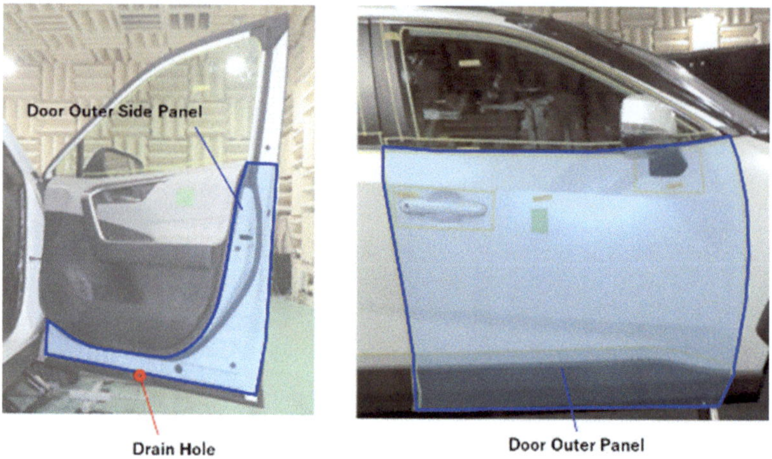


**Fig. 11.15.** Large contribution parts in the door outer panel area

3. Door inner panel area:

   Door inner panel surface, Speaker, service hole cover (See Fig. 11.16).

   (The multiple clip holes were small contribution.)

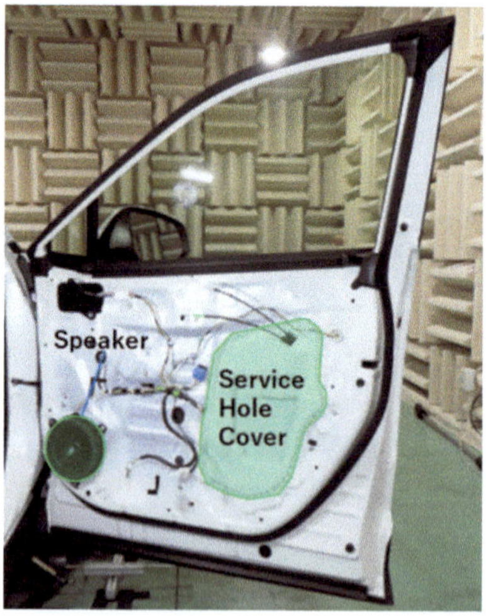


**Fig. 11.16.** Large contribution parts in the door inner panel area

4. Door trim area:

   Door trim surface, Inner belt molding, speaker grill (See Fig. 11.17). (The switch areas and the trim joints offered negligible contribution.)

**Fig. 11.17.** Large contribution parts in the door trim area

One of the highest contribution areas was the inner belt molding. As shown in Fig. 11.18, there are two possible routes to pass through this area: from the door inner space and from the door outer space. According to this investigation, although the energy level of the door inner space is lower than that of the door outer space, it was found that the influence from the door inner space was higher.

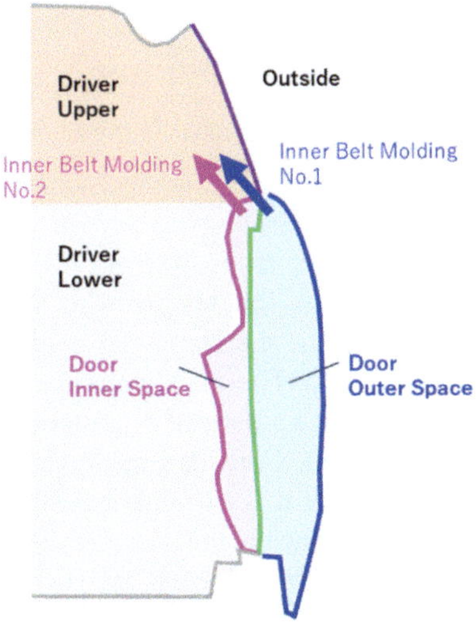


**Fig. 11.18.** Noise flow around inner belt molding in the door model

## 11.3 Simple Door Acoustic Performance Evaluation Tool

### 11.3.1 Making Simple Door Acoustic Performance Evaluation Tool

Based on the results of the investigation in Sect. 11.2, we created a door model that reproduces the acoustic power flows with the higher contribution (See Fig. 11.19).

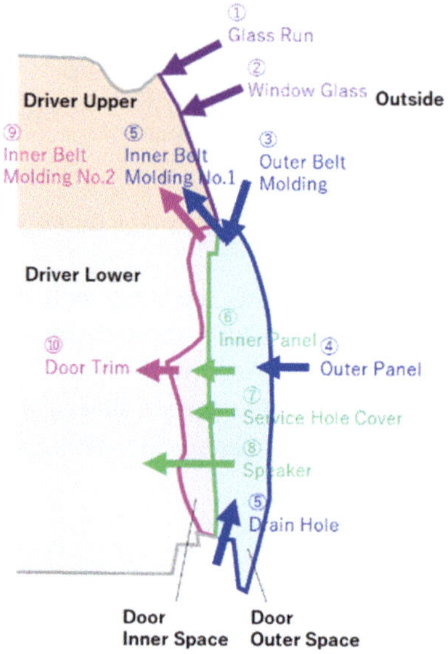


**Fig. 11.19.** Modeled acoustic power flows

The target frequency range in this study is from 400 Hz to 6500 Hz. It is also important that the tool should be easy for designers to access and does not require much time to learn. For these reasons, the statistical energy analysis (SEA) method was adopted as the calculation method. The SEA method targets high frequencies, and unlike the finite element method, the amount of calculations is very small, and even less for a small model such as a door. Considering the small amount of calculations required by the model and the fact that it is a tool that designers shall be familiar with, we chose to implement the tool using just a generic spreadsheet application.

By inputting geometric parameters such as cavity (acoustic space) and panel dimensions and material parameters required for the SEA calculations into a specific cell in the calculation sheet, the energy state of each acoustic space can be calculated.

In addition, the sound pressure level of the acoustic space (e.g. the sound pressure level in the upper driver space, which corresponds to the driver's ear position) and the acoustic power flow can also be visualized (See Fig. 11.20).

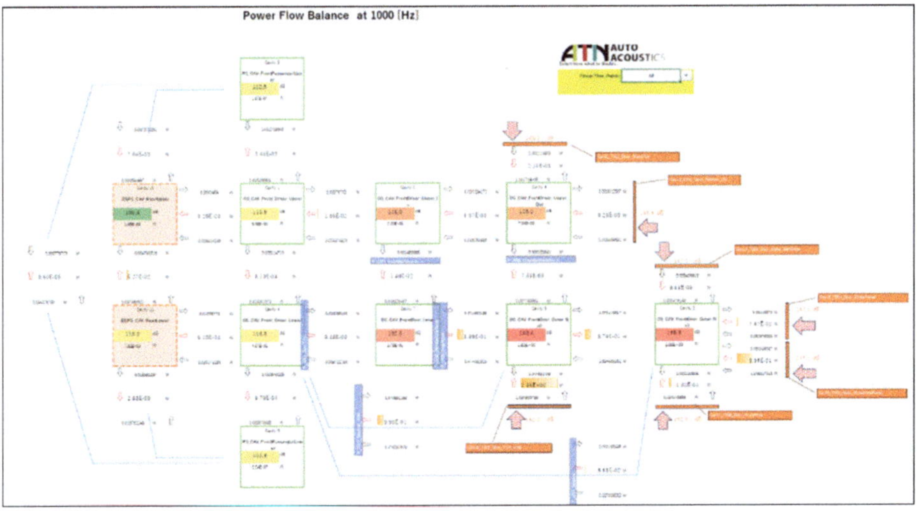


**Fig. 11.20.** Simplified Door Acoustic Performance Evaluation Tool

The tool calculates the energy state of each acoustic space (cavity), according to the SEA method. The coupling loss factors between cavities and the damping loss factors for each cavity are calculated, and the equilibrium relationship of the acoustic power flow between cavities is formulated and solved.

In addition, the transmission loss of SEA panels such as the window glass and the door outer panels made of steel, which is necessary for calculating the coupling loss factors is obtained from the mass law, while the sound absorption coefficient of glass and steel is set to a small value. The sound absorption coefficient of porous sound absorbing materials is simulated using measured Biot parameters. In addition, leakage from gaps such as door belt molding was determined by adjusting the transmission loss so that the coupling loss factor value matched the change in the sensitivity test results. (See Figs. 11.21 and 11.22). Similarly, adjustments are made to general surfaces based on all sensitivity test results to improve accuracy.

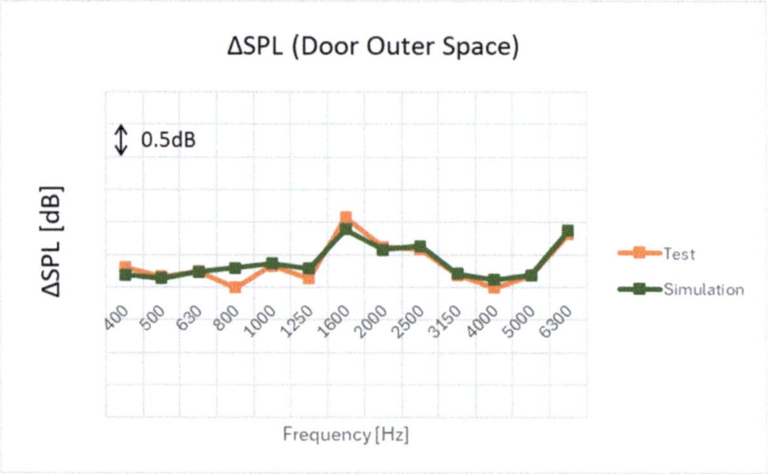


**Fig. 11.21.** Comparison of SPL change in door outer space between test and simulation (Outer belt molding)

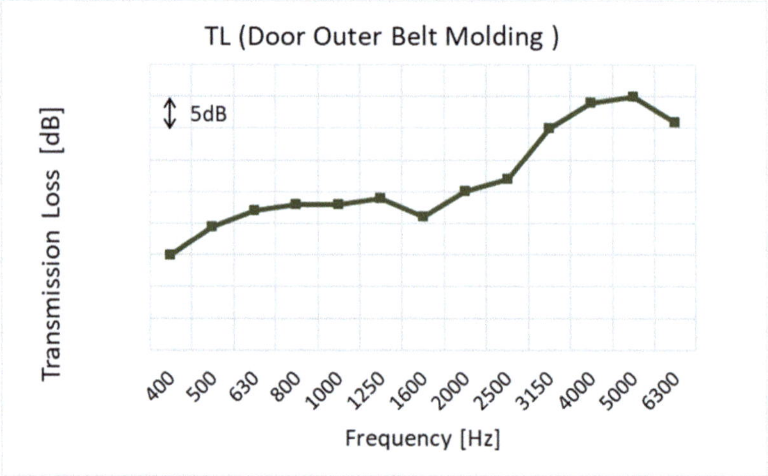


**Fig. 11.22.** Transmission Loss of outer belt molding after adjustment

We then verified whether the calculations in the created tool were correct by using the general-purpose SEA simulation software "REVAMP" (Autoneum). The results of calculations using the tool and REVAMP are shown in Fig. 11.23. The sound pressure levels in both spaces matched the results calculated using REVAMP.

**Fig. 11.23.** Comparison of SPL between Results Calculated by the spreadsheet-based Tool and REVAMP. (Top left: Driver Upper Space, Top right: Door inner space, Bottom right: Door outer space)

Next, the sound pressure levels in each space were compared between the experimental values and the results of the tool calculations (See Fig. 11.24).

It can be seen that the test results and the results of this tool are in good agreement.

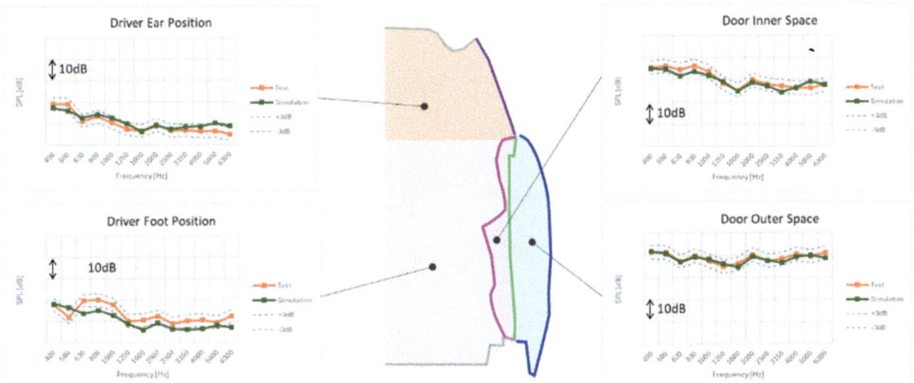


**Fig. 11.24.** Comparison of SPLs between experiment and result calculated by the tool

### 11.3.2 Evaluation of the Tool

To verify the validity of this tool, the sound absorption coefficient of the door inner space is changed, and the test sound pressure level at the driver's ear position is compared with the results of the tool's calculations.

Porous sound-absorbing material is placed behind the door trim so that it covers approximately 35% of the trim area. Note that the placement is made to avoid areas that would significantly change the transmission characteristics of areas such as speaker grilles (See Fig. 11.25). In the tool, the sound absorption of the door inner space, which is the space behind the trim, is changed based on the sound absorption coefficient and area of the sound-absorbing material.

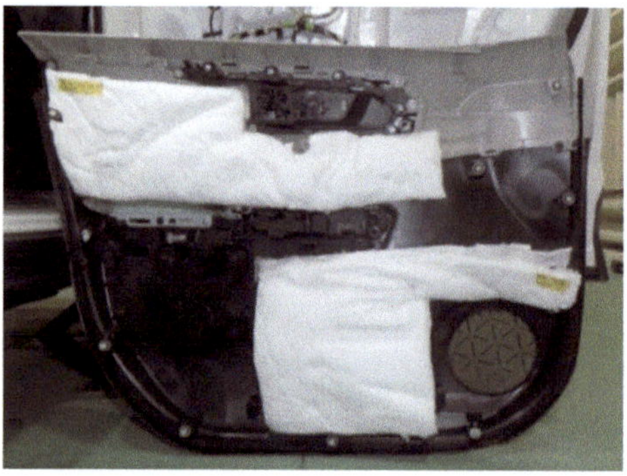

**Fig. 11.25.** Sound absorbing material placement on the back of the door trim

The test and simulation results are shown in Fig. 11.26. It was confirmed that the test trends were adequately captured.

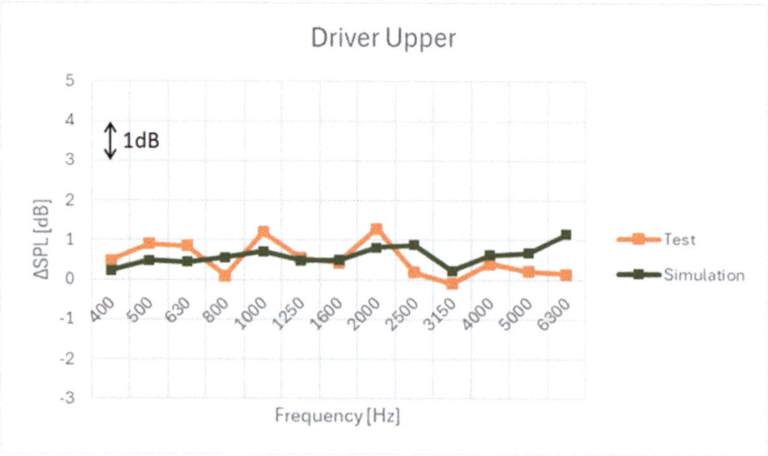


**Fig. 11.26.** Comparison of cavity SPL between the experiment and the simulation

Next, we focus on the acoustic power flow and the energy state of each space (See Fig. 11.27). From the calculation results of the tool, it can be confirmed that installing sound absorbing material reduces the energy in the door inner space, and that the acoustic power that flows from this space through the *Inner Belt Molding 2* to the *Driver Upper* space is also reduced. In Fig. 11.27, the red arrows indicate the power flow (watts) between different acoustic spaces, and *** in the boxes indicate the sound pressure levels (dB) in the same acoustic spaces. The diagram shows a 1/3 octave frequency band of 1 kHz.

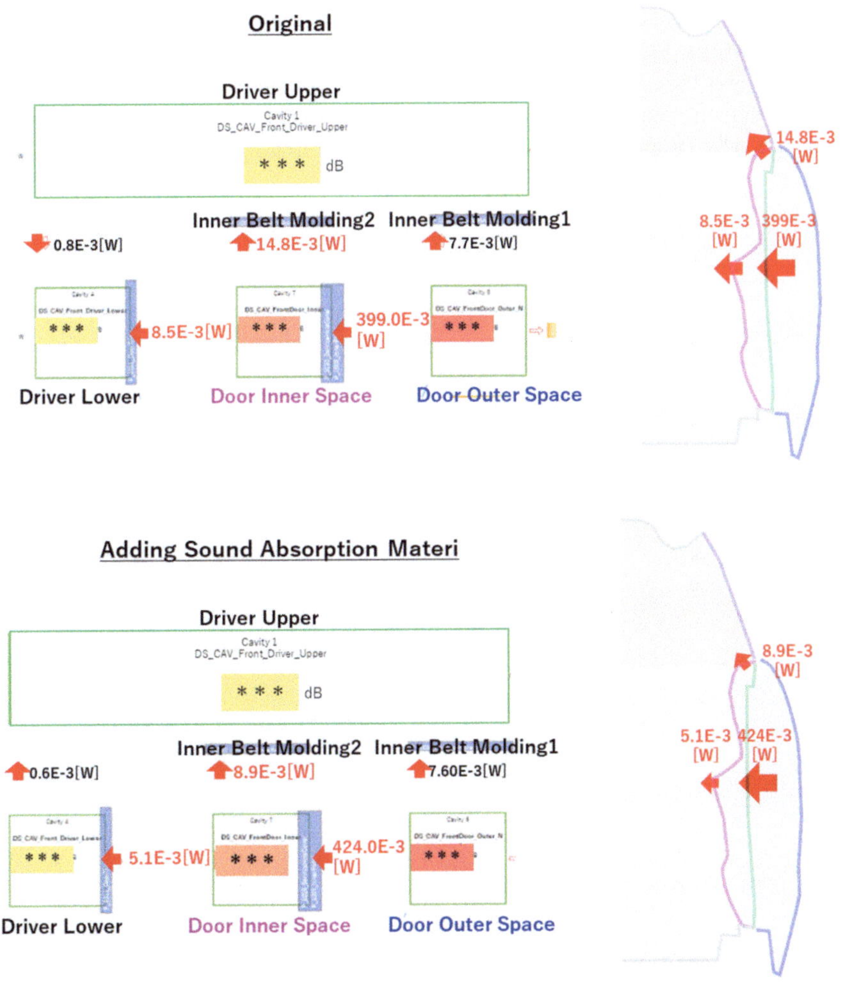


**Fig. 11.27.** Part of the acoustic power flow in the original state and after the application of sound absorption patch in the door inner acoustic space (as shown in Fig. 11.25)

## 11.4 Application to Conventional Vehicle SEA Models

In this section, we consider a method to confirm the effectiveness of acoustic countermeasures applied to a door for the entire vehicle, by first reflecting these acoustic countermeasures using the tool described in previous sections and then integrating the results obtained with the tool in a conventional SEA model that—in its original state—does not include the detailed description of the door structure.

The SEA method calculates the acoustic power transmitted between subsystems (cavities) and the acoustic power dissipated within each subsystem. If there is input power applied from the outside, this is also taken into account and the energy balance of the subsystems is calculated (See Fig. 11.28).

As shown in Fig. 11.28, acoustic power is exchanged between the subsystems, but if we focus only on the input/output *difference* of these acoustic powers, it appears that the sound is flowing from one subsystem to the other (for example, when $P_{12} > P_{21}$, it appears that *net* sound power is flowing from *Sub System1* to *Sub System2*).

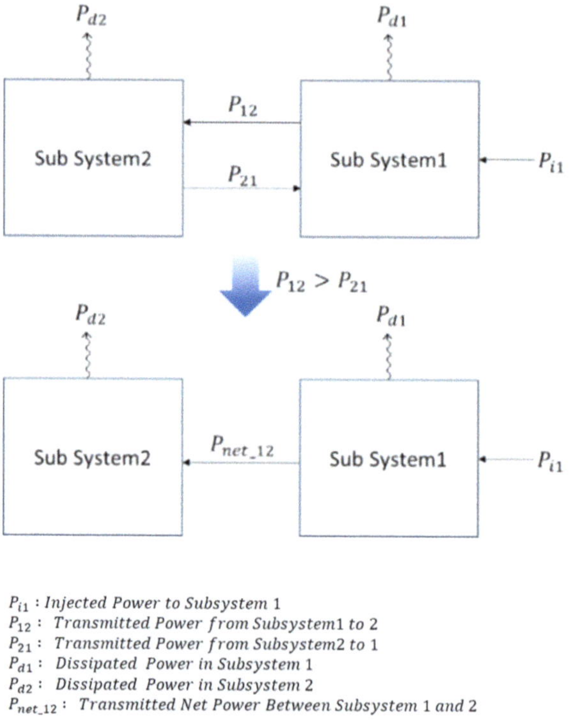


**Fig. 11.28.** Actual and Apparent Acoustic Power Flow

We try to apply this schematization to the door model. The door model of the developed tool has two spaces (cavity subsystems), with an inner panel and a service hole cover between these spaces. In conventional SEA model doors, all of these spaces and panels are represented as a single panel, which is meant to "condense" all the complex power flows through the vehicle door.

Therefore, to be consistent with this representation of a vehicle door normally used in conventional SEA models, the acoustic spaces and panels of the door model included in the developed tool are also considered as a single door panel (See Fig. 11.29). The acoustic power entering and leaving this simplified door panel consists only of the external input power $P_{i1}$ that is input from outside the vehicle to the door outer space, and the *apparent* power $P_{net\_23}$ that flows from the door inner space to the driver lower space.

From these acoustic powers $P_{i1}$ and $P_{net\_23}$, the acoustic characteristics of the *condensed* door panel normally used in conventional SEA models may be calculated as an *apparent* transmission loss.

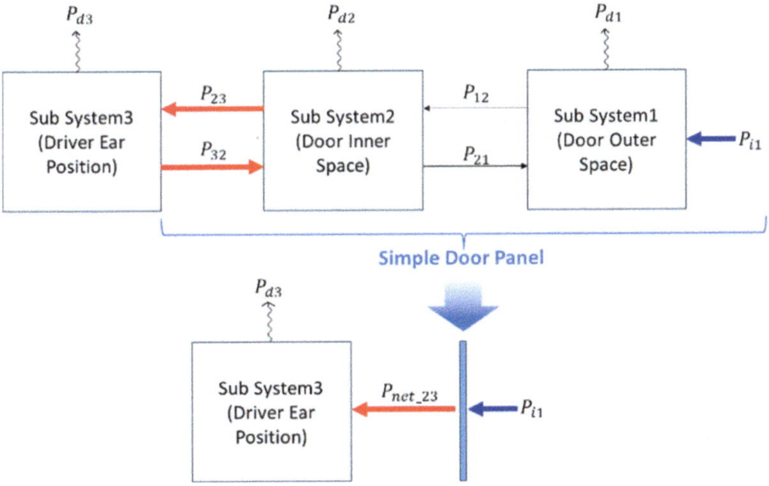


**Fig. 11.29.** Simplifying Complex Door Structure

The transmission loss is expressed by the following formula:

$$\text{Transimission Loss} = 10\,\log_{10}(\tau)^{-1} \tag{11.1}$$

$$\tau = \frac{I_t}{I_i} \tag{11.2}$$

where $\tau$ is the transmission coefficient, $I_i$, and $I_t$ are the acoustic intensities of the incident and transmitted waves, respectively.

The relationship between sound intensity and sound power is given as follows:

$$P = I \cdot S \tag{11.3}$$

where $P$ is the sound power, $I$ is the sound intensity, and $S$ is the area through which the sound power passes.

The areas of the incident and transmitted waves are assumed to be the same.

The sound intensity of the transmitted wave in the simplified door panel is calculated from the apparent sound power in the door area using Eqs. (11.1) to (11.3), and the sound intensity of the incident wave is calculated from the external input power.

The apparent door transmission loss calculated in this way was applied to the door of the conventional vehicle SEA model, in which the door was represented by a simple panel and the corresponding transmission loss was defined based on mass law, adjusted based on sensitivity tests as previously described.

Figure 11.30 shows the comparison between the sound pressure level at the driver upper space obtained with the conventional SEA model and the one obtained by replacing the transmission loss of the door in this same model with the one obtained from the tool described in previous sections, according to the procedure outlined here above.

SPLs compare well, demonstrating the validity of this method. (See Fig. 11.30).

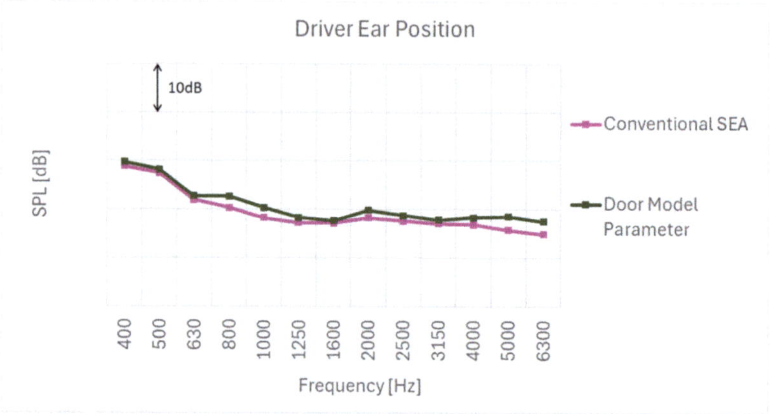


**Fig. 11.30.** Comparison of SPLs in driver upper space

## 11.5 Conclusion

A simple tool for the evaluation of door acoustic performance was created for studying the acoustic performance of vehicle doors at the upstream development stage.

To create the tool, the sensitivity of each part of the door was evaluated and noteworthy sound power routes were identified. A door model was created to reproduce these routes, and an spreadsheet-based tool that is easy for designers to access was created for allowing them to perform preliminary door acoustic performance evaluations.

The sound pressure levels in each space calculated with this tool were compared with the test results, and it was confirmed that the test results could be reproduced. The sound absorption characteristics inside the door were also changed, and the corresponding effect could be reproduced, confirming the validity of the tool. A method for applying the results of the evaluation using this tool to a conventional SEA model was also described.

**Acknowledgements.** In the writing of this paper, Autoneum provided us with appropriate advice throughout. We would like to express our deepest gratitude·

## References

1. Electric Vehicle Outlook 2024, 2–6: BloombergNEF (2024).
2. Global EV Outlook 2024, 102–117: IEA (2024).
3. Guo, R., Wei, X.-K, Gao, J. : Experimental NVH evaluation of a pure electric vehicle in transient operation modes. In: Proceedings of Vibroengineering Procedia, 10, 1(2016).
4. Lafont, T., Bertolini, C., Ronzio, F., Caprioli, D. : Application of Statistical Energy Analysis on a car : from the vehicle modeling to parts targeting. In: Proceedings of INTER-NOISE 2016, Hamburg, Germany (2016).
5. Nakajima, N., Fukumoto, I., Kokabu, S. : Study of effective interior noise reduction using detailed door model analysis technology development to visualize noise transfer flow by HSEA. In: Proceedings of the 2024 JSAE Congress (Spring), Yokohama, Japan (2024).
6. Cordioli, J. A., Calçada, M., Rocha, T., et al.: Application of the Hybrid FE-SEA Method to Predict Sound Transmission Through Complex Sealing Systems. SAE International Journal of Passenger Cars - Mechanical Systems (2011).

# Method for Determination of Psycho-Haptically Perceived Vibration in Steer-By-Wire Applications

Kenny Faske(✉) and Simon Dirauf

Schaeffler Technologies AG & Co. KG, Herzogenaurach, Germany
{faskeknn,dirausmo}@schaeffler.com

**Abstract.** The introduction of Steer-by-Wire marks the next stage in the development of steering systems in the automotive sector. This technology eliminates the conventional mechanical linkage, and the haptic feedback torque is provided by an electric feedback motor instead. This motor can introduce unwanted vibrations (e.g., motor controller, cogging torque) into the system, which potentially affects the steering feel negatively. So far, assessment methods and requirements have mainly been based on motor torque fluctuations evaluations. However, this approach does not fully correlate with users' subjective perceptions, as it does not consider humans frequency-dependent perception. This paper presents an analytical method validated through an experiment, which addresses this issue and objectively evaluates the subjective steering feel, providing a basis for specification and requirement development. Additionally, the limitations of the methodology are addressed, and further aspects for future improvement are discussed to establish requirement formulation standards as a target.

**Keywords:** Steer-by-Wire · psycho-haptic · hand · perception · vibration · steering · steer-feel · noise vibration harshness · NVH · automotive · algorithm · method

## 12.1 Introduction

### 12.1.1 NVH at Steer-By-Wire in Automotive and Its Challenges

The introduction of Steer-by-Wire (SbW) systems in the automotive sector marks a fundamental shift in steering technology. By eliminating the mechanical connection between the steering wheel and the wheels, new degrees of freedom arise in the design of steering feel, ergonomics, and safety functions. Furthermore, this technology not only enables a more flexible vehicle architecture but also serves as a prerequisite for highly automated driving.

The Schaeffler Group supports the introduction of this technology by developing innovative components and systems as well as controls for by-wire steering setups on


A. Heintzel (Ed.): Automotive Acoustics Conference 2025, Proceedings, pp. 176–188, 2026. https://doi.org/10.1007/978-3-658-51249-1_12

the front and rear axles. With its expertise in chassis mechatronics and drive-by-wire technology, Schaeffler is playing a significant role in driving the industry change necessary to achieve highly automated driving.

At the same time, moving away from mechanical components such as the steering column or rack-and-pinion steering introduces new challenges, particularly in the area of driver perception. In conventional systems, information about road surface structure, tire grip, or road irregularities is transmitted through natural feedback via the steering system. In SbW systems, these sensory inputs must be synthesized or derived from specific sensor data and appropriately modulated.

Haptic feedback, in particular, plays a central role in both the subjective driving experience and driving safety. Unwanted vibrations—caused, for example, by electronic or mechanical disturbances—can negatively affect the driving experience. Existing evaluation methods, such as measuring torque ripple, do provide objective metrics but often show insufficient correlation with subjectively perceived comfort or annoyance.

Against this background, there is a need for a new methodology to assess haptic feedback in SbW systems that systematically incorporates human perception. The aim of this paper is therefore to develop and validate a psycho-haptic evaluation method for steering wheel vibrations. The method is based on findings from human vibration perception and transfers principles from psychoacoustic research to the haptic domain. For the experimental implementation, a test bench environment equipped with a Magneto-Rheological Powder (MRP) brake system is used. This setup allows for flexible generation of steering forces and vibrations superimposed on steering movement, setting it apart from previous studies.

### 12.1.2 Status Quo of Steering Vibration Requirements

In practice, requirements for steering wheel vibrations in Steer-by-Wire systems are typically defined using metrics such as torque ripple, for example by specifying minimum and maximum peak value limits. However, this approach has been shown to lack sufficient correlation with subjectively perceived steering wheel vibrations and annoyance levels. As a result, threshold values in requirement specifications are often tightened in order to preemptively rule out potential comfort issues. This, however, can lead to over-dimensioning of system components or failure to meet requirements during system design. Therefore, it is essential to systematically consider human perception of mechanical vibrations at the steering wheel when defining requirements for steering wheel vibrations, in order to ensure a practical and cost-effective system design.

### 12.1.3 Human Sensation of Mechanical Hand Vibrations

The human perception of mechanical vibrations is highly frequency-dependent. Studies (including Morioka & Griffin [1]) indicate that the perception spectrum for vibrations on the finger or back of the hand typically ranges between 3 Hz and 300 Hz, with peak sensitivity between 80 and 160 Hz. This frequency dependency resembles the psychoacoustic perception of loudness, where the auditory system is particularly sensitive, but in a different frequency range [2, 3].

The perception of vibrations in the hand is mediated by specialized mechanoreceptors located in the skin, particularly within the dermis. There are several types of mechanoreceptors responsible for detecting different kinds of vibrations [4]:

- Pacinian Corpuscles: These receptors are located in the deeper layers of the skin (hypodermis) and within the connective tissues of the fingers and palms. They are particularly sensitive to high-frequency vibrations and respond quickly to changes in skin deformation. Their highest sensitivity is observed in the frequency range of 200 Hz to 300 Hz [20], although their entire perception range extends from 20 Hz to 1 kHz. They are primarily responsible for perceiving fast, high-frequency vibrations [9].
- Meissner's Corpuscles: These receptors are found in the upper skin layer and are also sensitive to vibrations. They respond particularly to low-frequency vibrations (approximately 10–50 Hz, but also from 5–100 Hz [20]) and are responsible for perceiving finer vibrations and touch [10].
- Merkel Cells: These receptors are associated with slow and continuous touches. They are particularly sensitive to small pressure changes and contribute to perceiving the shape and texture of surfaces. While they are not directly responsible for vibration detection, they play a role in perceiving static pressure [11].
- Ruffini Endings: These receptors respond to skin stretching and are specialized in perceiving slow, continuous movements and the positioning of limbs. They play a minor role in vibration perception but contribute to overall tactile sensation [12].

These receptors work in concert to convey the entire vibration experience, with their response varying across different frequencies and amplitudes. In particular, Pacinian Corpuscles are crucial for perceiving high-frequency vibrations, such as those that may occur in modern vehicle steering systems.

Furthermore, amplitude perception is also nonlinear: the perceived intensity of a vibration does not increase proportionally with the physical acceleration [5]. Therefore, converting it into a logarithmic relationship is appropriate.

## 12.2 Method for Determination of Psycho-Haptically Perceived Vibration in Steer-By-Wire Applications

### 12.2.1 Acoustic and Haptic Analogy

Established methods already exist for evaluating subjectively perceived airborne sound events using time-domain analysis. The most widely used and methodologically simplest approach is illustrated in Fig. 12.1. In this method, the sound pressure time signal is first weighted by frequency using a weighting function (e.g.: A-rating) that represents the subjective perception of intensity. Subsequently, a logarithmic level calculation is performed, considering a reference value corresponding to the auditory threshold—that is, the sound pressure level at which a sound becomes audible. The resulting time course of the sound level can then be analyzed with variable resolution, for example using (exponential) RMS averaging or appropriate settings for the Fast Fourier Transform (FFT).

This method allows for identifying the timing of specific sound levels or subjectively perceived intensities and drawing conclusions about potential sources of disturbance as well as suitable optimization measures.

This methodological approach also appears promising in the context of developing Steer-by-Wire systems. The following chapter therefore takes a closer look at the existing analogy between haptic and acoustic perception.

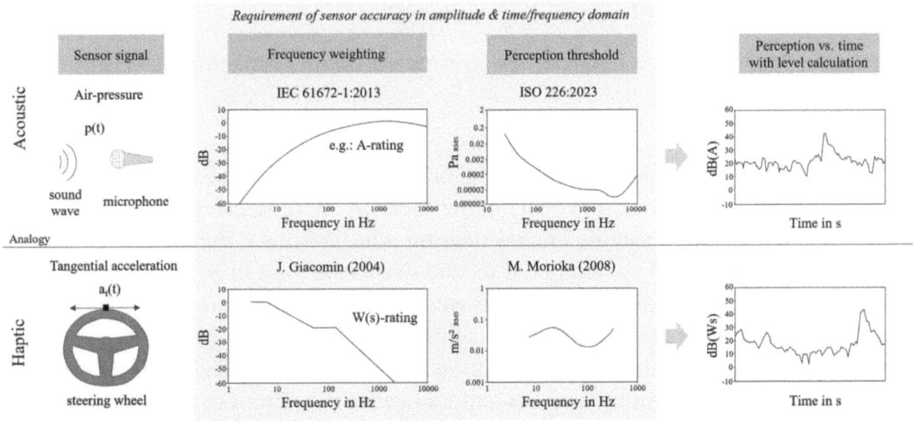


**Fig. 12.1.** Analogy between Acoustic and Haptic in the method for objectively assessing the subjectively perceived intensity.

Numerous studies on mechanical vibrations acting on the human body can already be found in the scientific literature. However, the application of these studies is primary on occupational safety, such as the Wh frequency weighting for hand-arm vibrations according to [6].

A more steering wheel application-orientated study of Giacomin [7] analyzed the perception of equal vibration intensities at different frequencies in the context of a steering system as conventionally used in automobiles. In this study a frequency-weighting curve Ws[1] for steering vibration was proposed. The key physical influencing factor in this case was the tangential acceleration at the radius of the hand position on the steering wheel.

Additionally, based on the research of Morioka [8], a perception threshold value can be derived, which defines the lower limit of vibrational intensity detectable by human senses. These investigations make it possible to calculate and evaluate subjectively perceived vibration intensities in the time domain.

The mentioned studies [7, 8] are based on the generation of vibrations using electrodynamic shakers. However, a potential influence of active steering rotation has not

[1] Ws curve defined as slope of 0 dB per octave from 3 to 6.3 Hz, −6 dB per octave from 6.3 to 50 Hz, 0 dB per octave from 50 to 160 Hz and −10 dB per octave from 160 to 315 Hz [7].

been considered so far. For this reason, the present paper follows a modified research approach, which is explained in more detail in Sect 12.2.3.

### 12.2.2 Sensor Requirements

The requirements for a sensor to determine the tangential steering wheel acceleration can be derived from Morioka [8] and Giacomin [7]:

- **Temporal resolution** to capture oscillations from 3 Hz up to at least 500 Hz
- **Amplitude resolution** to detect oscillations ranging from 0.02 m/s2 (perception threshold) to 20 m/s2 (dynamic range of 60 dB)

This paper presents two approaches for determining tangential steering wheel acceleration.

**Encoder**

A straightforward method to determine the tangential steering wheel acceleration is to use the built-in angle sensor (encoder) in the Steer-by-Wire steering system. Based on the steering wheel angle and the known steering wheel radius *r*, the tangential acceleration $a_t$ can be calculated by taking the second derivative with respect to time.

Encoders are discretized, for example with $N=2^{13}=8192$ increments per steering wheel revolution. According to [13], the sensor must be capable of resolving acceleration amplitudes from 1/10 of the perception threshold up to at least 500 Hz—that is, 0.002 m/s2 (RMS). Assuming a sinusoidal tangential acceleration, the following relationship can be derived (1) and plotted for common encoders, see Fig. 12.2:

$$a_{t,rms} = \frac{10 \cdot r \cdot (2\pi f)^2}{\sqrt{2} \cdot N} \tag{12.1}$$

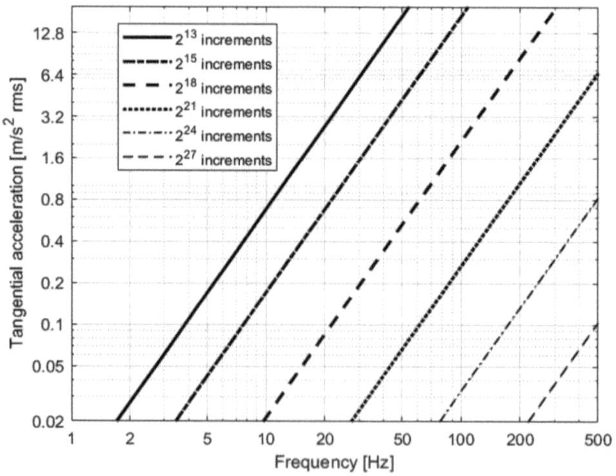


**Fig. 12.2.** Characteristic curves for common encoders to determine sinusoidal acceleration depending on frequency and amplitude.

From Fig. 12.2, it can be seen that with a typical rotary angle sensor offering $N = 8192$ increments per revolution, frequencies up to only 5 Hz can be analyzed at amplitudes up to ~0.17 m/s2 RMS. This frequency range is insufficient to meet the requirements for evaluating steering wheel vibrations. It would require $N \approx 7 \cdot 10^8$ increments per revolution to evaluate 0.02 m/s2 RMS up to 500 Hz. However, no encoder currently available—especially not in an automotive series application—can achieve such high angular resolution.

As an alternative, rotational rate sensors or gyroscopes could be used. These measure angular velocity directly, from which the required tangential acceleration can be calculated by a simple time derivative and multiplication by the radius. However, the extent to which these sensors can meet the requirements for evaluating steering wheel vibrations is not examined in this study.

**Acceleration Sensor**

An alternative method for capturing steering wheel vibrations is the direct measurement of tangential acceleration on the steering wheel using a suitable acceleration sensor, following the approach in [7]. For the present application, a piezoelectric accelerometer with a measurement range of ±5 g (corresponding to approximately ±49 m/s2 peak) proves to be suitable. According to the manufacturer's specifications, the broadband resolution of this sensor type is 0.001 m/s2 RMS in the frequency range from 1 Hz to 10,000 Hz. Such sensors are capable of reliably detecting vibrations in the range of approximately 1 Hz to 8000 Hz [14]. The Acceleration sensor can be easily mounted on the steering wheel and are negligible in terms of their impact on the moment of inertia in the steering system due to their low mass (<1%).

However, a disadvantage of this method lies in the requirement for a wired connection to the sensor, which can potentially interfere with the execution of steering maneuvers. Additionally, mechanical influences such as structure-borne noise during hand repositioning on the steering wheel may distort the measurement signal.

Since the chosen measurement method meets the requirements for detecting tangential acceleration, it was used in the present experimental study.

### 12.2.3 Evaluation Algorithm for Steering Vibrations

The method for evaluating subjectively perceived steering wheel vibrations is conceptually based on the approach for analyzing subjectively perceived airborne sound events described in Sect. 12.2.1. Initially, the time-discrete raw signal is transformed into the frequency domain by applying a Fourier Transform. Subsequently, a frequency dependent weighting of the amplitude spectrum is performed using the Ws-weighting curve. Since the Ws-curve does not define values for frequencies below 3 Hz, these are weighted in this study with a general damping value of −120 dB, in order to systematically suppress low-frequency signal components that are not perceived as relevant steering wheel vibrations.

The inverse Fourier transform is then used to convert the weighted spectrum back into the time domain, providing the basis for subsequent signal processing. To smooth the signal over time and perform an energy-based evaluation, a moving root mean square (RMS) is calculated. The averaging duration should be at least 0.33 s, which corresponds to a lower frequency limit of 3 Hz.

For the subsequent level calculation, the haptic perception threshold of 0.02 m/s2 RMS—as determined in [8]—can be used as a reference value across the entire frequency range. With this scaling, accelerations at the perception threshold correspond to a reference level of 0 dB, making the results more interpretable. Alternatively, the corresponding reference value of 1e-6 m/s2 according to ISO standards can also be used [15]. The basic steps for the calculation method are illustrated in Fig. 12.3.

The result of this evaluation algorithm is a time-dependent profile of the subjectively perceived steering wheel vibrations, which accounts for both frequency dependent sensitivity and dynamic amplitude behavior. This methodology enables the formulation of system-based requirements that can also be specified in detail at the subcomponent level.

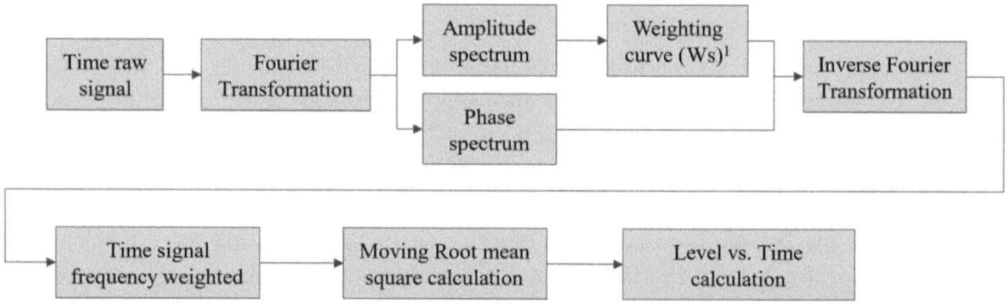


**Fig. 12.3.** Basic evaluation algorithm calculation steps. [1]Damping of −120 dB for frequencies at 0–3 Hz.

### 12.2.4 Experiment

As described in Sect. 12.2.3, the Magneto-Rheological Powder (MRP) brake was used in this study as new approach to vibration generation in order to account for the influence of superimposed steering motion. This system allows for the generation of nearly arbitrary time-dependent braking torques without disturbances known from electric motors as e.g. cogging. In the Steer-by-Wire application, steering resistance can be synthetically generated depending on the driving situation.

For this investigation, various constant target braking torques of 0.5 Nm and 2 Nm were each superimposed with sinusoidal signals at frequencies of 4 Hz, 8 Hz, 16 Hz, 32 Hz, and 64 Hz, with target torque amplitudes of 0.05 Nm, 0.1 Nm, 0.2 Nm, 0.4 Nm, and 0.8 Nm (at 2 Nm tests). Subjective ratings are typically conducted on a 1–10 scale. However, due to poor differentiation observed in preliminary tests for subjective evaluation, the scale was reduced to 1–3:

- Not perceptible
- Perceptible but acceptable
- Perceptible and unacceptable

Due to the wired sensor setup (see Sect. 12.2.2), the steering angle range was limited to ±90 degrees. However, the considered range of the steering angle also results from a typical steering gear ratio of 1:18, which occurs at a steering angle of 5°, defined as a typical range in vehicle dynamics [16].

All test participants were instructed to drive with both hands on the wheel and to intuitively grip it as if driving on a country road. A production steering wheel (VW-Arteon, without airbag) with a radius of 200 mm was used. A tangentially aligned acceleration sensor (PCB Piezotronics, Model 355B04, ±5 g) was firmly mounted on the wheel, oriented tangentially to the axis of rotation. The background noise of the accelerometer in the measurement chain was 0.02 m/s2 and was not perceptible by any participant.

Additionally, a torque measurement shaft with an integrated angle encoder (Kistler 4503B) was mounted between the MRP (prototype) and the steering wheel using torsionally rigid couplings (R+W, BKH 30). The tests were conducted using a laboratory test rig which simulated an automotive driving posture, see Fig. 12.4.

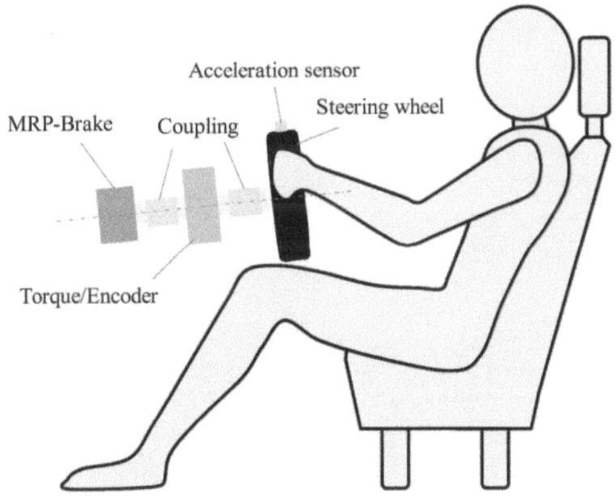


**Fig. 12.4.** Sensor and test setup for experiment.

### 12.2.5 Results

Figure 12.5 shows the respective levels at 0.5 Nm and 2 Nm load, overlaid for each investigated frequency, according to the calculation method presented in Sect. 12.2.3, illustrating the transition from "not perceptible" to "perceptible but acceptable" to "perceptible and unacceptable." Additionally, the corresponding transitions for the RMS values of the torques across the tested frequency range are displayed.

For the RMS values of the torques, the transitions are highly frequency-dependent, especially in the lower frequency range below 20 Hz. The spacing between the transitions is also not constant and varies with frequency. It can also be assumed, that the observed transitions are dependent on the moments of inertia in the steering system. However, the extent this influence has not been investigated in this experiment.

In contrast, for the levels of tangential acceleration that are subjectively perceived as "perceptible and unacceptable, not OK" a nearly frequency-independent relationship is observed. Similarly, the range from "not perceptible, OK" to "perceptible but acceptable, border" spans approximately 5 dB and also remains nearly constant across the frequency spectrum. Furthermore, no significant differences can be observed between the various load conditions of 0.5 Nm and 2 Nm.

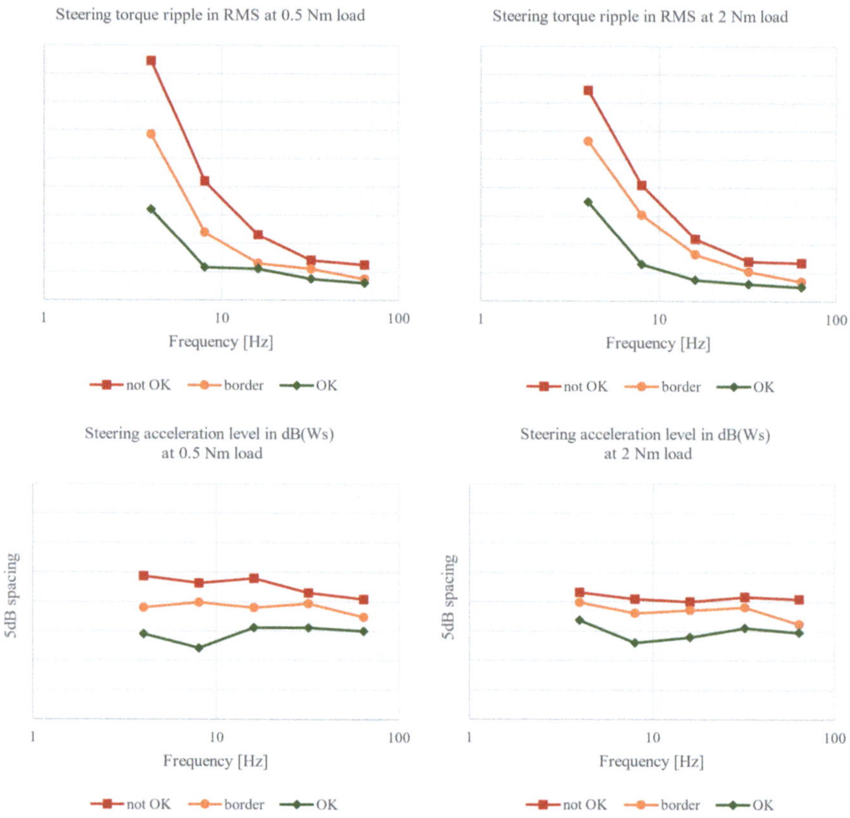


**Fig. 12.5.** Evaluation results for transitions of steer feel depending on frequency at different steering loads for torque ripple and tangential steering acceleration with Ws-frequency weighting curve (according to evaluation method proposed in this paper).

### 12.2.6 Examples

Based on the insights gained in Sect. 12.2.5, the proposed calculation method will now be validated using practical application examples:

**Inconsistency in Steering Feel During Steering Maneuver**

Using the methods presented in this study, an evaluation was conducted within a Steer-by-Wire system to assess a complaint related to a perceived inconsistency during steering. The analysis illustrated in Fig. 12.6 demonstrated that the threshold for "perceptible and unacceptable" vibrations was exceeded at the time of the short inconsistency in steering feel-event. Furthermore, the temporal progression of the signal allowed for the identification of a causal relationship, which could subsequently be addressed and resolved during the development phase.

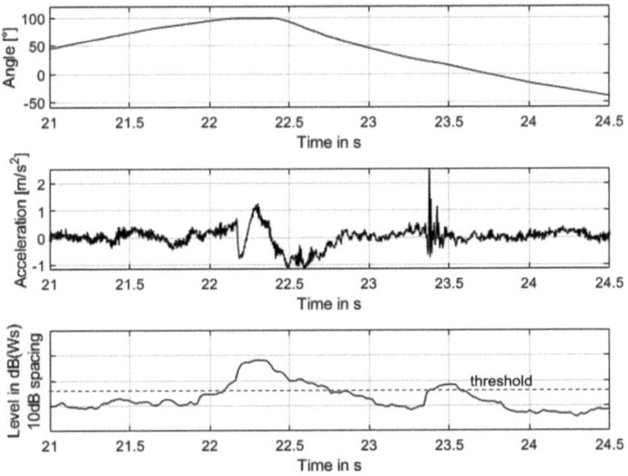


**Fig. 12.6.** Analysis plot for short inconsistency in steering feel that exceeds the threshold of "perceptible and unacceptable" at ~ 23.5 s apart from change of steering direction at ~22.2 s.

## Benchmark of Drive Motor

For a steer-by-wire system, two possible feedback motors were available. The regulated motors were compared with respect to their haptic feedback. Each motor was required to deliver an average torque of approximately 2 Nm, while the steering wheel was manually rotated at a constant speed of 50 deg/s. Although the second motor provides slightly increased torque, its amplitude is lower. This aligns with the subjective perception of slightly better haptic feedback from the second motor. An evaluation based only on torque is challenging. A difference in acceleration between the two motors is observable, but only at the amplitude are the occurring frequencies weighted and assessed according to human sensitivity, see Fig. 12.7. The presented method thus provided an objective assessment, on the basis of which a component could be selected.

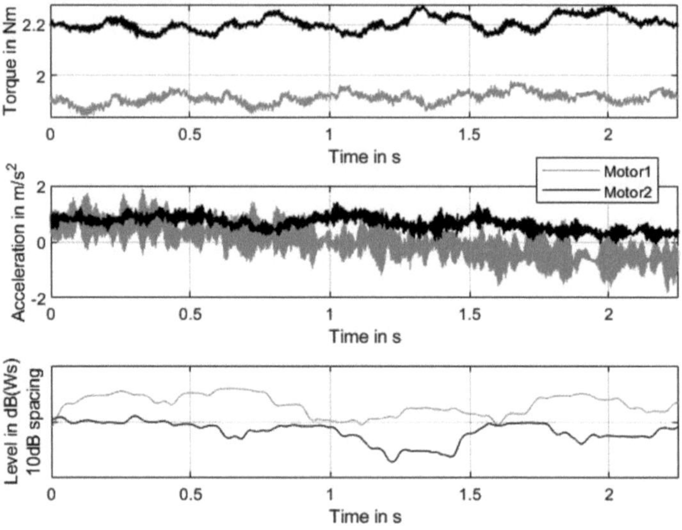


**Fig. 12.7.** Benchmark analysis of two electric motors using evaluation method proposed in this paper shows objective differences that correlates with subjective perception.

## 12.3 Challenges and Outlook

### 12.3.1 Challenges

Since the calculation method is based on the formation of RMS values and is thus related to an energetic metric, certain aspects of subjective perception cannot be adequately captured:

- Transient signal characteristics significantly influence vibrotactile perception. Impulses of different durations—even with identical RMS values—can be perceived very differently. This temporal dynamic is inadequately represented by purely energetic metrics [17].
- Masking effects also occur when vibrations consist of components that are close in frequency—for example, within the response range of Pacinian corpuscles (around 250–300 Hz). In such cases, stimuli may overlap in a way that impairs fine sensory resolution. An RMS-based evaluation tends to overestimate the perceived overall stimulus in these cases, as it merely sums the energy components without accounting for the sensory discrimination ability [18].
- In addition, for non-sinusoidal stimuli, such as noise-like vibrations, the presented calculation method reaches its limits: although identical RMS values may be calculated, literature shows that such stimuli are perceived differently than periodic sinusoidal signals [18, 19]. These differences can be attributed to the spectral structure and temporal irregularity of the signal and highlight the need for a more nuanced modeling of vibrotactile perception.

### 12.3.2 Outlook

The presented method enables a reliable prediction—particularly for sinusoidal vibrations—of whether a vibrotactile stimulus will be as perceivable by users of an electromechanical steering system. Due to its relatively low computational complexity, it appears well-suited for standardization within technical specifications and requirement documents.

The methodological approach is also fundamentally transferable to other input systems—for example, in Steer-by-Wire applications with steering knobs. However, this would require experimental redefinition of frequency-dependent perception thresholds and iso-sensitivity curves in order to accurately reflect the specific characteristics of such systems.

In the long term, the development of a model for vibrotactile perception—analogous to established loudness models in auditory perception [2]—should be pursued. However, this will require further psycho-haptic research to enable a more detailed representation of human perception of vibrotactile stimuli.

## References

1. Morioka, M.; Griffin, M. J. 2005 Measurement of human vibration perception J. Acoust. Soc. Am., 118(2), 1061–1071 https://doi.org/10.1121/1.1944566
2. Zwicker, E.; Fastl, H. 1990 Psychoacoustics: Facts and Models Springer ISBN: 978-3-540-65063-3
3. ISO 2017 ISO 532-1: Acoustics—Method for calculating loudness levels of steady sounds International Organization for Standardization
4. Griffin, M. J. 2012 Handbook of Human Vibration Academic Press ISBN: 978-0-12-303041-2
5. Stevens, S. S. 1957 On the psychophysical law Psychological Review, 64(3), 153–181
6. ISO 2001 ISO 5349-1: Mechanical vibration—Measurement and evaluation of human exposure to hand-transmitted vibration—Part 1
7. Giacomin, J., Shayaa, M. S., Dormegnie, E., & Richard, L. (2004). Frequency weighting for the evaluation of steering wheel rotational vibration. International Journal of Industrial Ergonomics, 33(8), 527–541. https://doi.org/10.1016/j.ergon.2003.12.003
8. Morioka, M., & Griffin, M. J. (2008). Absolute thresholds for the perception of fore-and-aft, lateral, and vertical vibration at the hand, the seat, and the foot. Journal of Sound and Vibration, 314(1–2), 357–370. https://doi.org/10.1016/j.jsv.2007.12.003
9. Mountcastle, V. B. 2005 The Sensory Hand: Neural Mechanisms of Somatic Sensation Harvard University Press ISBN: 978-0674019419
10. Johansson, R. S.; Flanagan, J. R. 2009 Coding and use of tactile signals from the fingertips in object manipulation tasks Nat. Rev. Neurosci., 10(5), 345–359 https://doi.org/10.1038/nrn2621
11. Halata, Z.; Grim, M.; Bauman, K. I. 2003 Funktionelle Morphologie der Merkel-Zellen Anat. Embryol., 207(6), 455–474 https://doi.org/10.1007/s00429-003-0346-5
12. Abraira, V. E.; Ginty, D. D. 2013 The sensory neurons of touch Neuron, 79(4), 618–639 https://doi.org/10.1016/j.neuron.2013.07.051
13. Berndt, G (1968). Funktionstoleranz und Messunsicherheit. Feinwerktechnik, 72(11), 441–444

14. PCB Piezotronics, Inc. (o. J.). Installation and Operating Manual—Model M355B04. Retrieved October 23, 2024, from https://www.manualslib.com/manual/2881802/Pcb-Piezotronics-M355b04.html
15. ISO 2017 ISO 8041–1:2017: Human response to vibration—Measuring instrumentation—Part 1
16. Klammerer, T. (2017). Potenziale einer Lenkmomentenempfehlung unter fahrdynamischen Gesichtspunkten (Dissertation, Technische Universität Dortmund). Eldorado. https://eldorado.tu-dortmund.de/bitstream/2003/34397/1/Dissertation.pdf
17. Hollins, M., Roy, E. A., Crane, S. A., & Dean, C. L. (2003). Vibrotactile adaptation on the hand: Neural vs. perceptual correlates. Somatosensory & Motor Research, 20(1), 33–43. https://doi.org/10.1080/0899022031000083827
18. Gescheider, G. A., Bolanowski, S. J., & Verrillo, R. T. (1996). Vibrotactile masking: Effects of one- and two-site stimulation. Perception & Psychophysics, 58(8), 1221–1232. https://doi.org/10.3758/BF03206837
19. Bolanowski, S. J., Gescheider, G. A., Verrillo, R. T., & Checkosky, C. M. (1988). Four channels mediate the mechanical aspects of touch. The Journal of the Acoustical Society of America, 84(5), 1680–1694. https://doi.org/10.1121/1.397184
20. Fleming MS, Luo W. The anatomy, function, and development of mammalian Aβ low-threshold mechanoreceptors. Front Biol (Beijing). 2013 Aug https://pubmed.ncbi.nlm.nih.gov/24376457/

# From Pulsations to Perfection: A Hybrid Approach for NVH Optimization in Automotive Fluid Systems

Jonas Holzbrecher[1,2,2(✉)] and Fadi Dohnal[3]

[1] Evamo – Pump Technology Solutions PS GmbH, Schwäbisch Gmünd, Germany
jonas.holzbrecher@pump-ts.com
[2] Institute of Measurement and Sensor Technology, UMIT Tirol, Hall in Tirol, Austria
[3] Research Center for Microtechnology, FHV—Vorarlberg University of Applied Science, Dornbirn, Austria
fadi.dohnal@fhv.at

**Abstract.** Noise and vibration challenges in automotive fluid systems arise primarily from pressure pulsations generated by displacement pumps. These pulsations, when reflected within the fluid system, can lead to resonances and unwanted acoustic phenomena. This paper introduces a hybrid approach combining experimental and simulation methods to model and mitigate such effects. A hybrid pump model is developed, incorporating both pulsation and impedance behavior, based on measurements from a reflection-free test bench and the Secondary Source Method. The model enables realistic simulation of pressure wave propagation and system resonances. By integrating the frequency-dependent pump impedance into a 1D system simulation model, resonance behavior can be analyzed and optimized virtually. Application to a commercial hydraulic steering system demonstrates the method's capability to reduce resonance amplification. The study highlights the critical role of accurate component modeling, especially for pumps and hoses, in predicting NVH phenomena in automotive fluid systems.

**Keywords:** Hydraulic system · Fluid-borne noise · Impedance

## 13.1 Introduction

Fluid power systems in vehicles serve key functions such as steering, cooling, and lubrication. Displacement pumps are widely used for their robustness but generate periodic flow pulsations [1], which propagate through the hydraulic lines and reflect at boundaries. If these pulsations coincide with system resonances, standing waves may form, increasing vibrations and noise, especially under low damping conditions. Engine-driven pumps intensify this risk due to their broadband excitation.


A. Heintzel (Ed.): Automotive Acoustics Conference 2025, Proceedings, pp. 189–206, 2026. https://doi.org/10.1007/978-3-658-51249-1_13

Conventional acoustic analysis detects noise effects but provides limited insight into the internal fluid dynamics. Since pressure sensors offer less data near pressure nodes, effective placement requires prior knowledge of wave propagation. To address this, 1D system simulation enables prediction of pressure pulsations, resonances, and mode shapes, provided components are accurately modeled and characterized independently.

While hoses and pipes are straightforward to simulate, pumps require a detailed acoustic model. This work proposes a hybrid pump model (HPM) based on experimental data, capturing both pulsation and frequency-dependent reflection via a simplified substitute model. Pump impedance is measured using the Secondary Source Method [2], and pulsation is measured under nearly reflection-free boundary conditions.

The methodology is applied to a fixed-displacement vane pump and a commercial steering system. It enables virtual evaluation of resonance behavior and supports early-stage optimization. The approach is broadly applicable to any fluid systems with pressure-induced NVH challenges.

## 13.2 Theoretical Background

### 13.2.1 Reflective Hydraulic System

In hydraulic line systems, pressure waves propagate as plane waves and reflect at termination components. The superposition of incident $p_i$ and reflected $p_r$ pressure waves creates interference patterns that represent the modal shape of the oscillating fluid column in the frequency domain [3]

$$p(x) = p_i e^{-\gamma x} + p_r e^{\gamma x} \tag{13.1}$$

Here, $\gamma$ denotes the complex wave propagation constant, which accounts for both phase shift and amplitude along the propagation path. The reflection coefficient $\Gamma$, representing the ratio of reflected to incident pressure wave, can be derived from two pressure measurements at different positions $x_1$ and $x_2$

$$\Gamma = \frac{p_r}{p_i} = \frac{p_2 e^{-\gamma x_1} - p_1 e^{-\gamma x_2}}{p_1 e^{\gamma x_2} - p_2 e^{\gamma x_1}} \tag{13.2}$$

Substituting Eq. (13.2) into Eq. (13.1) yields the total pressure distribution as

$$p(x) = p_i (e^{-\gamma x} + \Gamma e^{\gamma x}) \tag{13.3}$$

The acoustic impedance at position $x$ is defined as the ratio of pressure to volume flow $Q$ and can be expressed as

$$Z(x) = \frac{p(x)}{Q(x)} = Z_c \frac{e^{-\gamma x} + \Gamma e^{\gamma x}}{e^{-\gamma x} - \Gamma e^{\gamma x}} \tag{13.4}$$

Here, $Z_c$ is the characteristic impedance of the pipe

$$Z_c = Z_0 \xi = \frac{\rho c}{A} \xi \tag{13.5}$$

This quantity describes the wave resistance and links pressure to volume flow. It requires knowledge of the fluid density $\rho$, the cross-sectional area of the pipe $A$, and the speed of sound $c$. The correction factor $\xi$ accounts for viscous losses in unsteady laminar flow and is expressed as

$$\xi = \sqrt{-\frac{J_0(R^*)}{J_2(R^*)}} \approx \sqrt{1 + \frac{8\nu}{sR^2}} \tag{13.6}$$

In oscillatory pipe flow, the velocity profiles deviate significantly from the parabolic distribution known from steady-state Hagen–Poiseuille flow. These complex profiles can be described exactly using Bessel functions. The approximation on the right-hand side of Eq. (13.6) enables a simplified analytical solution. Here, $\nu$ denotes the kinematic viscosity, $R$ is the pipe radius, and $s$ is the Laplace variable.

Evaluating Eq. (13.4) at the termination ($x = 0$) yields the termination impedance

$$Z_T = Z_c \frac{1 + \Gamma}{1 - \Gamma} \tag{13.7}$$

Conversely, the reflection coefficient can be written in terms of the termination impedance as

$$\Gamma = \frac{Z_T - Z_c}{Z_T + Z_c} = \frac{R + jX - Z_c}{R + jX + Z_c} \tag{13.8}$$

The resistive component $R$ influences the magnitude of the reflection, while the reactive component $X$ determines the phase shift between incident and reflected waves. These expressions form the theoretical basis for modeling wave reflection and impedance matching in fluid-filled line systems. Like the reflection coefficient, acoustic impedance is frequency-dependent and serves as a key parameter for characterizing the reflective behavior of hydraulic components [5].

Typical boundary conditions can be derived from limiting cases: a closed end with $Z_T \gg Z_c$ results in $\Gamma = 1$, corresponding to total reflection without phase shift. An open end with $Z_T \ll Z_c$ yields $\Gamma = -1$, indicating total reflection with a 180° phase shift. Perfect impedance matching ($Z_T = Z_c$) leads to $\Gamma = 0$, i.e., no reflection.

While these idealized conditions provide a useful first approximation, they do not account for the complex, frequency-dependent reflection characteristics of real components. Accurate system modeling therefore requires consideration of the component impedance over a broad frequency range.

### 13.2.2 Transfer Matrix Method

The dynamic transfer behavior of a pipe with linear elastic material behavior can be described using a transfer matrix formulation

$$\begin{pmatrix} p_Y \\ Q_Y \end{pmatrix} = \begin{bmatrix} t_{11} & t_{12} \\ t_{21} & t_{22} \end{bmatrix} \begin{pmatrix} p_X \\ Q_X \end{pmatrix} \tag{13.9}$$

Here, the Laplace-domain pressure and volume flow at the inlet $X$ and outlet $Y$ form a state vector. Their relationship is governed by the transfer matrix **T**, based on four-pole network theory from electrical engineering (Fig. 13.1) [4].

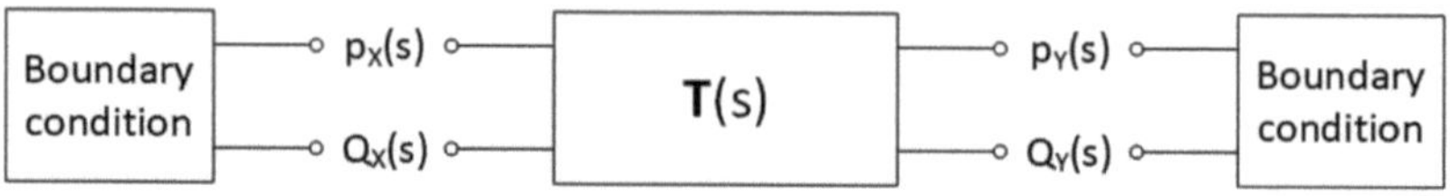


**Fig. 13.1.** Four-pole network of a pipe with pressure and volume flow at inlet X and outlet Y.

The entries of the matrix are derived from the wave equation, formulated as the superposition of forward- and backward-traveling waves

$$p(x) = C_1 e^{-\gamma x} + C_2 e^{\gamma x} \tag{13.10}$$

$$Q(x) = \frac{1}{Z_c}\left(C_1 e^{-\gamma x} - C_2 e^{\gamma x}\right) \tag{13.11}$$

By rewriting these expressions using hyperbolic functions, the system can be reformulated as

$$p(x) = D_1 \cosh\gamma x + D_2 \sinh\gamma x \tag{13.12}$$

$$Q(x) = \frac{1}{Z_c}(D_1 \sinh\gamma x - D_2 \cosh\gamma x) \tag{13.13}$$

Applying the boundary conditions at the pipe inlet ($x = 0$) yields

$$p(0) = D_1 = p_X \tag{13.14}$$

$$Q(0) = -\frac{D_2}{Z_c} = Q_X \rightarrow D_2 = -Q_X Z_c \tag{13.15}$$

Substituting Eqs. (13.14) and (13.15) into Eqs. (13.12) and (13.13) provides the pressure and flow at the pipe outlet ($x = l$)

$$p_Y = p_X \cosh \gamma l - Q_X Z_c \sinh \gamma l \tag{13.16}$$

$$Q_Y = \frac{p_X}{Z_c} = \sinh \gamma l - Q_X \cosh \gamma l \tag{13.17}$$

This leads to the final form of the transfer matrix for an unsteady laminar pipe flow

$$\begin{pmatrix} p_Y \\ Q_Y \end{pmatrix} = \begin{bmatrix} \cosh \gamma l & -Z_c \sinh \gamma l \\ -\frac{1}{Z_c} \sinh \gamma l & \cosh \gamma l \end{bmatrix} \begin{pmatrix} p_X \\ Q_X \end{pmatrix} \tag{13.18}$$

This derivation demonstrates that the Transfer Matrix Method (TMM) is fundamentally rooted in the physical principles of wave propagation within the pipe. In linear systems, two pressure measurements are sufficient to yield a solvable system of equations for the unknown volume flows $Q_X$ and $Q_Y$ at the inlet and outlet, respectively. [5].

The resulting formulation represents a distributed-parameter model, in which the fluid is treated as a continuous medium. This contrasts with lumped-parameter models, which approximate the system using discrete elements and become less accurate at higher frequencies due to their limited spatial resolution.

## 13.3 Acoustic and Hydraulic Pump Characterization

### 13.3.1 Pump Pulsation Measurement

Accurate characterization of pump pulsation is not possible in systems where reflections occur, as the superposition of incident and reflected waves creates location-dependent pressure distributions. Such measurements are therefore unsuitable for defining boundary conditions in time-domain system simulations. To obtain system-independent data, a reflection-free measurement environment is essential.

To enable this, a test bench with a reflection-less pipe termination is used. The setup consists of a pump connected to a measuring pipe with three pressure sensors, a hydraulic volume, and an orifice with adjustable diameter (Fig. 13.2). The hydraulic volume acts as a capacitance, attenuating low-frequency static pressure fluctuations. [6].

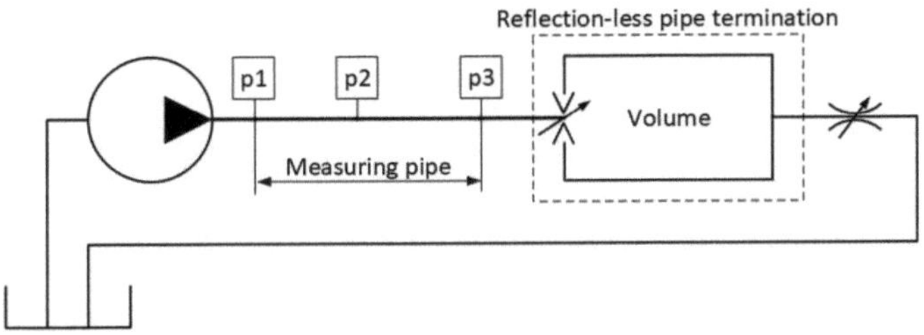


**Fig. 13.2.** Schematic representation of the reflection less-pipe termination.

Flow through the orifice follows the nonlinear orifice equation [7]

$$Q = A\sqrt{\frac{2\Delta p}{\rho}} \tag{13.19}$$

As the orifice cross-sectional area $A$ is reduced, the pressure drop $\Delta p$ increases, and the flow rate exhibits a nonlinear dependence on the pressure loss. To achieve an ideally reflection-less termination, the orifice impedance $Z_{RLPT}$ must match the characteristic pipe impedance, such that $Z_{RLPT} = Z_c$.

Because the orifice equation is inherently nonlinear, it must be linearized around the operating point to enable impedance matching. The resulting linearized impedance is given by the differential quotient

$$Z_{RLPT} = \frac{\mathrm{d}\Delta p}{\mathrm{d}Q} = \frac{1}{\alpha A}\sqrt{2\rho\Delta p} \tag{13.20}$$

This linearized form enables practical adjustment of the orifice to approximate the characteristic impedance and thus minimize wave reflections.

Impedance matching is performed automatically using in-pipe speed of sound measurements based on three pressure sensors [8]. These measurements allow continuous adjustment of the orifice diameter via pre-calibrated flow maps, ensuring that the orifice impedance matches the characteristic pipe impedance across varying operating conditions. The linearized orifice impedance at the operating point corresponds to the slope of the tangent to the nonlinear flow-pressure curve. As pump speed and flow rate vary, the control system iteratively adjusts the orifice cross-sectional area to maintain this match. While the termination behaves reflection-less over the relevant frequency range, unmodeled effects such as frequency-dependent viscous losses and flow separation prevent it from being strictly reflection-free.

Figure 13.3 shows pressure measurements from a vane pump operating at 750 rpm, 100 bar, and 60 °C. The signals from three sensors closely align, confirming proper tuning of the RLPT. Small deviations are due to propagation delay and damping. After time-delay compensation, the waveforms are nearly identical, with only minor amplitude differences resulting from varying sensor distances.

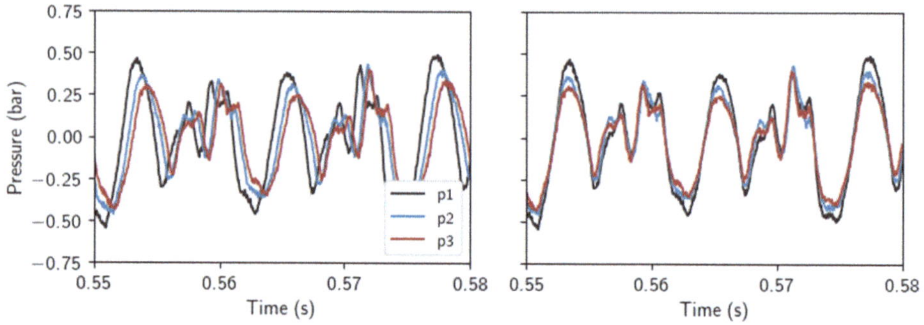


**Fig. 13.3.** Pulsation measurements with the reflection-less pipe termination.

In addition to the periodic pulsation, the signals contain uncorrelated disturbances such as noise and speed fluctuations. These are suppressed using Time Synchronous Averaging (TSA), in which signals are phase-aligned to a reference (e.g., an encoder) and averaged over multiple revolutions (Fig. 13.4). Windowing with overlap minimizes edge effects, particularly as the number of revolutions increases [9].

TSA is applied to each sensor and under different operating conditions. The resulting data is then used to calculate volume flow via the Transfer Matrix Method (TMM), enabling the creation of a boundary condition map for simulation. As a complete dataset was not yet available, a generic pulsation profile is used for subsequent analysis.

### 13.3.2 Pump Impedance Measurement

To determine the pump impedance, the standardized Secondary Source Method (SSM) is employed. Figure 13.5 shows the experimental setup: the pump is connected to a secondary pulsation source via a measuring pipe equipped with three pressure sensors positioned along its length.

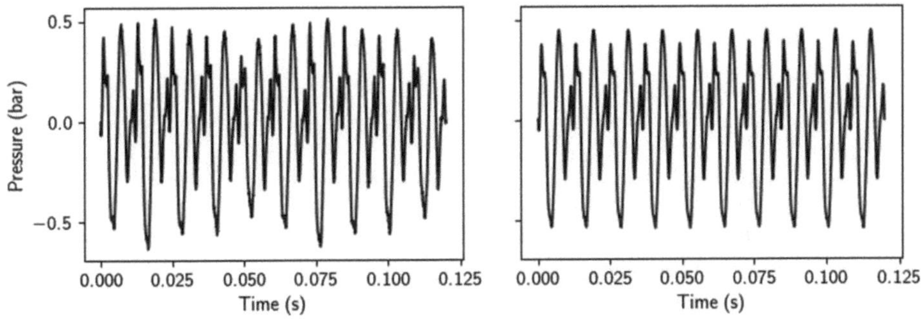


**Fig. 13.4.** Time synchronous averaging with different number of revolutions; left: 2 revolutions, right: 10 revolutions.

The pump is excited by the Secondary Source. Using the TMM, the volume flow at the sensor positions is calculated, which enables the determination of pressure and flow at the pump inlet, and thus the pump impedance. Only frequency components originating from the Secondary Source are evaluated, necessitating frequency-domain filtering. To achieve this, spectral separation between the pump and Secondary Source signals is required. A band-stop filter is applied to suppress the pump's pulsation components.

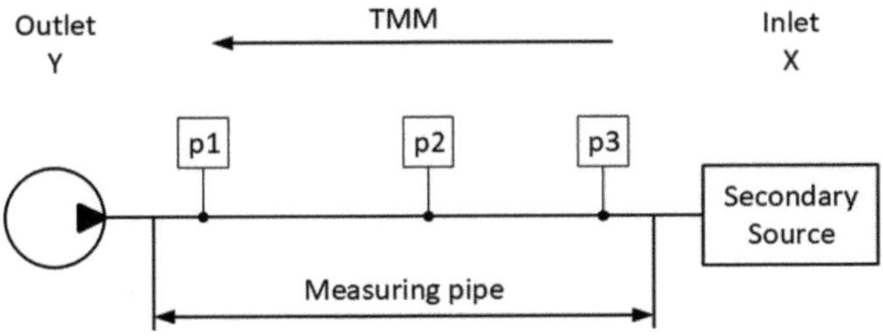


**Fig. 13.5.** Schematic representation of the Secondary Source Method.

Since real measurement data inevitably contain noise, the accuracy of the results is particularly affected at frequencies where the excitation from the Secondary Source is weak. This is especially critical near impedance minima, where small absolute deviations can lead to large relative errors. To improve robustness under such conditions, pressure transfer functions are estimated using the $H_1$ estimator [10]

$$H_1 = \frac{G_{YX}}{G_{XX}} \tag{13.21}$$

The $H_1$ estimator suppresses uncorrelated output noise, thereby improving the reliability of the impedance calculation. Accurate computation of the transfer function requires consistent input and output signal definitions, as shown in Fig. 13.5 Based on this configuration, the transfer matrix relating the pressure and flow transfer functions is given by

$$\begin{pmatrix} H_{pYpX} \\ H_{QYpX} \end{pmatrix} = \begin{bmatrix} \cosh \gamma l_{yx} & Z_c - \sinh \gamma l_{yx} \\ -\frac{1}{Z_c}\sinh \gamma l_{yx} & \cosh \gamma l_{yx} \end{bmatrix} = \begin{pmatrix} 1 \\ H_{QXpX} \end{pmatrix} \tag{13.22}$$

For the selected sensor pair, the volume flow transfer functions at the two sensor positions are computed as follows

$$H_{QXpX} = \frac{1}{t_{12}} H_{pYpX} - t_{11} \tag{13.23}$$

$$H_{QYpX} = t_{21} + \frac{t_{22}}{t_{12}} \left( H_{pYpX} - t_{11} \right) \tag{13.24}$$

The pressure $p_0$ and volume flow $Q_0$ at the pump inlet can then be calculated using

$$H_{p0pX} = t_{11} + t_{12} H_{QXpX} \tag{13.25}$$

$$H_{Q0pX} = t_{21} + t_{22} H_{QXpX} \tag{13.26}$$

The acoustic impedance at the pump inlet is subsequently obtained as the ratio of pressure to volume flow

$$Z_{SYX} = \frac{H_{p0pX}}{H_{Q0pX}} \tag{13.27}$$

Here, the subscript $S$ denotes the specific sensor combination used for the impedance calculation. To assess the quality of the measurement signals, the magnitude-squared coherence between the pressure signals is computed as

$$\gamma_{XY}^2 = \frac{G_{XY}}{G_{YY}} \tag{13.28}$$

where $G_{XY}$ is the cross-power spectral density and $G_{YY}$ the auto-power spectral density of the response signal [11]. High coherence values indicate reliable signal relationships, which are essential for accurate impedance estimation.

Figure 13.6 shows the SSM test bench configuration. The pump is connected to a tank on the suction side, while the high-pressure outlet is routed into a steel measuring pipe with constant diameter to ensure the applicability of the TMM. Piezoelectric pressure sensors (400 kHz natural frequency, 20 kHz sampling rate) are used to capture dynamic pressure pulsations along the pipe.

The Secondary Source is located at the opposite end of the pipe and is fully decoupled from the pump, allowing independent control of the excitation frequency. The setup is connected to the high-pressure port of the test bench via a T-junction, with the static system pressure regulated by a needle valve.

A directly actuated proportional valve serves as the Secondary Source and is controlled by an external signal generator. The valve piston oscillates with a stroke proportional to the input amplitude. Rapid modulation of the effective flow area—resulting from the opening and closing of the internal valve ports—leads to abrupt pressure drops and flow variations. These, in turn, generate pressure pulsations that propagate through the hydraulic system and excite the pump.

The fundamental frequency of the pulsation corresponds to the input signal, while higher harmonics arise as integer multiples. When operated with a frequency sweep, the valve produces broadband pulsations, making it well-suited for impedance analysis across a wide frequency range.

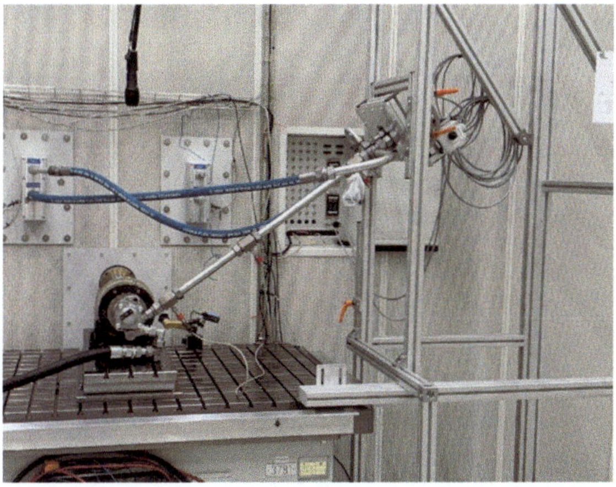

**Fig. 13.6.** Secondary Source Method test bench.

The following presents representative impedance measurements for a fixed-displacement vane pump. The dominant system resonance frequencies are primarily determined by the pipe geometry and the pump's frequency-dependent reflection behavior. For impedance analysis, only those pulsation frequencies generated by the pump itself are of interest. Accordingly, the target frequency range of the SSM is defined by the pump's maximum pulsation frequency.

For a vane pump with 10 displacement elements operating at 4000 rpm with ten displacement elements, the first harmonic occurs at approximately 666 Hz. From an energy point of view, the first three harmonics are most relevant, as higher-order components are significantly attenuated by system damping and contribute only marginally to excitation. As a result, the impedance analysis focuses on a frequency range up to approximately 2000 Hz.

In the test setup, the pump operated at 750 rpm, 100 bar, and 60 °C. The Secondary Source was actuated with excitation frequencies up to 300 Hz. Due to symmetric valve control across both control edges, two strokes per cycle were generated, effectively doubling the fundamental excitation frequency to 600 Hz and amplifying even-order harmonics. This enhances the excitation spectrum and enables reliable impedance measurements close to 2000 Hz.

In an extensive test campaign, the pump was characterized under varying speed, pressure, and temperature conditions. The results presented here represent a typical impedance curve used for parameter identification in the simulation model.

Figure 13.7 shows the averaged and interpolated results from three sensor combinations. The use of three pressure sensors enables the determination of the in-situ speed of sound and allows compensation for invalid sensor combinations. Although the upper frequency limit of 2000 Hz could not be reached due to limited excitation strength, a distinct impedance minimum with associated phase shift was reliably captured at 570 Hz, where the excitation amplitude was sufficient.

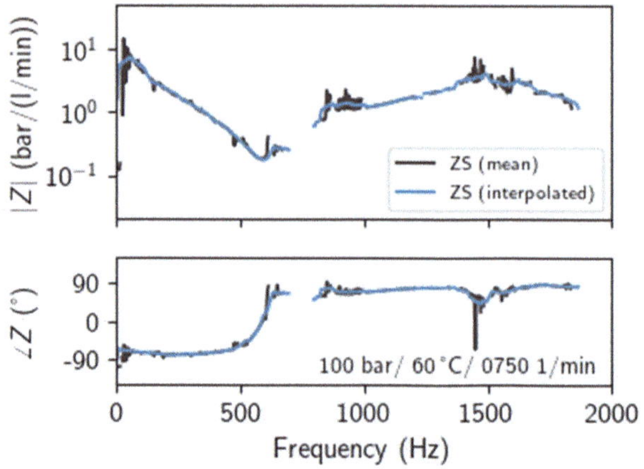


**Fig. 13.7.** Representative impedance measurement of a fixed-displacement vane pump.

## 13.4 Impedance Model

To represent the pump impedance, a geometric substitute model is used, which can be easily integrated into a 1D simulation. The model parameters are identified through curve fitting based on measurement data.

Impedance measurements indicate that the acoustic behavior of the pump can be well approximated by a Helmholtz resonator. Effective fitting requires an appropriate mathematical model. Since a Helmholtz resonator can be physically described as a spring-mass-damper system, its acoustic impedance can be derived directly from the equation of motion of a single-degree-of-freedom oscillator. [12] The resulting impedance equation of the resonator is given by

$$Z_{HHR} = j\omega m_a + \frac{1}{j\omega C_a} + d \tag{13.29}$$

where $m_a$ is the acoustic mass

$$m_a = \frac{\rho L_n}{A_n} \tag{13.30}$$

and $C_a$ is the acoustic compliance

$$C_a = \frac{V}{\rho c^2} \tag{13.31}$$

The geometric parameters relevant for fitting—neck length $L_n$, neck diameter $D_n$, cavity length $L_V$, cavity diameter $D_V$ and damping $d$—can be grouped in the vector

$$x = (L_n, D_n, L_v, D_v, d)$$

The objective of the optimization is to minimize the quadratic error function

$$\epsilon = \sum_{i=1}^{N} \left(Z_{\text{model,real}}(f_i) - Z_{\text{measured,real}}(f_i)\right)^2 + \sum_{i=1}^{N} \left(Z_{\text{model,imag}}(f_i) - Z_{\text{measured,imag}}(f_i)\right)^2 \tag{13.32}$$

The error function computes the difference between the measured and modeled impedance at each frequency. Since impedance is a complex quantity, the real and imaginary parts are treated separately. This ensures that both amplitude and phase information are accurately considered.

Since the underlying impedance model is a complex-valued function, any change in parameters affects both the real and imaginary parts. As a result, closed-form derivatives cannot be derived, making it difficult to define a clear search direction in a gradient-based optimization method. Therefore, parameter identification is performed using the gradient-free Powell optimizer [13]. Upon completion, the algorithm returns the optimal parameter set

$$\left(L_n^*, D_n^*, L_v^*, D_v^*, d^*\right) = \text{argmin } \epsilon$$

Figure 13.8 shows the measured impedance and the modeled impedance using the geometric parameters of the Helmholtz resonator identified through optimization. The gradient-free optimization method proves to be highly reliable. Both the anti-resonance frequency and the damping behavior are accurately captured by the model.

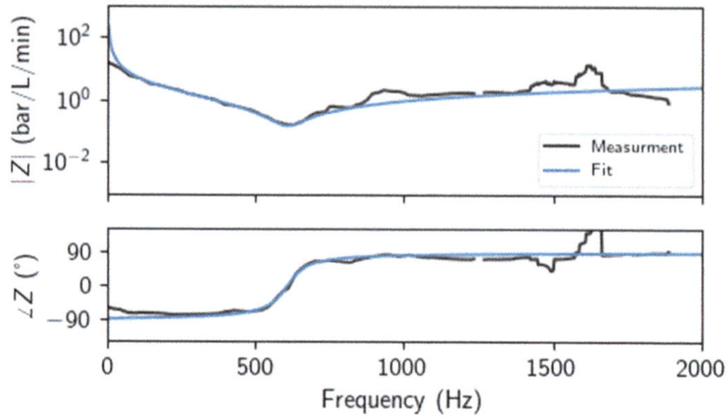


**Fig. 13.8.** Measured and fitted pump impedance, based on the Helmholtz resonator model.

## 13.5 Simulation Model

Figure 13.9 shows the simulation model, comprising the hybrid pump model HPM, the hydraulic line system, and a simplified steering model. The HPM includes the pump impedance, modeled as a Helmholtz resonator, and a signal generator for pulsation input. The simulation considers dynamic wave propagation only; convective effects and components like relief valves are neglected under steady-state conditions.

Pump volume flow is applied as a boundary condition. If measurement data is available, it is implemented as a pulsation map; otherwise, a generic sinusoidal excitation is used. A linear frequency sweep up to 2000 Hz is used. The excitation amplitude reflects the first harmonic typically observed in vane pumps.

The Helmholtz resonator consists of two pipe segments and an orifice, with geometry defined by parameter optimization. Since flow rate is prescribed, pressure drop is adjusted to match the target damping. The orifice damping is implemented via a linear resistance coefficient, avoiding the need for linearization at multiple operating points.

The steering system is simplified as an orifice and a hydraulic volume. The orifice represents inlet resistance, while the volume models the chamber behavior. Although the real steering gear has asymmetrical, load-dependent volumes, a constant-volume approximation is used here. Only the inlet bore is considered hydraulically relevant; other losses are neglected.

Two line configurations are simulated: one with rigid pipes, another with high-pressure hoses. While pipes are modeled with as ideal stiff, hoses exhibit viscoelastic, frequency-dependent behavior. Empirical data define hose parameters, allowing comparison of their effect on resonance behavior. If unknown, hose properties can be identified experimentally using the Two-Load-Two-Systems method, which accounts for frequency-dependent elasticity and installation effects. The resulting parameters ensure accurate dynamic modeling.

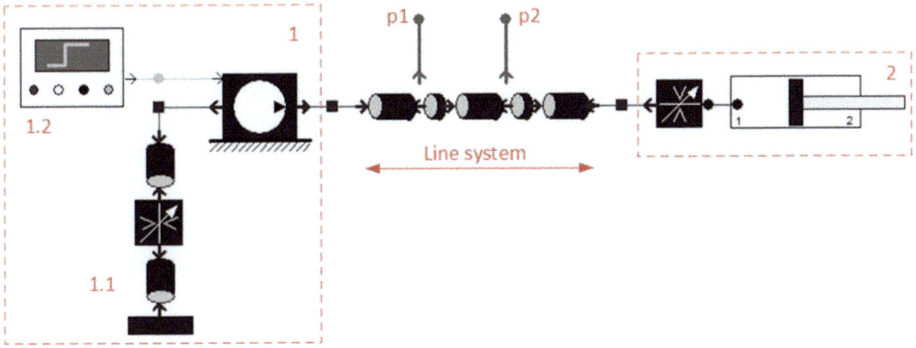


**Fig. 13.9.** Simulation model; 1) Hybrid pump model, 1.1) impedance model, 1.2) pump pulsation, 2) steering gear.

## 13.6 Results

In an initial simulation, the hydraulic line system is modeled entirely with rigid pipes. Wall compliance is neglected, meaning the speed of sound remains constant. A closed-closed boundary condition is assumed. This idealized configuration is commonly used in early-stage analyses of steering systems to estimate resonance locations and pressure amplitude distributions.

In the closed-closed setup, volume flow at both ends is zero due to blocked flow, leading to an increased pressure and the formation of antinodes at the boundary. The fundamental acoustic mode fits half a wavelength within the pipe, governed by

$$\lambda_n = \frac{2L}{n} \tag{13.33}$$

with the corresponding resonance frequencies

$$f_n = n\frac{c}{2L} \tag{13.34}$$

where $L$ is the pipe length, $c$ is the speed of sound, and $n$ is the mode number.

Figure 13.10 (left) shows the simulated resonance frequencies and pressure antinode locations along a 1.6 m pipe with closed-closed boundaries. The speed of sound is set to 1200 m/s, based on fluid compressibility and density. Using Eq. (13.34), the first resonance frequency is 375 Hz.

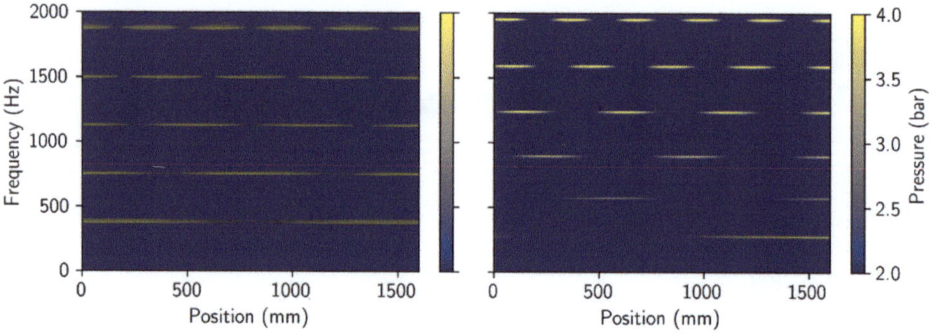


**Fig. 13.10.** Standing waves and resonance frequencies in a pipe system; left: closed-closed boundaries, right: pump-closed boundaries.

To account for realistic pump behavior, the impedance model is introduced in the simulation, incorporating the measured frequency-dependent pump impedance. The pipe end on the other side remains closed, maintaining the pressure antinode (Fig. 13.10, right).

On the pump side, the antinode shifts inward, indicating partial wave absorption within the pump. This increases the effective acoustic length and, according to Eq. (13.31), lowers the fundamental resonance frequency. With the real pump impedance, the first system resonance drops to 270 Hz, a reduction of 28% compared to the idealized case.

A similar effect is observed when the steering system is considered (Fig. 13.11 right). In this case, the pressure antinode at the end disappears, as fluid enters the attached volume. Although the orifice introduces damping, it does not act as an acoustically closed end. Consequently, the fundamental resonance frequency drops significantly to around 160 Hz, 57% lower than in the idealized closed-closed configuration.

The results underline the critical influence of termination impedance on the system's acoustic behavior. Realistic component modeling is essential for accurate prediction of resonance phenomena.

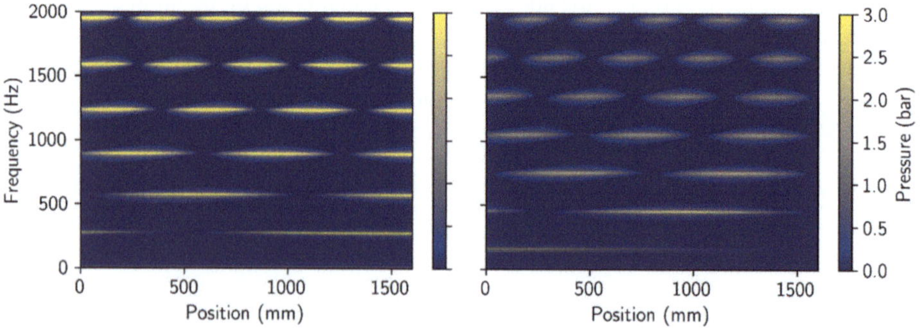


**Fig. 13.11.** Standing waves and resonance frequencies in a pipe system; left: pump-closed boundary configuration, right: pump-steering boundary configuration.

To complete the system analysis, the central pipe segment is replaced by a flexible hydraulic hose with high damping characteristics, chosen specifically to illustrate the frequency-dependent behavior of compliant line components. In order to accurately capture the hose dynamics, the simulation includes wall compliance, allowing for viscoelastic wall motion and its influence on bulk modulus and thus speed of sound.

Figure 13.12 compares the pressure transfer function between the inlet and outlet of the flexible hose to that of a rigid pipe segment, with both curves normalized to the fundamental resonance frequency. In the rigid pipe, sharp and narrow resonance peaks appear at integer multiples of the fundamental frequency, indicating low damping and the presence of well-defined standing wave modes.

This comparison clearly demonstrates how introducing a compliant, viscoelastic hose segment alters the acoustic behavior of the system. Compared to the rigid pipe, the hose attenuates pressure oscillations more effectively and suppresses higher-order resonances due to its increased damping. As a result, resonance peaks become broader and flatter, significantly reducing the quality factor. At higher frequencies, the peaks are no longer well-defined, which makes mode identification more complicated.

Additionally, the hose's viscoelastic and nonlinear properties introduce frequency-dependent propagation effects. Internal friction and material relaxation reduce the effective stiffness at higher frequencies, leading to a drop in the speed of sound and causing resonance frequencies to shift downward.

Spectrograms showing resonance frequencies and pressure antinode distributions (Fig. 13.13) confirm these findings. Since the boundary conditions at both ends remain unchanged, the qualitative distribution of standing waves also remains stable, as the effective acoustic length is preserved.

The influence of the hose's frequency-dependent propagation characteristics is evident in several aspects. A slight reduction in overall resonance frequencies is observed, attributed to the lower effective speed of sound.

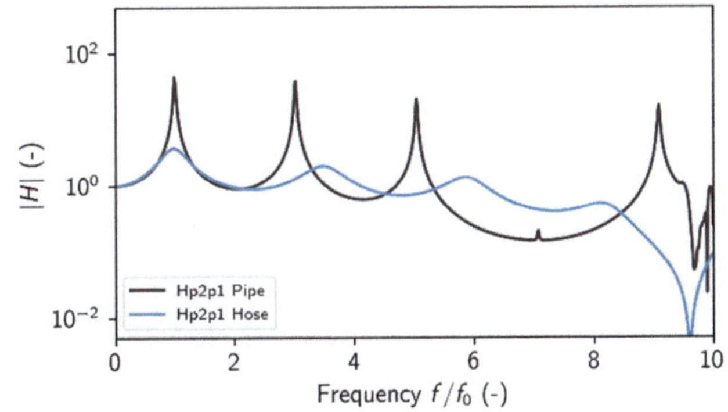


**Fig. 13.12.** Comparison of the pressure transfer function between a ideal stiff pipe and a high damping hose.

More notably, the spectrogram reveals the impact of increased damping: resonance peaks broaden, amplitudes diminish with frequency, and higher-order modes become indistinct or vanish entirely.

At higher frequencies, a shift in resonance peaks indicates weakly dispersive effects, suggesting that the hose alters phase velocity as a function of frequency. These dissipative and dispersive characteristics significantly affect the visibility and spacing of standing wave modes, highlighting the importance of incorporating realistic hose models in predictive simulations of pressure pulsation behavior.

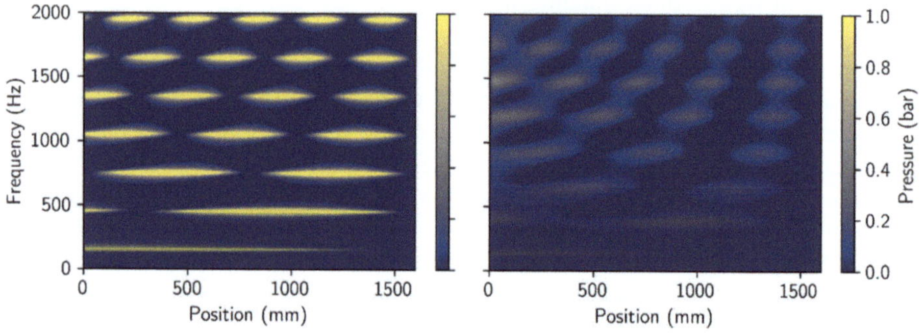


**Fig. 13.13.** Standing waves and resonance frequencies in a system with pump-steering boundary configuration; left: pipe system, right: system with highly damped hose.

A key factor of holistic system analysis is the pump's operating point: if its pulsation frequency aligns with a system resonance, targeted NVH measures are required to avoid amplified pressure pulsations or structure-borne noise.

Highly damped hoses can help, but their use is often limited by pressure, mechanical and packaging requirements. In practice, hoses are selected for routing flexibility, and steel-reinforced types offer merely little damping. Their effectiveness depends on both material and placement. In NVH-critical systems, mechanical coupling points and structure-borne paths must also be considered.

For existing systems, adjusting pipe length is a common tuning strategy. In one case, resonances at 160 Hz and 450 Hz coincided with pulsation frequencies during low-speed maneuvers (e.g., 1500 rpm engine speed, 10-vane pump, ratio $i = 1.8$). Because these conditions lack masking noise, they are prone to complaints. Shortening the pipe shifts resonances to higher, less critical frequencies, though this method is narrowband and may cause new resonances elsewhere.

In new designs, simulation enables early evaluation of resonance conditions using measured pump and steering impedances. One strategy is impedance matching, tuning the pump's impedance minimum to a system resonance (e.g., 450 Hz). At this point, the pump absorbs energy like a Helmholtz resonator, reducing pressure amplitudes.

Figure 13.14 shows how tuning eliminates the critical resonance. Though narrowband, this approach is highly effective when operating conditions are well defined, making impedance matching a powerful tool in early system development.

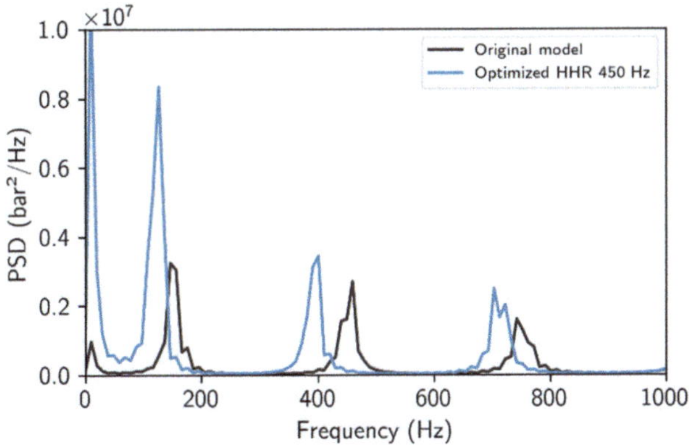


**Fig. 13.14.** System optimization with impedance matching.

## 13.7 Summary and Outlook

This work presents a hybrid methodology for analyzing and optimizing pressure-induced noise and vibration NVH in automotive fluid systems. A hybrid pump model was developed, combining experimentally obtained pump pulsation and impedance data. These parameters were integrated into a 1D simulation model, enabling realistic prediction of resonance frequencies and pressure wave behavior.

The approach avoids the need for full system measurements by characterizing key components independently. It captures the frequency-dependent reflection behavior of pumps and illustrates the influence of flexible hoses on wave propagation and damping. Application to a steering system demonstrates how acoustic resonances can be shifted or suppressed through impedance matching or line configuration changes.

Looking ahead, the method enables early-stage NVH assessment and design optimization with minimal experimental effort. Future work may include full system characterization, including the steering gear, using the principle of the TMM and SSM.

## References


1. **Stosiak, M., Karpenko, M., Skačkauskas, P., & Deptuła, A.** *Identification of pressure pulsation spectrum in a hydraulic system with a vibrating proportional valve.* Journal of Vibration and Control, 30(21-22), 4917–4930. (2024)
2. **Edge, K. A., & Johnston, D. N.** *The Secondary Source Method for the Measurement of Pump Pressure Ripple Characteristics Part 1: Description of Method.* Proceedings of the Institution of Mechanical Engineers, Part A: Journal of Power and Energy, 204(6), 33–40. (1990)
3. Wylie, E. B., & Streeter, V. L. Fluid Transients in Systems. Prentice Hall. (1993)
4. Munjal, M.L.: Acoustics of Ducts and Mufflers. 2nd edn. Wiley, Hoboken, NJ (2014)
5. Brumand-Poor, F., Kotte, T., Pasquini, E.G., Schmitz, K.: Signal Processing for Transient Flow Rate Determination: An Analytical Soft Sensor Using Two Pressure Signals. *Signals* **5**(4), 812–840 (2024)

6. Jarchow, M.: Maßnahmen zur Minderung hochdruckseitiger Pulsationen hydrostatischer Schrägscheibeneinheiten. PhD Technische Hochschule Aachen (1997)
7. Emerson Automation Solutions: The Engineer's Guide to DP Flow Measurement. 3rd edn. Emerson, Shakopee, MN (2019)
8. International Organization for Standardization (ISO): Hydraulic fluid power – Determination of the fluid-borne noise characteristics of components and systems – Part 2: Measurement of the speed of sound in a fluid in a pipe. ISO 15086-2:2000. ISO, Geneva (2000)
9. Bechhoefer, E., Kingsley, M.: A Review of Time Synchronous Average Algorithms. Proc. Annu. Conf. Progn. Health Manag. Soc. 1, 1–10 (2009)
10. Van den Hof, P.M.J., Schoukens, J., Pintelon, R.: Nonparametric data-driven modeling of linear systems. IFAC-PapersOnLine 50(1), 11779–11785 (2017)
11. Bendat, J.S., Piersol, A.G.: Random Data: Analysis and Measurement Procedures. 4th edn. Wiley, Hoboken, NJ (2010)
12. Mendes, B.O., Guimarães, G.P.: Numerical Study of the Helmholtz Resonators Behavior Using Impedance Functions. In: Proceedings of the 22nd International Congress on Sound and Vibration (ICSV22), Florence, Italy (2015)
13. Powell, M.J.D.: An efficient method for finding the minimum of a function of several variables without calculating derivatives. Comput. J. 7(2), 155–162 (1964)

# Experimental Determination of High Frequency System Sensitivities as Key Factor for Early NVH Optimization of Electrified Axles

Florian Löcken[1](✉), Jonas Richter[1], Hendrik Sell[1], and Enrico Kruse[2]

[1] Vibracoustic SE & Co. KG, Hamburg, Germany
{florian.loecken,jonas.richter, hendrik.sell}@vibracoustic.com
[2] Vibracoustic SE & Co. KG, Weinheim, Germany
enrico.kruse@vibracoustic.com

**Abstract.** Vibracoustic is a global supplier of NVH solutions such as subframe mounts and motor mounts for the automotive industry. As a prominent global automotive NVH development supplier, Vibracoustic has collaborated with OEMs for over twenty years to support their virtual NVH development and optimization. In the area of axle systems, Vibracoustic continually enhances its self-developed and unique NVH test facilities and methods to analyze and improve passenger car axles as subsystems, without the surrounding vehicle.

Over the past twenty years, several changes and trends have been observed in the automotive industry that have significantly influenced NVH development for new vehicles or platforms. A major change in recent years has been the transition from combustion engines to electrified drive concepts in battery electric vehicles (BEVs). As discussed in previous publications [1–3], the electric drive unit (EDU) in BEVs is highly integrated into the axle system. For example, in rear e-axles, the EDU is elastically suspended against the subframe, which is typically mounted to the vehicle body by elastic subframe mounts. The EDU, as a vibration source within the axle, generates structure-borne noise at higher frequencies compared to non-electrified axles.

Additionally, virtualization has become increasingly important in the early stages of NVH tuning. Effective simulations can substantially reduce time and cost-intensive hardware testing cycles by enabling early detection of design issues and assessment of countermeasures before physical prototypes are constructed.

This paper describes how experimental modal analyses, and the identification of high-frequency transfer functions are key factors for early NVH tuning of e-axles. It outlines how combining these methods assists in validating the basic modal alignment of isolation concepts and identifying potential


A. Heintzel (Ed.): Automotive Acoustics Conference 2025, Proceedings, pp. 207–219, 2026. https://doi.org/10.1007/978-3-658-51249-1_14

NVH weaknesses of the system early in the development process. Moreover, it demonstrates how both methods can be used to check, validate, and enhance FEA modeling at the OEM level.

**Keywords:** e-axle NVH · structure borne noise · subsystem high frequency transfer behavior

## 14.1 Introduction

The analysis and optimization of Noise, Vibration, and Harshness (NVH) in vehicle axle systems are integral components of Original Equipment Manufacturer (OEM) vehicle or platform development. The reduction of vibrations and the enhancement of road noise isolation are inherent to every axle development process. Typically, OEMs employ a well-established combination of proven methods and extensive experience derived from previous similar axle systems.

With the advent of electrified axles, new NVH challenges have emerged. As outlined in our publications [1–3], the electric drive unit (EDU) in Battery Electric Vehicles (BEVs) is significantly integrated into the axle system. For instance, in rear e-axles, the EDU is elastically suspended against the subframe, which is usually mounted to the vehicle body utilizing elastic subframe mounts. Consequently, the EDU acts as a vibration source within the axle, generating structure-borne noise relevant to comfort at higher frequencies compared to non-electrified axles. In this context, we as Vibracoustic therefore also speak of "e-NVH".

Virtualization plays a significant role in the early stages of NVH tunings. By using effective simulations, time and cost-intensive hardware testing cycles can be reduced through early detection and resolution of design issues and assessment of countermeasures before new physical prototypes are built. In virtual NVH analyses of axle systems, structural dynamic FEA models are typically used to predict vibration transfer and identify potential weaknesses in structural parts such as the subframe. To increase confidence in simulation results, it is necessary to validate simulation models in the early development phases. For NVH assessments, often two key questions must be addressed by validation:

- What is the frequency range within which the current simulation model is valid?
- Does the model cover the main vibration phenomena that need to be considered?

This paper explains how experimental modal analyses and the identification of high-frequency transfer functions can become key factors for early NVH tuning of e-axles. It describes how combining these methods aids in validating the basic modal alignment of isolation concepts and identifying potential NVH weaknesses early in the development process. Additionally, it shows how both methods can be used to verify, validate, and improve FEA modeling at the OEM level.

## 14.2 Background and Motivation

Vibracoustic has been conducting optimization projects on axles in cooperation with its customers for over 20 years and has analyzed more than 100 axles from different vehicle classes. Figure 14.1 provides an overview of various test setups developed at Vibracoustic throughout the years. These NVH axle system analysis methods were created to target different frequency bands for investigation, addressing multiple NVH critical vibration phenomena. The core of these NVH analyses involves connecting the axle system to multi-axis load cells located at the interfaces between the axle and the body, which measure the internal forces between them against the test rig.

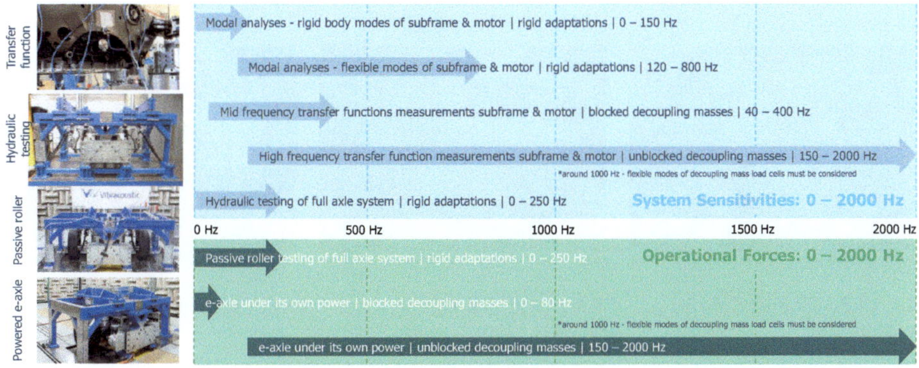


**Fig. 14.1.** Overview of Vibracoustic's Axle Systems NVH Investigation Methods and their target frequency ranges

As detailed in the publication [3], in many test setups, the axle is rigidly attached to the load cells in a specifically designed frame for axle analyses. This direct attachment method is exemplarily illustrated in Fig. 14.2 on the left and is specially designed and manufactured for the investigated axle. Through further development of the frame and the adaptation design, the test setups remain free from eigenmodes up to 100–150 Hz while maintaining maximum rigidity. The rigidity of the attachment is also experimentally determined by measuring the "input stiffness" in advance of each axle analysis. These measures extend the confidence range for axle analyses with this attachment method up to approximately 250 Hz.

To extend the application of the test rig for "e-NVH" measurements up to 2 kHz, it is necessary to ensure that the eigenmodes of the test rig do not interfere with the force measurement. Therefore, the concept of a rigid connection of the axis has been replaced by an elastically designed mounting with high-frequency decoupling elements using rubber mounts.

As described in detail in [1, 2], the decoupling is designed so that the rigid-body modes of the elastically supported mass fall within a narrow and low frequency band, ensuring that vibrations from the test rig are isolated by the inertial properties of the spring-mass system. The force measurement is performed using piezoelectric load cells, which are integrated into the decoupling mass. In the final test application, partial masses and stiffnesses (such as subframe mounts) of the attached axle interact with the spring-mass system of the decoupling element. Considering this interaction, the lower frequency limit for measurements with the decoupling elements is below 150 Hz.

The elastic mounting of the decoupling elements can be bridged by additional side plates. This enables the axle system to be measured against a rigid connection without disassembling the entire attachment to the test rig. Figure 14.2 illustrates the three connection variants that can be generally represented in the test setups for force measurements.

**Fig. 14.2.** Connection variants of the axle that can be represented in general on the NVH test methods [1]

As also illustrated in first Figure (see Fig. 14.1), the experimental setups are categorized into two main focuses: system sensitivity and operational forces, according to the type of excitation.

To analyze the transmission sensitivity to vibrations, the axle system is intentionally excited at specific critical points using actuators that generate sinusoidal signals across varying frequencies. By combining these excitations with the force measurements at the axle's connection interfaces, the sensitivities can be evaluated through transfer functions. For example, in the paper [3], NVH axle analyses with hydraulic excitation at the wheel carrier are described in detail. Such investigations often serve as fundamental references for NVH analysis and the optimization of axle systems in collaborative projects between Vibracoustic and an OEM.

To measure the operational forces transmitted by the axle system, the test rig including the axle with mounted wheels can also be operated on a dyno test bench. The test rig is designed in such a way that the experimental setup from the hydraulic measurements can be directly transferred without having to remove the axle or the adaptation with the force measurement from the test rig. The roller can either drive or "drag" the axle, or in the case of electrically driven axles, it can also function as a brake under load. As described in [1, 2], operating an e-axle outside an actual vehicle environment requires several system components and parameters that are not initially provided. This mainly includes the high voltage power supply, liquid cooling, CANbus communication for drive signals, warning and error messages, and the residual bus simulation.

Figure 14.3 resents the experimental setup and a typical example of an investigation where dynamic force is input into the test rig's frame at one of the subframe points in the X direction. The figure illustrates how the transmitted force over frequency varies with the applied drive torque. Depending on the frequency bands, either the maximum recuperation torque or the maximum acceleration torque induces the highest dynamic forces.

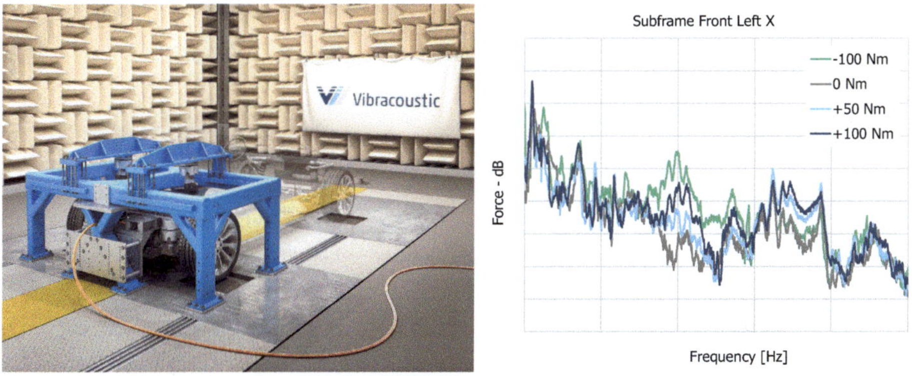


**Fig. 14.3.** Experimental setup to study operational forces under operating conditions and measured operational forces at subframe mount location under different driving torques [1]

## 14.3 Basic Modal Alignment of Motor and Subframe

Every powertrain, whether internal combustion based or electric, on its elastic mounting system will exhibit rigid as well as flexible eigenmodes. When internal or external forces or torques fall in line with one of these modes, the incoming vibrations can be amplified and lead to NVH issues across the vehicle. The continued operation of such a system near one of its resonances can also lead to permanently high dynamic loads and ultimately to durability issues.

Hence, it has been standard practice to conduct a modal analysis and optimization of the entire powertrain mounting system, starting with the powertrain's own six rigid body modes. If the powertrain is double-isolated through an elastically suspended subframe, all twelve rigid body modes of both objects (powertrain and subframe on their respective mounts) must be considered simultaneously. As shown in Fig. 14.4, double isolation can also be provided through e. g. suspended brackets or cross-members—either way it is highly recommended to always modally optimize the assembled double-isolated system in its entirety.

Modal targets consist of frequency ranges and coupling limits for each eigenmode. They are usually derived from several sources, such as:

- operational conditions and excitations of the mounted objects, e.g. the engine or motor, the gearbox and the power electronics
- excitations from other vehicle elements, such as suspension or driveline components
- potentially overlapping sensitivity ranges of other vehicle components, such as the vehicle body or the steering system
- frequency ranges with increased human body sensitivity.

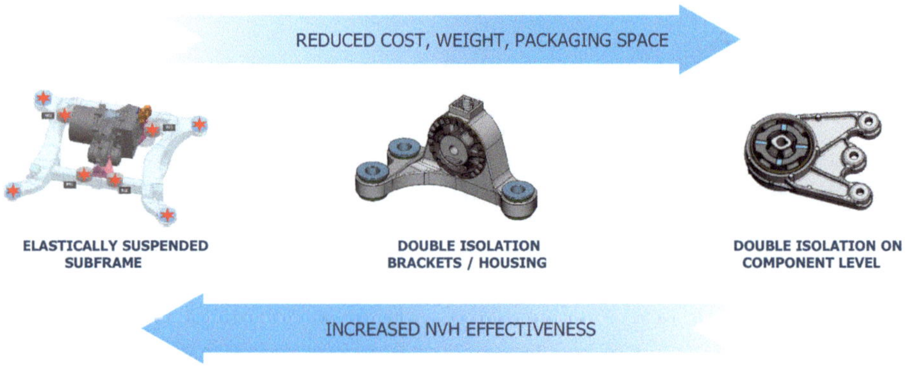


**Fig. 14.4.** General benefits and disadvantages of various double-isolation solutions [4–6]

When setting modal targets for an electric drive unit, its specificities are considered:

- lighter weight compared to a combustion engine
- no start/stop vibration
- no idle excitation
- (usually) no gear shifting

Figure 14.5 shows an overview of the default modal targets for a double-isolated electric drive unit. A typical sign for the modal alignment of an electric drive unit is the significantly higher bounce frequency at 20 Hz and above: there is no need to isolate an idling combustion engine anymore, so initial mount stiffness can be increased, pushing the bounce mode safely away from the human body sensitivity range of 4–8 Hz. This also helps in better separating motor and subframe modes from each other, thus avoiding unwanted modal couplings.

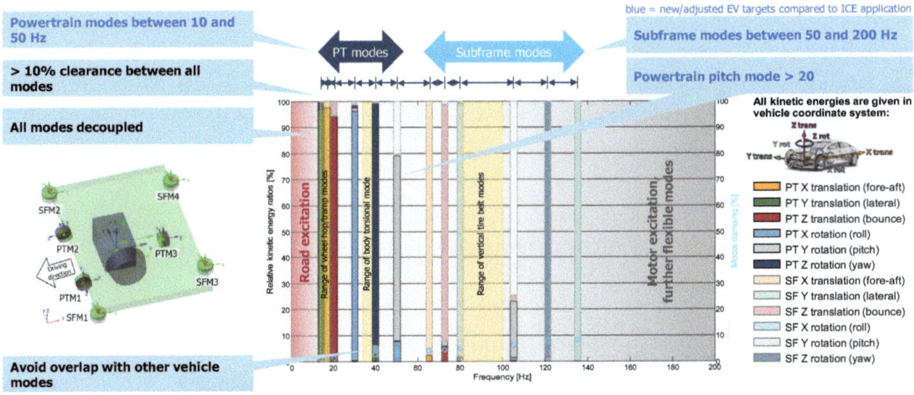


**Fig. 14.5.** Default targets for the rigid body modes of a double-isolated electric drive unit [4–6]

To achieve these targets for a given mounting system, linear dynamic stiffnesses of all mounts in all directions are the main tuning parameters. In some cases, mount positions and orientation, or even the number of mounts, can be varied, too. Figure 14.6 shows sketches of typical mounting system architectures for electric powertrains. In the upper row (examples 1 through 3) all bushings have the same orientation, generally leading to a soft system tuning in the axial direction of the bushings and a stiff system tuning in the radial direction of the bushings. In contrast, the architecture variants in the lower row (examples 4 through 6) feature a mixed orientation of bushings, leading to a more balanced overall system stiffness in all directions.

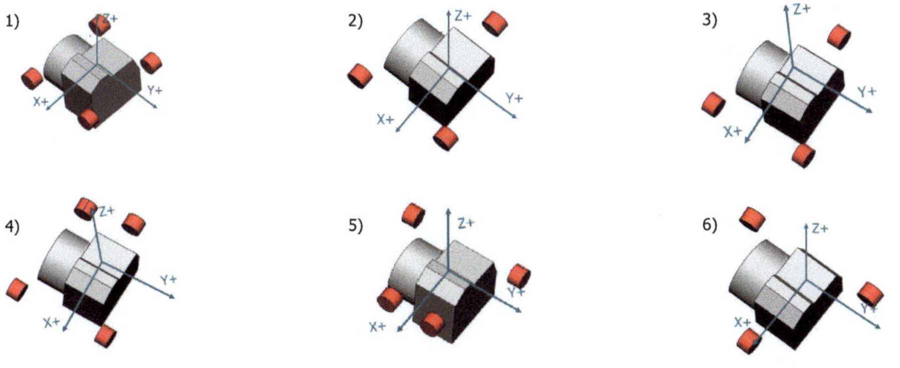


**Fig. 14.6.** Examples of typical 3-point and 4-point electric powertrain mounting systems [4–6]

After this initial linear system tuning is finished, the full nonlinear static force-deflection-curves of all mounts are being optimized, aiming to meet isolation and maximum travel targets under specified g-force and torque conditions. All these investigations can be carried out early in the development project using multi-body simulation software such as Adams or Vibracoustic's proprietary layout tools.

We consider this initial quasi-static and low-frequency dynamic system tuning indispensable, as it lays the groundwork for a well-balanced NVH performance. The further investigation and optimization of higher-frequency NVH phenomena, for instance involving flexible modes of the subframe or continuum modes of the mounts' rubber elements, are building on it.

## 14.4 Experimental Determination of Rigid and Flexible Mode Shapes

Modal analysis is a fundamental technique in the study of structural dynamics, particularly for understanding how mechanical systems are sensitive to vibrational excitation by identifying the natural frequencies, mode shapes, and modal damping characteristics of a system.

Rigid body modes are motions in which the components of an assembly move without deformation, typically observed at lower frequencies. Looking at the subsystem of subframe and e-motor with 12 degrees of freedom, its rigid body modes generally lie within the range of 10 to 150 Hz, see Fig. 14.5. These modes are significant in evaluating the vehicle's modal alignment of the motor mount system and are also important in primary ride characteristics, such as aftershake. Accurate identification of these modes helps to prevent resonance interactions with suspension and drivetrain frequencies.

In addition, flexible modes occur at higher frequencies and involve internal deformation of the structure or its components. These modes can significantly influence the vibrations transmitted into the vehicle body, directly affecting passenger comfort.

Modal analysis provides the dynamic characteristics of a structure—natural frequencies, damping, and mode shapes—by evaluating its response to hammer or shaker excitation. Frequency Response Functions (FRFs), measured at key locations, form the basis for identifying both rigid and flexible modes that are critical for NVH evaluation and simulation model validation. Accurate identification of flexible modes depends heavily on proper sensor placement. It is important to place sensors at locations with significant modal displacement rather than at nodal points, where the structure exhibits little to no motion for certain modes. Even preliminary mode shapes from an unvalidated FE model can aid in optimal accelerometer placement by highlighting regions of higher structural displacement. Once FRFs are acquired, an Auto-MAC (Modal Assurance Criterion) analysis should be performed to ensure correct mode shape interpretation and avoid misidentification due to spatial aliasing or closely spaced modes.

The experimentally derived modal parameters can be used to calibrate and validate the multi-body simulation (MBS) model of the subframe and electric motor, ensuring that the simulated dynamics align with real-world behavior. Understanding these characteristics is essential for verifying the intended modal alignment of the system.

Figure 14.7 illustrates a modal analysis test setup of a subframe/e-motor subsystem mounted on concrete foundations with rubber-isolated feet and interface adaptations. The measured results are used to calibrate the MBS model by tuning key parameters such as the masses and inertias of the subframe and motor, as well as the stiffness characteristics of the mounts at defined operating points.

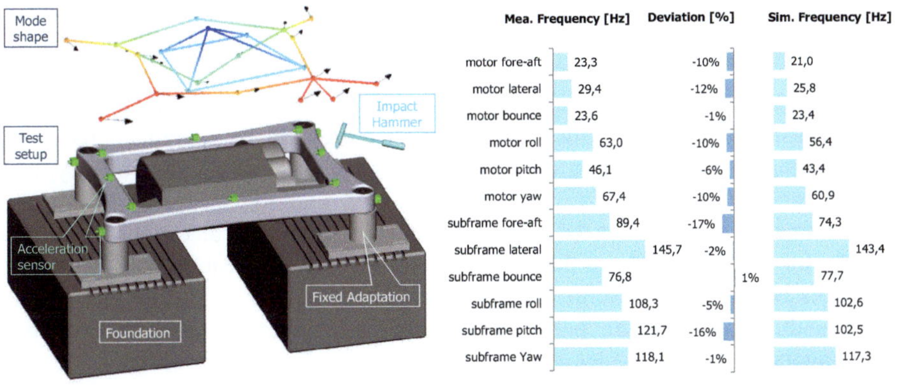


**Fig. 14.7.** Test Setup for Modal Analysis and Comparison of Experimental vs. MBS Simulated Rigid Body Modes

At frequencies beyond the rigid body range, structures such as subframes begin to exhibit flexibility. These flexible modes significantly affect the NVH transmissibility of the subframe and require a finite element (FE) model supported by detailed modal analysis for accurate characterization. Precise mode identification necessitates a higher density of acceleration sensors to effectively capture complex deformation patterns. A finer spatial resolution is typically required around structural joints or regions of expected high modal curvature. Based on the FE model, the resulting data can be used to refine parameters such as mass distribution and stiffness along the structure, particularly relevant for subframes manufactured from cast aluminum, where material variation can influence dynamic behavior.

While Auto-MAC compares mode shapes derived from experimental data, MAC analysis is also used to assess the correlation between measured and final simulated mode shapes. Values close to 1.0 indicate a strong match, confirming the validity of the FE model. Significant deviations may reveal local stiffness inaccuracies or insufficient sensor coverage during testing.

## 14.5 High Frequency Structure Borne Transfer Measurements for Early System Validation

### 14.5.1 Test Setup for Structure Borne Transfer Measurements

To gain understanding about the transmission of vibrations from the e-motor to the vehicle body, a series of high-frequency transfer measurements is performed. Instead of fixed adaptations, decoupling elements are placed between each subframe mount core and the foundation. Each decoupling element is equipped with a dynamic load cell that measures the interconnecting forces as described in Sect. 14.2.

For excitation, an inertia shaker is used to apply a sine sweep or noise excitations ranging from 40 to 2000 Hz. Both the excitation force and the resulting acceleration at the same location are measured. By reusing the accelerometer setup from the modal analysis, various transfer functions e.g. force-force, acceleration-force, force-acceleration, and acceleration-acceleration FRFs, can be derived from the acquired data. These transfer functions provide a comprehensive view of the system's behavior and can be compared with existing data, such as vehicle measurements, which typically represent acceleration data.

The small form factor of the shaker also enables excitation at various locations of interest, for example, close to the motor mounts to identify location-specific sensitivities. This is particularly useful for motor assemblies still in the design phase, where understanding local dynamic behavior can help guide structural improvements before finalizing hardware.

Using an artificial excitation signal, rather than the operating motor, enables the identification of the dynamic sensitivity rather than recording the transmitted forces during operation. This approach supports a well-targeted and effective tuning of the system properties. Moreover, it is well-suited for early development phases when an operable motor unit is not yet available.

Depending on the specific setup, the decoupling elements show a flexible mode around 1000 Hz, due to limited stiffness of the load cell. To ensure a valid FE-model, Vibracoustic provides surrogate stiffness values that customers can implement in their own simulations. These surrogate parameters represent the frequency-dependent behavior of the decoupling elements and can be included to enhance model accuracy.

As described in Sect. 14.3, double-isolated systems are proven to significantly enhance NVH performance, particularly at higher frequencies. Using the applied test methodology in combination with high-frequency decoupling elements, the effectiveness of double isolation can be both verified and quantified.

Figure 14.8 illustrates the resulting sum of forces in the Z direction at the subframe mounts, measured in response to a center-of-gravity (COG) excitation in the Z direction of the e-motor. At 700 Hz, the transmissibility drops by 60 dB, demonstrating the high-frequency isolation achieved through the double isolation concept. This attenuation level demonstrates the system's ability to suppress structure-borne noise transmission into the vehicle cabin in critical high-frequency ranges.

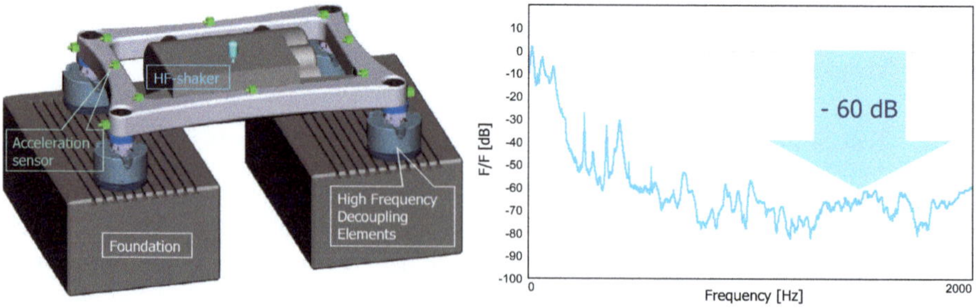


**Fig. 14.8.** Sum of transferred forces in the Z-direction at the subframe mounts, showing the transmissibility drops by 60 dB above 700 Hz

### 14.5.2 Test Method Applications in Early NVH Optimization

The described combination of modal analysis and high-frequency transfer measurements provides a valuable framework for early-stage NVH optimization. To minimize the transmission of vibrations from the EDU to the vehicle body, modal and transfer methods work in tandem. The transfer measurements offer insight into the frequency bands where the system is most sensitive to excitation, indicating which ranges require attenuation. Modal analysis delivers the structural understanding necessary to guide targeted design changes, such as adjusting local structure stiffnesses, mass distribution, isolation via the rubber mounts or mass damper application. These modifications are first implemented virtually in the FE model, validated through modal analysis, and subsequently confirmed by repeated transfer measurements, ensuring a reliable and efficient optimization process. Even at an early development stage with no running motor, using just a mass representation, this closed-loop methodology enables refinement of subsystem dynamics before integration into the full vehicle. This can reduce late-stage development iterations and accelerate NVH maturity. This approach reflects again the V-shaped development approach as described in [3], where measurement-driven model validation and optimization are followed by hardware refinement and system-level verification.

The methodology proves effective in supporting both optimization of structural suspension parts as well as rubber mounts. An illustrative example can be seen in frequencies, which can become critical due to excitations generated by the gearset of electric drive units. Improving the transmissibility of the axle system within this range is therefore essential for reducing structure-borne noise. As shown in Fig. 14.9 for an exemplary subframe, the application of the described methodology led to a reduction in RMS transmissibility by about 3 dB.

The methodology is equally suited for supporting mount development in early design phases. By combining modal analysis with high-frequency transfer measurements, the effects of modified mount characteristics can be evaluated even in the absence of an operable drive unit. In the example shown in Fig. 14.9 on the left, an increase in overall motor mount stiffness led to a 10 dB rise in force transmissibility in the investigated frequency range. Modal analysis revealed a shift in rigid body modes, while the flexible modal behavior remained unchanged. This highlights the methodology's capability not only to guide mount tuning and isolation strategy development, but also to predict and understand the effects of design changes across the entire NVH-relevant frequency spectrum.

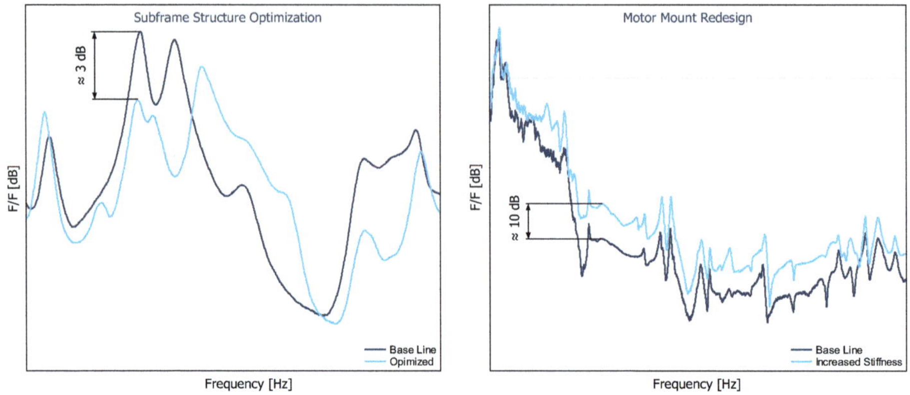


**Fig. 14.9.** Sum of transferred forces at the subframe mounts, showing the effect of modifying the subframe structure (left) and motor mount stiffness (right)

## 14.6 Summary and Outlook

This paper describes how Vibracoustic has improved its experimental capabilities to study high-frequency structure-borne noise transfer of e-driven axles at a subsystem level. By focusing on transfer sensitivities and using the e-motor as a passive mass, it is well-suited for early development phases when an operable motor unit is not yet available. It explains how experimental modal analyses and the identification of high-frequency transfer functions work together, and how combining these methods helps in validating the basic modal alignment of isolation concepts and identifying potential NVH weaknesses early in the development process. Additionally, it was discussed how both methods assist OEMs in validating and enhancing their FEA modeling.

The continuous extension and improvement of experimental capabilities and simulation methods enable concurrent virtual NVH optimization at Vibracoustic and the OEM.

A major challenge in optimizing and predicting the noise, vibration, and harshness (NVH) transmissibility of electrically driven axles is the complex, multi-dimensional response exhibited by elastic mounts and their structural components (such as housing, core, and sleeve) under high-frequency excitation. Consequently, Vibracoustic is currently developing new high-frequency part tests and mount representations in full six degrees of freedom, including transfer, input, and output stiffness characteristics, to facilitate the creation of "best virtual prototypes" for fully coupled simulations.

## References

1. Sell, H., Löcken, F., Kruse, E.: NVH optimization of electrified axles new "front-loading" approaches. In: Heintzel, A. (eds) Automotive Acoustics Conference 2021. Proceedings, pp. 82–96, Springer Nature, Wiesbaden (2021)

2. Kruse, E., Sell, H., Löcken, F.: Electric Motor Mounting System Validation with Subsystem and Powered e-Axle Tests. In: SAE Technical Paper 2023-01-1097—Noise and Vibration Conference and Exhibition (2023)
3. Sell, H., Löcken, F., Kruse, E., Reinauer, A.: Holistic approach to axle NVH assessment and optimization. In: 12th International Munich Chassis Symposium 2021. Proceedings, pp. 624–642, Springer Nature, Wiesbaden (2021)
4. Roth, D., Kruse, E.: Dedicated Powertrain Mounting Systems for Electric Vehicles. In: Automotive NVH Comfort Conference, Le Mans, France (2021)
5. Kruse, E.: Partnerships on NVH foster mutual learning. In: ATZ worldwide 122, pp. 22–25, Springer, Heidelberg (2020)
6. Kruse, E., Chauvet, L.: Adding Comfort to Mobility—Mount Solutions for Electric Motors. In: Aachen Acoustic Colloquium, Aachen, Germany (2019)

# Characterization of the Electromagnetic Noise Generated by Pulse Width Modulation in Electric Motors

Salomé Wanty[1,2](✉), Martin Glesser[1], Etienne Parizet[2], and Nicolas Totaro[2]

[1] Dassault Systèmes, SIMULIA Manatee, Villeneuve d'Ascq, France
{salome.wanty,martin.glesser}@3ds.com
[2] Laboratory of Vibration and Acoustics, INSA-Lyon, Villeurbanne, France
{etienne.parizet,nicolas.totaro}@insa-lyon.fr

**Abstract.** Pulse Width Modulation (PWM) is used to supply electric motors with variable signals from a DC battery. This process generates high-frequency harmonics in the electromagnetic forces that produce vibrations and noise radiation of the stator. These harmonics create a symmetrical distribution of frequencies around the switching frequency and its multiples. The resulting component of electromagnetic noise is called PWM noise, which varies depending on the operating point. From a sound quality perspective, the typical frequency signature of PWM noise is known to produce tonality, roughness, or fluctuations. However, in some cases, PWM noise also includes some periodic impulsive noise. In the present study, a perceptual experiment is conducted to characterize the perception of this particular noise and initiate the development of a specific sound quality index. Encouraging results suggest that this "periodic impulsiveness" is related to the amount of close prominent peaks in the signals.

**Keywords:** Sound Quality · Electric Motor · Pulse Width Modulation · Electrical Rattle

## 15.1 Introduction

The acoustic behavior of electric motors differs from that of combustion engines. Although they are quieter, electric motors radiate a noise composed of many harmonics, which can be unpleasant. These harmonics can be separated into "low frequency" noise mainly produced by slotting effects (influenced by slot/pole combination) [1], and "high frequency" noise influenced by the Pulse Width Modulation (PWM) supply technique [2]. This present study focuses on PWM noise, which is characterized by the frequency signature in Fig. 15.1. PWM harmonics are distributed symmetrically around multiples of the switching frequency $f_{swi}$, with a deviation depending on the electrical frequency $f_e$ that controls the rotation speed of the motor [3]. These


A. Heintzel (Ed.): Automotive Acoustics Conference 2025, Proceedings, pp. 220–228, 2026. https://doi.org/10.1007/978-3-658-51249-1_15

harmonics can induce a strong tonality, but also some modulation effects. It is therefore necessary to perform specific experiments to know the relation between PWM harmonics and perceived sound attributes.

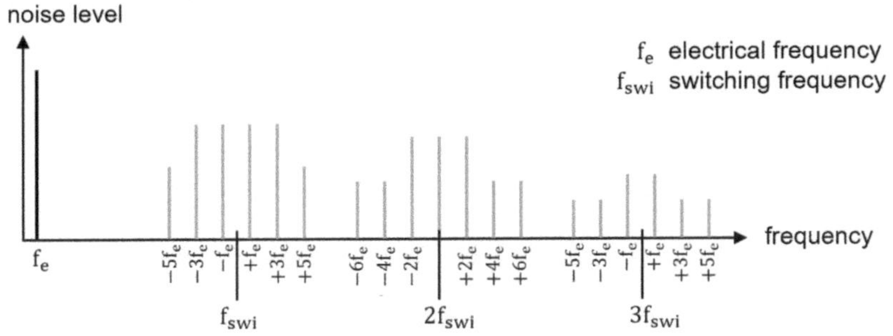


**Fig. 15.1.** Characteristic noise spectrum of PWM noise.

A previous study was performed on several PWM strategies [4]. Pairwise comparisons analysed by Individual Difference Scaling provided a 2D perceptual space, revealing two main sound dimensions. Listening to the stimuli in regards to their position in the perceptual space revealed the significance of some periodic impulsive noise, called "electrical rattle" (e-rattle) in the following. However, no existing index was able to describe this effect. A perceptual experiment was therefore carried out to characterize the perception of e-rattle noise. The identification of the relationship between the subjective responses and the spectral properties of the stimuli paves the way to the development of a new sound quality index.

In the following, e-rattle noise is described in part 2. The perceptual experiment is presented in part 3, and the results are discussed in part 4.

## 15.2 Electrical Rattle Noise Characterization

The term "rattle" is usually employed to describe a rapid and regular succession of impulsive sounds, from quickly repeated impacts between two parts of a system. A similar type of sound was detected in electric motors noise at specific operation points, in terms of rotation speed and torque. It cannot be considered as roughness [5] because the modulation's pulsation is too low. Nor it can be characterized by the fluctuation strength [6] because it implies a specific timber. Even if its periodicity is close to fluctuation strength, it has a more percussive nature.

The terms electrical rattle (e-rattle) and mechanical rattle are employed in the following to refer to the repetitive impulsive aspect of the sounds, while differentiating its source. Even if e-rattle can be related to mechanical rattle in terms of perception, they differ in several aspects. First, the mechanical rattle results from a mechanical excitation with contact [7] (it is impact-induced), whereas e-rattle comes from a purely electromagnetic excitation, without contact. Then, their frequency signature is different, because PWM harmonics inducing e-rattle follow a known structure, while the mechanical rattle is characterized by a richer spectrum. Finally, mechanical rattle is generally introduced by components inside the vehicle's interior, while e-rattle emerges from the electric motor. Mechanical rattle is often associated with Buzz and Squeak (BSR). Although e-rattle differs from BSR, the way the latter is characterized can be inspiring for the development of an e-rattle index. BSR is usually characterized by measurements [8, 9]. An index was proposed to characterize the impulsivity of BSR, based on a combination of the signal's kurtosis and its loudness [10]. Machine learning was also used to create a detection tool from measurements [11].

Unlike mechanical rattle, which is caused by time-varying phenomena, e-rattle noise seems to result from frequency modulations. The spectral analysis of the stimuli involved in the first perceptual experiment [4] revealed a link between e-rattle and the presence of many prominent peaks in a narrow frequency band. This concentration of close peaks creates a high-density zone in the noise spectrum, which leads to a superimposition of modulation effects. Unlike existing impulsivity indices, e-rattle noise could thus be characterized from the frequency spectrum of the signals, as the sum of peaks prominences. A perceptual experiment was performed to characterize the human perception of e-rattle noise and verify the relevance of the proposed index.

## 15.3 Perceptual Experiment

### 15.3.1 Participants

Fifty-four people (31 men, 22 women, aged between 18 and 43, mean age = 23) participated in the experiment. They were 41 students from the Mechanical Engineering Department of INSA Lyon, and 13 members of SIMULIA Manatee team. None of them reported hearing impairment. The experiment took place in a soundproof booth. Sounds were presented through headphones (Sennheiser HD 650). The output level was calibrated to 80 dBA thanks to a binaural head.

### 15.3.2 Noise Stimuli

The sound set was composed of 60 synthetic stimuli whose structure, shown in Fig. 15.2, followed the characteristic structure of PWM noise. The stimuli were composed of 3 groups of 20 harmonics, centered around the first 3 multiples of the switching frequency $f_{swi}$. The amplitude envelope of the groups were controlled by a Gaussian distribution curve.

Two parameters, presented in Fig. 15.3, defined the harmonic structure within each of the three groups. The 20 peaks were equally distributed, according to the distribution frequency. The standard deviation of the Gaussian distribution set the envelope width at −3 dB, to determine the amplitude of the peaks. The envelope width also has an impact on the number of peaks, since harmonics at the edge of the group could be cancelled.

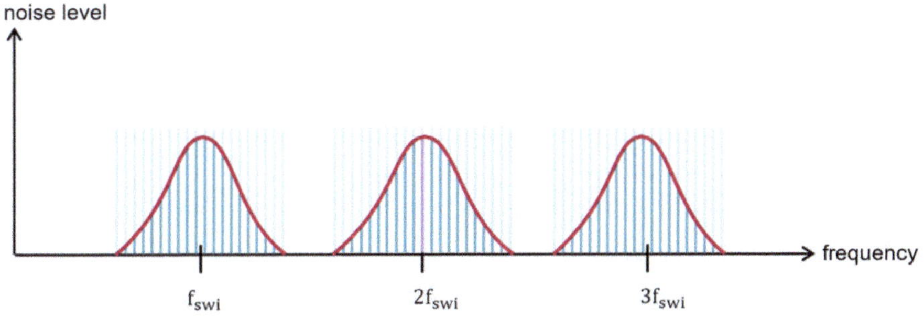


**Fig. 15.2.** Frequency and amplitude structure of the stimuli.

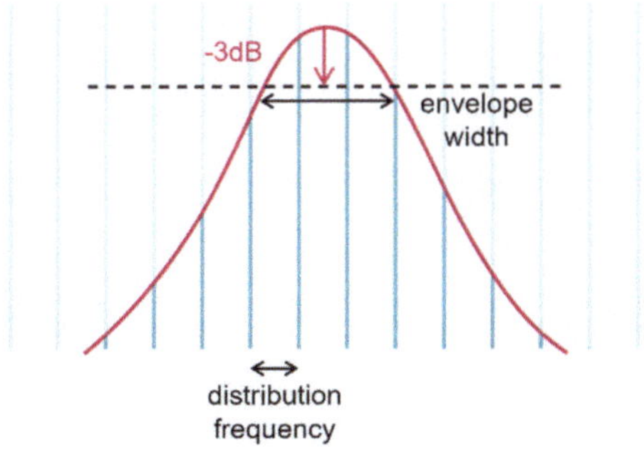


**Fig. 15.3.** Parameters used for the creation of all stimuli.

Two center frequencies ($f_{swi} = 500$ Hz or 1 kHz) were used to generate the stimuli, to verify the impact of pitch on the phenomenon. Thirty stimuli were generated per center frequency, with the distribution parameters given in Table 15.1. The example of nine spectra centered on the first group of harmonics are shown in Fig. 15.4. All stimuli lasted 3 s, with a 0.2 s linear fade-in and -out. They were presented with an equalized loudness level.

**Table 15.1.** Harmonic distribution parameters for the 30 stimuli of each switching frequency involved in the experiment.

| | | Distribution frequency | | | | | |
|---|---|---|---|---|---|---|---|
| | | 5 Hz | 10 Hz | 15 Hz | 20 Hz | 25 Hz | 30 Hz |
| Env. Width | 20 Hz | s1 | s6 | s11 | s16 | s21 | s26 |
| | 50 Hz | s2 | s7 | s12 | s17 | s22 | s27 |
| | 100 Hz | s3 | s8 | s13 | s18 | s23 | s28 |
| | 200 Hz | s4 | s9 | s14 | s19 | s24 | s29 |
| | 400 Hz | s5 | s10 | s15 | s20 | s25 | s30 |

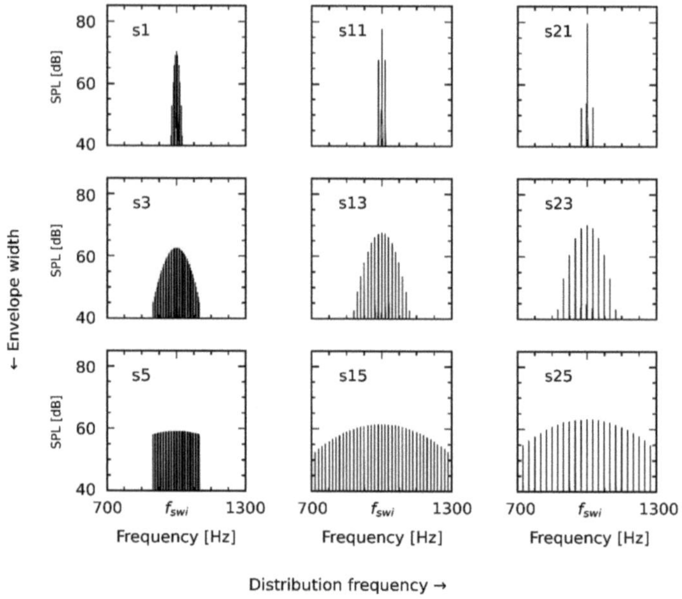


**Fig. 15.4** Nine example spectra from the sound set, centered on the first group of harmonics, with a switching frequency of 1 kHz.

### 15.3.3 Methodology

The experiment consisted of two parts, separated by a short discussion between participant and experimenter.

**Presentation of e-rattle Noise to Participants**

The first part of the experiment aimed at introducing rattle noise to the listeners. No difference between mechanical rattle and e-rattle was explained to the listeners, because even if their source differs, the resulting perceived sounds are comparable. Since rattle noise can be difficult to recognize, there was a need to clearly define it to the listeners before they evaluated the stimuli, to differentiate it from fluctuation or roughness. To do so, the introductory page in Fig. 15.2 was designed with examples of real and synthetic signals. Since the experiment was performed in France, the interface was written in French. The first four signals were recordings of a train with a flat wheel, a truck at constant speed, a tractor gearbox in reverse position, and a washing machine in spin cycle. All these signals presented some rattle (from mechanical sources), at different pitches and with different pulsations. They came from common real-life situations that the listeners may have already heard. This was intended to reassure them and put them in a good position for the rest of the experiment.

The next five signals were synthetic. The first was white noise, the second was a pure tone (440 Hz). The third presented some fluctuation (two sines at 440 Hz and 444 Hz), and the fourth presented some roughness (two sines at 440 Hz and 470 Hz). The fifth was a e-rattle noise example from the set of stimuli described in paragraph 3.2. It was mentioned that only the last one presented some rattle noise, according to the experimenters' definition.

The last four signals came from the previous perceptual experiment [4]. These were recordings of an electric motor powered by four different PWM strategies, with varying degrees of rattle. This was intended to show listeners that there can be different levels of rattle (Fig. 15.5).

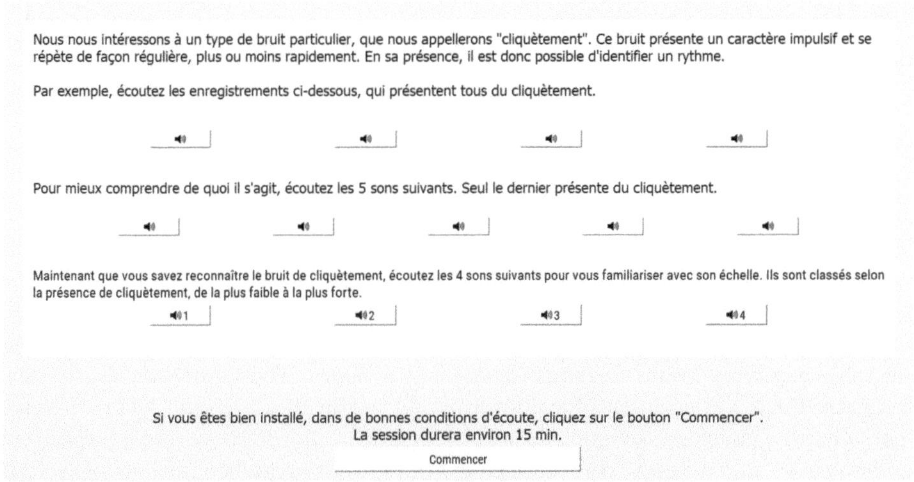


**Fig. 15.5.** Introductory page from the perceptual experiment interface.

### Discussion with the Listeners

A short discussion time (3 min max.) took place afterwards, to answer any questions from the participants. The instructions for the second phase of the test were then delivered to the listeners, using written material.

### Sound Evaluation Method

Listeners were asked to estimate the presence of rattle in the stimuli they heard. They estimated the sensory magnitude related to rattle by adjusting the position of a slider on a non-numerical continuous scale ranging from "very weak" to "very strong" ("très faible" and "très forte" in French). The definition of the scale with words instead of numerical values eliminates the eventual bias due to the tendency of listeners to use some numbers more frequently than others [12]. The test interface was designed so that participants could not move on to the next stimulus until they had listened to the stimulus at least once, and moved the slider on the scale at least once.

No turning back was possible between the different stimuli, to avoid lengthening the test duration. A random order of stimuli was used among all participants to eliminate eventual order effects (Fig. 15.6).

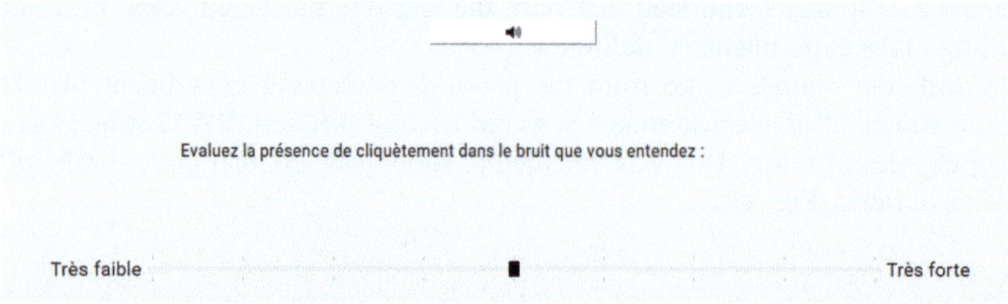


**Fig. 15.6.** Rattle noise assessment page from the perceptual experiment interface.

## 15.4 Results

The classification of the participants answers with Euclidean distances did not show significant groups. All evaluations are presented in Fig. 15.7, with the corresponding stimuli's parameters. The results show an increasing tendency with respect to the envelope width. This confirms the link between e-rattle and the amount of close prominent peaks, because the wider the envelope the more peaks there are. The distribution frequency only seems to impact low e-rattle values. The higher the distribution frequency, the lower e-rattle ratings for envelope widths of 20 Hz and 50 Hz. Indeed, when a narrow envelope is combined with a high distribution frequency, the spectrum has few peaks, and e-rattle does not appear. On the contrary, when a narrow envelope is combined with a low distribution frequency, the spectrum has more peaks and e-rattle may appear.

No significant difference can be seen between the stimuli centered on 500 Hz multiples and 1 kHz multiples, with an average relative difference of 4% and a maximum difference of 10%. This means that pitch does not impact the perception of e-rattle, but it would be necessary to test more different frequencies to confirm this.

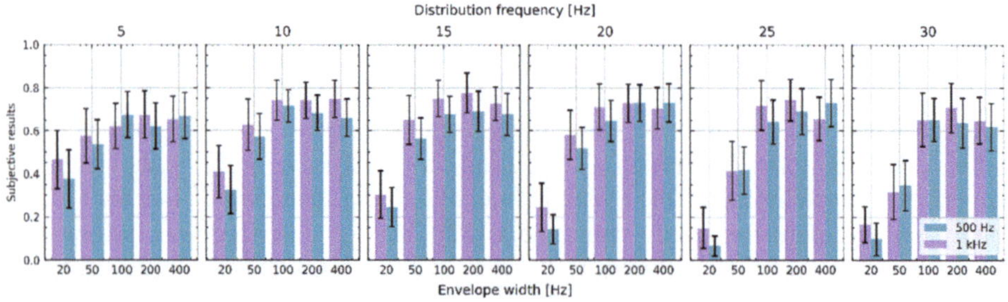


**Fig. 15.7.** Average answers from the listeners for all stimuli, grouped by synthesis parameters, with standard errors of the means.

All subjective answers are presented in ascending order in Fig. 15.8. Two distinct areas can be seen. In the first part, e-rattle ratings evolve with a steeper slope than in the second part, which represents medium to strong e-rattle assessments. This curve confirms the relevance of studying rattle perception as a continuous progressive scale, rather than a binary notion. Listening to the stimuli in regards to their position confirms the tendency perceived by the participants. On the left part of the curve are positioned stimuli with increasing fluctuation, progressively becoming e-rattle. On the right part of the curve, all stimuli present e-rattle, but they evolve in terms of pulsation and timber. The perception of e-rattle then seems to be a combination of these two notions.

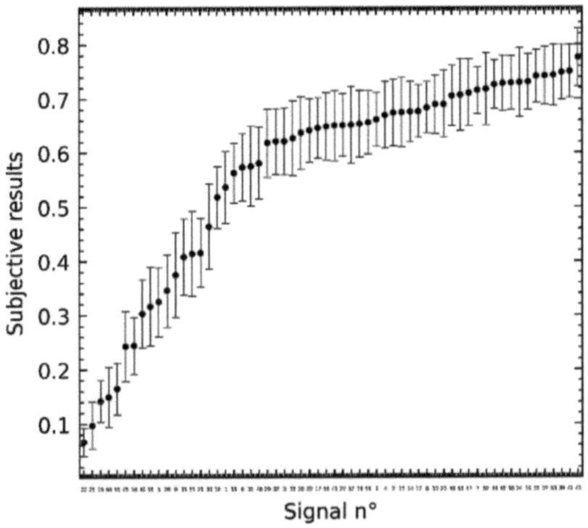


**Fig. 15.8.** Subjective answers ranked in ascending order.

Studying the stimuli ranking in regards to their spectrum parameters confirmed that there is a link between the number of prominent peaks and the perceived e-rattle. Further results will be discussed during the presentation at the conference. A complete index will be needed in the future to represent this phenomenon.

## 15.5 Conclusion

This study shows the need for new sound quality metrics specific to electric motor sounds, with a perceptual test intended to characterize e-rattle noise. The stimuli listened by the participants were synthesized according to the characteristic PWM spectrum. Two harmonic distribution parameters (distribution frequency and envelope width) were used to vary both the frequency and the amplitude of the peaks. Observing the stimuli ranked in ascending order seems to confirm the relevance of the sum of close peaks prominence to characterize the presence of e-rattle in a noise. Further studies will be necessary to develop a complete and robust index.

## References

1. G. Verez, G. Barakat, Y. Amara, and G. Hoblos, "Impact of Pole and Slot Combination on Vibrations and Noise of Electromagnetic Origins in Permanent Magnet Synchronous Motors," *IEEE Trans. Magn.*, vol. 51, no. 3, pp. 1–4, 2015, https://doi.org/10.1109/TMAG.2014.2354019.
2. A. Andersson, D. Lennström, and A. Nykänen, "Influence of Inverter Modulation Strategy on Electric Drive Efficiency and Perceived Sound Quality," *IEEE Transactions on Transportation Electrification*, vol. 2, 2016, https://doi.org/10.1109/TTE.2015.2514162.
3. M. Glesser, S. Ni, K. Degrendele, S. Wanty, and J. L. Besnerais, "Sound quality analysis of Electric Drive Units under different switching control strategies," presented at the SIA Automotive NVH Confort Conference, Le Mans, 2021, p. 8.
4. S. Wanty, J. Le Besnerais, E. Parizet, and N. Totaro, "Simulation of pulse width modulation strategies: validation based on sound quality evaluations," in *INTER-NOISE and NOISE-CON Congress and Conference Proceedings*, Nantes, France: Institute of Noise Control Engineering, 2024, pp. 5162–5167. https://doi.org/10.3397/IN_2024_3555.
5. *Psychoacoustic metrics for ITT equipment—Part 2 (models based on human perception)*, ECMA-418-2, Genève., 2022. [Online]. Available: https://www.ecma-international.org/wp-content/uploads/ECMA-418-2_2nd_edition_december_2022.pdf?trk=organization_guest_main-feed-card_reshare-text.
6. *acousticFluctuation.* (since R2020b). [MATLAB Audio Toolbox]. Available: https://fr.mathworks.com/help/audio/ref/acousticfluctuation.html.
7. W. Deng and S. Zuo, "Electromagnetic Vibration and Noise of the Permanent-Magnet Synchronous Motors for Electric Vehicles: An Overview," *IEEE Transactions on Transportation Electrification*, vol. 5, no. 1, pp. 59–70, 2019, https://doi.org/10.1109/TTE.2018.2875481.
8. F. Kavarana and B. Rediers, "Squeak and Rattle—State of the Art and Beyond," presented at the Noise & Vibration Conference & Exposition, 1999, pp. 1999-01–1728. https://doi.org/10.4271/1999-01-1728.
9. S.-H. Shin and C. Cheon, "Experimental characterization of instrument panel buzz, squeak, and rattle (BSR) in a vehicle," *Applied Acoustics*, vol. 71, no. 12, pp. 1162–1168, 2010, https://doi.org/10.1016/j.apacoust.2010.07.006.
10. G. Cerrato-Jay, J. Gabiniewicz, J. Gatt, and D. Pickering, "Automatic Detection of Buzz, Squeak and Rattle Events," presented at the SAE 2001 Noise & Vibration Conference & Exposition, 2001, pp. 2001-01–1479. https://doi.org/10.4271/2001-01-1479.
11. J. Nam, S. Kim, and D. Ko, "Quantitative Prediction and Analysis of Rattle Index Using DNN on Sound Quality of Synthetic Sources with Gaussian Noise," *Sensors*, vol. 24, no. 16, p. 5128, 2024, https://doi.org/10.3390/s24165128.
12. B. Giordano, P. Susini, and R. Bresin, *Experimental methods for the perceptual evaluation of sound-producing objects and interfaces*, vol. 1. Oxford University Press, 2001. https://doi.org/10.1093/gmo/9781561592630.article.42130.

# Advanced Battery Representations for Optimal Body Integration

Marc Brughmans[1(✉)], Mahmoud Elkafafy[1], Andrey Hense[1,2], and Elke Deckers[2,3]

[1] Siemens Industries Software, NV, Leuven, Belgium
{marc.brughmans,mahmoud.el-kafafy,andrey.hense}@siemens.com
[2] Department of Mechanical Engineering, KU Leuven, Diepenbeek, Belgium
elke.deckers@siemens.com
[3] Flanders Make @KU Leuven, Leuven, Belgium

**Abstract.** Predicting NVH performance of an electric vehicle (EV) at full vehicle level with data-based modeling techniques requires a lot of data to be created for the training of AI/ML models. Hybrid system modeling techniques provide an ideal framework to generate the required data in a trustworthy way. In this environment the best of CAE and Test worlds are combined: accurate prediction of system response is realized through a test-based modeling approach in which a CAE based component is integrated. The parametric nature of the CAE model enables the NVH expert to generate and analyze many system variants.

The first section of the paper is dedicated to the process of building a hybrid system model, connecting a vehicle body (CAE component) with an EV Battery (Test component). Special attention will be given to estimate the global and local dynamic behavior of the battery at the coupling interface. The test-based battery model is further completed with rotational DoFs to extend the frequency range of applications. This section illustrates the environment for a body designer to perform body optimization without being concerned with the accuracy of the complex battery subsystem.

The second section has the focus on the generative aspect of the hybrid system modeling approach. For this purpose, a hybrid system model is created with a parametric representation of the battery component and its connection to the body. The focus here is not on the test representation of the body but on the application of AI/ML methods in the optimization of the battery design itself.

**Keywords:** Electric vehicle · MLMM · Modal parameters · Hybrid modelling


A. Heintzel (Ed.): Automotive Acoustics Conference 2025, Proceedings, pp. 229–245, 2026. https://doi.org/10.1007/978-3-658-51249-1_16

## 16.1 Accurate Battery Representation in the Vehicle Body System

### 16.1.1 Introduction

The automotive industry's transition towards Electric Vehicles (EV) presents novel engineering challenges in vehicle body development. The integration of battery packs fundamentally alters both the structural integrity and dynamic characteristics of vehicles. This paradigm shift necessitates a comprehensive redesign of the vehicle body, incorporating structural enhancements to effectively accommodate the heavy battery structure.

For precise development, engineers require accurate simulation models for both the vehicle body and battery components. While automotive body simulation benefits from decades of expertise and refined methodologies, less experience exists in battery structural modeling. When developing a comprehensive FE model of the integrated body-battery system, there's a significant risk that less sophisticated battery FE representations could compromise the overall simulation accuracy. This could potentially lead to misguided body design optimizations based on inaccurate battery behavior predictions.

Consequently, in scenarios where the battery design is already finalized and physical components are available, body design engineers would achieve more reliable results by integrating a test-based representation of the battery in the overall vehicle system model.

### 16.1.2 Challenges with Test Based Representations for Battery Structures

Creating a test-based modal model begins with measuring Frequency Response Functions (FRFs) at all battery-to-body interface points. These measurements can be directly incorporated into Frequency Based Substructuring (FBS) methodologies to construct a comprehensive vehicle model [1]. However, this approach faces significant challenges due to the extensive interface between body and battery structures. The high number of interface degrees of freedom (DoFs) necessitates processing a very large number of FRFs which creates substantial computational complexity, particularly in the matrix inversion operations of the FBS coupling operator. Controlling the numerical condition of the inversion is indeed challenging due to the presence of measurement noise, phase inconsistencies, non-reciprocities and large response level ranges.

In this paper, we propose an innovative approach to achieve more robust coupling conditions by utilizing a high-fidelity modal model that provides a refined, smoothed approximation of the measured FRFs. However, several technical challenges must be addressed to develop such a modal model effectively. The primary complexity stems from processing numerous FRFs with potentially disparate magnitude scales.

Sophisticated processing algorithms are required to handle magnitude variations while maintaining accuracy across the entire frequency spectrum.

The next section discusses the developed solutions from a theoretical point of view. Afterwards, the developed solutions are applied in a CAE modeling environment to validate the workflow and provide a proof of concept. Finally, the new developed techniques are demonstrated using frequency response functions (FRFs) measured on a battery structure provided by Hyundai Motor Company (HMC).

### 16.1.3 Innovative Workflow for Efficient Creation of High-Fidelity Modal Model

The proposed workflow aims to address two key aspects: 1) ensuring reasonably accurate modal synthesis for all FRFs, regardless of their magnitude level, and 2) executing the modal estimation process in a time-efficient manner. The different blocks of the proposed workflow are illustrated in Fig. 16.1. As shown in the figure, the workflow consists of two main steps: 1) estimating a reduced-size, in terms of reference DoFs, modal model, and 2) expanding the reduced-size modal model to the full original-size FRFs matrix.

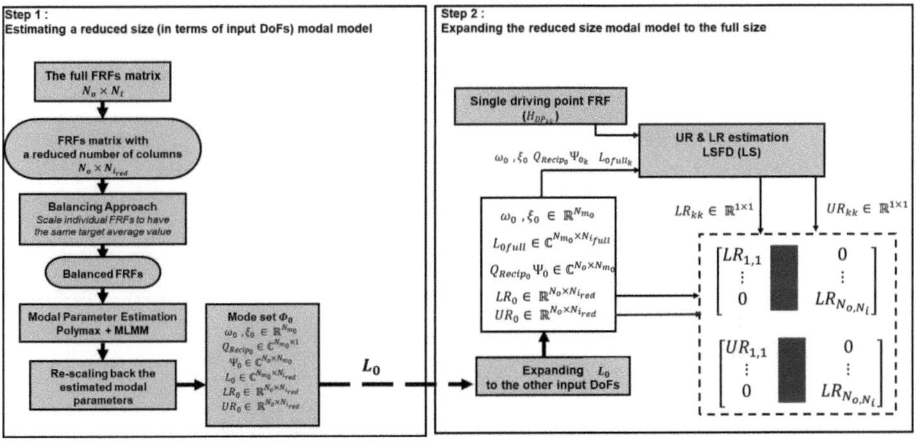


**Fig. 16.1.** Workflow of the enhanced modal parameter estimation approach.

To estimate a reduced-size modal model, a selection of a few columns (for example 4 columns) from the full-size FRFs matrix must be made. After selecting the reduced number of columns, the first step in the workflow, estimating reduced-size participation factor vectors, consists of three main sub-steps. In the first sub-step, a magnitude-based balancing approach is applied to the reduced FRFs matrix to equalize the different FRFs in terms of magnitude. In the second sub-step, the modal parameters of the component under test (CUT) are estimated in the frequency band of interest by applying the well-known Polymax and MLMM (Maximum Likelihood estimation of a Modal Model) estimators [2, 3] to the balanced reduced-size FRFs matrix. Finally, in the third sub-step, the estimated mode shape vectors, participation factor vectors, and elements of the lower and upper residual terms matrices are rescaled to make them consistent with the original magnitude levels of the processed FRFs. Once the reduced-size modal model is estimated, a full-size modal model is retrieved by extending the estimated participation factor vectors, the lower residual matrix, and the upper residual matrix in the second step of the proposed workflow [4].

### 16.1.4 Workflow Validation

In this section, the workflow is validated using a dataset consisting of Frequency Response Functions (FRFs) extracted from a finite element model (FEM) of an EV battery, which contains 927712 nodes (see Fig. 16.2, left). The FRFs were simulated at 26 body-to-battery connection points (see Fig. 16.2, right), each with 3 translational degrees of freedom (DoFs) in the x, y, and z directions, resulting in a square FRF matrix with 78 rows (outputs) and 78 columns (inputs) at each frequency line. Figure 16.3 shows some typical FRFs and illustrates the wide range in response level. In the workflow validation study, the frequency range was limited to 25–160 Hz, as the FEM had been experimentally calibrated up to this frequency.

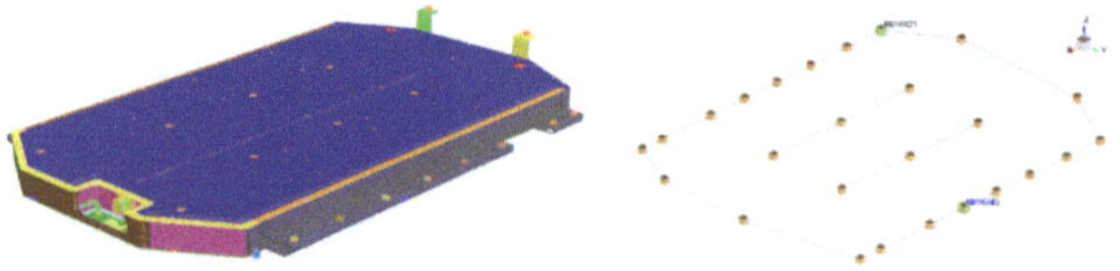

**Fig. 16.2.** The battery pack FEM (left) and Body-to-Battery connection wireframe (right).

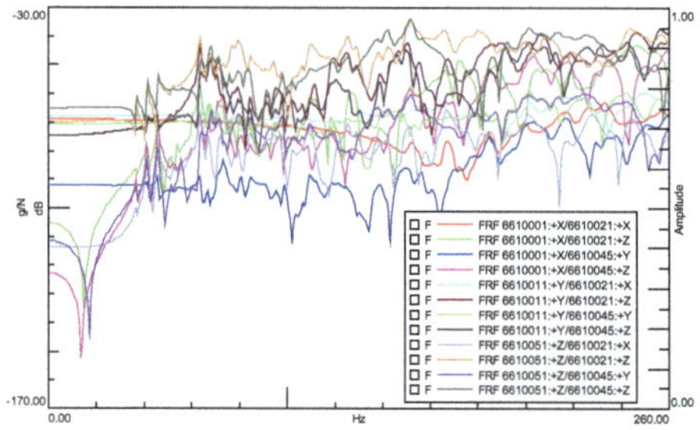


**Fig. 16.3.** Typical FRFs at connection points.

Starting from the 78×78 FRF matrix extracted from the battery model, the goal is to identify a modal model that captures both the global flexible modes and the local flexibility at the connection DoFs in this frequency band. The identification of that modal model must ensure both accuracy and relatively fast computation. The challenges arise from two main factors: 1) the number of flexible modes should be minimized while maintaining a good fit, and 2) local flexibility must be included for all the coupling degrees of freedom (DoFs) and this requires including all 78 columns of the FRFs matrix in the modal fitting process, which is computationally impractical.

The standard workflow for modal parameter estimation involves the complete 78*78 FRF matrix to identify a modal model that captures both the global flexible modes and the local (residual) flexibility at the connection points. The modal fitting process becomes very challenging and computationally impractical.

The new workflow brings down the complexity by separating the identification of the global flexible modes from the calculation of the local flexibility. By breaking up the process in two phases, the analysis can be drastically speed-up while not giving in on the accuracy as each phase has its own dedicated focus.

During the first phase, the global modes of the battery are identified. For this purpose, a few columns are selected from the full-size FRFs matrix. Indeed, there is no need to involve all 78 inputs for the excitation of the global modes. By reducing the number of columns, the modal estimation process is significantly speed up without loss of accuracy. In this case four columns are selected, two corresponding to each of the two excitation locations shown in Fig. 16.2-right (X,Z in loc.1 and Y,Z in loc. 2). This results in a $78 \times 4$ FRFs matrix, which is then processed through the implemented FRFs balancing technique to equalize the magnitude levels of the 312 FRFs.

Having balanced the FRFs matrix, the combined Simcenter Polymax- MLMM modal modal solutions are then applied to the balanced matrix to optimize the modal model imposing the FRFs reciprocity, real mode shapes, and the negative mass sensitivity (NMS) constraints on the identified model. As a result, an accurate modal model with 48 global modes was extracted from the FRFs.

In the second phase of the process, the focus is shifted to the extraction of the local residual flexibility at the connection points. In the first phase, the information about the local flexibility was only available for 4 connection DoFs. So, in the second part of the workflow, the driving point FRFs of the other connection points are processed to extend the set of residual flexibility vectors. As only the diagonal of the FRF matrix is involved, this process is executed in a very efficient way.

The final mode set with 48 global modes and 78 residual flexibility vectors was obtained with the new workflow in very fast, but accurate way. Compared to the standard modal fitting process that took 2.5 h to complete, the new workflow was executed in just 45 s. Figure 16.4 illustrates the good accuracy of the obtained modal model by comparing the FRFs synthesized with the modal model against the true FRFs. The figure shows examples for FRFs with low and high magnitude level. The virtue of FRF balancing can be well observed in the low magnitude level FRF.

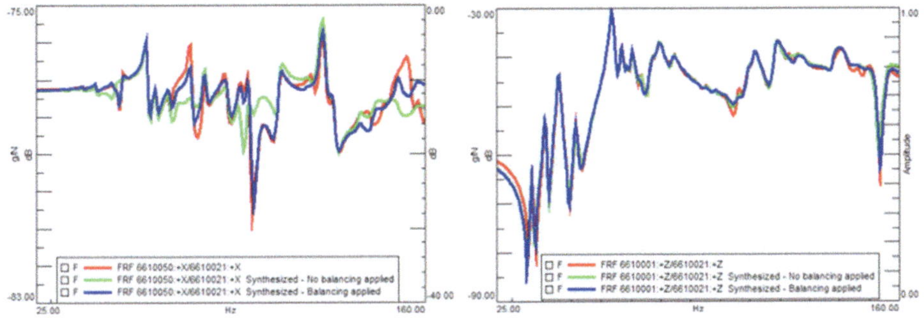


**Fig. 16.4.** Modal synthesis for low magnitude level FRF (left) and high magnitude level FRF (right).

As a final validation step in this section, Fig. 16.5 illustrates the body-to-battery coupling results for selected target DoF FRFs when the MLMM-based identified battery modal model, is coupled to the vehicle body modal model (blue curve in Fig. 16.5). These results are compared to the reference case modal system assembly (red curve in Fig. 16.5), which utilizes the full mode set identified by FEM for both the body and battery within the frequency band of interest.

Two key conclusions can be drawn from the results. First, the figure clearly demonstrates that the battery significantly impacts the NVH characteristics of the vehicle body. The system represented by the green curve in Fig. 16.5 (body without battery) exhibits a notably different response compared to the system represented by the red curve (battery coupled to the body with the full mode set). This highlights the importance of considering body-to-battery coupling when assessing the NVH performance of an electric vehicle (EV). Second, the coupling result obtained using the MLMM-based reduced-order modal model, considering both global and local flexibilities, albeit with only 48 flexible modes, closely matches the result from the full system assembly. These findings validate that the proposed workflow effectively delivers a reduced-order modal model capable of adequately capturing the contributions of both global flexible modes and local flexibilities at the coupling degrees of freedom (DoFs).

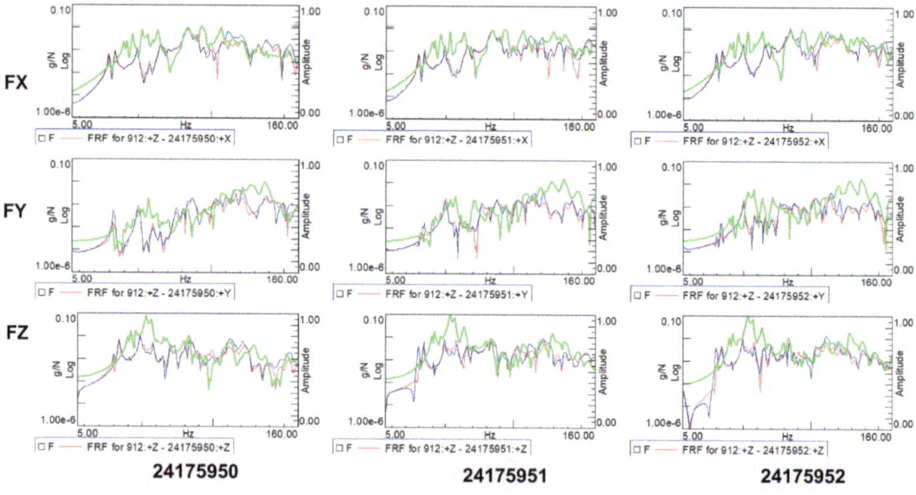


**Fig. 16.5.** Body-to-battery coupling I Reference assembly (red curve) versus the MLMM-based modal system assembly (blue curve). Body without battery (green curve). Note: The shown result is for the seat rail vibration as a body target response (912: +Z) and unit body load at the rear engine mount locations (24175950, 24175951, 24175952 +X, +Y, and +Z).

### 16.1.5 Application of the Workflow to Create a Test Based Modal Model of an EV Battery

Having validated the process in the lower frequency range, the developed workflow is now applied to an extended frequency range using test data that were measured during a modal test of the battery structure. In this modal test, the battery structure was free-free suspended in bungees (cf. Fig. 16.6) and excited up to 2 kHz at 78 coupling DoFs (26 body connection points in x, y and z- directions) using the Simcenter Qsources hardware integral shakers (Qish-Integral shaker) mounted on Qtrx excitation adaptor. Frequency response functions were measured consecutively under different multiple angles as shown in Fig. 16.6. The angled FRFs set is processed to reconstruct the FRFs with the orthogonal inputs of the global axis system. The acceleration responses were simultaneously acquired at the 78 coupling DoFs using modal accelerometers PCB 356A15. This results in a 78×78 FRFs matrix at each frequency line in the excited frequency range. Some typical measured FRFs are shown in Fig. 16.6.

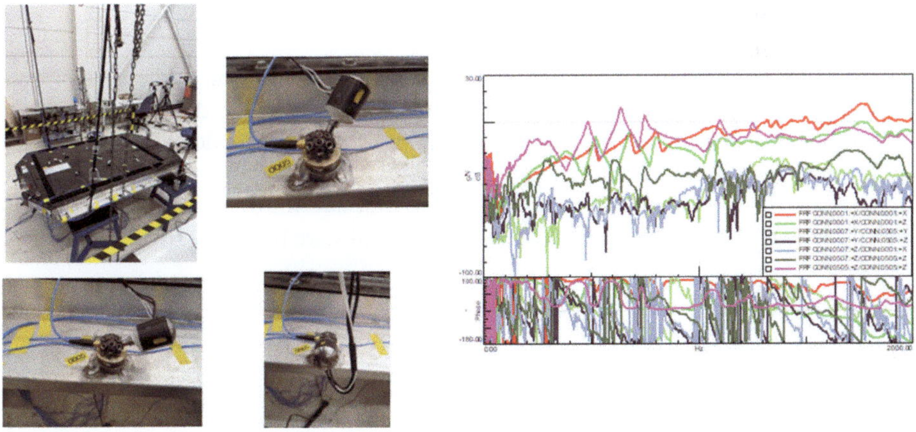

**Fig. 16.6.** The battery test set-up and some typical measured FRFs at connection points.

In our analysis, we have targeted our frequency application range [100–1000 Hz] to start from the validated FE battery upper frequency limit (<160 Hz) and go up to the mid frequency range (1 kHz) where CAE body models can still be represented by FEM. Using the developed workflow, an optimized modal model was obtained with 47 global modes (excited by 4 inputs) and 78 residual flexibility vectors. The objective of the modal fitting process was to obtain excellent accuracy fit for the high-level magnitude FRFs and to capture well the level and shape of the low level magnitude FRFs. Good results were obtained with the developed workflow as can be seen in the modal synthesis validation shown in Fig. 16.7.

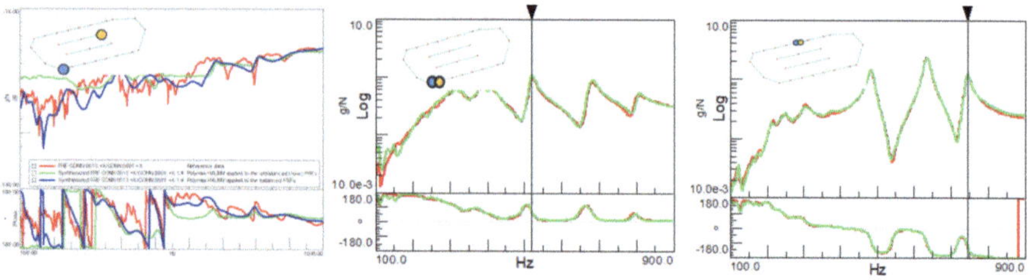


**Fig. 16.7.** Modal synthesis for low magnitude level FRF (left) and high magnitude level FRFs (mid and right). Measured curve (red), Modal synthesis with unbalanced FRFs (green), Modal synthesis with balanced FRFs (blue).

### 16.1.6 Completing the Test Based Modal Model with Rotational DoFs

With the extended frequency range for the battery-body assembly, the need for involving rotational DoFs in the connection of the subsystems arises. Indeed, imposing only translational continuity in the coupled assembly state is too weak when higher order component shapes are involved.

Using FE based shape functions the deformation pattern around the battery connection points was reconstructed for each mode shape. From the deformation pattern the rotational displacement was extracted and inserted in the mode shape, preserving the original modal parameters (frequency, damping, scaling). Figure 16.8 illustrates this process for the mode shapes related to the first and third high peak in the FRFs shown in the mid and right window of Fig. 16.7 (peaks are indicated by the cursor).

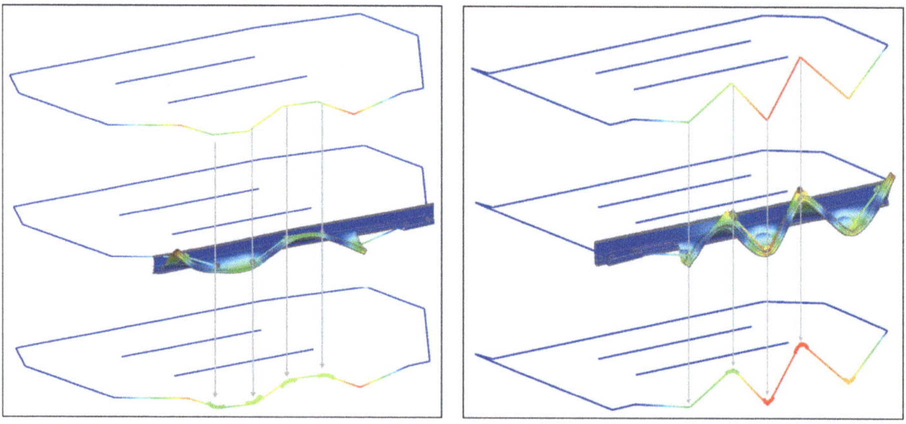

**Fig. 16.8.** Examples of Test mode completion with rotational DoFs.

### 16.1.7 Hybrid Body (CAE)—Battery (Test) Modal Assembly

The test based modal model of the EV battery developed in the previous section was processed as a modal superelement for Nastran and coupled to the body FEM. With a modal forced response solution sequence (SOL111), FRFs were calculated from 20–1000 Hz in a very efficient way without any frequency-dependent stiffness properties, unlike FRF-based substructuring methods. Furthermore, the modal based coupling scheme embedded in the modal forced response solution enables to identify high frequency battery component modes that play an important role in the assembly and body countermeasures can be more efficiently devised. Figure 16.9 shows in the top window the acceleration response at the seat rail position to a unit force input at one of the rear engine mounts. In the bottom part of Fig. 16.9 the modal participations of the 47 battery component modes are shown and two of them are highlighted corresponding to the modes illustrated in Fig. 16.8.

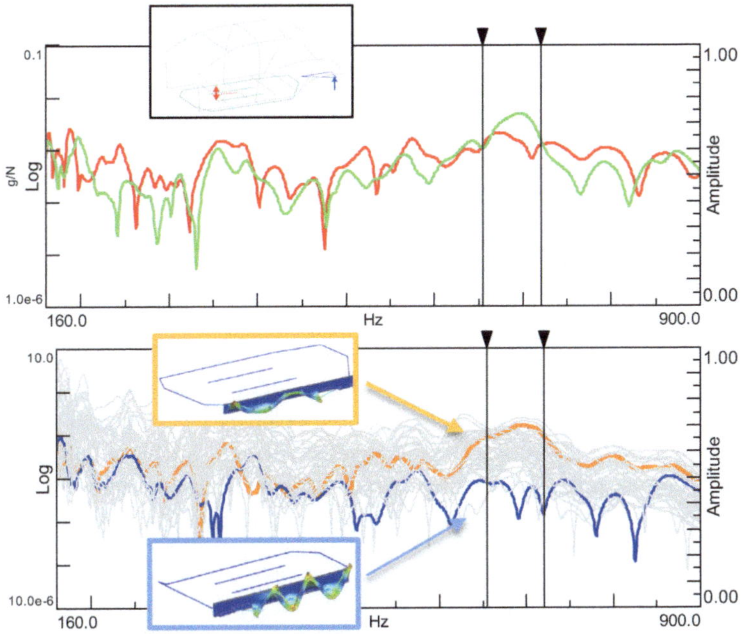


**Fig. 16.9.** Top Window: Body FRF, green curve: body only, red curve: body+battery Bottom Window: Modal participation of 47 battery component modes in the assembly FRF (body+battery).

In the frequency range indicated by the double cursor in Fig. 16.9, the battery component mode corresponding to the orange curve is very active in the assembly. Completing this mode with rotational DoFs is important to impose deformation conformity in the assembly. This is evidenced in Fig. 16.10 where the effect of the presence of rotational DoFs in the battery mode shapes on the body-battery coupling is shown.

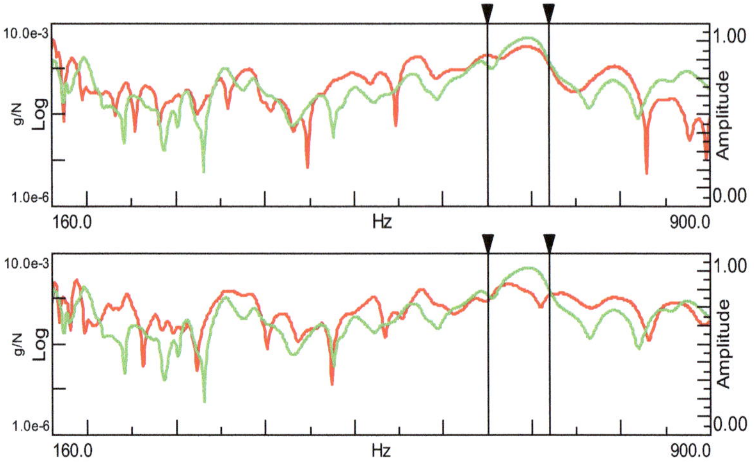


**Fig. 16.10.** Body FRF without (Top Window) and with (Bottom Window) rotational DoFs involved in the coupling (green curve: body only, red curve: body+battery).

## 16.2 Parametric Battery Representation in the Vehicle Body System

While in the first section the attention is given to developing an accurate and efficient experimental representation of the battery for a body designer to perform body optimization, in the current section the perspective changes to the battery pack design, optimization and integration. The battery pack's inherent stiffness can contribute to overall body rigidity, potentially reducing the need for additional reinforcement structures and thereby offsetting some of the battery weight penalty.

However, determining the optimal integration strategy requires a modeling approach that can accurately predict system-level NVH performance across various design configurations efficiently. From the battery pack development perspective, the vehicle body design can be fixed and reduced to a modal model representation, also known as Component Mode Synthesis (CMS) [1], while a parametric concept battery model is developed to reduce computation time and enable comprehensive design space exploration. For the current low frequency analysis, the vehicle body modal model is simulation-based, though as previously demonstrated, it can be experimentally established for a higher frequency range.

Using the integrated FE model, our investigation focuses on evaluating AI/ML methods on the creation of surrogate models to accelerate optimization analysis. These surrogate models enable rapid design optimization across multiple objective functions and constraints, facilitating efficient exploration of the complex design space without the computational burden of full finite element analyses for each design iteration.

The previous mentioned approach to efficiently simulate vibration responses of the integrated battery and body is presented in [5]. The results demonstrated that machine learning models can accurately predict vibration levels at the seat rail position based on battery pack design parameters, particularly at lower frequencies.

The integration strategy of the battery pack with the vehicle body represents another critical aspect of EV design with significant implications for NVH performance. The number, and location of connection points between the battery and body structure are as important as the stiffness distribution to determine the system-level dynamic behavior. The current investigation focuses on optimizing the connection layout configuration between the battery pack and vehicle body using the same coupled parametric FE model developed for the sizing study.

Building upon previous experience with machine learning techniques, we apply similar surrogate modeling approaches to the connection layout problem. The ML models are trained in data generated from the parametric FE simulations with varying connection configurations. The workflow of the analysis is summarized in Fig. 16.11.

Design exploration starts with a baseline configuration in which all structural connections between the battery pack and vehicle body are active (Fig. 16.2). Design variations are achieved by deactivating, in a symmetric manner, pairs of connections from the left and right sides of the battery pack. This bilateral deactivation pattern maintains the structural balance of the battery-body system. As a constraint in the design space a minimum of four connections is considered to maintain essential structural integrity.

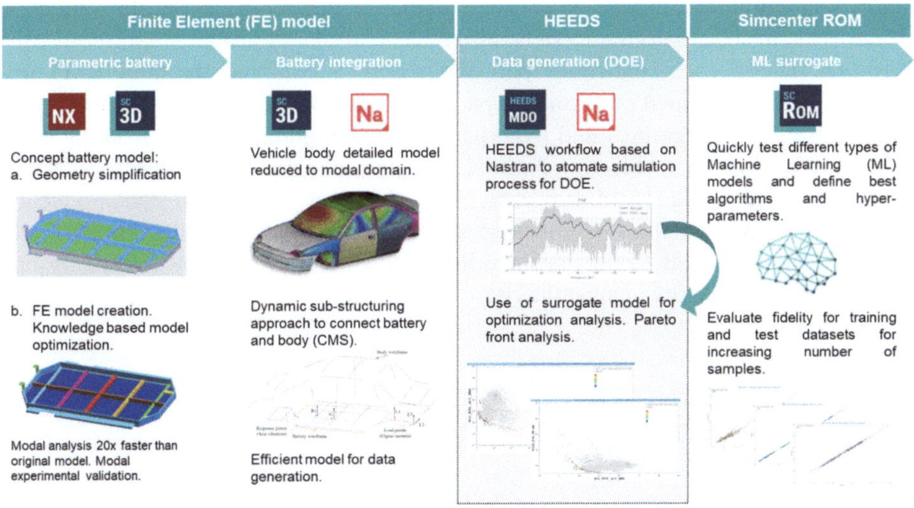


**Fig. 16.11.** Workflow to evaluate connection distribution.

The investigation evaluates the feasibility of minimizing connection points while maintaining acceptable performance across multiple criteria: (i) acceleration response at seat rail position (Fig. 16.9), (ii) average battery acceleration at connection interface, and (iii) maximum element strain energy (ESE) across battery-body connecting elements. For quantitative assessment, root mean square (RMS) values are calculated for these parameters within the frequency range of 0.5–80 Hz. The ESE metric serves as an indicator of load distribution and junction reliability. As illustrated in Fig. 16.12, the distribution of maximum ESE values demonstrates an inverse relationship with connection quantity, where fewer connections generally correspond to increasing strain energy. Nevertheless, the graph reveals reduced-connection configurations that maintain ESE values comparable to the baseline design.

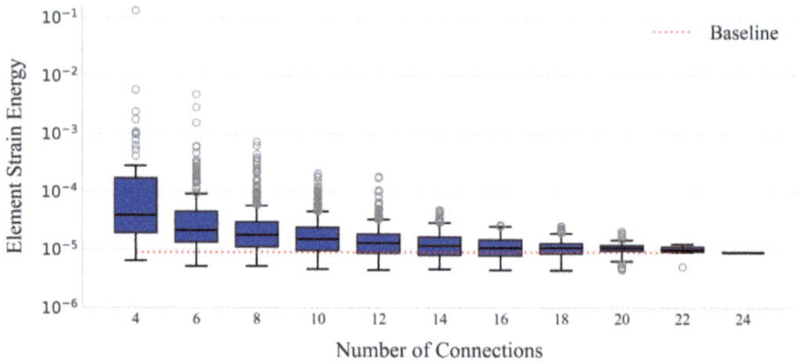


**Fig. 16.12.** Element strain energy boxplot per number of connections for a unit force input at one of the rear engine mounts in vertical direction.

Reduced order models are generated using Simcenter ROM, enabling rapid evaluation of multiple ML models to enhance predictive fidelity [6] across the validation dataset. The machine learning models utilize 12 binary input features representing the activation state of each connection pair, while outputs comprise the three RMS values of the aforementioned performance parameters. Model fidelity varies substantially across different architectures and hyperparameter configurations, necessitating comprehensive evaluation to identify optimal surrogate models for the specific dataset. As demonstrated in Fig. 16.13, Bayesian Neural Network (BNN) and standard Neural Network (NN) architectures demonstrate superior performance for the current application. Relatively small models achieve high validation accuracy by effectively balancing model complexity against the risk of overfitting the training data.

| | bnn | rom_4 | rom_6 | rom_5 | rom_1 | rom_2 | nn | rom_7 | rom_8 | rom_9 | rom_3 | rom_10 | rom_11 |
|---|---|---|---|---|---|---|---|---|---|---|---|---|---|
| Type of model | Bayesian Neural Network | Neural Network | Neural Network | Neural Network | Neural Network | Neural Network | Neural Network | Gaussian Process | Gaussian Process | Response Surface Model | Neural Network | Gaussian Process | Response Surface Model |
| Training fidelity [%] | 95.518 | 95.555 | 94.480 | 94.792 | 97.678 | 97.533 | 96.721 | 93.794 | 93.356 | 92.063 | 96.879 | 90.668 | 83.442 |
| Validation fidelity [%] | 93.632 | 92.765 | 92.454 | 91.813 | 91.668 | 91.236 | 90.488 | 89.489 | 89.485 | 89.167 | 87.695 | 87.226 | 82.305 |
| Total trained epochs | 10 | 50 | 50 | 50 | 50 | 50 | 100 | | | | 50 | | |
| Number of layers | 2 | 2 | 1 | 1 | 2 | 3 | 2 | | | | 3 | | |
| Layer types | Dense<br>Dense | Dense<br>Dense | Dense | Dense | Dense<br>Dense | Dense<br>Dense<br>Dense | Dense<br>Dense | | | | Dense<br>Dense<br>Dense | | |
| Number of cells | 20<br>20 | 32<br>13 | 27 | 25 | 36<br>17 | 39<br>16<br>9 | 32<br>13 | | | | 41<br>18<br>11 | | |
| Activation types | relu<br>relu | tanh<br>tanh | tanh | tanh | tanh<br>tanh | relu<br>relu<br>relu | relu<br>relu | | | | relu<br>relu<br>relu | | |
| Training duration [s] | 87.0424 | 5 | 2 | 2 | 7 | 10 | 14 | 127 | 9 | 0 | 12 | 6 | 0 |
| Training method | | Levenberg-Marquardt | Levenberg-Marquardt | Levenberg-Marquardt | Levenberg-Marquardt | Levenberg-Marquardt | Levenberg-Marquardt | | | | Levenberg-Marquardt | | |
| Max order of inputs | | | | | | | | | | 3 | | | 2 |
| Max order of cross terms | | | | | | | | | | 3 | | | 2 |
| Number of coefficients | | | | | | | | | | 1365 | | | 273 |
| Minimal size | | | | | | | | | | No | | | No |
| Fixed point | | | | | | | | | | No | | | No |
| Kernel | | | | | | | | Gaussian | Matern type 5 | | | Exponential | |
| Kernel variance | | | | | | | | 0.879043 | 1.08016 | | | 13.8851 | |
| Length scale | | | | | | | | 2.31698 | 3.48078 | | | 1526.04 | |
| Noise level | | | | | | | | 0.01 | 0.01 | | | 1E-08 | |
| Fitting method | | | | | | | | subset_of_regre... | subset_of_regre... | | | subset_of_regre... | |

**Fig. 16.13.** Simcenter ROM detailed information of model sweep for 1494 samples in the training set.

| Number of Samples | | Validation Fidelity | | | |
|---|---|---|---|---|---|
| Training | Validation | Acc. 912 | Acc. Avg. Bty. | Max. ESE | Overall |
| 3065 | 766 | 99 | 95.5 | 85.7 | 93.4 |
| 1992 | 1839 | 99 | 95.7 | 85.7 | 93.4 |
| 1494 | 2337 | 98.9 | 95 | 87 | 93.6 |
| 996 | 2835 | 98.7 | 95.2 | 82.45 | 92.1 |
| 498 | 3333 | 98.3 | 92.9 | 81.2 | 90.8 |
| 268 | 3563 | 97.2 | 91.6 | 78.1 | 89 |

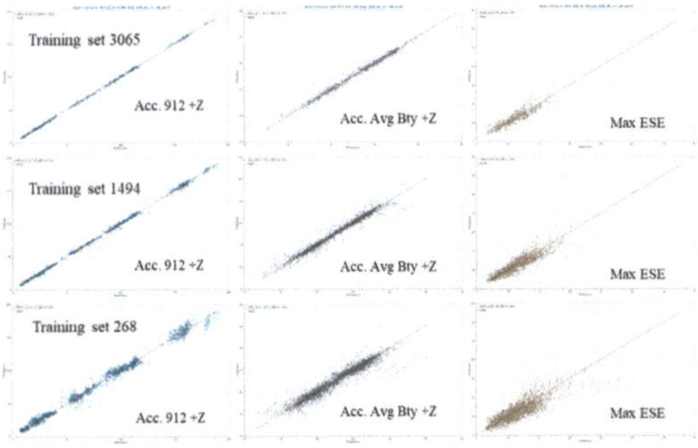


**Fig. 16.14.** BNN fidelity for increasing number of samples in training set (left table) and specific output parameter fidelity (right).

The BNN emerged as the most accurate predictive model across various training set sizes, with the additional advantage of quantifying prediction uncertainty. While BNN training requires more computational resources than standard neural networks, this overhead remains negligible relative to the dataset generation timeframe. Figure 16.14 illustrates BNN fidelity as a function of training set size, confirming the expected positive correlation between data volume and model accuracy. An optimal balance between data generation effort and model performance is achieved with approximately 1000 samples, although smaller datasets already demonstrate satisfactory representation of response trends. Notably, prediction fidelity varies across output parameters; models trained with just 268 samples adequately predict seat rail acceleration for novel designs, whereas element strain energy prediction requires more training data due to its nonlinear behavior across connection configurations and presence of statistical outliers. The implementation of feasibility constraints to narrow the design space would enhance model accuracy while reducing the data requirements for developing robust surrogate models.

High-fidelity surrogate models facilitate efficient multi-objective optimization processes, enabling rapid evaluation of several objective functions and constraints. These models provide invaluable real-time feedback to project engineers, accelerating design space exploration, expediting assessment of alternative project scenarios, and enhancing system-level understanding.

Despite accuracy limitations, models trained on reduced datasets can significantly accelerate the design process. Hereafter the analysis focuses on a model trained on just 268 simulations to explore design insights efficiently. The ML model can predict all 4096 possible connection combinations faster than executing a single FE simulation. Since a bigger simulation dataset is available, the complete simulation data is used as reference to precisely quantify prediction errors. The analysis proceeds by clustering designs that demonstrate improvements across all three key performance indicators, thereby reducing the design space to 346 configurations. Figure 16.15 illustrates these selected designs in a scatter plot representation. A distinctive advantage of the BNN approach is its inherent uncertainty quantification for each prediction, providing critical context for decision-making. To visualize uncertainty trends, Fig. 16.15 includes an average uncertainty line, revealing higher uncertainty in response regions with sparse training data while confirming model accuracy within the regions of primary interest.

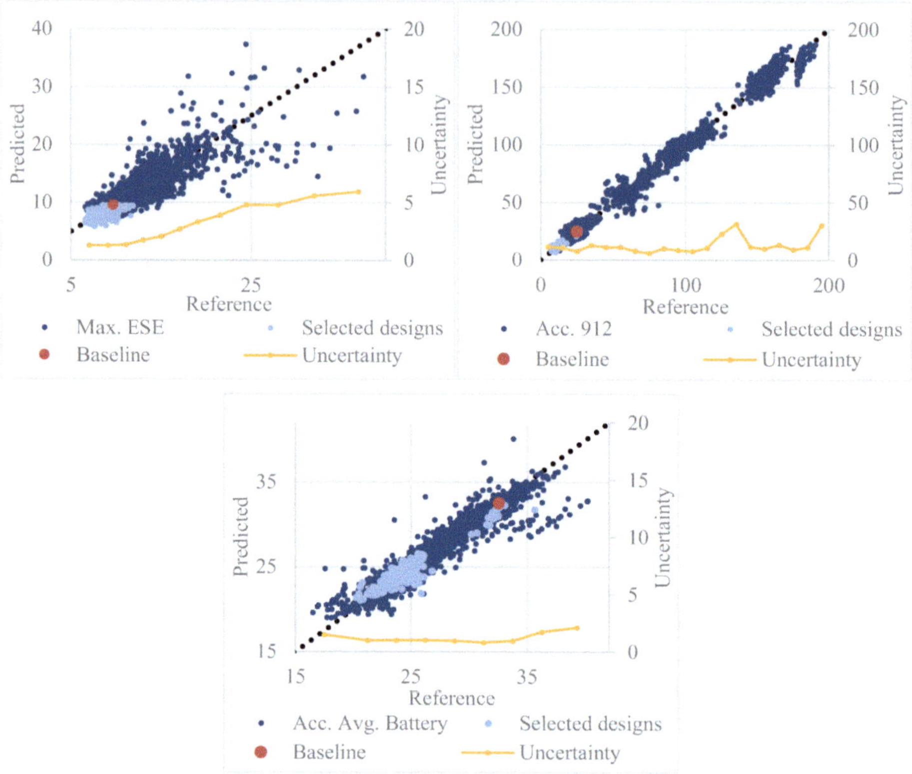


**Fig. 16.15.** Scatter plot of reference (simulation data) with predicted value of ML model created with small dataset.

The inherent uncertainty in predicted values means that optimization processes using this data would identify solutions approximating, rather than precisely achieving, theoretical optimum. Analysis of the clustered design configurations reveals critical insights regarding the relative importance of specific connection points for performance optimization. Figure 16.16 quantifies the activation frequency of each connection pair across the improved design cluster. Connection pairs consistently present in superior designs, such as C1 and C3, exhibit activation frequencies approaching 1, indicating their essential contribution to system performance. Conversely, connections with low activation frequencies, exemplified by C4, are predominantly deactivated in high-performing configurations, suggesting their minimal or potentially detrimental impact on the system's dynamic behavior.

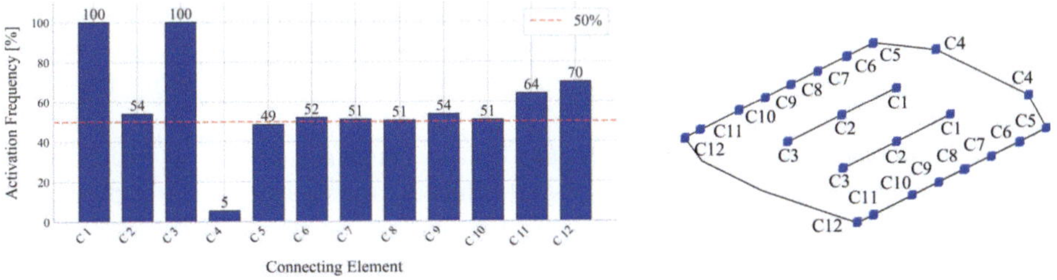


**Fig. 16.16.** Frequency of active connections in the cluster of improved designs (left) and connections numbering on the right.

To elucidate the observed trends in connections C1, C3, and C4, both the Boruta feature selection method [7] and Sobol indices [8] can be applied, yielding convergent conclusions. The global sensitivity given by the Sobol indices correlate the importance of each connection to each of the response's variance. Connection C1 emerges as the most influential feature affecting seat rail acceleration; its absence permits the development of a localized vehicle floor mode that significantly amplifies vibration. Connections C3 and C4 predominantly influence battery average vibration, albeit through different mechanisms—vehicle body mode shapes induce elevated battery vibration when C4 is connected, whereas C3 connections enhance stiffness in the battery's central region, effectively attenuating vibration. Additionally, connections C11 and C12 demonstrate moderate influence on battery vibration while substantially impacting maximum ESE. Elevated ESE manifests in configurations characterized by low battery natural frequencies, where mode shapes reveal the battery essentially suspended from limited connection points. A notable outlier exhibiting extreme strain energy occurs when only connections C1 and C6 remain active, resulting in pronounced battery rotation around these connection points. This phenomenon exemplifies on a smaller scale the higher ESE in configurations lacking C11 and C12 connections.

Connections positioned along the side frame (C5, C6, C7, C8, C9, and C10) demonstrate comparatively limited influence on the three key performance indicators. This reduced impact suggests that acceptable vibration performance can be maintained with a strategic reduction in these peripheral connection points, offering potential simplification of the battery integration design without compromising dynamic performance characteristics.

## 16.3 Conclusions

In this paper, we introduced a novel workflow to create a test-based battery representation that can be easily integrated with a CAE body model to provide an accurate body development framework for body optimization in an extended frequency range. After validation of the workflow, the process was applied to an EV battery using test data that were measured during a modal test. Specific methods were developed to extract successfully global and local dynamic properties from the large amount of test data. With the extended frequency range for the battery-body assembly, comes the need to involve rotational DoFs in the coupling definition. A new method using FE based shape functions was introduced to provide the missing rotational information.

The second section of the paper focused on the interface layout between battery and body. A modeling methodology was introduced capable of efficiently predicting system-level NVH performance across diverse battery design configurations using high-fidelity surrogate models. The predictive accuracy increases proportionally with training data volume. However, it was shown that even surrogate models with moderate accuracy can provide valuable insights into fundamental system behavior patterns.

For the datasets examined, Bayesian Neural Networks emerged as the optimal machine learning approach, with their inherent uncertainty quantification offering critical context for analysis reliability and confidence assessment within regions of interest. The implementation of result clustering, as opposed to individual design evaluation, effectively identifies broader performance trends when combined with global sensitivity analysis, while demonstrating reduced susceptibility to model accuracy limitations.

While this investigation presents the analyses of stiffness distribution and connectivity layout as discrete studies, it is important to acknowledge their inherent interdependence. The dynamic behavior of the integrated battery-body system emerges from the complex interaction between these design domains. Future research will address this limitation by implementing a comprehensive approach that simultaneously evaluates both parameter sets, enabling the exploration of potential synergistic effects and more holistically optimized design solutions.

**Acknowledgements.** The authors gratefully acknowledge the support of VLAIO (Flemish Innovation & Entrepreneurship) through the O&O project ADRENALINE (HBC.2022.1027). The authors gratefully acknowledge the support of Hyundai Motor Company (HMC) both for providing the battery hardware and the FE model. The work was supported by the project IN-NOVA: Active reduction of noise transmitted into and from enclosures through encapsulated structures, which has received funding from the European Union's Horizon Europe program under the Marie Sklodowska-Curie grant agreement no. 101073037.

## References

1. D. D. Klerk, D. J. Rixen and S. N. Voormeeren, "General framework for dynamic substructuring: history, review, and classification of techniques," *AIAA Journal,* vol. 46, no. 5, pp. 1169-1181 https://doi.org/10.2514/1.33274, 2008.
2. B. Peeters, H. Van der Auweraer, P. Guillaume and J. Leuridan, "The PolyMAX frequency-domain method: a new standard for modal parameter estimation?," *Shock and Vibration,* vol. 11, pp. 395-409, 2004.
3. M. Elkafafy, B. Peeters, P. Guillaume and T. De Troyer, "Constrained maximum likelihood modal parameter identification applied to structural dynamics," *Mechanical Systems and Signal Processing,* Vols. 72-73, pp. 567-589, 2016.
4. M. Elkafafy and M. Brughmans, "Optimizing NVH Performance in Electric Vehicles Through Enhanced Modal Parameter Estimation Techniques," in *IMAC-XLIII, International Modal Analysis Conference*, 2025.

5. A. Hense, F. Chauvicourt, M. Brughmans, S. van Ophem and E. Deckers, "Evaluation of machine learning techniques for the NVH design of electric vehicle battery," in *International Conference on Noise and Vibration Engineering (ISMA2024)*, Leuven, Belgium, 2024.
6. "Simcenter Reduced Order Modeling 2310 | User Guide," Siemens Digital Industries Software, 2023.
7. M. B. Kursa and W. R. Rudnicki, "Feature Selection with the Boruta Package," *Journal of Statistical Software,* vol. 36, no. 11, 2010.
8. I. M. Sobol, "Global sensitivity indices for nonlinear mathematical models and their Monte Carlo estimates," *Mathematics and Computers in Simulation,* vol. 55, p. 271–280, 2001.

# Optimization of Mass-Spring Noise Control Treatment Using Permeable Heavy Layer

Alexander Peiffer(✉) and Marc Burghardt

Audi AG, Ingolstadt, Germany
{alexander.peiffer,marc.burghardt}@audi.de

**Abstract.** This paper is concerned with a new set-up of mass-spring noise control treatment. The double wall system established by the mass-spring-system added to the car body structure is the most efficient set-up for noise control due to the double wall effect. However, at the double wall resonance such systems have a lower performance as the single wall configuration. The classical countermeasure against this is to increase the damping of the spring material, e.g. by using viscoelastic foams. The increase of damping is often costly and restricted to polyurethane foams. This prohibits the use of nonwoven materials that cannot achieve similar damping due to the dominance of the fluid phase wave propagation in the fiber material. In contrast to this fiber spring materials provide a lower spring stiffness leading to better high frequency performance. In this paper a method is presented to use a permeable mass layer that allows control of the resonance drop with eligible loss of high frequency performance. This paves the road for environmentally friendly and high performing mass spring systems using polyester fiber material.

**Keywords:** Noise control treatment · Transmission loss · Sound-reduction index · Mass-spring systems

## 17.1 Introduction

For the path towards a more sustainable automotive industry the main technology is the development of batterie electric vehicles (BEV). In addition, there is an urge to increase the recycling rate of the noise control package. One candidate for this is the mass-spring system applied as acoustic barrier at firewalls or car floors. In premium cars a layup of polyurethan (PUR) foam in combination with a thermoplastic heavy layer (HL) highly filled with inorganic material is used. The heavy layer is sometimes additionally treated with a carpet cover. Spring layers of visco-elastic foams with high damping provide high performance in the double wall resonance range and the high frequency regime.

However, the PUR-HL combination cannot be recycled or produced form recycled material. Thus, Audi is investigating new mass-spring concepts using polyethylene (PET) as spring and heavy layer material.


A. Heintzel (Ed.): Automotive Acoustics Conference 2025, Proceedings, pp. 246–260, 2026. https://doi.org/10.1007/978-3-658-51249-1_17

## 17.2 Foundations of Mass Spring Noise Control Systems

The sections deal with the principal effect of double wall systems and the properties of foam and fiber spring materials. It presents an acoustic model of a permeable heavy layer and its effects on the double wall transmission loss (TL) also called sound reduction index (R).

### 17.2.1 Dynamics of Foam and Nonwoven Fiber Material

The wave propagation in fluid saturated porous material was first described by Biot [1, 2]. In these papers Biot describes the coupled propagation of elastic and fluid waves. At that time only basic models regarding fluid wave propagation in porous material were available. Decades later the fluid wave propagation in the skeleton of the porous absorber was developed by Johnson [3] and extended by Champoux and Allard [4]. This equivalent fluid model of wave propagation is called Johnson-Champoux-Allard (JCA) model. The full theory is presented in Allard's textbook [5]. In the context of this paper the porous material theory is presented only to that extend that the general differences between foam and fiber material can be understood. Please refer to [5] for the detailed theory.

The PUR foam material consists of an open cell foam with specific elastic wave properties usually assumed to be isotropic. In vacuo the spring stiffness is determined by the complex Young's modulus of the skeleton $\boldsymbol{E} = E(1 + j\eta)$. $E$ is the real valued Young's modulus and $\eta$ the damping loss factor of the material. Typical values of viscoelastic PUR foams in vacuo are $E = 90 \ldots 150$ kPa and $\eta = 20\%$. When the pores are filled with air there is an additional stiffness in parallel to the foam skeleton. The effective spring modulus is in the range of $E_{\text{eff}} = E + K_{air} \approx 200–260$ kPa (Fig. 17.1).

In contract to this the fiber material is modelled as an equivalent fluid and the spring properties are determined by the acoustic properties of air and the parameters of the JCA model. The key parameter is the bulk modulus $K$ given by:

$$K(\omega) = \frac{\frac{\rho_0 c_0^2}{\Phi}}{\kappa - (\kappa - 1)\left[1 + \frac{8\eta_d}{j\Lambda'^2 \,\text{Pr}\, \omega\rho_0}\sqrt{1 + j\rho_0 \frac{\omega\, \text{Pr}\, \Lambda'^2}{16\eta_d}}\right]^{-1}} \overset{\omega \to 0}{\longrightarrow} \frac{\rho_0 c_0^2}{\Phi\kappa} = 110 \text{ kPa} \tag{17.1}$$

With $\rho_0$, $c_0$, $\kappa$, Pr and $\eta_d$ being the density, speed of sound, ratio of specific heat, the Prandtl number and the dynamic viscosity of air, respectively. $\Phi$ and $\Lambda'$ are the volume porosity and the thermal characteristic length. These are the only constants related to the skeleton. From this model two consequences follow:

1. The spring stiffness of fiber material is given by an isothermal (denoted by the $1/\kappa$ term) air spring and therefore lower than the PUR + air spring stiffness.
2. The options for introducing damping in Eq. (17.1) are limited.

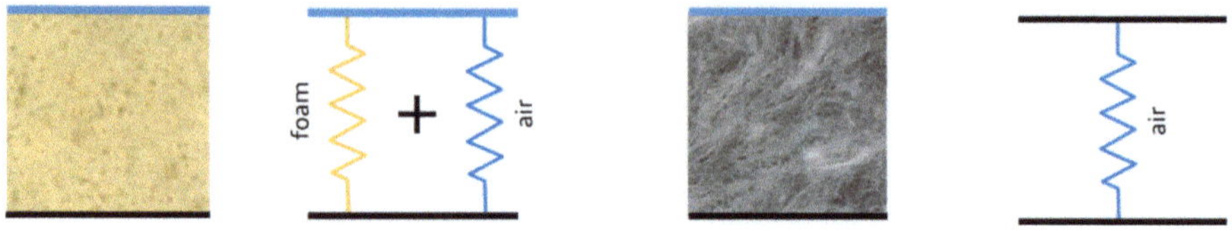


**Fig. 17.1.** Material structures and stiffness equivalent circuit of foam and fiber spring layer

To conclude, due to the different physics of wave propagation both spring materials exhibit different stiffness properties. The foam with higher stiffness and damping, the fiber with less stiffness and damping.

### 17.2.2 Acoustic Transmission Through Double Wall Structures

The double wall or double leaf structure is the most efficient passive means of noise insulation [6]. With similar mass a much higher transmission loss compared to single walls is achieved above the double wall frequency.

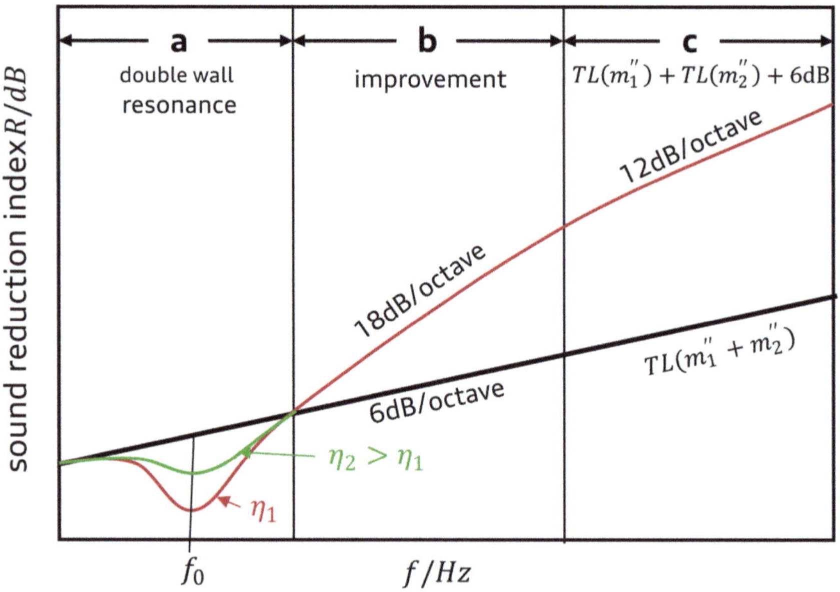


**Fig. 17.2.** Transmission loss of single and double wall systems

In Fig. 17.2 a typical transmission loss of a double wall set-up is compared to the curve of a single wall with same overall weight. Above the double wall resonance $f_0$ the mass-spring system performs better than a single wall. The resonance frequency for perpendicular sound waves is estimated by [8].

$$f_0 = \frac{1}{2\pi}\sqrt{\frac{s''}{\mu''}} \text{ with } \mu'' = \frac{m_1''m_2''}{m_1''+m_2''} \underbrace{\approx}_{} m_1'' \gg m_2''m_2'' \tag{17.2}$$

$s''$ and $\mu''$ are the stiffness per area and the effective mass per area, respectively. $m_1''$ and $m_2''$ are the masses per area of both layers of the double wall. The double prime denotes a quantity per unit area. The mass of the spring is neglected in this treatment. In case of elastic materials, the spring stiffness is determined by the young modulus $E$ and the bulk modulus $K$, in case of fiber material that is modelled as equivalent fluid the stiffness results from the bulk modulus $K$ of the air spring and the thickness $t$ of the spring material.

$$s'' \approx \frac{E+K}{t} \text{ or } s'' = \frac{K}{t} \approx \frac{\rho_0 c_0^2}{\kappa t} \tag{17.3}$$

At the resonance frequency the mass-spring system performs less than the single wall. The countermeasure to the drop in transmission loss is to introduce damping into the spring layer. The higher the damping the less serious the resonance drop.

In case of viscoelastic PUR foams this drop is reduced by the damping in the solid phase of the foam. For fiber material this is not an option because of the discussions in Sect. 17.2.1.

### 17.2.3 Comparison of PUR and Fiber Mass-Spring Systems

A first comparison is performed based on transmission loss tests according to ISO 15186 [10]. In Fig. 17.3 the transmission loss of mass spring noise control treatments with two heavy layers of different weight and different spring material, PUR and fiber, are compared. Note the lower double wall resonance of both fiber NCT leading to better high frequency performance in combination with a strong resonance drop. In contrast to this, the resonance of the PUR NCT occurs at higher frequencies but with moderate drop and lower performance at high frequencies.

The options to increase the damping in the nonwoven material via fiber diameter variations and porosity is investigated at three different version (Table 17.1) as shown in Fig. 17.4. The strong variation of fiber properties has low impact on frequency and shape of the resonance frequency drop. Thus, the discussions in Sect. 17.2.1 are verified and the shows that this is not an option for efficient resonance drop mitigation.

Beside the superior high frequency performance of fiber material, the degradation at the double wall resonance is too strong for automotive applications

**Table 17.1.** Fiber properties of spring material in Fig. 17.4

| Material | Porosity | Mass per length/dtex |
|---|---|---|
| Extra fine PES | 0.999 | 0.2 |
| PES Fiber 200 g/m$^2$ | 0.97 | 1.87 |
| PES Fiber 500 g/m$^2$ | 0.925 | 4.2 |

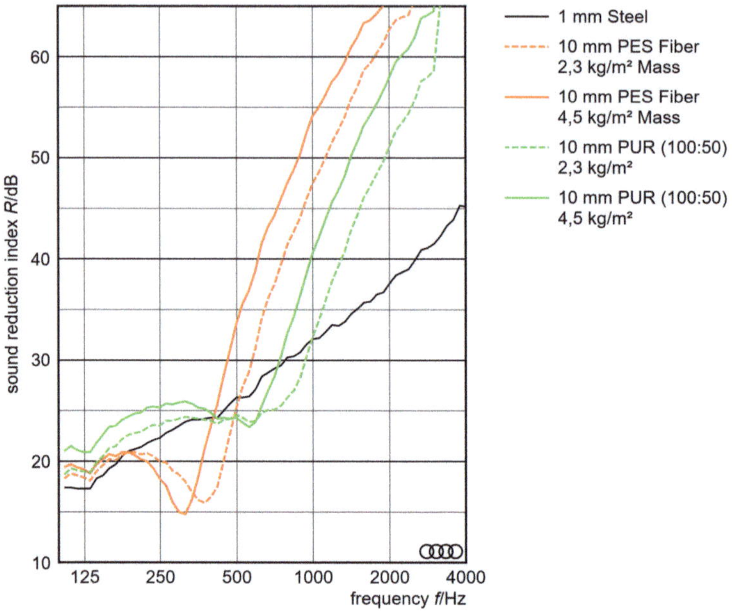


**Fig. 17.3.** Comparison of PUR and fiber mass-spring systems

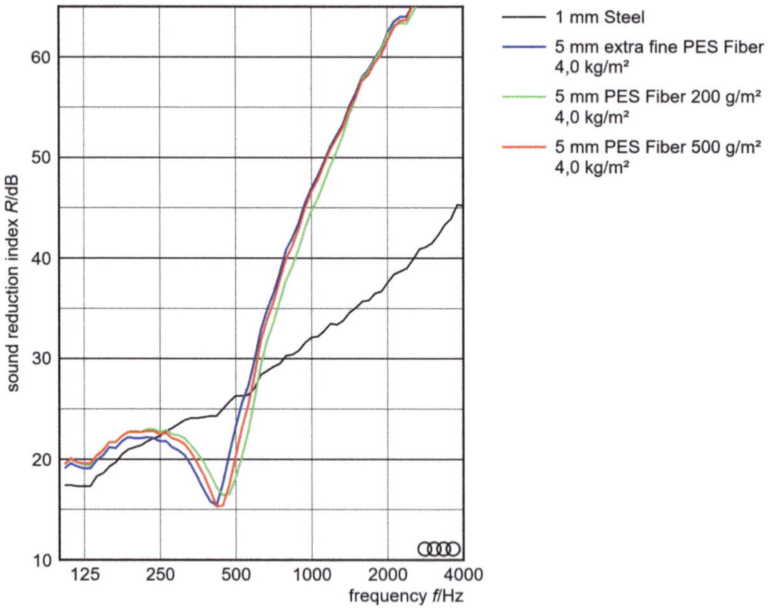


**Fig. 17.4.** Transmission loss of three different mass-spring NCT using different fiber materials

### 17.2.4 Permeable Heavy Layer

One option to overcome this dilemma is to use a heavy layer with customized flow resistance. This paper concerns a selected permeability that reduces the double wall drop. The resistance values required for this purpose are much higher than those for surface absorption designs. In the following the theory and basic principle of this mass control is presented and compared to test results of PUR and nonwoven noise control packages.

### 17.2.5 Transfer Matrix Modelling

The performance of noise control treatments such as mass-spring systems applied to steel plates is modelled by acoustic material models of porous and elastic materials as summarized in the book of Allard [5]. The wave propagation through a layup of multiple infinite layers of such materials is simulated with the transfer matrix method (TMM) also described in [5] or in a simpler version by Peiffer [6] using pure equivalent fluid degrees of freedom: pressure and velocity. All cases are simulated using the open-source *python toolbox for vibroacoustics* (pyva™) [7]. This toolbox implements the TMM from [5] and the hybrid method [6] in Python and allows simulating the transmission coefficient or transmission loss (TL) of plates and noise control treatments. In case of hybrid FEM/SEA simulation the plate is modelled as a deterministic finite element model (FEM) using the hybrid coupling loss theory of statistical energy analysis (SEA).

### 17.2.6 Modelling of Porous Heavy Layer

The plate layer without shear stress contact to the spring material is modelled by the transfer impedance $z_t$ using the degrees of freedom of the equivalent fluid model: pressure $p_i$ and the normal velocity per unit area $v = v_1 = v_2$ (Figs. 17.5 and 17.6).

$$z_t = \frac{p_1 - p_2}{v} = \frac{\Delta p}{v} \tag{17.4}$$

For the illustration of the general principle, it is sufficient to model the heavy layer as an ideal limp layer with the mass transfer impedance $z_m$.

$$z_m = j\omega m_2'' \tag{17.5}$$

Considering a heavy layer that is permeable the velocity per unit area is shared between heavy layer velocity $v_m$ and the fluid velocity per unit area through the pores $v_{air}$, hence $v = v_{air} + v_m$.

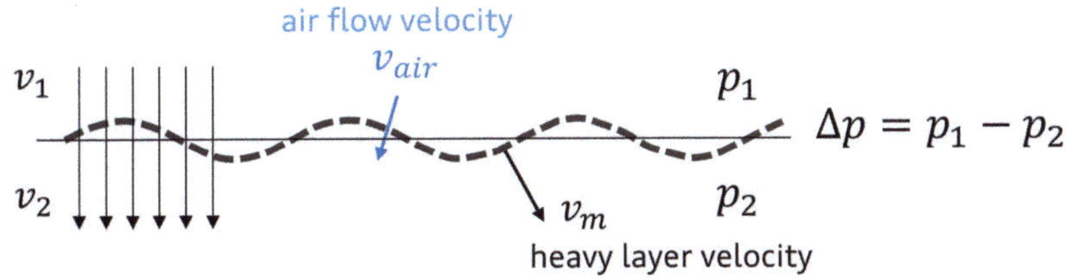


**Fig. 17.5.** Pressure and velocity convention of a perforated heavy layer

The permeability is determined by the specific flow resistance $R_s$ that is considered as frequency independent for the sake of simplicity. The transfer impedance of the pervious layer follows from the parallel arrangement of resistance and mass impedance as shown in Fig. 17.6. The effective transfer impedance of a perforated layer reads as:

$$z_{t,\text{eff}} = \frac{j\omega m_2'' R_s}{R_s + j\omega m_2''} \tag{17.6}$$

Obviously, for high resistivities $R_s \gg j\omega m_2''$ the transfer impedance is equal to the mass transfer impedance $z_{t,\text{eff}} \rightarrow z_m$. For low resistance $R_s \ll j\omega m_2''$ the air flow dominates the impedance $z_{t,\text{eff}} \rightarrow R_s$.

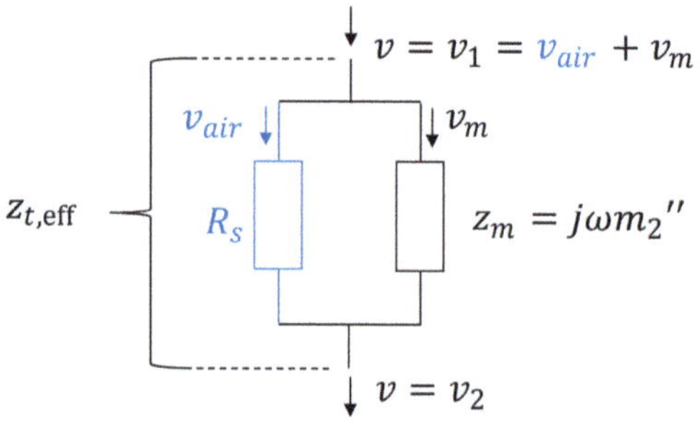


**Fig. 17.6.** Equivalent circuit of the perforated heavy layer.

One may define the frequency $\omega_{1/2}$ at which half the mass is active. This frequency is found when the transfer reactance equals half the reactance of the impermeable heavy layer:

$$\frac{m_2''}{2}\omega_{1/2} = Im\{z_{t,\text{eff}}(\omega_{1/2})\} \Rightarrow \omega_{1/2} = \frac{R_s}{m_2''} \tag{17.7}$$

One consequence of this mass reduction is that the double wall resonance is reduced according to Eq. (2) when the heavy layer is permeable. This mass control feature is the key to the double wall drop reduction for mass-spring systems with fiber springs. Normalizing Eq. (17.5) with reduced frequency $\omega' = \omega/\omega_{1/2}$ reads:

$$z_{t,\text{eff}} = \frac{j\omega' R_s}{1 + j\omega'} \text{ with } \omega' = \frac{\omega}{\omega_{1/2}} \tag{17.8}$$

In Fig. 17.7 the slope of the transfer impedance normalized with respect to $R_s$ is shown revealing that the maximal effective reactance occurs at $\omega_{1/2}$.

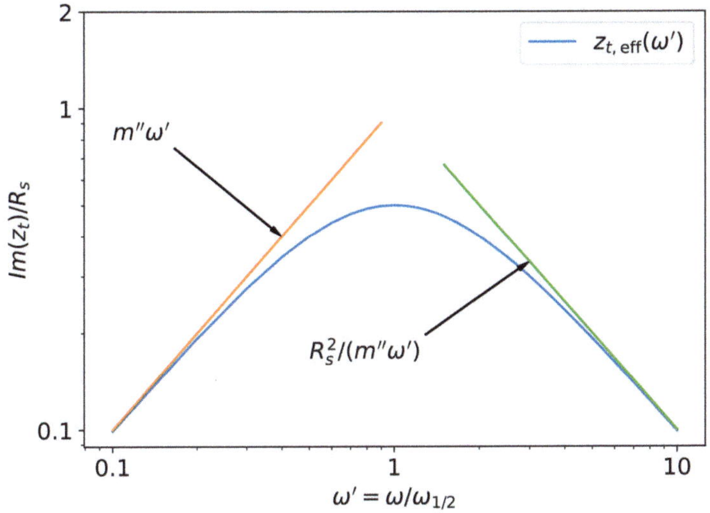


**Fig. 17.7.** Imaginary part of permeable heavy layer

## 17.3 Parameter Study

In the following the principal correlation between resistance, effective mass and double wall resonance with respect to the resulting transmission loss is derived. From these considerations a clear design rule is established enabling high performance mass spring systems with nonwoven fiber spring material.

### 17.3.1 Simulation Results

The transmission loss is investigated based on a set-up as show in Fig. 17.8.

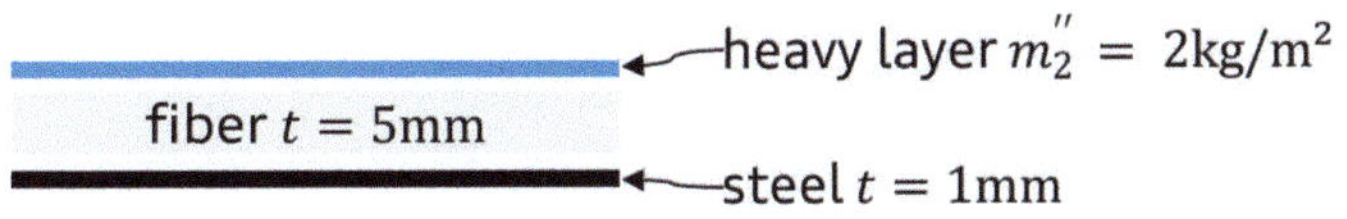


**Fig. 17.8.** Set-up of simulation configuration

The fiber layer is modelled as an equivalent fluid using the limp version of the Johnson-Champoux-Allard (JCA) model [3, 4, 9]. The transmission loss is calculated using angular averaging [1, 6] of plane wave transmission. With the given parameters and an estimated stiffness of the fiber bulk modulus of $K \approx \frac{\rho_0 c_0^2}{\Phi \kappa} = 110$ kPa the following double wall resonance frequency is given.

$$\omega_0 = \sqrt{\frac{\rho_0 c_0^2}{t\mu''}} \approx 2\pi\, 600\, \text{s}^{-1} \tag{17.9}$$

Note, that this estimation is only correct for normal incidence. The diffuse field double wall frequency is supposed to be lower.

**Table 17.2.** Parameters of fiber material

| Parameter | Symbol | Value | Unit |
|---|---|---|---|
| Porosity | $\Phi$ | 0.92 | - |
| Flow resistivity | $\sigma$ | 45000 | Pa s m$-^2$ |
| Bulk density | $\rho_{bulk}$ | 50.0 | kg m$^{-3}$ |
| Tortuosity | $\alpha_\infty$ | 1.5 | - |
| Viscous characteristic length | $\Lambda$ | 56 | μm |
| Thermal characteristic length | $\Lambda'$ | 122 | μm |
| Fluid | | Air | |

The diffuse field transmission loss results are depicted in Fig. 17.9. As reference the transmission loss of the single 1mm steel plate is denoted by the dotted line. The impermeable line reveals the known shape of fiber systems, and the transmission loss is nearly 10dB lower at resonance than the single plate.

The relationship between the required resistivity and the mass per area and the double wall resonance is supposed to be linear.

$$R_s = a_s m_2'' \omega_0 \tag{17.10}$$

Numerical studies reveal that the best compromise is at $a_s = 2$ but a range of $a_s \in [1.5, 3.0]$ is also a good compromise between reduced double wall transmission loss drop and acceptable loss in high frequency performance. The resulting transmission loss of this range also shown in Fig. 17.9. Note that with this rule the efficient mass $m''_{\text{eff}}$ follows from the imaginary of $z_{t,\text{eff}}$:

$$m''_{\text{eff}}\omega_0 = Im\{z_{t,\text{eff}}(\omega_0)\} = \frac{m''_2\omega_0}{1 + \frac{1}{a_s^2}} \tag{17.11}$$

The efficient mass that is operative at the original resonance frequency $\omega_0$ is at 70–90% of $m''_2$. This generates a shift of the new double wall resonance to slightly lower frequencies. Since the trough is less steep this frequency cannot be well identified. In this case this leads to the resistance values of $R_s = 15079 \ldots 22619\,\text{Pa}\,\text{s}\,\text{m}^{-1}$ . With this resistance range the resonance collapse is less than 5 dB at an acceptable reduction in high frequency transmission loss.

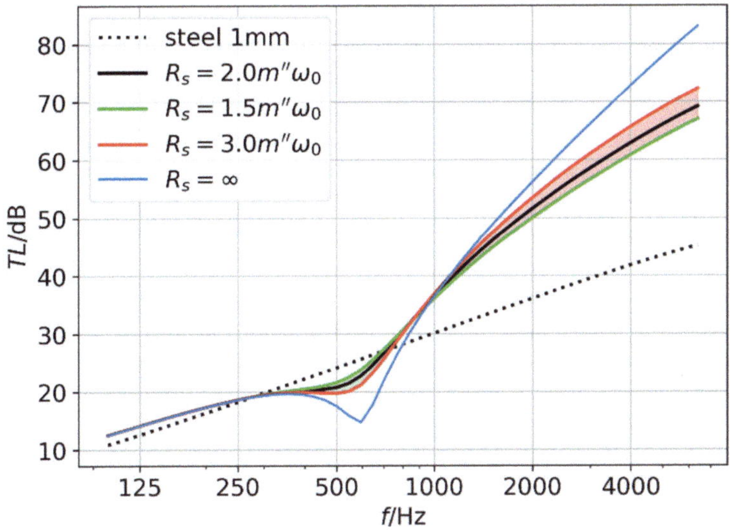


**Fig. 17.9.** Transmission loss of 1mm steel pure and with various mass-spring systems.

To conclude, carefully controlling the heavy layer mass by means of flow resistance replaces the spring damping and enables the use of environmentally friendly fiber material. It is important to understand, that the resistivity values are in completely different value ranges compared to absorption designs. Typical values for absorber applications are $R_s = 600 \ldots 1200\,\mathrm{Pa\,s\,m^{-1}}$. In addition, and due to the lower spring stiffness the mid and high frequency performance is better than the PUR foam transmission loss.

### 17.3.2 Experimental Results

To prove the concept experimentally two permeable mass layers are tested in the transmission loss test bench at Audi.

**Slitted Heavy Layer**

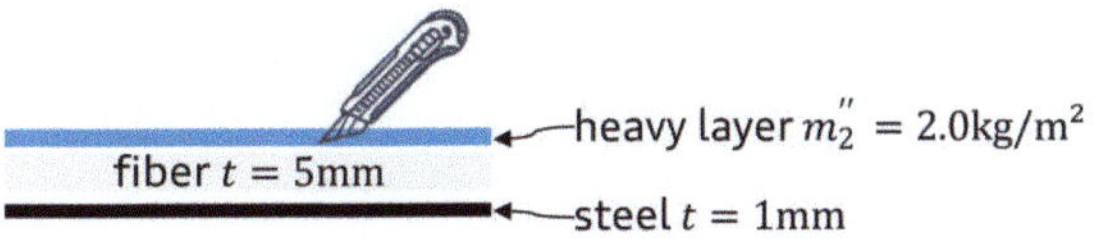


**Fig. 17.10.** Picture and Layup of slitted heavy layer

A conventional heavy layer is modified by several cuts as depicted in Fig. 17.10. The closure of the cuts is prevented by small paper pieces. The estimated resistance due to the slits is $R_s \approx 20\,\text{kPa}\,\text{s}\,\text{m}^{-1}$. The fiber properties are given in Table 17.2. In Fig. 17.11 the measured and simulated transmission loss is shown. The poor agreement at low frequencies follows from the fact that the implemented transfer matrix method does not consider the aperture of the specimen (1.2 x 1.0 $\text{m}^2$). Even though the resistance is estimated, and fiber material parameters were not precisely diagnosed, the high frequency simulation and test results agree well. This further confirms the fact, that the fiber spring properties are mainly determined by the air properties.

The permeable heavy layer effect is confirmed as the resonance drop is reduced without losing too much high frequency performance. According to Eq. (17.10) the selected resistance corresponds to $a_s = 2.6$ which is in the proposed resistance interval.

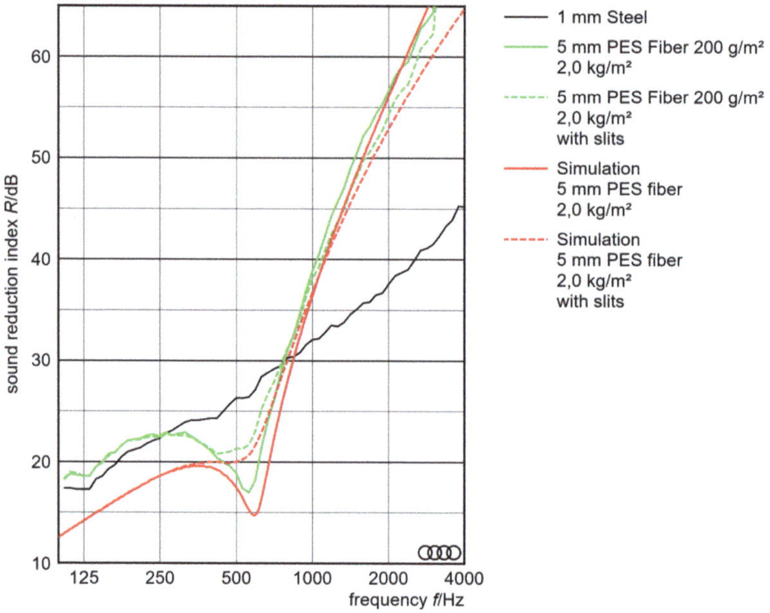


**Fig. 17.11.** Measured and simulated transmission loss of slitted mass layer mass spring system

**Fig. 17.12.** Test specimen of compressed PES fiber layup (by courtesy of Autoneum)

**Compressed PES Heavy Layer**

An industrial solution was created by a compressed PES layer that serves as heavy layer added to a 20mm spring layer (Figs. 17.12 and 17.13). By using different compression rate three resistance values are realized: $R_s = 6080$, $9825$ or $85000\,\mathrm{Pa\,s\,m^{-1}}$. The resistance was measured using a flow resistance tester according to ISO 9053-1 [11]. In this case the hybrid FEM/SEA method [6] was used to calculate the transmission loss. Thus, the plate is modelled as a finite element system, the noise control treatment by the transfer matrix method. The transmission loss is calculated using the hybrid formulation of coupling loss factors as described in detail in Sect. 11.3 of [6]. With this approach, the aperture of the test specimen is included.

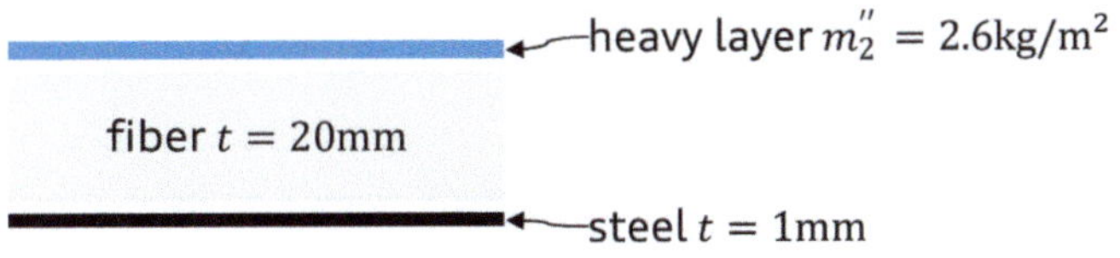


**Fig. 17.13.** Layup of PES heavy layer configuration

With the given parameters the double wall resonance is at $\omega_0 = 2\pi\,270\,s^{-1}$. The optimal resistance according to Eq. (17.10) would be $R_s = 8820\,\mathrm{Pa\,s\,m^{-1}}$. The given values correspond to $a_s = 1.4$, $2.2$ or $19.3$. The first two are in the required range.

The solid lines in Fig. 17.14 show the simulation results and the dotted the test results. The high frequency results agree very well showing that the reduction of transmission loss in this regime is precisely calculated. In the frequency regime near the resonance the test and simulation results don't coincide perfectly, but the fundamental effect is well represented and in case of the test results the positive impact of the resistance control is even better than predicted.

One reason for the difference may be that the plate is modelled as FEM system, but the mass spring system is described by a discrete but infinite representation [6].

In general, this proves that a practical implementation fulfils exactly the expected and intended result. The results for the high resistance value illustrate that high resistance layer are like impermeable layers as expected.

The resistance values of the heavy layer are too high for highly efficient absorption applications. However, an impermeable heavy layer has no absorption, and the permeable design has one further positive impact by providing absorption of $\alpha_s = 22\ldots32\%$ (Fig. 17.15).

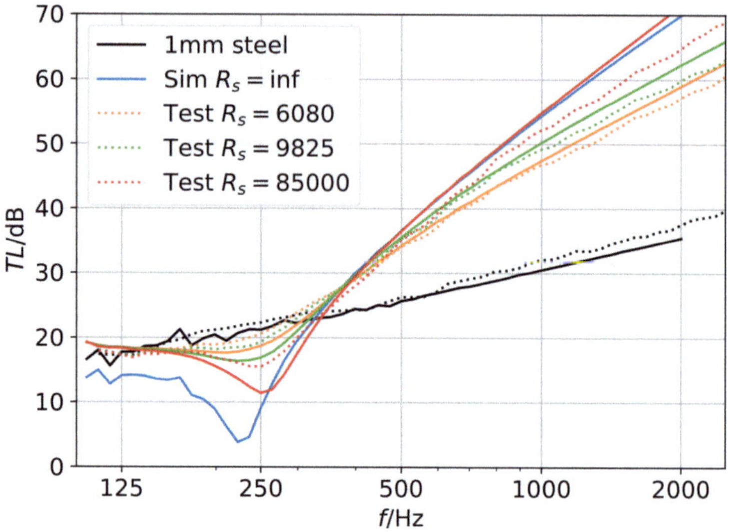


**Fig. 17.14.** Simulated and measured transmission loss of PES heavy layer systems

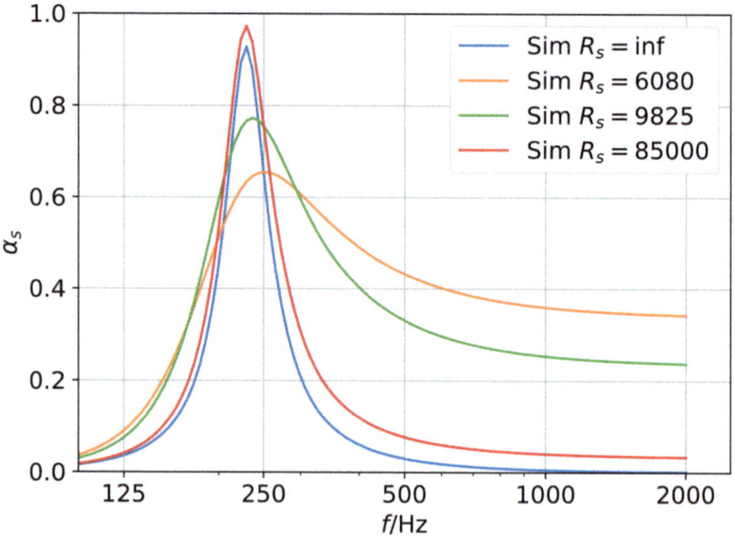


**Fig. 17.15.** Simulated diffuse field absorption of PES heavy layer systems

### 17.3.3 Automotive Application Context

The derivation of the preceding sections is based on ideal and flat configuration without leakages and grommets. Whether the permeable layer concept is working in real car applications, for example a firewall mass spring NCT, was further investigated by Venor and Burghardt [12]. In Fig. 17.16 the insertion loss of a firewall NCT under real world excitation is shown. The PUR NCT shows the medium performance at the double resonance and high transmission loss at high frequencies. The green curve depicts the fiber version with impermeable heavy layer. The resonance drop at lower frequency with better mid frequency is clearly visible. However, the resonance drop is too strong. In contrast to this, the proposed permeable heavy layer designs provide excellent low and mid frequency performance sacrificing 3dB at high frequency and worst case (FR min).

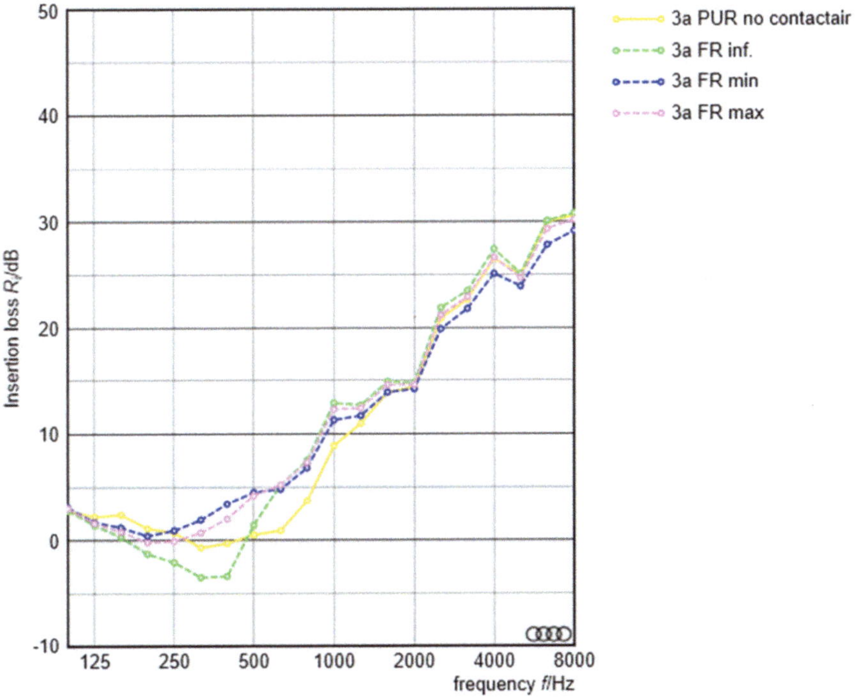


**Fig. 17.16.** Difference of pressure level with PUR and fiber NCT [12]

## 17.4 Conclusions

With permeable heavy layers the performance of premium PUR based mass-spring systems can be achieved or even overachieved due to the lower stiffness of fiber spring systems. This paves the road for environmentally friendly, single material fiber systems with high performance at same or lower weight. The optimal range for the required resistance is usually higher than for absorption layouts but keeping a minimum absorption as positive side effect.

The tolerance interval of the designed resistance is relatively high and does not establish a too strict accuracy. This simplifies industrial solutions and avoids expensive precision requirements.

However, this technique is not restricted to fiber springs but can also improve the transmission loss of PUR or other mass-spring systems. The patent for this permeable design is pending with application number **DE 10 2024 117 099 A1**.

## References


1. M. A. Biot, "Theory of Propagation of Elastic Waves in a Fluid-Saturated Porous Solid. II. Higher Frequency Range", *The Journal of the Acoustical Society of America*, (28), Nr. 2, p. 179–191, 1956.
2. M. A. Biot, "Theory of Propagation of Elastic Waves in a Fluid-Saturated Porous Solid. I. Low-Frequency Range", *The Journal of the Acoustical Society of America*, (28), Nr. 2, p. 168–178, 1956.
3. D. Johnson, J. Koplik, und R. Dashen, „Theory of dynamic permeability and tortuosity in fluid-saturated porous media", *Journal of Fluid Mechanics*, vol. 176, p. 379–402, 1987.
4. Y. Champoux und J.-F. Allard, „Dynamic tortuosity and bulk modulus in air-saturated porous media ", *Journal of Applied Physics*, vol. 70, Nr. 4, p. 1975–1979, Aug. 1991.
5. Allard, J.-F., Atalla, N., Propagation of Sound in Porous Media, 2nd ed. Wiley, UK (2009).
6. Peiffer, A., Vibroacoustic Simulation: An Introduction to Statistical Energy Analysis and Hybrid Methods. JOHN WILEY, New York, USA (2022).
7. Pyva homepage, https://pyva.eu
8. Fahy F., Sound and Structural Vibration: Radiation, Transmission and Response. London, The United Kingdom: Academic Press (1985).
9. Johnson, D., Koplik, J. und Dashen, R., Theory of dynamic permeability and tortuosity in fluid-saturated porous media, Journal of Fluid Mechanics, Bd. 176, p. 379–402, 1987
10. Champoux, Y., Allard, J.-F., Dynamic tortuosity and bulk modulus in air-saturated porous media, Journal of Applied Physics, Bd. 70, Nr. 4, p. 1975–1979, Aug. 1991
11. Doutres, N. et al., Validity of the limp model for porous materials: A criterion based on the Biot theory, Journal of the Acoustical Society of America, (122), Nr. 4, p. 2038–2048, 2007
12. ISO 15186-1:2000 Acoustics—Measurement of sound insulation in buildings and of building elements using sound intensity, Part 1: Laboratory measurements.
13. ISO 9053-1:2018 Acoustics—Determination of airflow resistance—Part 1: Static airflow method
14. Venor, J., Burghardt, M., Reduced flow resistivity in vehicle noise control treatment, to be published.

# Comfort Evaluation of the Introduction of Mono-Material Floor Insulator in D-segment Vehicles

Keun-Young Kim[1], Théophane Courtois[2](✉), Jeongbin Heo[1], Hyundae Cho[1], Kichang Jo[1], Dongju Yoon[1], and Andrew Beam[3]

[1] Hyundai Motor Company, Hwaseong-si, South Korea
{kky816,huio0868,hyundae,dudreamck, toktik}@hyundai.com

[2] Autoneum Management AG, Winterthur, Switzerland
theophane.courtois@autoneum.com

[3] Autoneum North America, Bloomsburg, USA
andrew.beam@autoneum.com

**Abstract.** In this article, we present the results of an exhaustive evaluation on comfort and weight, when replacing a multi-material carpet by a 100% polyester carpet. These results leverage on the outcome of a vehicle sound package optimization prototyping activity, which proves the possibility of reducing the overall weight of the vehicle while improving its NVH performance, by introducing a new polyester fiber-based floor insulator concept. This new concept features a newly developed fiber decoupler, which allows the substitution of conventional spring-mass systems based on a polyurethane foam decoupler and a mass layer, by an entirely polyester fiber-based solution, including a suitable polyester film. The concept is validated by means of flat samples characterization, both for the acoustic insulation and absorption performance and for static compression. In a second step, a part prototyping activity has been performed in order to assess the industrial feasibility of the concept for an existing part already in mass-production for a mid-size vehicle. We present the challenges of the prototyping activity with respect to manufacturing, weight and geometrical considerations, as well as the verification of the carpet characteristics, such as 3d-static compression. The part is finally installed in the mass-production vehicle in order to verify the geometrical and overall comfort quality, including NVH performance, brought by the part. Despite some weight and manufacturing limitations inherent to prototyping constraints, the conclusion of the article will pave the path to industrial implementation thanks to comprehensive weight and scenarios breakdowns.

**Keywords:** Mono-material · Polyester · End Life Vehicle · Vehicle Acoustics · Static compression


A. Heintzel (Ed.): Automotive Acoustics Conference 2025, Proceedings, pp. 261–276, 2026. https://doi.org/10.1007/978-3-658-51249-1_18

## 18.1 Introduction

In order to reach the objectives of climate neutrality in the automotive industry, the governments are already now setting time objectives in terms of post-consumer recycled content in the production of vehicles. For example, the European Green Deal, as part of the vision to achieve the carbon neutrality by 2050, is setting an objective of 25% of postconsumer recycled content to be introduced as the minimum weight of recycled content of plastics in vehicles by 2030. To reach this target, the car manufacturers must already plan the next generation of vehicle development, to be ready with parts, in addition to the battery and motor, that can be dismantled and reused at End-of-Life of vehicle, such as recyclable mono-material trim parts. The carpet and floor insulators systems are considered challenging candidates, due to their multi-material content, and their large weight contribution to overall vehicle weight, while the replacement of such systems must imperatively be done without compromising with weight, comfort and cost in mass production of mid-size vehicles.

The number of polymers used in the construction of vehicle parts must be minimized to allow for the feasibility of the End-of-Life recycling. In the field of sound package parts, the automotive industry is already exploiting or investigating the possibilities offered by the post-industrial polyester recycling loop, as illustrated by the Autoneum Video [1], leveraging on the potential presence of polyester components in a car, such as underbody parts, wheel outer liner parts or trunk trims parts, which can all contribute, together with other post-consumer polyester supplies, to the volumes required for an economical automotive polyester recycling industry. In this article, we investigate the impact on acoustics and static compression of a 100% polyester carpet prototype part in comparison to the corresponding mass-production multi-polymer carpet part. This research activity is a prerequisite to enable the polyester based automotive recycling industry strategy. This article does not deal with the feasibility of recycling at product end of life.

The Table 18.1 summarizes the material composition of the investigated carpet floor insulator. Each layer serves a distinct function: the carpet pile and coating enhance appearance, durability, and cleanability; the substrate felt compressed in a range of 2-3mm adds stiffness and acoustic performance; and the decoupler provides static stability and noise insulation.

**Table 18.1.** Composition of the investigated carpet/floor insulator.

| | Mass production part composition | | Prototype part composition | |
|---|---|---|---|---|
| Bill of Materials | Polymer | Recycled content [%] | Polymer | Recycled content [%] |
| Carpet pile | Polyester | 0% or 70% | Polyester | 70%-80% |
| Carpet coating | Latex and PE | 0% | Polyester | 0% |
| Needled felt | Polyester | 80% | Polyester | 80% |
| Barrier film | PA/PE | 0% | Polyester | 0% |
| Decoupler | Polyurethane foam | 0% | Polyester fibers | 80%-90% |

### 18.1.1 Introduction of a Polyester-Based Carpet Concept for Vehicle NVH

Ensuring vehicle interior NVH is crucial when introducing new floor insulator materials. In a preliminary vehicle optimization study [2], it was shown by means of panel contribution analysis, that it is possible to introduce new sustainable and light materials keeping vehicle NVH targets under control. In this preliminary study, the methodology has been used to define and prototype manually an optimal weight versus performance sound-package including a mono-material polyester carpet. The contribution of the dash and floor silencer carpet handmade prototypes on test vehicle acoustic transfer functions (ATF) was of 0.4 dB improvement from tires averaged excitation to driver ears and 0.9 dB from rear tires to rear right passenger ears without need for material weight increase. The graph Fig. 18.1 shows the overall vehicle NVH performance improvement as a conclusion of this optimization activity.

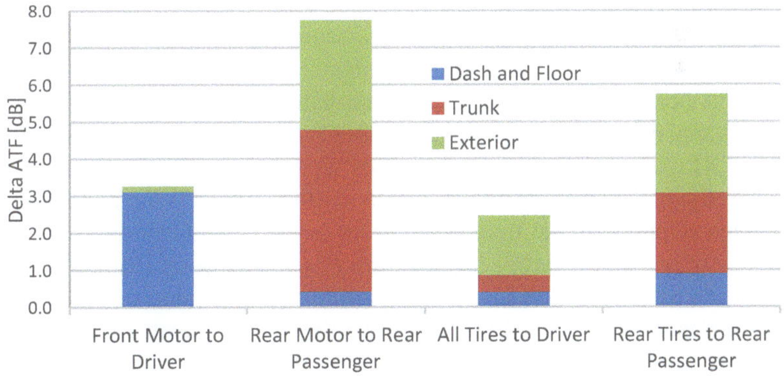


**Fig. 18.1.** Vehicle acoustic transfer functions improvement (Delta ATF) as result of sound-package optimization including a mono-material carpet floor silencer. Tested delta ATF handmade prototype versus original sound package (no weight increase compared to original).

The floor silencer concept used in this preliminary vehicle optimization study featured the substitution of Polyurethane foam decoupler by a polyester air-lay felt decoupler, and the mass backed carpet by a polyester compressed felt backed carpet. Figure 18.2 illustrates the schematic substitution, where the "hard felt" substitutes the mass by means of a needled compressed felt that achieves a minimum stiffness, and where "Decoupler felt" substitutes the foam by means of a soft molded fibers that meets stiffness requirements. All other ingredients of this system, such as the film or the carpet itself and/or the carpet coating must be made of polyester.

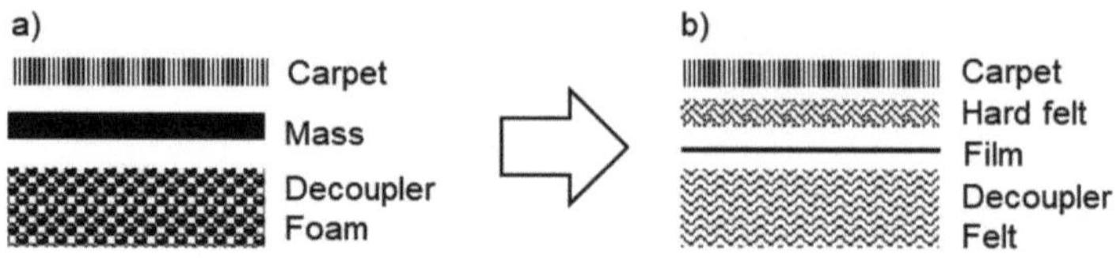


**Fig. 18.2.** Schematic substitution of a foam-mass spring floor carpet system (a) by a fiber-based system (b), allowing for the usage of polyester fiber-based materials.

The fundamental concept of this substitution is based on two key principles: the relative softness of the felt decoupler in comparison to the foam decoupler, and the minimal stiffness of the hard felt coupled with a backing film.

**Felt Decoupler Stiffness** The felt decoupler plays the role of the spring in a so-called spring-mass insulation system, which is a major vehicle floor carpet property to meet the vehicle NVH requirements. A felt in a range of approximately 1000 g/m$^2$ features a lower dynamic youngs Modulus than a viscoelastic foam of approximately 70 kg/m$^3$ typically used for floor applications. Consequently, a floor insulator using felt exhibits a lower Frequency spring-mass resonance and has correspondingly a superior insulation performance compared to one using foam, assuming the top mass remains unchanged. This must be done without reducing static stiffness, which usually declines when dynamic stiffness decreases. Vehicle manufacturers specify a minimum value for static stiffness. A too low static stiffness impacts the quality perception for passengers stepping on the floor. The curves (a), (b), and (c) in Fig. 18.3 show the insertion loss of a 3 kg/m$^2$ heavy layer on top of a viscoelastic foam decoupler having a dynamic stiffness of 285 kPa (curve (a)) compared to that obtained by replacing the viscoelastic foam with a Soft High Resilient "HR" foam decoupler typically used for dash applications with a dynamic stiffness of 55 kPa (curve (b)) and with that obtained by replacing the viscoelastic foam with a felt in this example having a dynamic stiffness of approximately 30 kPa (curve (c)). The spring-mass frequency decreases and the insulation performance improves in higher frequency ranges as the dynamic stiffness decreases.

**Hard felt stiffness** The hard felt utilized as a carpet substrate can substitute the mass in a spring-mass insulation system, provided that the hard felt meets a minimum stiffness for a specific area weight and thickness, and that a film is placed between the hard felt and the decoupler to ensure the necessary spring boundary condition. In the article [3], it is demonstrated how the insertion loss of a spring-mass system featuring a hard felt mass behaves according to a 12 dB/octave in frequency ranges below the quarter wave-length resonance frequency $f_{1/4}$ of the hard felt, according to the formula (18.1), where $E$, $AW$ and $t$ are respectively the compression dynamic modulus, the area weight and the thickness of the hard felt. This formula applies without the presence of carpet on top, but the principle is still valid for the hard felt present in the carpet backing.

$$E = 4 \cdot {f_{1/4}}^2 \cdot AW \cdot t \tag{18.1}$$

Good insulation at high frequency is achievable thanks to proper compression of the hard layer. Figure 18.3 curve (d) shows the insertion-loss of a 2 kg/m$^2$ carpet with hard felt and film on top of the above-described felt decoupler with a dynamic stiffness of about 30kPa. The insertion loss curve shows a 12 dB/octave behavior up to a frequency of about 6300 Hz and it shows levels comparable to those obtained with the HR soft foam decoupler with 3kg/m$^2$ mass layer on top. In summary, using a soft felt decoupler increases the insertion loss more effectively than foam, while compensating for high-frequency loss inherent to the reduced area weight of the carpet hard felt. Considering the sensitivity of vehicle NVH to insertion loss in the range of 1000 Hz to 4000 Hz, this fiber-based concept allows for vehicle NVH improvement at same or lower weight, while using polyester fibers.

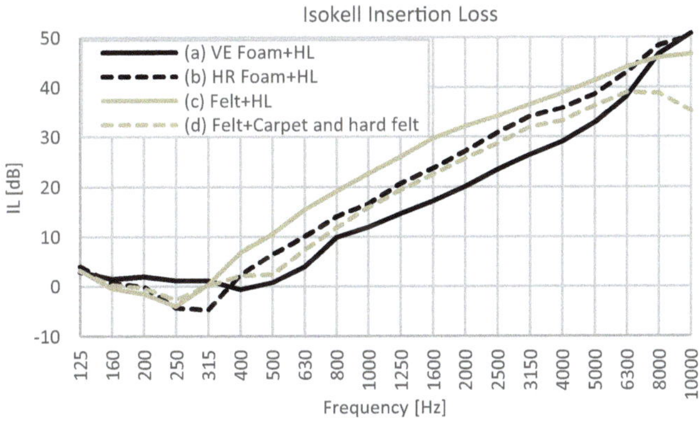


**Fig. 18.3.** Insertion loss performance of 20 mm thick flat samples for different decoupler stiffness and top mass. (a) VE Foam E = 285 kPa and Heavy-layer 3 kg/m$^2$; (b) HR Foam E = 55 kPa and Heavy-layer 3 kg/m$^2$; (c) Felt 1000 g/m$^2$ and E = 30 kPa and Heavy-layer 3 kg/m$^2$; (d) Felt 1000 g/m$^2$ and E = 30 kPa and 750 g/m$^2$ carpet, 750 g/m$^2$ hard felt and 400 g/m$^2$ impervious backing.

## 18.2 Presentation of the Case Study of a D-segment Vehicle Floor Carpet

This paragraph describes the case study, namely the comparison of the mass-production carpet of the test vehicle versus polyester prototype carpet. The test vehicle is a D-segment vehicle in production in North America. The mass-production carpet is in production also in North America and combines a carpet blank made of a tufted carpet with latex adhesive and PE coating, laminated to a needled felt and a non-polyester film, together with a decoupler made of injected polyurethan foam. Figure 18.4 shows the layout of the carpet blank and the foam decoupler. One can see the areas of the carpet blank, which features back-injected foam. The part is assembled in the vehicle with the foam surface laying on the vehicle floor body-panel.

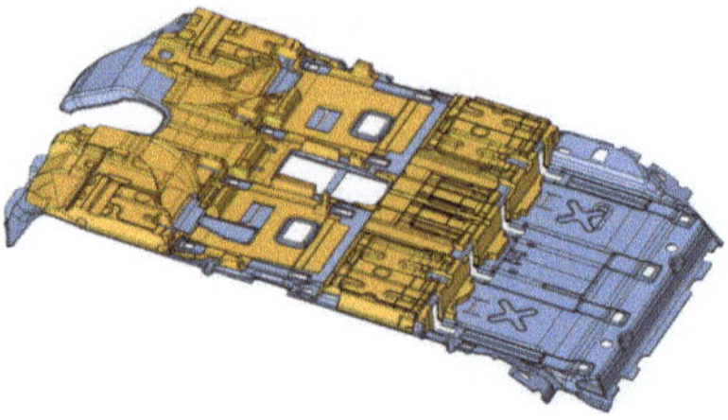

**Fig. 18.4.** D-segment test vehicle's floor carpet layout. The carpet blank is represented in blue, and the foam is represented in orange.

The prototype polyester component features a comparable layout, utilizing the fiber-based concept described in the introductory paragraph. The carpet blank consists of a tufted carpet with co-polyester adhesive laminated to a needled felt and a polyester film. The decoupler includes a type of a novel polyester felt referred to in this study as "High Resilient Felt" (HR felt), which is made from a specific fiber blend and features fibers randomly oriented in a "three-dimensional manner", while traditional standard air-lay felts (due to their web-forming process) feature fibers that are predominantly "in-plane-oriented", i.e., fibers which predominantly lie a plane perpendicular to the thickness direction. This material allows for excellent molding shape-ability, conforming to similar geometry targets as the foam. In addition to the superior shape-ability compared to an air-lay felt type, the so-called HR felt provides higher static compression performance at lower area-weight. Indeed, in this study, the relevance of the weight is crucial. The polyester prototype has been planned with various weight scenarios, as shown in Fig. 18.5. In this study, (a) refers to the mass-production carpet comprising a tufted yarn of 240 g/m$^2$ and using latex and PE backing technology. The polyester material lay-up (b) used for the prototype comprises a 360 g/m$^2$ tufted yarn with coPET (co-polyester) backing technology and 750 g/m$^2$ hard felt (called "FBR" in what follows) available from production. This corresponds to a nominal weight of 1550 g/m$^2$, which is heavier than the reference (a). Meanwhile, the polyester concept (c) assumes the usage of the same tufted yarn FBR as in mass production and a slightly heavier FBR, all in all resulting in the same carpet blank weight as reference (a). While the concept (c) is technically feasible, it has been prototyped using the same carpet blank as reference (a) for practical reasons, rather than employing a coPET backing technology as initially planned. In all versions, including the serial production one, the so-called FBR corresponds to a compressed polyester needled felt. To demonstrate the impact of the low weight HR felt decoupler versus air-lay felt type, the concept (b) has been prototyped with two different decouplers, respectively HR felt 900 g/m$^2$ and air-lay polyester felt 1200 g/m$^2$.

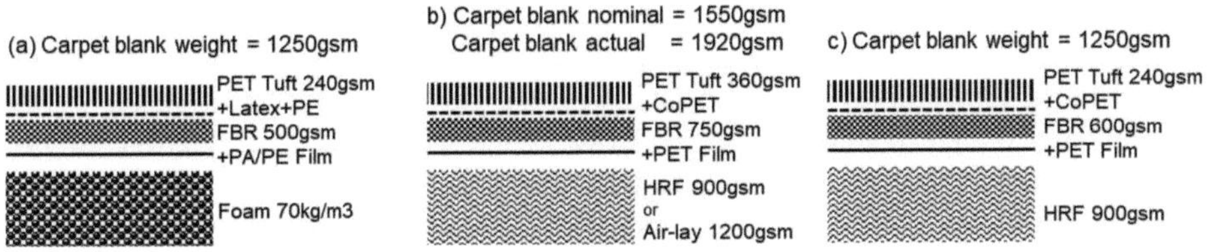


**Fig. 18.5.** A comparative analysis of the composition of mass-produced floor carpet with the material lay-up (a) against two polyester material lay-ups: lay-up (b) corresponds to the polyester prototype; lay-up (c) corresponds to the iso-weight polyester concept.

## 18.3 Flat Sample Performance

### 18.3.1 Acoustic Performance

To evaluate the acoustic performance of the polyester concept compared to the mass-production reference, several flat samples have been produced. All flat samples have been manufactured using a non-industrial process. The carpet blanks and decouplers were molded separately and then bonded together with spray-applied pressure-sensitive adhesive. The estimated added weight of the spray adhesive is 100 g/ $m^2$. In the reference sample (Fig. 18.5(a)), the foam decoupler was made using the production foam recipe, which has a density of 70 kg/m3 and a dynamic Young's modulus of about 150 kPa. Table 18.2 summarizes the tested flat samples and their main properties. In the prototypes sample (Fig. 18.5(b)), all layers are effectively made of mono-material polyester, but with higher weight than reference. In the polyester concept sample (Fig. 18.5(c)), we utilized a mass-produced carpet to illustrate the effect of maintaining the same weight.

**Table 18.2.** List of flat samples tested for acoustic performance. The blank weights are actual weights, taking into account additional prototype adhesive, and differ from nominal weights.

| | | Mass production | Polyester proto with Air-lay | Polyester proto with HRF | Iso-weight |
|---|---|---|---|---|---|
| Sample | Total weight | 2500 g/$m^2$ | 3300 g/$m^2$ | 3000 g/$m^2$ | 2300 g/$m^2$ |
| | Total thickness | 21.5 mm | 22 mm | 22 mm | 21.5 mm |
| Carpet blank | Blank weight | 1240 g/$m^2$ | 1920 g/$m^2$ | 1920 g/$m^2$ | 1240 g/$m^2$ |
| | Tuft | 240 g/$m^2$ | 360 g/$m^2$ | 360 g/$m^2$ | 240 g/$m^2$ |
| | Backing | Latex | coPET | coPET | Latex |
| | FBR | 500 g/$m^2$ | 750 g/$m^2$ | 750 g/$m^2$ | 500 g/$m^2$ |
| Decoupler | Material | Foam | Air-lay | HRF | HRF |
| | Thickness | 15 mm | 15 mm | 15 mm | 15 mm |
| | Weight | 70 kg/$m^3$ | 1250 g/$m^2$ | 950 g/$m^2$ | 950 g/$m^2$ |
| | Dynamic stiff | ~150 kPa | ~30 kPa | ~85 kPa | ~85 kPa |

Figure 18.6 illustrates the effect that the carpet blank layer has on the absorption coefficient of various samples measured in the Alpha-Cabin reverberant room. The absorption of the samples mainly depends on the carpet blank that is on top of the sample. Two sample results are presented here: one is the mass production sample, and the other is a polyester prototype with air-lay felt decoupler, featuring respectively latex-PE and coPET backing. For coPET backing, absorption above 1000 Hz is higher because the coPET adhesive allows acoustic energy to be absorbed through both the tuft and FBR layers. In contrast, with latex-PE backing, acoustic energy is dissipated only through the tuft. Therefore, the mono-material polyester carpet demonstrates better absorption compared to the conventional one. Below 1000 Hz, the absorption is driven by the spring-mass effect interference and has lower relevance for the final vehicle NVH.

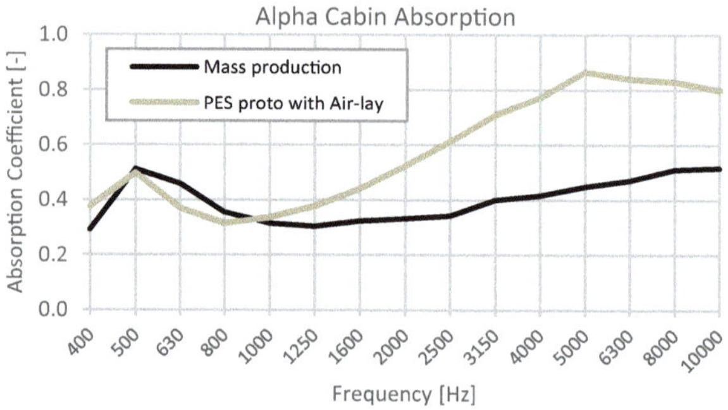


**Fig. 18.6.** Alpha Cabin absorption of flat samples featuring the mass production latex carpet and the mono-material polyester carpet.

Figure 18.7 presents the insertion-loss of the different flat samples, measured in the Isokell double room. All samples show an insulation slope of about 12 dB/octave above the spring-mass resonance, due to proper FBR compression. This high frequency limit is lower in case of 750 $g/m^2$ FBR (~6300 Hz) compared to 500 $g/m^2$ due to the higher compression requirements to comply with Eq. (18.1). The iso-weight sample utilizes the same carpet blank as the mass production sample. Nonetheless, due to the inclusion of the HR felt decoupler, its insulation performance is enhanced by up to 10 dB compared to the mass production sample. In the case of the two mono-material polyester prototype samples, the better insulation is, in part, the result also of higher carpet weight. Overall, this suggests that a sample with the same carpet blank as the polyester prototype, but with same weight as the mass production carpet, and an HR felt decoupler would perform similarly to the tested iso-weight sample. Below 315 Hz, the shift of the spring-mass resonance to lower frequencies translates into lower insulation. Considering that structure-borne noise is predominant in vehicle NVH within this frequency range, the performance of the polyester sample should be assessed using a suitable method for evaluating the structure-borne noise performance of insulation components, as outlined in the article [4].

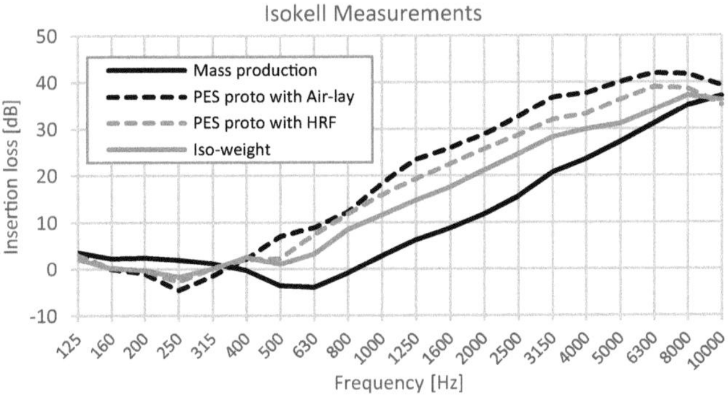


**Fig. 18.7.** Isokell measurement of flat samples featuring the mass production latex carpet and the mono-material polyester carpet, as well as foam and felt decoupler (according to Table 18.2).

### 18.3.2 Static Compression Stiffness Performance

An essential comfort function that should be maintained with the introduction of a mono-material floor carpet system is the static compression stiffness performance, as described in the introduction. This is a challenge since felt materials, with lower dynamic stiffness than foam for acoustic reasons (paragraph 3.1), may result in statically softer materials. Compared to standard air-lay felts, the polyester HR felt of this study enhances the resistance to compression through its unique fiber orientation structure, while substantially preserving the dynamic spring effect required for effective acoustic wave dissipation. Figure 18.8 illustrates examples of those fiber orientation structures. Compared to other fiber structures, the specific HR felt structure of this study is advantageous because it can be manufactured by exploiting existing air-lay felt production processes already in place for conventional cotton felts.

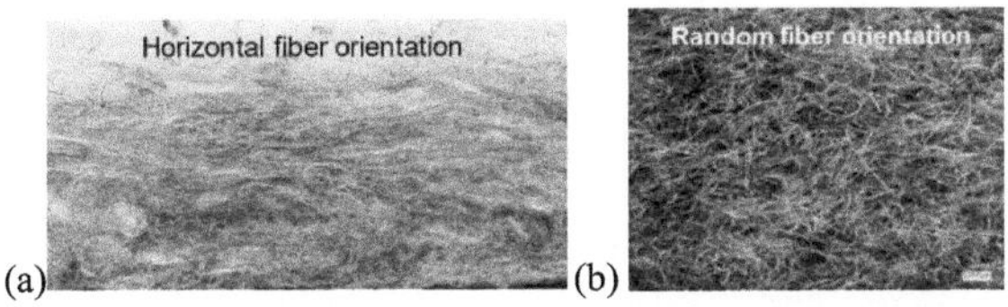


**Fig. 18.8.** Example of polyester decoupler felt structures. The left image (a) is representative of so-called polyester air-lay felt, also used as prototype in this study. The right image (b) is representative of random fiber orientation, such as the HR felt of this study.

Car manufacturers typically assess carpet system quality by measuring the static stiffness of the decoupler layer beneath the carpet, which significantly influences static compression. The most commonly used test standard, referred to as CLD40 (Compression Load Deflection), evaluates both the stiffness and the resilience effect after multiple compression cycles, as detailed in the ISO standard [5]. Figure 18.9 illustrates a comparative analysis of the CLD40 values for polyester HR felt at varying densities in comparison to commonly used foam in floor carpet applications, including the specific type of foam employed in the mass production component of this study. It is interesting to remark that the CLD40 values of the felt at densities with an area weight of 900 g/m2 compressed from 35 mm to 20 mm match those of viscoelastic foams. A typical air-lay felt would be able to match those values at much higher densities, and therefore at higher area weight. This indicates that HR felts, thanks to their specific fiber orientation and structure, may be used to replace foams with weight advantage compared to air-lay felts.

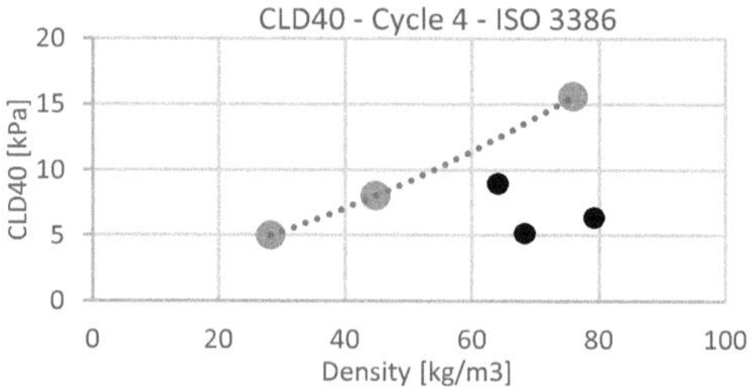


**Fig. 18.9.** Values and trend-line of CLD40 (according to standard ISO 3386 [5]) for the High Resilient Felt (●) in function of the felt density, in comparison to a sampling of several typical viscoelastic foams (•).

## 18.4 Part Performance

To evaluate the feasibility and comfort performance of the mono-material polyester concept in the D-segment mass production vehicle, prototype parts have been manufactured. The material compositions of the different prototypes are identical to those of the flat samples, as described in Table 18.2. The prototypes have been manufactured by using the mass-production carpet tooling with some modifications of the molding process to accommodate for the change of materials of the carpet blank. In that case, the carpet blank comprises the full pre-laminated carpet tuft with backing, FBR hard felt and film. A proto tool has been designed and produced to accommodate felt molding for the decoupler. Similar to the flat samples, the molded decoupler has been manually applied to the molded carpet blank using sprayable pressure sensitive adhesive. The approximate weight of the sprayed adhesive is 100 g/m$^2$. It is crucial to note that while this method is suitable for prototyping, it does not reflect an actual efficient process, which would be typically used in conditions of industrial mass-production but is sufficient for the purpose of the study.

Table 18.3 summarizes the weight variation of the different prototypes versus serial part weight, including carpet blank and decoupler. As stated in paragraph 2, the polyester prototype carpet is heavier than the mass-produced carpet. This difference in weight is not only attributable to the available materials sourced from existing production but also due to prototyping constraints. Specifically, an additional 300 g/m$^2$of adhesive was used for the pre-lamination of film to the FBR hard layer though this wouldn't be needed in conditions of industrial mass production. The nominal versus actual weight indicated in the Table 18.3 reflects this difference in weight. The decoupler coverage also affects the part weight under prototyping conditions. Figure 18.10 illustrates the prototype's felt covering the entire carpet surface, in contrast to the foam used in mass production carpets that have "no-decoupler" areas and where the felt is fully compressed. This setup may be useful in industrial conditions for NVH performance to offset acoustic leaks in areas without decouplers. It is also possible to pre-trim the molded felt with a water jet to fit the serial foam design.

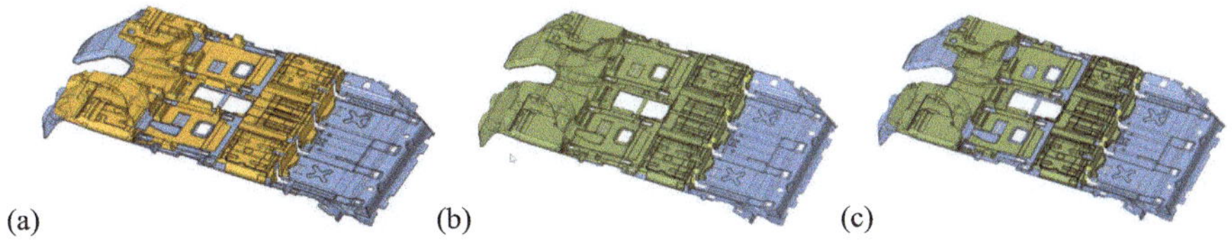


**Fig. 18.10.** Decoupler coverage. (a) Foam serial coverage; (b) Felt proto coverage (better NVH); (c) Feasible felt coverage (optimal weight).

**Table 18.3.** Weight breakdown of the manufactured floor carpet prototypes (estimated excl. Accessories, incl. Adhesive). ***Tested weight.*

| Weight in [kg] | | Mass production | Polyester proto with Air-lay | Polyester proto with HRF | Iso-weight |
|---|---|---|---|---|---|
| Carpet | Actual | 4.2 | 6.4 | | 4.2 |
| | Nominal | | 5.4 | | |
| Adhesive | Actual (spray) | - | | 0.3 | |
| Decoupler | Actual | 2.5 | 3.0 | 2.3 | 2.2 |
| | Optimal | | 2.2 | 1.7 | 1.7 |
| Total | Actual | **6.7**** | 9.7 | **9.0**** | **6.7**** |
| | Nominal/ Optimal | | 7.6 | 7.1 | 5.9 |

In conclusion, the prototype incorporating the HR felt is 34% heavier than the serial part, whereas it is expected to be nominally only 6% heavier. Using an iso-weight carpet ensures the prototype's weight matches the serial part and can even offer a 12% weight advantage with optimal decoupler coverage, due to the felt area weight compared to foam density.

### 18.4.1 3D Static Compression Performance

The prototypes have been used to perform static compression performance evaluation at part level. The evaluation detailed in paragraph 3.2 indicates that the HR felt decoupler exhibits comparable or superior CLD40 performance when compared to the foam material, making it a viable material solution for substituting the foam. Moreover, it has also already been demonstrated that not only the decoupler, but also the presence of the carpet and eventually its stiffness, as well as the 3D configuration with boundary conditions contribute to the overall part compression performance [6]. In fact, certain automobile manufacturers already evaluate static compressions, in vehicle test conditions, for all newly developed components.

The testing method applied in this study is an adaptation of the part specification test [7]. The floor carpet component lays on the vehicle body panel. A compression displacement is applied by means of an 80 mm diameter indenter, which simulates foot pressure on the carpet, with force measurements being monitored throughout the process. The test is conducted by placing the indenter in a location where the decoupler measures 35 mm in thickness. A single compression cycle up to 40% is considered. In this setup, the result depends on both the decoupler compression and the carpet deformation. For testing the prototypes, a 3D-printed test jig featuring the geometry of the front left floor panel of the test car has been manufactured and installed on a ZwickRoell displacement-strength testing machine, as shown Fig. 18.11. The testing setup has also been modelled by means of non-linear Finite Element Modelling (see Fig. 18.11) and correlated against the test. It is worth mentioning that the simulation requires the input of both the decoupler characterization in compression and the carpet hard layer in tensile and bending.

**Fig. 18.11.** Static compression assessment setup for prototype parts. (a) 3D-printed testing jig; (b) part testing; (c) FEM simulation model.

The results of the test and simulation are presented in Fig. 18.12. Figure 18.12(a) compares the mass production part with the two polyester prototypes featuring respectively air-lay and HR decoupler felts (see Table 18.3). The prototype with HR felt shows a high correlation between test and simulation results, which confirms the good accuracy of the simulation. The HR felt at 900 g/m$^2$ area weight can double the compressional stiffness of the part in comparison to the 1200 g/m$^2$ polyester air-lay, due to its unique fiber orientation structure.

It is supposed that the carpet hard layer (the molded blank) has an impact on the result. The prototype carpet is heavier and stiffer compared to the mass production carpet, which contributes to improved performance. Figure 18.12(b) shows the impact of the carpet on the static compression performance in case of HR felt decoupler. With the same carpet layer stiffness, the prototype with HR felt performs same as the serial part with foam. This is to confirm that HR felt can replace the serial foam for part-level static compression, up to a decoupler thickness of about 35 mm, with a felt weight of 900 g/m$^2$. This substitution applies if the carpet is replaced with a similar polyester carpet of the same weight and stiffness.

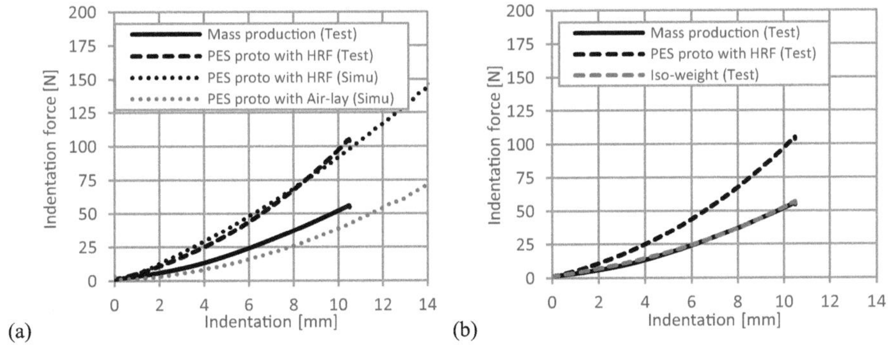


**Fig. 18.12.** Part static compression performance from test and simulation. Indenter is 80 mm diameter. Decoupler at indentation location is 35 mm thickness. (a) Mass production compared to polyester prototypes; (b) Effect of carpet change.

### 18.4.2 Vehicle Acoustic Performance

The components were installed in the vehicle to assess noise reduction and performance during driving. The prototype parts and accessories fitted well in the D-segment test vehicle, showing that the felt decoupler meets space and stability requirements.

In this study, one focuses on the reduction of the harmonic acoustic power, that is transmitted through the walls of the car, especially through the floor, where the floor carpet component is installed. For this, one uses the so-called Power Based Noise Reduction (PBNR), as described in the article [8]. The PBNR concept is used by car manufacturers to set vehicle acoustic specifications for sound package design, as it indicates how many dB the noise is reduced when it passes through the vehicle's sound package. A higher PBNR (in dB) indicates better noise reduction performance. More precisely, the PBNR, by using an acoustic power-based approach, refers to the ratio of a transmitted acoustic power to incident acoustic power. In very simple terms, it quantifies the power-flow through the vehicle structure.

In this study, the PBNR is sequentially measured with the mass production and the prototype part. In one case, the acoustic power input is considered at tire positions. In that condition, the noise is "travelling" through multiple paths from the tires to the interior passenger ears, such as through the floor, but also through the trunk. In another case, the acoustic power input is placed outside the car, just below the floor panel. In that case, the PBNR depends mainly on the reduction of acoustic power flowing though the floor area. Figure 18.13 illustrates the comparison of the PBNR results measured in the vehicle with the production part against two polyester prototypes, which feature air-lay and HR decoupler felts respectively (refer to

Table 18.3). Figure 18.13(a) illustrates the PBNR between the tire location and the average listening position of all passengers, whereas Fig. 18.13(b) depicts the PBNR between the underfloor location and the average listening position of all passengers. The noise reduction difference between the parts falls within the test approximation range. Therefore, despite the polyester prototypes being heavier and having better acoustic performance at the sample level (see paragraph 3.1), they contribute equally to noise reduction through the floor. This is due to the 3D construction of the parts, with varying thickness, making all three parts perform similarly. In the case of PBNR from floor, both prototype curves are systematically slightly higher than the mass production curve above 400 Hz. It is unclear how the higher weight of polyester prototypes affects vehicle acoustic performance. To address this, the iso-weight prototype in Table 18.3 has been measured. Results will be shared in a future version of this article.

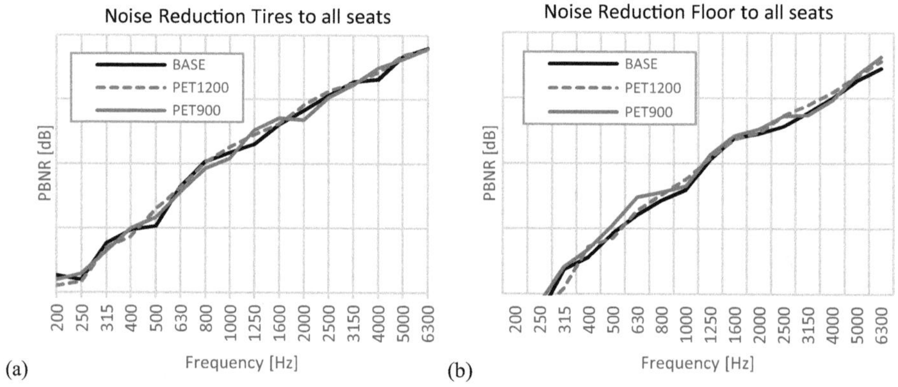


**Fig. 18.13.** Tested PBNR in a serial vehicle with different floor carpet components in place. Higher is better. Y-axis scale is 10dB/division. (a) is the PBNR with input power at tire location; (b) is the PBNR with input power under the floor. "BASE" means "with mass production part"; "PET1200" means "Polyester prototype with Air-lay felt decoupler" and "PET900" means "Polyester prototype with HR felt decoupler" having weights according to Table 18.3.

Finally, the performance of the parts has been evaluated in driving conditions. Figure 18.14 presents the average sound pressure levels at all passengers' listening positions in constant speed driving conditions of 80 kph (with all-wheel drive powertrain) with the different floor carpet components installed. Like for the PBNR tests, only the mass production part, as well as the two polyester prototypes with respectively air-lay felt and HR felt decoupler have been measured. The iso-weight part result will be provided in a later version of this article. Figure 18.14(a) and (b) respectively show the results on rough and smooth road conditions. Under rough road conditions, the results are within the test uncertainty range, with slightly lower noise levels for the prototype with air-lay felt in the 1600–4000 Hz range. This difference may arise from testing approximation; however, it appears precisely in the frequency range where the flat sample performance discussed in paragraph 3.1 is expected to affect vehicle performance. The same observation applies to smooth roads, where both polyester prototypes exhibit lower sound pressure levels between approximately 1600 Hz and 4000 Hz third octave bands.

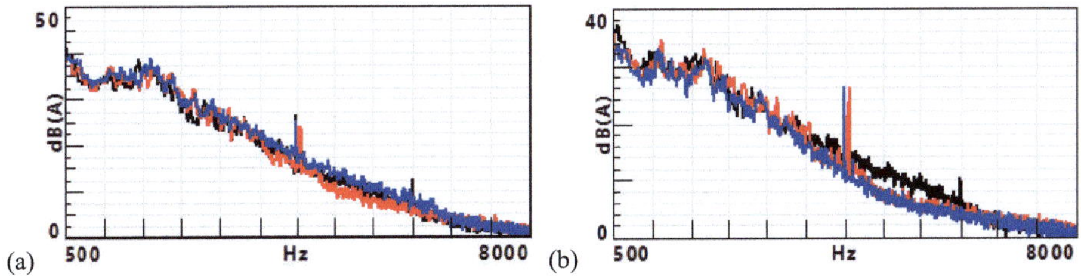


**Fig. 18.14.** All seats average sound pressure level in driving conditions at 80 kph constant speed. (a) rough road; (b) normal (smooth) road. Black curve: mass production part; blue curve: polyester prototype with HR felt decoupler; red curve: polyester prototype with air-lay felt decoupler.

## 18.5 Conclusion

In summary, we have presented a 100% polyester material lay-up to enable a mono-polymer floor carpet component while preserving key comfort performance aspects, including static compression and acoustic properties. The material lay-up is based on the control of several design variables: the carpet hard felt substrate weight and compression, the presence of a backing film between the carpet blank and the decoupler, and finally the exploitation of a novel felt decoupler featuring combined low dynamic but high static stiffness. Both comfort functions have been investigated and validated on samples, and then confirmed by means of part assessment in the vehicle. Although the mono-material demonstration was conducted with a carpet that is heavier than the reference, and although the iso-weight performance demonstration utilized the serial carpet layer, it is possible to conclude that the 100% polyester material lay-up can replace the mass production part without increasing weight.

The primary objective of this study is to facilitate a recycling loop that includes the End-of-Life recycling of products. This approach aims to enhance the long-term recycling content of polyester products by increasing the use of post-consumer recycling materials sourced from used floor carpets, in addition to the bottle flakes already obtained from the PET-bottle recycling market. Even in the absence of an End-of-Life recycling loop, whose implementation will require at least one full vehicle life-cycle time, it is necessary to track the recycling content of mono-material parts placed on the market today. The Table 18.4 provides an overview of recycling content for an iso-weight 100% polyester part produced today, based on the material proposal of this study in comparison to the mass-production part. The main recycled content comes from fibers used in the decoupler and the carpet hard felt. For the carpet, even with the shift to recycled tuft filament from bottle flakes, the mono-material blank is currently limited to up to 59%. This limitation is mainly due to the virgin coPET used for the back-coating and backing film of the mono-material carpet and will be increased in the future.

**Table 18.4.** Recycling content of both mass production part and mono-material component at equal weight, expressed as a percentage of the part weight (excl. Accessories).

| | Mass production part with virgin PET BCF | | Mass production part with recycled PET BCF | | 100% polyester concept part | |
|---|---|---|---|---|---|---|
| | Weight | Recycled content | Weight | Recycled content | Weight | Recycled content |
| Carpet | 4.2 kg | 32% | 4.2 kg | 52% | 4.2 kg | 59% |
| Decoupler | 2.5 kg | 0% | 2.5 kg | 0% | 2.2 kg | 90% |
| **Part (excl. Accessories)** | 6.7 kg | **20%** | 6.7 kg | **32%** | 6.4 kg | **69%** |

The material solution proposed in this study can be used for producing mono-material parts for future vehicle developments. However, the feasibility of recyclability at the End-of-Life stage must be investigated before these parts can be included in a recycling loop. To achieve this, several aspects must be investigated. One aspect resides in the exploitation of existing loops for the re-use of production waste through washing, shredding, granulation, and filament re-spinning, like processes already established for certain polyester materials [1]. Another aspect involves adapting and validating recycling processes for materials in aged condition, given that traceability and collection processes are established.

## References

1. Autoneum Website Video: https://www.autoneum.com/wp-content/uploads/2020/04/Reclaiming-and-recycling-of-Ultra-Silent_final_small.mp4.
2. Ki-Chang Jo, Ki-Sang Chae, Hyun-Woo Kim, Jung Jang, T. Courtois, T. Lafont, "Noise Transmission Path Analysis and Sound Package Optimization for Electric Vehicle using Numerical Simulation", Acoustic Society of Korea (ASK) conference, Korea, 2024.
3. E., Negro, A., Marino, G. Boreanaz, "Optimisation of a Dash "Hybrid" Insulator-Absorber Based on Pareto Analysis of SEA Simulation Results", Proceedings of 1st International ATZ Conference on Vehicle Acoustics, Zurich, 2011.
4. T. Courtois, M. Seppi, F. Ronzio, L. Sangiuliano, T. Yano, "Importance of the evaluation of the structure-borne NVH performance for light weight trim design", Proceedings of 3rd International ATZ Conference on Vehicle Acoustics, Zurich, 2011.
5. DIN EN ISO 3386-1, "Determination of stress-strain characteristics in compression—Part1".
6. TO BE INCLUDED: Impact of carpet on static compression.
7. PV 3919-B, "Faservlies- und PUR-Schaumstoff-Dämpfungen. Bestimmung der Stauchhärte", Volkswagen Konzernnorm.
8. J. Zhu, D. Hammelef, M. Wood, "Power-Based Noise Reduction Concept and Measurement Techniques", SAE 2005 Noise and Vibration Conference and Exhibition, US, May, 2005.

# Mono-Material TPU Foam and Heavy Layer with Improved Acoustic Performance and Recyclability

Mark Brennan(✉), Annelies Vandevelde, Gilles Geumez, Xiaodong Guo, and Kristof Verniers

Huntsman Polyurethanes, Tienen, Belgium
{mark_brennan,annelies_vandevelde,gilles_geumez,xiaodong_guo,kristof_verniers}@huntsman.com

**Abstract.** The versatility of polyurethane materials makes them useful in a variety of automotive noise and vibration applications. To meet the growing demand for more sustainable solutions, and the increasing sustainability targets set by Automotive OEMs, a novel foam and heavy layer concept, based on thermoplastic polyurethane, has been developed.

In this work, we introduce this low-density thermoplastic polyurethane foam, made in-situ in a standard way with a physical blowing agent. It is developed to have the right elasticity and damping characteristics and foam morphology for automotive sound insulation applications. It can be further combined with a heavy layer with the same or similar polymer backbone as the foam. This combination facilitates recovery of the materials at end of life and avoids mixed plastic streams which is common for these parts currently. We will present both experimental test results and simulation results of this concept and compare it to traditional polyurethane foams used in acoustic applications.

**Keywords:** Mono-material · Sustainability · Recyclability · TPU Foam

## 19.1 Introduction

The versatility of polyurethane (PU) materials can enable their use in a wide range of automotive applications, including sound absorption, sound insulation, and vibration damping [1]. However, these materials have traditionally been challenging to recycle. To meet the growing demand for more sustainable materials and the increasing sustainability targets set by automotive original equipment manufacturers (OEMs), new types of PU foam materials must be developed. This involves addressing recyclability issues at the end of their life cycle while ensuring that the materials maintain their intended performance.


A. Heintzel (Ed.): Automotive Acoustics Conference 2025, Proceedings, pp. 277–286, 2026. https://doi.org/10.1007/978-3-658-51249-1_19

PU foams, which are mostly thermoset foams, can be recycled chemically using glycolysis [2]. However, this process requires cost-effective collection methods and large-scale recycling plants to be economical. On the other hand, thermoplastic polyurethane (TPU) polymers, used in automotive parts, wires and cables, sports equipment, and textile coatings, can be recycled mechanically. This involves heating and melting the polymer and then injection molding it into new parts [3].

In this paper, we present a low-density TPU foam material that retains an open-cell foam structure and possesses the necessary viscoelastic properties for automotive acoustic applications. For dash and under-carpet applications, PU foam materials are often combined with a heavy layer, usually composed of filled ethylene propylene diene monomer (EPDM) rubber. This combination further complicates the recyclability of the entire part. By replacing EPDM with a TPU polymer, we can create a mono-material of TPU polymer and foam, which will simplify the recycling process.

## 19.2 Thermoplastic Polyurethane Foam Development

### 19.2.1 Physical Crosslinking in Thermoplastic Polyurethane

Unlike thermoset polyurethanes, thermoplastic polyurethanes (TPUs) become pliable when heated, allowing them to be reshaped multiple times without significant degradation. TPUs are composed of linear segmented block copolymers with alternating hard and soft segments. The hard segments provide toughness and rigidity, while the soft segments offer flexibility and elasticity.

Selecting the appropriate hard segments can lead to the formation of a semi-crystalline polymer backbone. This refers to the degree to which the polymer chains are ordered in a regular, repeating pattern. These hard segments, typically composed of diisocyanates and chain extenders, tend to form crystalline regions due to their ability to pack closely together. Higher crystallinity generally enhances durability and increases the melting points of TPUs, thereby improving their thermal stability and physical crosslinking.

The ability of TPUs to be reshaped multiple times without significant degradation makes them highly versatile for various applications. This property is particularly beneficial in industries where materials need to be recycled or repurposed, such as automotive, sports equipment, and textile coatings. The combination of toughness, flexibility, and recyclability positions TPUs as a promising material for sustainable development.

Moreover, the adaptability of TPUs can allow for their use in innovative applications, such as in the development of advanced automotive components that require both durability and recyclability. This makes TPUs an attractive option for manufacturers aiming to meet stringent sustainability targets while maintaining high performance standards (Fig. 19.1).

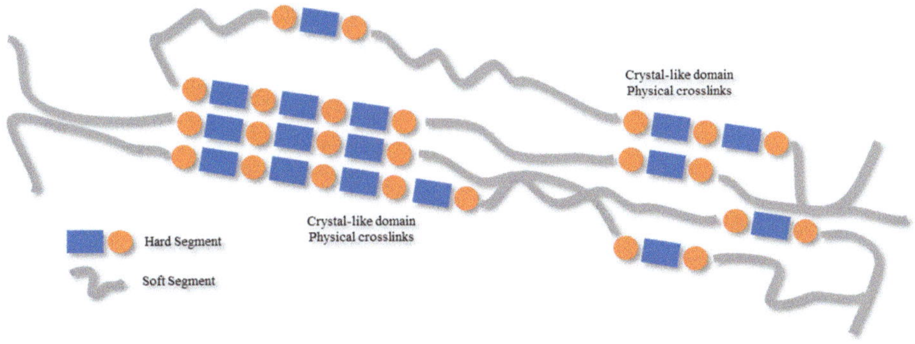


**Fig. 19.1.** Illustration of the semi-crystalline polymer backbone structure of a TPU.

### 19.2.2 Making a TPU Foam

Water-blown flexible PU foams are created using water, which reacts with isocyanates to form $CO_2$ as the primary blowing agent. This process results in low-density, open-cell foam with a polyurethane-polyurea polymer backbone. Increasing the water content lowers the foam density but also creates more polyurea, eventually leading to a brittle and friable foam. The foam softens at higher temperatures and degrades before melting due to its mixed polyurethane-polyurea backbone and inherent cross-linking.

Making TPU foam requires different blowing agents instead of water. For high-density foam (>200 kg/m3), gases such as nitrogen, carbon dioxide, or hydrocarbons (e.g., pentane) are physically mixed into the TPU polymer melt at high pressure. The pressure drop is carefully controlled to form nucleates of gas, and the thermoplastic melt is cooled to control bubble size.

For low-density foam (<200 kg/m3), the physical blowing agent is mixed with the TPU building blocks before the TPU is formed. The physical blowing and polyurethane reaction occur simultaneously while the thermoplastic polymer and foam cellular structure are being created. Careful selection of catalysts and surfactants ensures the formation of open-cell, low-density foam.

### 19.2.3 Mechanical Properties of TPU Foam

The load-bearing properties of flexible foams are crucial to their function and are typically measured using a simple compression test. The figure below shows typical results for a physically blown low-density TPU foam, which are comparable to water-blown PU foams. From the compression test, the hardness of the foam at higher strains (typically 40%) and an estimate of the Young's modulus at low strains can be recorded.

For applications such as sound insulation and vibration damping in automotive settings, dynamic electromagnetic shaker tests are performed for varying loads. In these tests, the foam acts as the spring and damper in a mass-spring-damper system. The figure below illustrates a forced vibration measurement of a TPU foam between two masses of different values. From this type of measurement, the dynamic modulus and loss factor can be estimated. TPU foams, like thermoset PU foams, can be developed to be highly resilient (loss factors in the range of 0.1–0.2) or viscoelastic (loss factors >0.3) (Fig. 19.2).

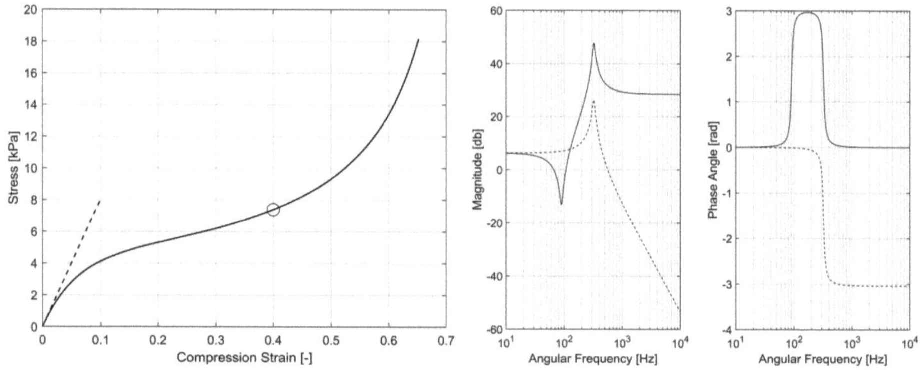


**Fig. 19.2.** Example stress-strain curve of a low density TPU foam with a Young's modulus of 80 kPa and a compression hardness of 7.4 kPa (left) and corresponding forced vibration transfer functions of the same foam under load of 2.8 kPa (right).

The hardness, Young's modulus, and damping properties of TPU foams can be controlled by adjusting the amount of physical blowing agent used and/or by modifying the TPU building blocks. This control can be achieved independently, unlike in water-blown thermoset PU foams. The figure below shows typical properties of low-density foams made with two different TPU polymer backbones. All foams exhibit durability due to the semi-crystalline nature of the TPU backbone, with a low compression set (<10%) achievable through careful polymer design (Fig. 19.3).

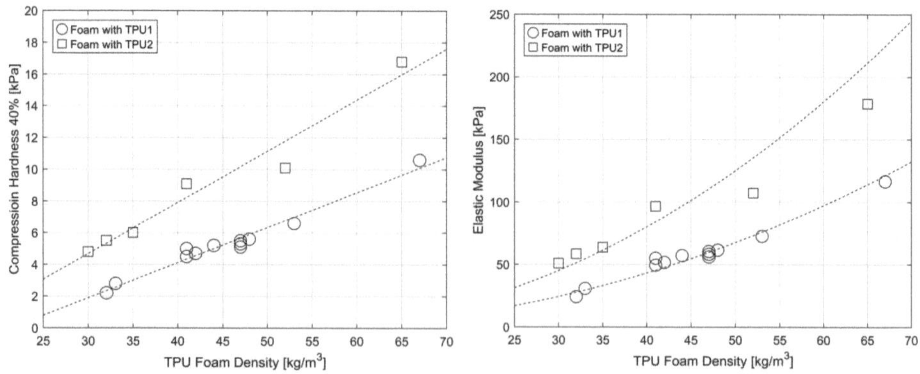


**Fig. 19.3.** Compression hardness at 40% strain of TPU foams with different polymer backbones in a density range of 30–70 kg/m3 (left) and estimated elastic modulus for the same foams (right).

### 19.2.4 Cellular Morphology of TPU Foam

The foam cell morphology of low-density TPU foam is comparable to open-cell thermoset PU foam in terms of cell sizes and degree of openness. This similarity can allow for the development of lightweight, flexible absorbers using TPU foam. The figure below shows an example of the foam morphology reconstructed from X-ray microCT, a powerful imaging technique used to visualize the internal structure of materials in high resolution. Furthermore, sound absorption measurements, using an impedance tube, of two foams made at different densities. The values are typical for flexible PU foam absorbers (Fig. 19.4).

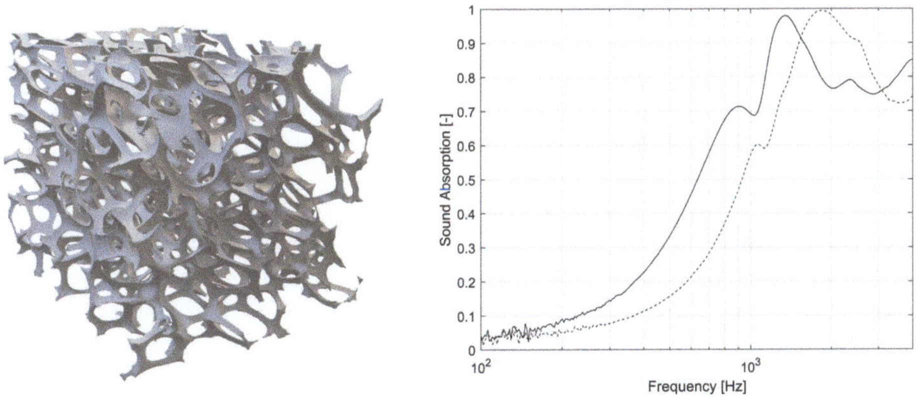


**Fig. 19.4.** Typical foam cell morphology of a low density TPU foam (left) and sound absorption of TPU foams measured in an impedance tube (right).

### 19.2.5 Recyclability of TPU Foam

Since the polymer backbone is made from TPU, once the material is collected for recycling purposes, it can be melted like any other TPU polymer and reshaped into new parts. The melting temperature of a thermoplastic polymer is measured using shear rheology. The melting temperature is defined at the point where the storage modulus and loss modulus are equal, indicating the transition from solid-like to liquid-like behavior. Thermoset polymers do not exhibit such a crossover before thermal degradation. The figure below contrasts the melt rheology of TPU and PU foam, showing a clear 'crossover' melting point in TPU. PU foam softens but never achieves a 'crossover' point before thermal degradation (Fig. 19.5).

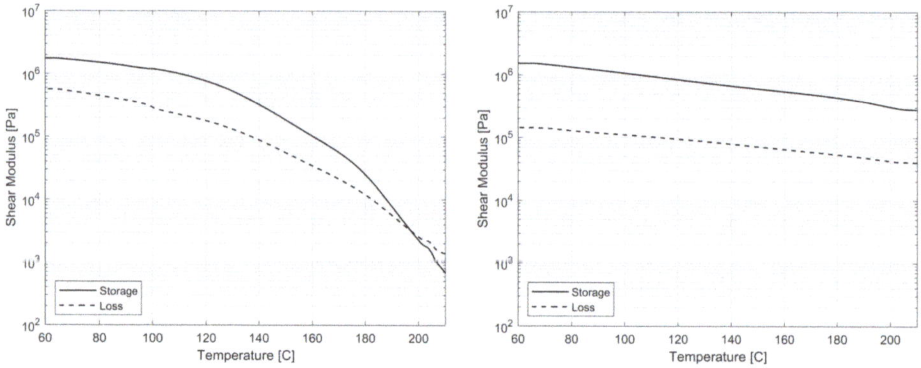


**Fig. 19.5.** Melt rheology of TPU foam shows thermoplastic behavior (left), a clear transition from elastic to viscous behavior, e.g. crossover at 190° C, in contrast, water-blown thermoset PU foams (right) soften but do not melt before polymer degradation.

Once melted, the TPU polymer can be reshaped into new forms typical for TPU applications. Below are examples of TPU foam recycled into new forms, such as partially compression molded parts, pallets, and wire (Fig. 19.6).

**Fig. 19.6.** Some examples of TPU foam reshaped into new parts.

## 19.3 Integrating a TPU Heavy Layer and TPU Foam

In automotive sound insulation packages, PU foam is often combined with a filled ethylene propylene diene monomer (EPDM) rubber heavy layer to create a mass-spring-damper system. In this system, the EPDM heavy layer acts as the mass, while the PU foam provides elasticity and damping. This multi-material layup complicates the recyclability of the part, as the materials need to be separated and sorted after dismantling for efficient recycling.

A TPU polymer can be used as an alternative to the EPDM heavy layer, either filled or unfilled, depending on the requirements. Using the same TPU polymer for both the foam and the heavy layer can further aid recyclability. Typical properties for TPU include a density of approximately 1150 kg/m3, Shore A hardness of 50–100, tensile strength of 32–56 MPa, and elongation of 300–900%. Below is an illustration of two comparable foam-heavy layer concepts with the same mass and foam elasticity. The TPU example assumes the same or similar TPU polymer backbone in both the foam and heavy layer, creating a mono-material system (Fig. 19.7).

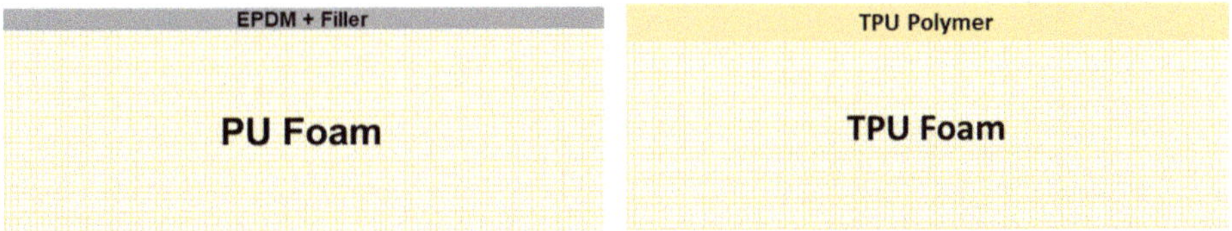


**Fig. 19.7.** Mono-material concept consisting of TPU heavy layer and foam as an alternative to the reference insulator, EPDM and PU foam.

## 19.3.1 Sound Transmission Loss Predictions and Measurements

Using transfer matrix methods (TMM), it is possible to predict the performance of mono-material TPU foam and heavy layers compared to traditional EPDM-PU foam layups. Based on the typical mechanical properties of PU foam and the known achievable properties of TPU foam and polymer, predictions of sound transmission loss for the layups are shown in the figure below. Comparable performance is achievable with the TPU mono-material even without the addition of fillers. Further improving the sound absorption properties of the TPU foam will enhance the overall performance.

Sound transmission loss measurements in an impedance tube are also performed on TPU mono-material foam and polymer. Below is a sample prepared by back foaming a TPU foam onto a heavy layer made from the same material. The sound transmission loss measurements are comparable to TMM predictions for a TPU poro-elastic foam and TPU heavy layer (Figs. 19.8 and 19.9).

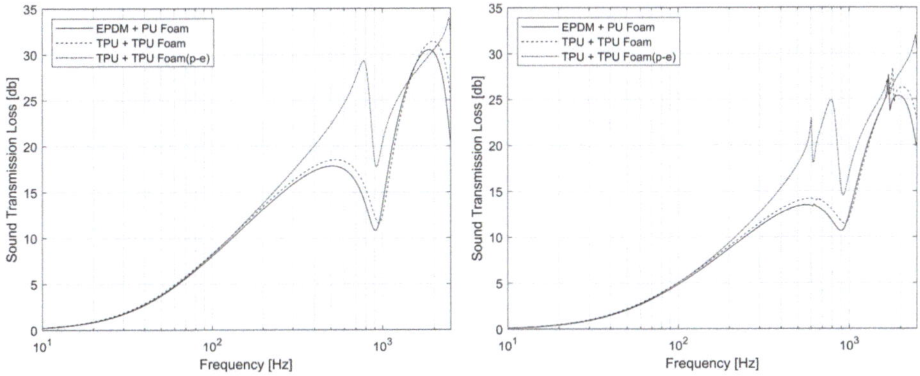


**Fig. 19.8.** Prediction of airborne sound transmission loss for a plane wave (left) and diffuse field (right) for three different configurations (i) filled EPDM heavy layer of 2.5 kg/m$^2$ and PU foam (ii) unfilled TPU heavy layer of 2.7 kg/m$^2$ and TPU foam (iii) unfilled TPU heavy layer of 2.7 kg/m$^2$ and sound absorbing TPU foam.

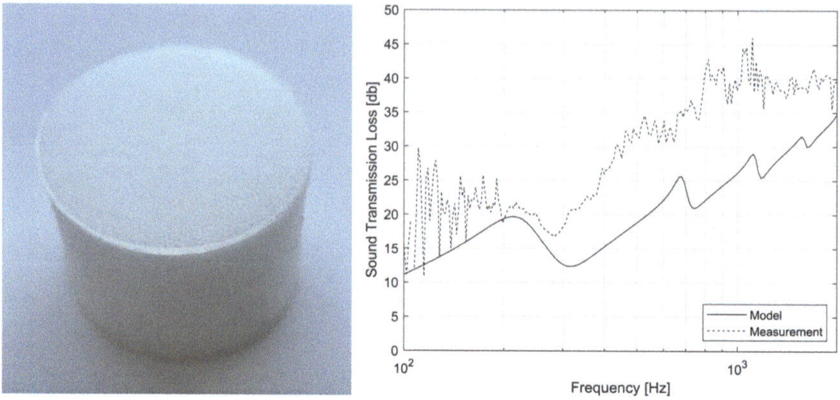


**Fig. 19.9.** Sample of mono-material (left) of TPU foam back foamed onto unfilled TPU heavy layer, made from the same polymer, and cut for impedance tube measurement of sound transmission loss (right) compared to model prediction.

### 19.3.2 Prototype Parts Made with TPU Foam and Heavy Layer

Automotive parts, such as dash insulators and under-carpet padding, can be manufactured using TPU foam and polymer. These parts can be produced using the same equipment currently used for making similar components, with some additional processing capabilities to incorporate physical blowing agents. This can allow for the creation of lightweight, durable, and recyclable parts.

Below is an illustration of a prototype carpet made from TPU foam, showcasing its lightweight structure and effective sound insulation properties. This part demonstrates the potential of TPU foam in automotive applications, providing a more easily recyclable and high-performance alternative to traditional materials (Fig. 19.10).

**Fig. 19.10.** Prototype part for under carpet applications (intentionally blurred to protect design) made with TPU foam, TPU heavy layer and TPU inserts.

## 19.4 Conclusions

The new TPU foam matches the performance of thermoset PU foam in terms of hardness, Young's modulus, and damping properties. This makes it potentially suitable for applications requiring sound insulation and vibration damping, such as automotive parts.

The ability to melt and reshape TPU foam offers significant advantages in terms of recyclability. Unlike thermoset PU foam, which cannot be melted and reshaped, TPU foam can be recycled into new parts, reducing waste and supporting circular economy initiatives.

Using the same TPU polymer for both the foam and heavy layer creates a mono-material system, which further enhances recyclability. This approach addresses the industry's sustainability challenges by simplifying the recycling process and reducing the environmental impact of automotive sound insulation parts.

A prototype part has been presented, demonstrating the feasibility of using TPU foam and polymer in automotive sound insulation applications. This prototype shows a potential solution for the industry to meet its sustainability goals. By adopting TPU foam, manufacturers can produce high-performance, recyclable parts that contribute to a more sustainable future.

## References[1]

1. Randall, D., Lee S.: The Polyurethanes Handbook, Wiley, London (2002).
2. Markovic, V., Hicks, D. A.: Design for Chemical Recycling, Philosophical Transactions: Mathematical, Physical and Engineering Science 355, no. 1728 (n. d.): 1415–25 (1997).
3. Wölfel, B., Seefried, A., Allen, V., Kaschta, J., Holmes, C., Schubert, D.W.: Recycling and Reprocessing of Thermoplastic Polyurethane Materials towards Nonwoven Processing. *Polymers* 2020, 12, 1917.

[1] While all the information in this publication is to the best of our knowledge and belief accurate at the date of publication, NOTHING HEREIN IS TO BE CONSTRUCTED AS A WARRANTY, EXPRESS OR OTHERWISE. Huntsman warrants only that its products meet the agreed specifications. Typical properties, where stated, are to be considered as representative of current production and should not be treated as specifications. In all cases, it is the responsibility of the user to determine the applicability of such information and the suitability of any product for its own particular purpose. The sale of products referred to in this publication is subject to the general terms and conditions of sale of Huntsman International LLC or of its affiliated companies.

# Vehicle Vibration Insulation Assessment Through Multi-Domain Substructuring

Alessandro Zucchini[1,2](✉) and Frank Naets[1,3]

[1] Department of Mechanical Engineering, KU Leuven, Heverlee, Belgium
alessandro.zucchini@bmw.de, frank.naets@kuleuven.be
[2] BMW Group, München, Germany
[3] Flander's Make@KU Leuven, Leuven, Belgium

**Abstract.** This paper presents a comprehensive framework for evaluating insulation performance in complex mechanical assemblies using a multi-domain dynamic substructuring approach. Building on the dual assembly formulation, we extend classical frequency-based substructuring by integrating physical, modal, and frequency domain representations. This hybrid formulation enables the coupling of experimentally characterized subcomponents with numerically reduced models, preserving interpretability and computational efficiency. A central contribution is the formulation of insulation as a frequency-dependent transfer response functions between interface degrees of freedom, allowing the identification of anti-resonance-driven isolation behavior. Using Lagrange Multipliers Frequency-Based Substructuring (LM-FBS), we establish a consistent methodology for assembling reduced and full-order models while enforcing interface compatibility. The method is demonstrated on a vehicle subsystem consisting of an electric drive unit, rubber mounts, and a rear axle carrier. Results illustrate how insulation trends can be explained through modal contributions and response shape functions, enabling targeted structural modifications. The proposed approach bridges experimental and numerical modeling techniques to provide design engineers with actionable insights into high-frequency vibration insulation.

**Keywords:** NVH · Vibration insulation · Substructurings

## 20.1 Introduction

Dynamic substructuring has its roots in domain decomposition techniques, which aim to improve the efficiency of solving large scale finite element models. Traditionally, problems were decomposed into subdomains to facilitate iterative interface coupling—this is often referred to as a second level of decomposition. A first level typically refers to discretization strategies such as finite element methods.

In the context of dynamics, efficiency can be further improved by representing substructures not through detailed discretization but by their generalized dynamic responses. These responses can be handled in different domains—physical, modal,


A. Heintzel (Ed.): Automotive Acoustics Conference 2025, Proceedings, pp. 287–300, 2026. https://doi.org/10.1007/978-3-658-51249-1_20

or frequency—and can be transformed between one another through appropriate operations.

We propose to extend the dual substructuring formulation to enable the coupling of models represented in different domains: physical, modal, and frequency. This allows for a more flexible and comprehensive analysis across modeling paradigms. Next, we integrate the dynamically characterized components of a car assembly—initially using simplified representations and later using full-fidelity models—to build a realistic coupled structure. This includes the integration of a parametric reduced-order model (ROM) of the rear axle carrier.

The resulting coupled system enables the computation and interpretation of insulation curves, derived from ratios of Frequency Response Functions (FRFs). Particular attention is given to the physical interpretation of trends in these curves, providing insights for engineers seeking to understand and improve the dynamic performance of vehicle assemblies.

Further, we demonstrate how structural modifications guided by high-frequency insulation principles can be applied to the complete vehicle model. These modifications successfully achieve targeted changes in the insulation behavior, highlighting the role of anti-resonances and associated structural parameters in governing insulation effectiveness.

## 20.2 Limitation in Existing Dynamic Substructuring

For evaluating the insulation, a crucial role is played by the anti-resonances of a system and by the structural parameters related to them. The roots of dynamic substructuring (DS) can be found in the field of domain decomposition as a second level of decomposition [1]. Here, complex problems are separated into subdomains. These are iteratively solved for the interface coupling as their complexity is reduced by representing substructures using their general responses. DS can be performed in different domains and conceptually same information is contained in all different representation and it is possible to switch from one domain to the other following particular transformations [2]. Figure 20.1 shows a brief overview of these different analysis domains.

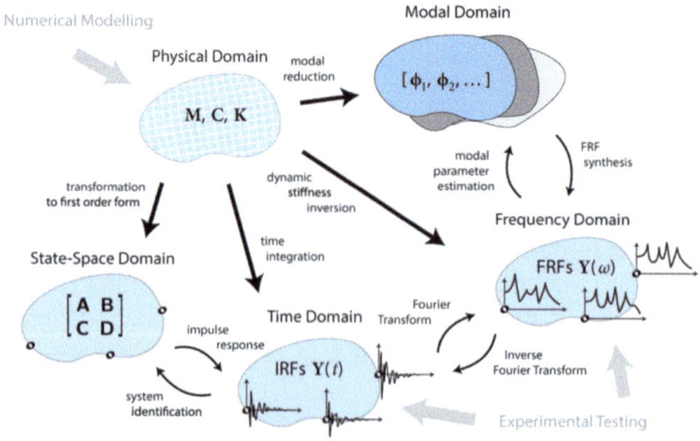


**Fig. 20.1.** Various analysis domains for dynamic substructuring (ref. [1]).

Because of the ease of expressing the solution of a differential equations of a dynamical structural system by means of its frequency response functions (FRFs), the frequency domain is the domain of interest for most of the projects and the most commonly used for substructuring. This type of approach presents a series of advantages:

- dynamic models are reduced at their boundary interface points, therefore they are quite compact and allow for fast computation.
- the DOFs are physical, leaving an easy interpretation of the coupling for the user.
- models can be easily derived from measurements or numerical models from FEM.

However, there is still a challenge in interpreting these FRFs as a design engineer, especially when trying to assess which structural parameter affects which natural (anti)-resonance. Moreover, using the physical interface coordinates in the frequency domain:

- limits the observability of the results only to a few connection points to which the models are reduced.
- having only the measured or simulated nodes available, only a direct connection to those few points is possible.
- does not allow for any structural parameter's change since there is no longer a direct relation between FRFs and structural parameters.

As engineers we do not merely want to observe the insulation behavior, but we want to explain the reason behind a certain peak; to do so, the best way is to look at the Response Shape Functions (RSF) of a system. Looking at RSF means displaying the deflection shape of a system under a certain excitation for a specific frequency. By doing so, one can better understand which structural parameters are responsible for that specific deflection shape, and target the right structural changes in order to modify is response if desired.

The reader can easily imagine that having only the connection's DOFs available, it is not possible to:

- analyse how the entire geometry deflects.
- change local parameters in terms of mass and stiffness for specific locations.

For these reasons we propose to use a dual formulation to perform the substructuring, rather than the more common primal formulation. This enables a more direct interpretation of the contributions of the response of the various subdomains.

## 20.3 Dual Assembly Framework for Insulation Evaluation

Evaluating insulation performance—particularly at higher frequencies—requires a careful analysis of the system's anti-resonances and the structural parameters influencing them. This is done by assembling structures from the frequency domain—such as measured FRF matrices—with numerical models derived in the physical and modal domains. Physical and modal domain models offer a clear link between system parameters and dynamic response, which is essential for understanding the root causes of insulation trends—particularly resonance and anti-resonance features. We propose a modified dual assembly method that combines domains, allowing us to retain the practical benefits of frequency based modelling while gaining interpretability and control from physical and modal representations.

### 20.3.1 Insulation Definition

Considering a system with two degrees-of-freedom of interest, after applying an excitation F causing a displacement X we can define the receptance as the ratio between the displacement and force $H(j\omega)=X/F$. Here j is the imaginary unit number and $\omega$ is the angular frequency. The response of this system can then be obtained as:

$$\begin{bmatrix} X_1 \\ X_2 \end{bmatrix} = \begin{bmatrix} H_{11} & H_{12} \\ H_{21} & H_{22} \end{bmatrix} \begin{bmatrix} F_1 \\ F_2 \end{bmatrix} \tag{20.1}$$

The insulation from DOF 1 to 2 can be evaluated as the ratio of their responses:

$$G_{1\to 2}(j\omega) = \frac{X_1(j\omega)}{X_2(j\omega)} = \frac{H_{11}(j\omega)F_1 + H_{12}(j\omega)F_2}{H_{21}(j\omega)F_1 + H_{22}(j\omega)F_2} \tag{20.2}$$

And considering a only an input load on DOF this can be further simplified to:

$$G_{1\to 2}(j\omega) = \frac{X_1(j\omega)}{X_2(j\omega)} = \frac{H_{11}(j\omega)}{H_{21}(j\omega)} \tag{20.3}$$

It is indeed possible to express the insulation between two points as the ratio of the forced responses or as the ratio of FRFs. More insights are available in [3, 4, 5 and 6].

### 20.3.2 Dual Assembly Approach Through Lagrange Multipliers

Lagrange Multipliers Frequency Based Substructuring (LM-FBS) is a substructuring technique used for computational simulations of coupled mechanical systems [7, 8, 9 and 10]. It is a frequency domain technique that allows for efficient modelling of complex systems by dividing them into smaller, more manageable substructures. Those sub-models can be obtained from information available in the physical, modal, state-space or time domain and then transformed into the frequency domain. This method uses Lagrange multipliers to enforce the conformity along the subsystem interfaces, and obtain the response of the entire structure. More is explained inside [7, 8 and 11]. The LM-FBS method is particular interesting for obtaining the dynamic behaviour of substructures, like rubber mountings coupled with a fixed preload, where the measurement of the substructure alone is difficult to perform and the influence of preloads, temperature or other effects might not be known in advance.

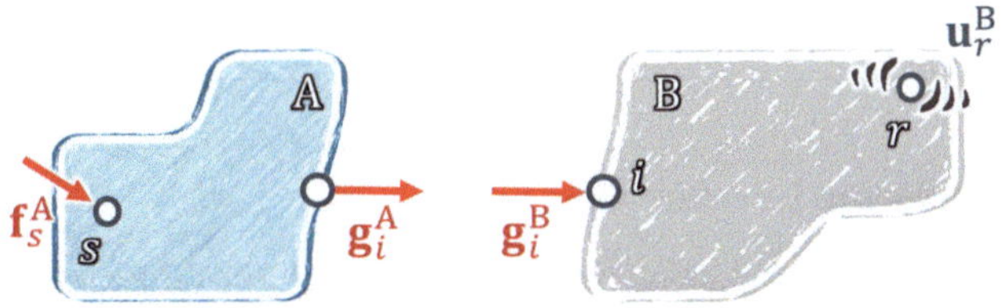


**Fig. 20.2.** Interface interaction when coupling substructures A and B into the assembly AB (ref. [2]).

Figure 20.2 illustrates two coupled substructures and their interaction along the interface point *i*.

The matrix of the coupled system is assembled in a dual manner (also called an augmentation formulation in the field of multibody dynamics). The compatibility condition on the interface will be juxtaposed to the subsystem dynamic equilibrium equation. Additional unknown interface forces are then introduced to enforce the compatibility constraints so that the local equilibrium equations can be written as:

$$\boldsymbol{Z}^{bl}\boldsymbol{u} + \boldsymbol{B}^{T}\boldsymbol{\lambda} = \boldsymbol{f} \tag{20.4}$$

where $\boldsymbol{Z}^{bl}$ is the dynamic stiffness matrix of the assembly in block-diagonal form, the operator $\boldsymbol{B}^{T}$, defines the DOFs to which the equivalent connecting forces $\boldsymbol{\lambda}$ will be applied. $\boldsymbol{\lambda}$ are in fact the Lagrange multipliers associated with these compatibility constraints.

Combining the dynamic equilibrium equations with the compatibility equations, the behaviour of the complete system with rigid coupling can be described as:

$$\begin{bmatrix} \mathbf{Z}^{\mathrm{bl}} & \mathbf{B}^{\mathrm{tot}^{T}} \\ \mathbf{B}^{\mathrm{tot}} & 0 \end{bmatrix} \begin{bmatrix} \mathbf{u} \\ \lambda \end{bmatrix} = \begin{bmatrix} \mathbf{f} \\ 0 \end{bmatrix} \tag{20.5}$$

For which the solution can be obtained as:

$$\begin{bmatrix} \mathbf{u} \\ \lambda \end{bmatrix} = \begin{bmatrix} \mathbf{Z}^{\mathrm{bl}} & \mathbf{B}^{\mathrm{tot}^{\mathrm{T}}} \\ \mathbf{B}^{\mathrm{tot}} & 0 \end{bmatrix}^{-1} \begin{bmatrix} \mathbf{f} \\ 0 \end{bmatrix} = \mathbf{Z}^{\mathrm{tot}-1} \begin{bmatrix} \mathbf{f} \\ 0 \end{bmatrix} \tag{20.6}$$

To evaluate the insulation between two points that have a unique source of excitation, we only need the transfer and driving FRF between these points. For this reason and for computational convenience, we do not need to evaluate the FRFs between all the possible DOFs but one can impose a unit excitation for each DOF s of interest, and obtain the response at these nodes. In this way, we will obtain the unit response or receptance $\mathbf{H}$ (plus the Lagrange multipliers) of all DOFs after a unit excitation in a specific set of them:

$$\begin{bmatrix} \mathbf{H}_{i1} & \dots & \mathbf{H}_{\mathrm{is}} \\ \lambda_1 & \dots & \lambda_{\mathrm{s}} \end{bmatrix} = \begin{bmatrix} \mathbf{Z}^{\mathrm{tot}-1} \begin{bmatrix} \mathbf{f}_1 \\ 0 \end{bmatrix} & \dots & \mathbf{Z}^{\mathrm{tot}-1} \begin{bmatrix} \mathbf{f}_{\mathrm{s}} \\ 0 \end{bmatrix} \end{bmatrix} \tag{20.7}$$

Here, $\mathbf{f}_s$ is a zero-vector with the dimensions of the total amount of DOFs and only a unit value in element $s$.

### 20.3.3 Substructuring Interface Description

From a finite element model, it is possible to extract full-order mass and stiffness matrices. In order to describe the physical connections between the various connecting substructures, we typically need to project the displacements and forces from a set of nodes related to the interface, to a single reference interface node [12 and 13]. To do so, we "reduce" the number of DOF by projecting the total amount of displacements, given by the sum of the dependent and independent displacements, over the independent DOFs through a projection matrix **A**. Such projection matrix **A** presents therefore the size of the total number of DOFs times the number of independent DOFs:

$$\mathbf{u}_{\text{tot}} = \begin{bmatrix} \mathbf{u}_{\text{IN}} \\ \mathbf{u}_{\text{DN}} \end{bmatrix} = \mathbf{A}\mathbf{u}_{\text{IN}} \tag{20.8}$$

Keeping in mind these 'spiders' (see e.g. Fig. 20.3) present one independent node in the middle of the spider and dependent nodes (DN) around it connected with the rest of the mesh. Here the independent nodes are also the nodes where the excitations are applied. The displacement of the dependent nodes are calculated with the equations of a rigid free-body from the displacements of the single independent nodes (IN). To properly couple a full-order numerical model, one should first eliminate of the DOFs associated with the dependent nodes describing their motion as a projection of the motion of the independent nodes. One can construct the projection matrix for a rigid spider connection (also denoted as RBE2 in Nastran) as follows:

$$\mathbf{A} = \begin{bmatrix} \begin{bmatrix} \mathbf{I} & 0 \\ 0 & \mathbf{I} \end{bmatrix} \\ \begin{bmatrix} \mathbf{I} & \mathbf{r}_{\text{ID}1} \\ 0 & \mathbf{I} \end{bmatrix} \\ \vdots \\ \begin{bmatrix} \mathbf{I} & \mathbf{r}_{\text{ID}i} \\ 0 & \mathbf{I} \end{bmatrix} \end{bmatrix} \tag{20.9}$$

where $\mathbf{r}_{\text{ID}i}$ is the skew-symmetric matrix of the relative position going from the independent node to the i-th dependent node. Independent nodes should not be reduced, therefore a unitary value appears in the projection matrix. Figure 20.3 shows an example of a rigid spider projection on a simple hexagonal body.

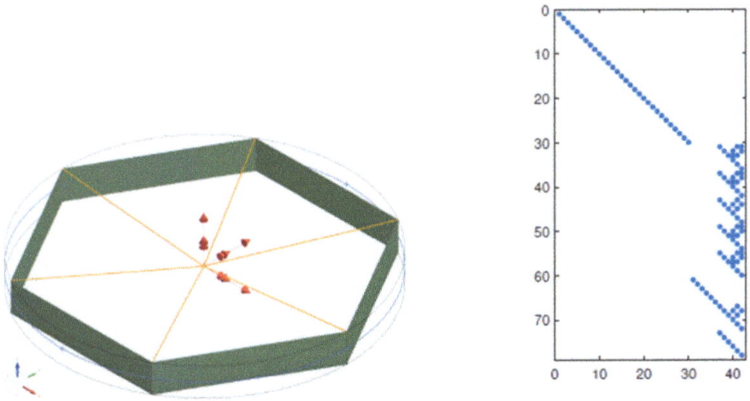


**Fig. 20.3.** Rigid spider (RBE2) element representation.

In such an example, one independent node in the middle of the spider is rigidly connected to six dependent nodes around it. Six other nodes are part of the mesh and they figure as independent not having any RBE connection.

Alternatively one can define an averaging connection, where the central interface node represents the average motion of all surrounding nodes. In Nastran these 'RBE3' connections are defined by a single dependent node located at the center of the spider and multiple independent nodes (IN) distributed around it (see Fig. 20.4). One can construct the projection matrix for these interface connections as follows:

$$\mathbf{A} = \begin{bmatrix} \mathbf{I} & 0 & \cdots & 0 \\ \vdots & \mathbf{I} & \vdots & \vdots \\ \begin{bmatrix} \mathbf{I} & \mathbf{P}_{\mathrm{w1}} \\ 0 & \mathbf{I} \end{bmatrix} & \begin{bmatrix} \mathbf{I} & \mathbf{P}_{\mathrm{w2}} \\ 0 & \mathbf{I} \end{bmatrix} & \cdots & \begin{bmatrix} \mathbf{I} & \mathbf{P}_{\mathrm{wi}} \\ 0 & \mathbf{I} \end{bmatrix} \end{bmatrix} \tag{20.10}$$

where $\mathbf{P}_{\mathbf{wi}}$ is the weighted projection of the forces and moments acting on the center node as a result of the independent nodes. Like for the rigid spider, independent nodes should not be reduced, therefore they are related with a unit value inside $\mathbf{A}$.

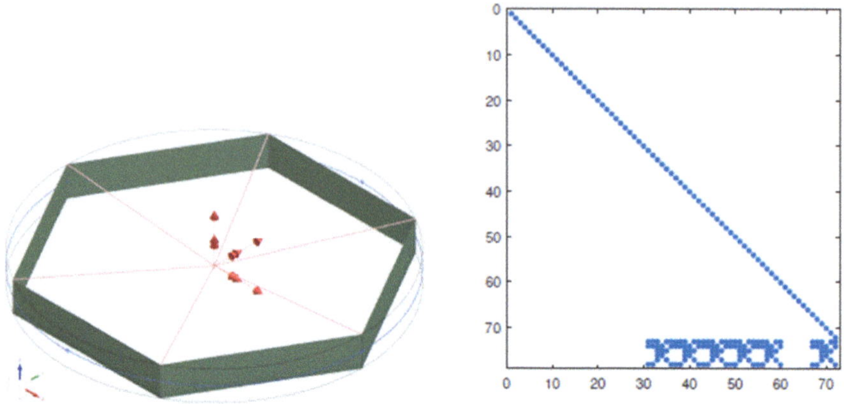


**Fig. 20.4.** Flexible spider (RBE3) representation.

### 20.3.4 Frequency-Modal Domain Coupling

For the coupled model, we will assume a system with two components, a first component which is characterize through a full FE model (FE), a second model representing experimentally characterized coupling components like bushings (EX), and a third model consisting of a modally reduced FE model (RE).

For the FE model, the appropriate projections need to be performed to account for the equivalent interfaces:

$$\mathbf{M}_{\mathrm{FE}}{}^{(f)} = \mathbf{A}^T \mathbf{M}_{\mathrm{FE}}{}^{(g)} \mathbf{A}, \mathbf{K}_{\mathrm{FE}}{}^{(f)} = \mathbf{A}^T \mathbf{K}_{\mathrm{FE}}{}^{(g)} \mathbf{A} \tag{20.11}$$

And the corresponding full component impedance can be obtained as:

$$\mathbf{Z}_{\mathrm{FE}}{}^{(f)} = -\omega^2 \mathbf{M}_{\mathrm{FE}}{}^{(f)} + \mathbf{K}_{\mathrm{FE}}{}^{(f)} \tag{20.12}$$

Equivalently, for the reduced order model, assuming a reduction basis $\mathbf{V}$ the impedance can be evaluated as:

$$\boldsymbol{Z}_{\boldsymbol{RE}} = -\omega^2 \mathbf{V}^{\mathbf{T}} \mathbf{M}_{\mathrm{FE}}{}^{(f)} \mathbf{V} + \mathbf{K}_{\mathrm{FE}}{}^{(f)} \tag{20.13}$$

For this case the block-diagonal impedance matrix becomes:

$$\boldsymbol{Z}^{bl} = \begin{bmatrix} \boldsymbol{Z}_{\boldsymbol{FE}} & 0 & 0 \\ 0 & \boldsymbol{Z}_{\boldsymbol{EX}} & 0 \\ 0 & 0 & \boldsymbol{Z}_{\boldsymbol{RE}} \end{bmatrix} \tag{20.14}$$

Here the full state, load and Lagrange multiplier vector are

$$\boldsymbol{u} = \begin{bmatrix} u_{FE} \\ u_{EX} \\ u_{RE} \end{bmatrix}, \boldsymbol{f} = \begin{bmatrix} \boldsymbol{f}_{\boldsymbol{FE}} \\ \boldsymbol{f}_{\boldsymbol{EX}} \\ \boldsymbol{f}_{\boldsymbol{RE}} \end{bmatrix}, \boldsymbol{\lambda} = \begin{bmatrix} \boldsymbol{\lambda}_{\boldsymbol{FE-EX}} \\ \boldsymbol{\lambda}_{\boldsymbol{EX-RE}} \end{bmatrix} \tag{20.15}$$

The full coupling matrix **B** can then be represented as:

$$\mathbf{B} = \begin{bmatrix} \boldsymbol{B_{FE}} & \boldsymbol{B_{EX}} & \boldsymbol{B_{RE}} \end{bmatrix} \tag{20.16}$$

This system can then be solved in accordance to the methods outlined in Sects. 3.1 and 3.2 to perform the insulation analysis on a system level, as will be illustrated in the next section.

## 20.4 System-Level Validation: Engine with Rear-Axle-Carrier and Car-Body

As an example of this approach, we will analyze the behavior of an (electric) drive unit (EDU) being connected through a set of rubber mounts (experimentally characterized [14, 15, 16 and 17]) to a rear axle carrier (RAC) which is also suspended to the car body (CAR) as shown schematically in Fig. 20.5. The EDU, RAC and CAR are here represented by (reduced order) finite element models. We will focus only on the first layer of insulation, ignoring the influence of the second layer of rubber mounts and the car body. The RAC is represented as a FEM model, which can be reduced to a smaller equivalent model through a project on the dominant eigenmodes in the frequency range of interest. For this mesh:

- Most elements are shell elements, meaning they provide a dynamic description both for the translational and rotational DOFs.
- the nodes at the interface are connected with the rest of the body through Nastran RBE3 elements.

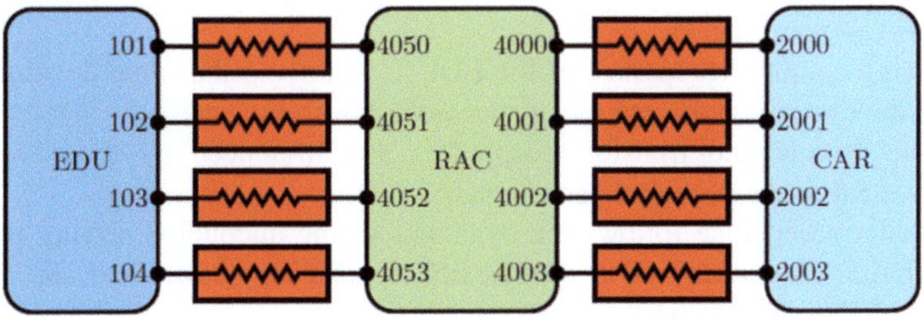


**Fig. 20.5.** Conceptual overview of electric drive unit connected through an axle carrier to a vehicle body.

The dynamic interface of the EDU and of the rubber mounts is not rigidly "connected" to the interface nodes of the RAC, but rigidly projected to the corresponding independent nodes.

Once again, having the full-order model of the RAC available, allows us to perform a reduction in a modal sense and obtain a lighter model. Figure 20.6 shows the six rigid body modes and the first 9 flexible modes with global body deflections. In fact, there are some local modes corresponding with resonances of non-interesting subcomponents such as appendices, holes, ribs or other particular connections, which are not relevant for the insulation we intend to evaluate and which are omitted from the model. For this specific application, we chose a total of 30 modes to describe the dynamics of the RAC.

Figure 20.7 shows the topology representation of the dynamic stiffness matrices assembled in block-diagonal form and their corresponding Boolean matrices. The first consideration is that the system of equations which is being solved is relatively small (order of 150 equations) with respect to the size of the full-order model of the RAC ($n \approx 5 \cdot 106$). Also interesting to note here is that the Boolean matrix of the RAC is not a simple identity matrix but is a dense matrix coming from the projection matrix $\mathbf{A}$ and pre-multiplied for the reduction basis $\mathbf{V}$ coming from the modal selection of eigenvectors.

Once we obtain the coupled receptance of the assembled system, following the same procedure as before we can plot the insulation between two DOFs as a ratio of receptances. In Figs. 20.8 and 20.9 one can see:

- in light blue the uncoupled driving receptance of the EDU;
- in orange the driving and transfer receptance of an uncoupled rubber mount;
- in green the uncoupled driving receptance of the RAC;
- in dashed red the coupled driving and transfer receptances between the two interface points and the consequent insulation given by a direct evaluation of Eq. (20.3).
- in solid blue there are the coupled receptances and the corresponding insulation obtained with the proposed multi-domain FBS.

When considering the insulation behavior in Fig. 20.9, we can observe the following behavior which provides critical insights for the designer which needs to determine the desired behavior of various components and substructures:

- as one would expect, the insulation curve is changing in correspondence with certain anti-resonance peaks;
- the driving coupled receptance of the EDU is following the same dynamic behaviour of the uncoupled driving receptance at higher frequencies;
- both the driving and transfer receptances are influenced by the resonant and anti-resonant peaks of the rubber mounts;
- the transfer receptance follows the dynamics of the uncoupled driving receptance of the RAC. This allows us to spot flexible modes very easily and observe if and how much they shift in frequency.

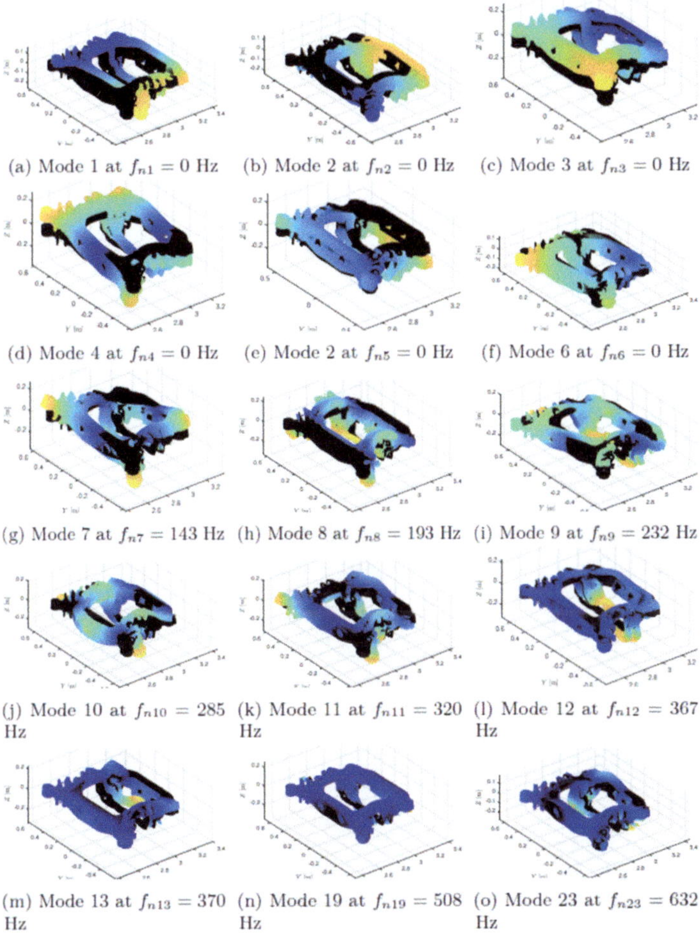


**Fig. 20.6.** Dominant modes for reduction of rear-axle-carrier model (ref. [13]).

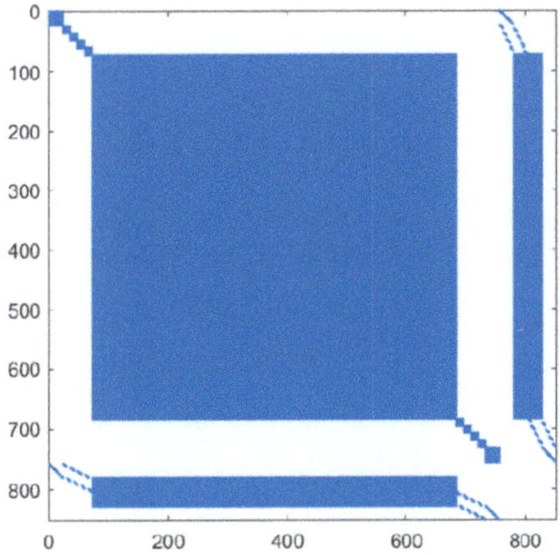


**Fig. 20.7.** Topology for assembled EDU-RAC impedance matrix (ref. [13]).

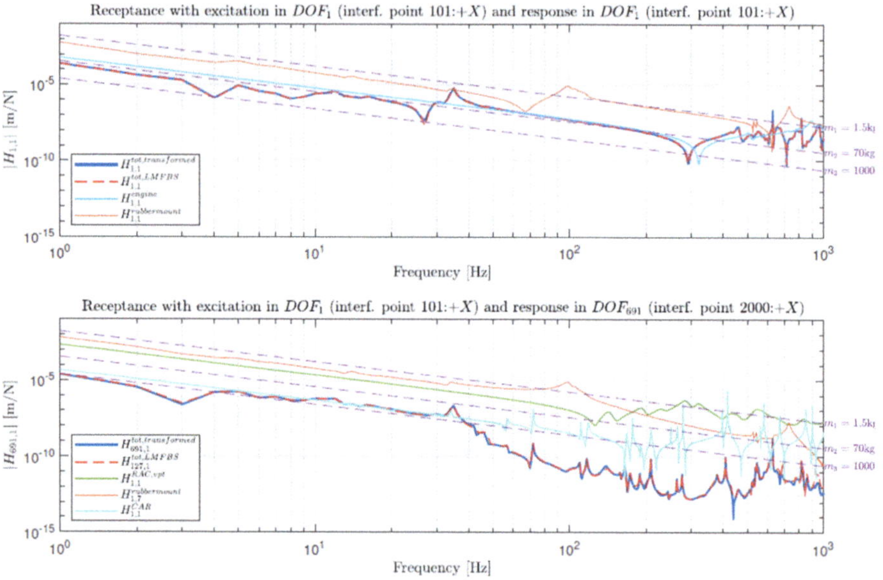


**Fig. 20.8.** Receptances of the structures constituting the EDU-RAC system (ref. [13]).

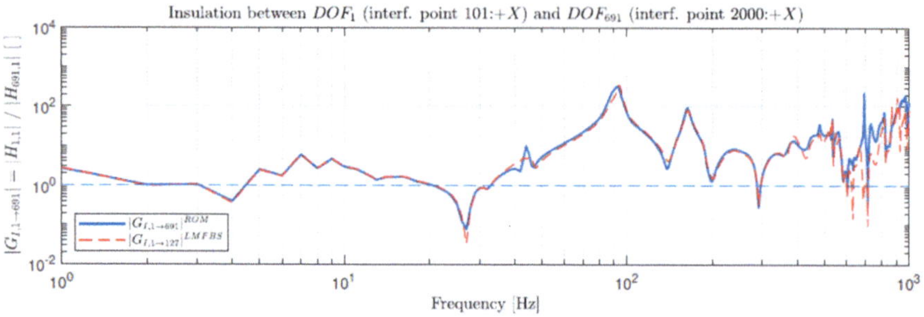


**Fig. 20.9.** Insulation between interface nodes 101 and 2000 before and after the rubber mounts in X-global direction (ref. [13]).

## 20.5 Conclusion

This work has demonstrated a versatile and interpretable framework for dynamic substructuring aimed at insulation evaluation in complex mechanical assemblies. By extending the dual assembly formulation to support coupling across physical, modal, and frequency domains, we enable the integration of detailed numerical models with experimentally characterized subcomponents. This hybrid approach retains the computational advantages of frequency-domain modeling while enhancing the physical interpretability essential for design-driven applications.

The use of Lagrange multipliers within the frequency-based substructuring (LM-FBS) context allows for accurate and efficient assembly of reduced-order and full-order models. The formulation of insulation as a ratio of receptances provides a clear physical interpretation of dynamic interactions across subsystems, particularly highlighting the role of anti-resonances and their dependence on structural parameters.

The methodology was successfully applied to a representative vehicle subassembly, where insulation behavior between an electric drive unit and a rear axle carrier was analyzed. The results reveal how flexible modes and interface dynamics influence insulation, and how targeted structural modifications can be guided by modal insights.

Overall, the proposed framework serves as a powerful tool for engineers seeking to not only quantify but also understand and optimize insulation characteristics in complex structural systems. Future work will focus on extending this approach to nonlinear and time-varying systems, and integrating optimization techniques for automated design improvements.

## 20.6 CRediT Authorship Contribution Statement

**Alessandro Zucchini**: Conceptualization, Methodology, Software, Investigation, Writing—Original draft preparation, Writing—Review & Editing, Visualization.

**Frank Naets**: Conceptualization, Resources, Supervision, Writing—Review & Editing, Project administration, Funding acquisition.

**Acknowledgements.** The European Union is also gratefully acknowledged for their support of the research of A. Zucchini under the ECODrive research project (GA 858018). Views and opinions expressed are however those of the authors only and do not necessarily reflect those of the European Union. The European Union cannot be held responsible for them. Internal Funds KU Leuven are gratefully acknowledged for their support.

## References

1. D. Rixen, "Dynamic substructuring," *Wikipedia,* 01 2024.
2. M. S. Allen, D. Rixen, M. van der Sejis, P. Tiso, T. Abrahamsson and R. L. Mayes, Substructuring in Engineering Dynamics: Emerging Numerical and Experimental Techniques, Springer International Publishing: CISM International Centre for Mechanical Sciences, 2019.
3. A. Zucchini, F. Naets und A. Hülsmann, "Transmissibility and insulation evaluation inside a multi-degrees of freedom system," in *Forum Acusticum 2023,* Turin, Italy, 2023.
4. M. Maia, P. V. Urgueira und R. A. B. Almeida, "Whys and Wherefores of Transmissibility," *Vibration Analysis and Control,* 01 2011.
5. A. M. R. Ribeiro, J. M. M. Silva und N. M. M. Maia, „On the generalisation of the transmissibility concept," *Mechanical Systems and Signal Processing,* Bd. 1, Nr. 14, pp. 29-35, 2000.
6. N. M. M. Maia, J. M. M. Silva und A. M. R. Ribeiro, „The transmissibility concept in multi-degree-of-freedom systems," *Mechanical Systems and Signal Processing,* Bd. 1, Nr. 15, pp. 129-137, 2001.

7. D. Klerk, D. Rixen und J. Jong, "The Frequency Based Substructuring Method reformulated according to the Dual Domain Decomposition Method," in *Proceedings of the XXIV International Modal Analysis Conference*, St. Louis, MO, 2006.
8. D. Klerk, D. Rixen und S. Voormeeren, „General Framework for Dynamic Substructuring: History, Review, and Classification of Techniques," *AIAA Journal - AIAA J,* Bd. 46, pp. 1169–1181, 2008.
9. M. van der Sejis, "Experimental dynamic substructuring," TU Delft, https://doi.org/10.4233/uuid:28b31294-8d53-49eb-b108-284b63edf670, 2016.
10. J. Ortega Almirón, F. Bianciardi, P. Corbeels, R. Ullmann und W. Desmet, "Mount characterization analysis in the context of FBS for component-based TPA on a wiper system," in *Forum Acusticum 2020*, Lyon, France, 2020.
11. M. Haussler, D. C. Kobus und D. Rixen, „Parametric design optimization of e-compressor NVH using blocked forces and substructuring," *Mechanical Systems and Signal Processing,* Bd. 150, pp. 107-217, 2021.
12. A. Ruiz de Aguirre Malaxetxebarria, "How RBE1 RBE2 and RBE3 work?," AERSYS KNOWLEDGE UNIT, 2014.
13. A. Zucchini und F. Naets, A dual substructuring formulation for electric vehicle drivetrain vibrations, Leuven, Belgium: KU Leuven, 2025.
14. A. Zucchini, F. Naets und A. Hülsmann, "Comparison of free-free and fixed boundary conditions for non-linear rubber mount characterization," in *ISMA—30th International Conference on Noise and Vibration Engineering and USD—International Conference on Uncertainty in Structural Dynamics*, Leauven, Belgium, 2022.
15. A. Zucchini, F. Naets, A. Hülsmann und W. Desmet, "Characterizing non-linear elements with free-free and fixed boundary conditions for dynamic substructuring," in *NOVEM—7th International Conference on Noise and Vibration emerging methods*, Auckland, New Zealand, 2023.
16. A. Zucchini, F. Naets und A. Hülsmann, „Characterisation of rubber mounts through Virtual Point Transformation using different boundary conditions in the context of Dynamic Substructuring," in *IMAC XLI—A Conference and Exposition on Structural Dynamics*, Austin, Texas (USA), 2023.
17. A. Zucchini, A. Hülsmann und F. Naets, "A methodology for the experimental characterization of rubber isolators with frequency and preload dependency," *Mechanical Systems and Signal Processing,* Nr. 225, p. 112225, 2025.

# Hybrid Virtual Prototyping for Accelerated Development of Road Noise Performance

Domenico Minervini[1](✉), Junmin Park[2], Heungsuh Park[3], and Tom Dirickx[1]

[1] Siemens Digital Industries Software, Leuven, Belgium
{domenico.minervini,tom.dirickx}@siemens.com
[2] Hyundai Motor Europe Technical Center, Rüsselsheim Am Main, Germany
jmpark@hyundai-europe.com
[3] Yundai Motor Company, Hwaseong-si, Republic of Korea
11010@hyundai.com

**Abstract.** To address the growing complexity in vehicle variants, the automotive industry is focusing on characterizing the NVH (Noise, Vibration, and Harshness) profile of vehicles using virtual assemblies of subcomponents. With the rise of electrification, tire/road noise has become a critical factor in cabin noise, making the representation of tire dynamics crucial. To achieve a comprehensive virtual vehicle noise performance assessment, this study implements an innovative methodology combining experimental and simulation techniques. Initially, through Frequency-Based Substructuring (FBS) decoupling a representative Frequency Response Function (FRF) model of the tire under operational boundary conditions is extracted. Secondly, a hybrid Virtual Prototype Assembly (VPA) model is developed by integrating a test-based trimmed body and tire model with a simulation-derived suspension model. Wheel center blocked forces, experimentally identified, are applied to this hybrid-VPA model to predict virtual vehicle road noise performance. Finally, a multi-level transfer path analysis (TPA) is utilized to precisely identify body receiver contact loads derived from wheel blocked forces and virtual assembly dynamics. This approach enables a detailed classical transfer path analysis that can strategically guide development decisions in body design, bushing stiffness optimization, and chassis configuration.

**Keywords:** Road Noise · TPA · C-TPA · FBS · Decoupling · Tire

## 21.1 Introduction and Motivation

In the last years, among the many challenges brought by the growing diversity of powertrains, electrification and hybridization contributed to a significant increase in available variants for each vehicle. This represents a very important challenge on the accessibility of the full-vehicle prototypes to individually test at the end of the cycle. Therefore, automotive Original Equipment Manufacturers (OEMs) are exploring


A. Heintzel (Ed.): Automotive Acoustics Conference 2025, Proceedings, pp. 301–315, 2026. https://doi.org/10.1007/978-3-658-51249-1_21

different techniques to allow system testing performances without the need of a real assembled vehicle. The so-called "virtual prototyping" enables us to assess a large number of variants and actively control the NVH performances via a modular representation and combination of subcomponents. It allows early detection of possible issues and system optimization at a stage where the impact of the cost and time of making modification is minimal. On the component level, electrification and the consequently phasing out of internal combustion engines highlighted the role of the tire/road noise as one of the main contributors to the full-vehicle noise profile [1].

Among the different technical solutions to this problem Component-Transfer Path Analysis (C-TPA) represent the most promising one [2, 3]. This set of techniques allows an invariant source characterization independently from the vehicle availability, allowing possible exchanges between the full-vehicle engineers and the component (tire, suspension, etc.) experts, but also from supplier and car manufacturer. A crucial point in the C-TPA, as a building block approach, is to represent and combine all the different components in their realistic operational conditions. For tires it is widely known [1, 4, 5] how the boundary conditions such as preload, inflation and rolling speed affect the modal behavior, thus having a realistic test-based representation of the tire subcomponent is of primary importance.

The adoption of component-based TPA methodologies in industry has evolved from proof-of-concept projects toward design exploration and Virtual Prototype Assembly (VPA), with the final goal being integration within actual vehicle development processes. This evolution requires combining various modeling approaches, connecting test and simulation data in a unified framework for comprehensive NVH assessment.

The current challenge in the industry is in the identification of the building blocks for C-TPA, being these representative component models, in their realistic operational boundary conditions. Such components must fulfill the FBS requirement of being free at the connection point with the inclusion of all the phenomena occurring in realistic condition such as preload, stick-slip and other nonlinearities. For simple components, for example a lower control arm, all these requirements are easily satisfied by measuring in free-free boundary conditions. Other components, for instance shock absorbers or tires, cannot be directly characterized, thus requiring indirect approaches like Frequency-Based Substructuring (FBS) Decoupling [5–7].

In this paper, a novel hybrid approach combining experimental and simulation techniques is presented. FBS Decoupling is utilized for extracting realistic tire dynamics under operational conditions, which are then integrated with simulation-based suspension models and experimental body models in a Virtual Prototype Assembly framework. This methodology fulfills the need for integration between experimental and numerical domains in virtual prototyping workflows. The hybrid-VPA model, combined with experimentally identified wheel blocked forces, enables accurate road noise prediction while the multi-level transfer path analysis identifies critical paths and provides design guidance. This approach is valuable in the early stages of vehicle development, where design changes are still possible with minimal impact on cost and schedule.

## 21.2 Theory

This section presents the theoretical background of the hybrid methodology proposed in this paper. Component-based TPA and Virtual Prototype Assembly, FBS Decoupling, and blocked forces methodology are briefly described. This section aims at giving the reader a solid base before diving into the application and workflow. The existing literature is referred where needed.

### 21.2.1 Component-Based TPA

Component-based Transfer Path Analysis (C-TPA) is a set of techniques that allows the NVH performance of an assembly to be predicted using independent component characterizations [2, 3]. Source components are characterized by invariant loads (Blocked Forces) and impedances at their output connection interfaces. Receiver components are characterized by impedances and transfer sensitivities between their input and output connection interfaces.

In the framework of C-TPA, any component can be reduced to a comprehensive model using only characteristic data at its connection interfaces. Different types of component models can be used depending on the function of a component in its assembly [3]. Source components are mainly characterized by operational loads and impedance FRFs, passive components by impedance and transfer FRFs, mounts by stiffnesses, and receivers by impedance and target FRFs.

Despite their simplicity, collaborating using component models can be challenging in practice, as there is no unique way to define a component model for a given component. Standardized component templates are crucial to facilitate the exchange of component models. Such templates prescribe the input and output interfaces, naming conventions, coordinate systems, and operational dependencies.

Virtual Prototype Assembly (VPA) is the process of building and analyzing virtual prototypes starting from component characterizations. The key advantage of VPA is the possibility to evaluate NVH performance before the first physical prototype is available. Component characterizations from any origin (test-based and/or CAE-based) are published to a standardized component library. This capability to seamlessly integrate experimental and numerical components represents one of the major strengths of the VPA approach, enabling hybrid models that leverage the best aspects of both domains. Next, virtual prototypes are unambiguously assembled as block diagrams using standardized components from the library. Finally, the NVH performance of the virtual prototypes is predicted and assessed from objective and subjective point of view.

For the actual coupling of components, Frequency Based Substructuring (FBS) is employed. In the framework of FBS coupling, all components are represented by FRF matrices and combined following a specific nomenclature. The A-structure represents the first structure (e.g., tire), the B-structure is the second structure (e.g., body), and the assembly of the two is called AB-structure (e.g., full-vehicle). In the framework of FBS decoupling, B-structure is usually the unknown structure, but for sake of consistency, in this manuscript it follows the naming in Fig. 21.1, thus it will be A. [5–7]. In this manuscript, the Degrees of Freedoms (DoFs) are categorized so that 1 represent the structural internal DoFs of the A-Structure, 2 represent the interface DoFs of the A-Structure, 3 is the interface DoFs of the B-Structure, 4 is the structural DoFs of the B-structure (e.g. knuckle), 5 is the acoustic DoFs of the B-structure (e.g., cabin pressure). The compliances are denoted by H. For a rigid connection case, such as a tire coupled to a vehicle, the dynamics of the assembly can be described by solving simultaneously the equations of the individual subsystems together with the equilibrium and compatibility equations, as in Eq. (21.1).

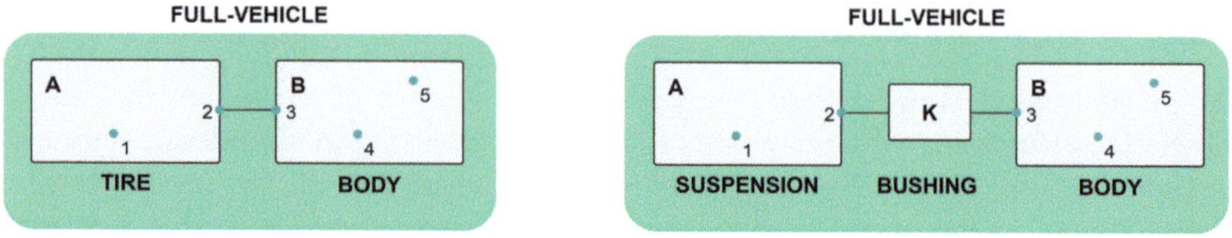


**Fig. 21.1.** Substructuring representation of full-vehicle assemblies: (left) rigid connection between tire and body; (right) suspension and body substructures connected via soft connection characterized by bushing stiffness K.

In this formulation, the core of the coupling process is at the interface where displacement compatibility and force equilibrium are enforced.

$$\boldsymbol{H}^{AB} = \left( \begin{bmatrix} \boldsymbol{H}^{A}_{11} & \boldsymbol{H}^{A}_{12} & 0 \\ \boldsymbol{H}^{A}_{21} & \boldsymbol{H}^{A}_{22} & 0 \\ 0 & 0 & \boldsymbol{H}^{B}_{44} \end{bmatrix} - \begin{bmatrix} \boldsymbol{H}^{A}_{12} \\ \boldsymbol{H}^{A}_{22} \\ -\boldsymbol{H}^{B}_{42} \end{bmatrix} \left( \left[ \boldsymbol{H}^{A}_{22} \right] + \left[ \boldsymbol{H}^{B}_{33} \right] \right)^{-1} \begin{bmatrix} \boldsymbol{H}^{A}_{12} \\ \boldsymbol{H}^{A}_{22} \\ -\boldsymbol{H}^{B}_{43} \end{bmatrix}^{T} \right) \tag{21.1}$$

For a soft connection case, such as suspension components connected via bushings, an additional flexibility term is included to represent the dynamic behavior of the connecting element, as in Eq. (21.2).

$$\boldsymbol{H}^{AB} = \left( \begin{bmatrix} \boldsymbol{H}^{A}_{11} & \boldsymbol{H}^{A}_{12} & 0 \\ \boldsymbol{H}^{A}_{21} & \boldsymbol{H}^{A}_{22} & 0 \\ 0 & 0 & \boldsymbol{H}^{B}_{44} \end{bmatrix} - \begin{bmatrix} \boldsymbol{H}^{A}_{12} \\ \boldsymbol{H}^{A}_{22} \\ -\boldsymbol{H}^{B}_{42} \end{bmatrix} \left( \left[ \boldsymbol{H}^{A}_{22} \right] + \left[ \boldsymbol{H}^{B}_{33} \right] + \boldsymbol{K}^{-1} \right)^{-1} \begin{bmatrix} \boldsymbol{H}^{A}_{12} \\ \boldsymbol{H}^{A}_{22} \\ -\boldsymbol{H}^{B}_{42} \end{bmatrix}^{T} \right) \tag{21.2}$$

This expanded formulation allows for the inclusion of mount or bushing dynamics in the FBS coupling process, creating a more realistic representation of vehicle subsystem interfaces [8]. The ability to handle both rigid and flexible connections makes the FBS methodology particularly suitable for automotive applications, where components are connected through various types of interfaces.

### 21.2.2 FBS Decoupling

Creating good assemblies requires capturing all components under realistic operating conditions. For frequency-based substructuring, source and receiver impedance FRFs with appropriate boundary conditions are needed. However, this creates a conflict: FBS requires unconstrained component models at connection interfaces, while realistic component dynamics need proper operational conditions [5–7].

FBS decoupling is a methodology to identify an unknown component FRFs by removing the influence of the secondary structure from the assembly of the two. The unknown subsystem-A dynamics can be derived by (21.3).

$$\boldsymbol{H}^{A} = \left(\boldsymbol{H}^{AB} \text{-} \boldsymbol{H}^{AB}\boldsymbol{E}^{AB^{T}}\left(\boldsymbol{C}^{AB}\boldsymbol{H}^{AB}\boldsymbol{E}^{AB^{T}} - \boldsymbol{C}^{B}\boldsymbol{H}^{B}\boldsymbol{E}^{B^{T}}\right)^{+}\boldsymbol{C}^{AB}\boldsymbol{H}^{AB}\right) \tag{21.3}$$

where $\boldsymbol{H}^{AB}$ is the FRF matrix of the assembly, $\boldsymbol{H}^{B}$ is the FRF matrix of the known substructure, and $\boldsymbol{E}$ and $\boldsymbol{C}$ are respectively the equilibrium and compatibility Boolean matrices. Depending on the choice of equilibrium and compatibility equations, different decoupling formulations can be derived [7]. In classical formulation (C-C method), equilibrium and compatibility are solved only at the interface points. Alternative formulations can be derived when solving the equations at non-interface DoFs, as in the A-A method, which is particularly useful when the interface is not accessible [6]. These methods differ in accuracy and efficiency depending on the specific measurement conditions. From experimental point of view, the success of the decoupling is strictly connected by the quality of information provided in the core matrix. Challenges include accessibility, consistency, repeatability, noise, damping estimation and positioning errors. In the process, the inversion of the core matrix amplifies errors, making a strategic approach essential [6].

### 21.2.3 Blocked Forces

In the context of C-TPA, source characterization refers to the quantitative identification of operational conditions when a source is coupled to a receiver [3]. Blocked forces are the forces that the source would apply on a fully rigid receiver. Blocked forces are invariant with respect to the receiver, making them valuable for transferring source characterization between different assemblies.

Blocked forces can be identified in-situ without the need to detach the source from the passive receiver structure, according to (22.4).

$$\boldsymbol{f}^{A}_{2,bl} = \boldsymbol{H}^{AB^{\dagger}}_{42}\boldsymbol{a}^{AB}_{4} \tag{21.4}$$

In the equation, $\boldsymbol{f}_{2,bl}$ are the blocked forces, $\boldsymbol{H}_{42}$ are the transfer functions between interface excitation and indicators responses, and $\boldsymbol{a}_{4}$ are the operational responses at the receiver. This approach is particularly useful when the connection interface has complex geometry or is difficult to access. In the road noise context, blocked forces represent the wheel center excitation caused by tire-road interaction. For accurate characterization, measurements should ideally capture the rolling condition of the tire [5]. The importance of the rolling condition increases in the 80–100 Hz range, where the difference between static and rolling tire dynamics becomes significant [9].

While blocked forces are invariant source characteristics, contact forces represent the actual forces at the interface during operation. For a given interface, contact forces can be calculated from blocked forces, as highlighted by (22.5).

$$f_{3,int}^{AB} = \left[H_{22}^{A} + H_{33}^{B} + K^{-1}\right]^{-1} H_{22}^{A} f_{2,bl}^{A} \tag{21.5}$$

In Eq. (21.6) the acoustic pressure at target locations can be calculated using either blocked forces or contact (also called "interface") forces:

$$p_5 = H_{53}^{B}\left[H_{22}^{A} + H_{33}^{B} + K^{-1}\right]^{-1} H_{22}^{A} f_{2,bl}^{A} = H_{52}^{AB} f_{2,bl}^{A} = H_{53}^{B} f_{3,int}^{AB} \tag{21.6}$$

Contact forces will play a key role in the multilevel TPA approach presented later, providing insights into the actual force transmission paths through various interfaces (wheel to suspension to body), critical for identifying design modifications.

## 21.3 Workflow and Subcomponents

This section describes the hybrid approach workflow (see Fig. 21.2) and the different subcomponents used in the proposed methodology. The integration of experimental and numerical components requires specific characterization techniques for each component to ensure proper interfacing within the virtual assembly. Each subcomponent presents different challenges in terms of capturing operational conditions and interface dynamics.

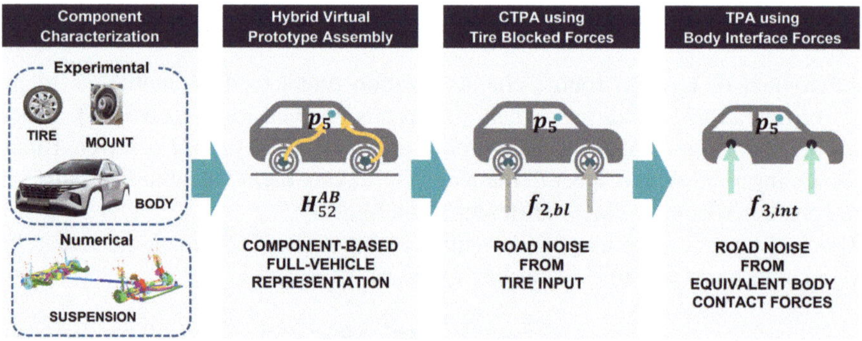


**Fig. 21.2.** Workflow representation. Firstly, the numerical and experimental subcomponent are characterized. Secondly, the VPA methodology is validated on full-vehicle measurements. Successively, the cabin pressure is calculated with a tire blocked-forces approach. Finally, the same pressure is validated using multilevel TPA.

The overall workflow consists of four main steps. First, each subcomponent (tire, body, suspension, bushings) is characterized independently using either experimental or numerical approaches. Second, the characterized subcomponents are assembled in a virtual prototype using FBS coupling techniques. Third, operational loads (blocked forces) identified on a tire test rig are applied to the virtual assembly to predict road noise performance. Finally, a multi-level TPA is performed to identify critical paths and components for targeted design improvements.

### 21.3.1 Tire—Experimental Subcomponent

For researching the best possible model, two different tire model have been measured, the first one through direct measurement on a lightweight test rig, the second one through an indirect method making use of two consecutive measurement of the vehicle with and without tire. A comparison between the two methods will follow in the next chapter.

**Interface—Direct Measurement**
The tire substructure is characterized experimentally using a direct measurement approach. The tire is mounted on a static fixed-free setup where the wheel hub is suspended using elastic bungees to approximate free-free boundary conditions at the interface, while the contact patch is fixed on the ground. Vehicle preload is applied to ensure realistic tire deformation. Virtual Point Transformation and shaker testing is exploited to reduce the dynamics at the interface [5, 9].

**Interface—FBS Decoupling**
Alternative to direct measurement, FBS decoupling is applied to extract the tire dynamics in operational conditions. For this application, the vehicle itself has been used as a test rig for the decoupling operation. The tire interface is measured twice: once with the tire mounted on the vehicle (HAB) and once without it (HA). This approach allows the tire to be characterized under realistic preload conditions while maintaining the necessary unconstrained interface required for FBS coupling [5].

**Blocked Forces—in Situ TPA**
Blocked forces at the wheel center are identified using an in-situ TPA approach. Instead of using a specialized test rig, the vehicle itself is used for operational measurements on a single road surface type. The tire is rolling on the actual road while responses are measured at various locations on the vehicle. Blocked forces are calculated by inverting the measured transfer functions from wheel hub to a set of indicator positions on the vehicle structure. These blocked forces represent the invariant source characteristics that can be transferred to different vehicle configurations in the virtual assembly.

### 21.3.2 Body—Experimental Subcomponent

The vehicle receiver structure is characterized for two conditions. The first condition is the vehicle without wheels, considering the four wheel centers (as the connection interface). This is used for the evaluation of the virtual assembly of the vehicle body and suspension by comparison with experimental data. The second condition is the characterization of the vehicle with suspension receiver for coupling the tires at the wheel centers.

**Vehicle Without Wheel**
The vehicle component is characterized experimentally with the wheels removed but suspension attached. Shaker testing is performed at wheel hubs and combined with VPT to identify the dynamics at the interface and the Noise Transfer Functions (NTFs) to the target locations. The resulting FRFs capture the dynamics of the body including the suspension-to-body interface, which is essential for accurate road noise transmission prediction.

**Vehicle Without Suspension**
The vehicle body is also characterized with the entire suspension removed, the so-called "trimmed body condition", to enable an assembly with the numerical suspension model. This provides clean interfaces at the suspension attachment points on the body side. Shaker testing is performed at these interfaces, and transfer functions to interior noise target locations are measured.

### 21.3.3 Suspension—Numerical Subcomponent

The suspension subcomponent is derived from detailed finite element models. Modal reduction techniques are applied to extract the interface dynamics while maintaining computational efficiency. The suspension model includes all major components (knuckle, strut, subframe) with appropriate constraints and connections. Special attention is given to the interface dynamics at both the wheel hub and the body attachment points to ensure compatibility with the experimental subcomponents.

### 21.3.4 Bushing—Numerical and Experimental Subcomponent

Bushings represent a critical component in the road noise transmission path. Their characterization combines numerical models with experimental validation. Frequency-dependent stiffness and damping properties are extracted from material tests and component-level dynamic tests. The bushing models include preload effects and amplitude-dependent behavior based on experimental identification. For the virtual assembly, bushings are represented as flexible connections in the FBS coupling formulation as described in the theory section.

As consistency and repeatability is key [6], all the measurements have been acquired using a Qsources Integral shaker combined with the in-plane excitation adapter Q-TRX. Simcenter Scadas are used as acquisition systems. The data processing has been performed in Siemens Simcenter Testlab. The numerical simulation has made use of Simcenter 3D.

## 21.4 Results and Validation

This section presents the validation of the proposed methodology. Two applications are considered. First, the improvement in tire model accuracy obtained through FBS decoupling is evaluated. Second, the validity of the hybrid C-TPA methodology for full-vehicle road noise prediction is assessed.

### 21.4.1 Application 1: Improved Tire Model Through FBS Decoupling

At component level, the interface dynamics obtained through direct fixed-free measurement are compared with those extracted using FBS decoupling. Figure 21.3 shows a comparison between the translational DoFs of the $\boldsymbol{H}^{A}_{22}$ matrix. Direct measurements and FBS decoupling demonstrate similar behavior up to approximately 230 Hz across all directions, with critical differences emerging at higher frequencies.

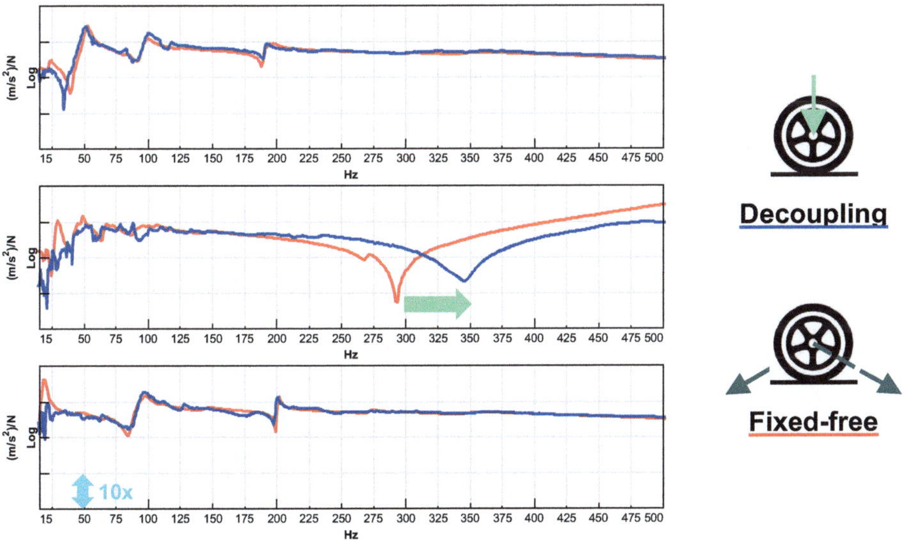


**Fig. 21.3.** FBS Decoupling comparison at component level. The reduced FRFs at the wheel center in lateral, axial and vertical direction are shown. In blue, the decoupled tire FRFs and in red the fixed-free tire FRF.

The flexible tire modes and cavity modes are well captured. In direct measurements, the test-rig influence remains negligible up to 220 Hz, beyond which FBS decoupling becomes necessary to isolate component dynamics from the support. The shift visible in the second subplot, can be linked to the influence of the bungees on the flexible mode of the rim disk and spokes.

Figure 21.4 shows the validation at the assembly level, in this case the cabin noise pressure is compared. The large overestimations caused by the direct method are connected to the influence of the test rig in the component measurements. Conversely, the decoupled tire subcomponent, shows a very good fit, as the antiresonance and the high-frequency tire dynamics are isolated from any unwanted dynamics of the test setup.

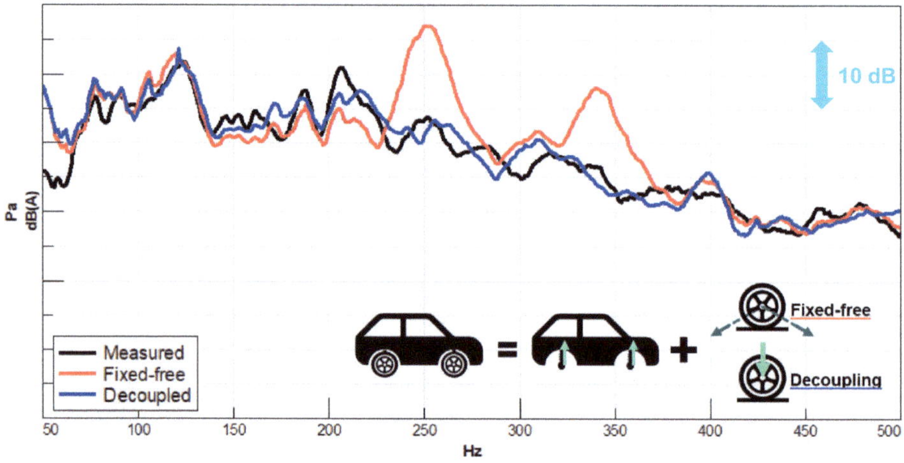


**Fig. 21.4.** FBS Decoupling comparison at assembly level. Cabin pressure predictions obtained by coupling different tire models to the vehicle model: fixed-free tire measurement prediction (red), and FBS decoupling prediction (blue). The reference (black) is measured in assembly-conditions.

The FBS decoupling approach captures the true component dynamics under operational conditions by extracting the component model from an actual assembly state. This provides enhanced boundary condition representation, particularly valuable for components like tires where preload significantly influences dynamic behavior. Between dominant tire modes such as rim twist and spoke bending, the decoupling approach reveals more accurate anti-resonance behavior that fixed-free setups fail to capture. As the frequency range of interest in NVH analysis has expanded from the historical 50–250 Hz to 500 Hz in current applications, this improved accuracy becomes increasingly relevant. Thus, the decoupled tire model is used for the remaining steps of this work.

### 21.4.2 Application 2: Hybrid CTPA

The hybrid CTPA integrates experimental and simulation components for road noise prediction, combining test-based trimmed body, simulated suspension, and test-based tire models extracted via FBS decoupling. Components were assembled using

Eq. (21.2), with bushings modeled as flexible connections with frequency-dependent properties. Validation using wheel center blocked forces shows good correlation between predicted and measured interior noise levels (Fig. 21.5). The model captures both booming phenomena (around 110 Hz) and tire cavity resonance (210 Hz) critical for road-noise applications.

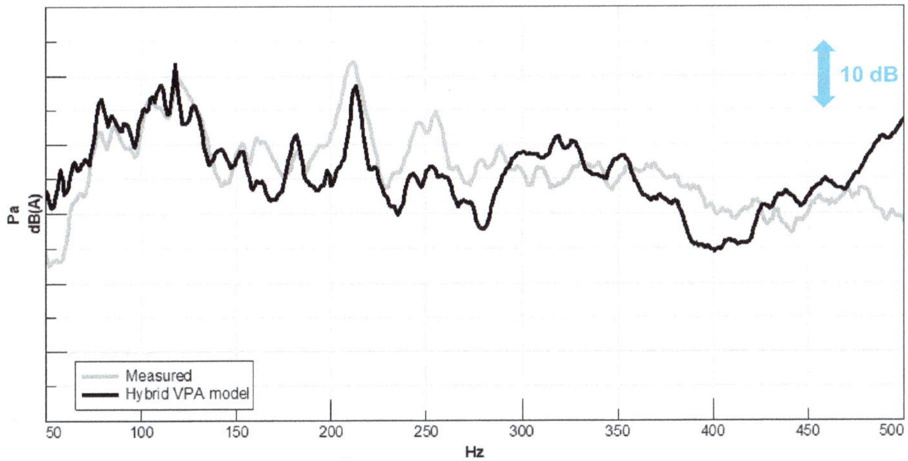


**Fig. 21.5.** Validation of Hybrid VPA approach. In grey, the reference cabin pressure measured in full-vehicle condition. In black, the cabin pressure synthesized by the virtual assembly of both numerical and experimental subcomponents.

## 21.5 Design Exploration

By simulating NVH performance predictions in the early stages of vehicle development using Component-TPA and Virtual Prototype Assembly, NVH development goals and system-level KPIs (Key Performance Indicators) can be set, and systematic development can be conducted by establishing system unit level targets.

### 21.5.1 Multi-Level Transfer Path Analysis

Multi-level Transfer Path Analysis allows to identify and quantify the paths through which vibrations and noise travel from their sources to the receiver.

Based upon the calculated dynamics of the virtual assembly, the wheel center blocked forces (at the source) are propagated to trimmed body interface/contact forces at the receiver component. Detailed knowledge at the receiver interface on contact forces and single path contributions allows the identification of critical suspension transfer paths exciting the body structure and definition of adequate countermeasures.

From the virtual assembly the transmissibility $\boldsymbol{T}_{32}$, from source blocked forces to receiver interface forces can be calculated from Eq. (21.5), as shown in Eq. (21.7).

$$\boldsymbol{T}_{32} = \boldsymbol{f}^{AB}_{3,int} \cdot \left(\boldsymbol{f}^{A}_{2,bl}\right)^{-1} = \left[\boldsymbol{H}^{A}_{22} + \boldsymbol{H}^{B}_{33} + \boldsymbol{K}^{-1}\right]^{-1} \boldsymbol{H}^{A}_{22} \tag{21.7}$$

With the interface forces known at the receiver side, a detailed "classical" Transfer Path Analysis can be performed, calculating single path contributions. The resulting contribution map supports the identification of critical contribution paths, as shown in Fig. 21.6. The single path contributions are grouped per connection point location, combining all 3 DoFs translational contributions.

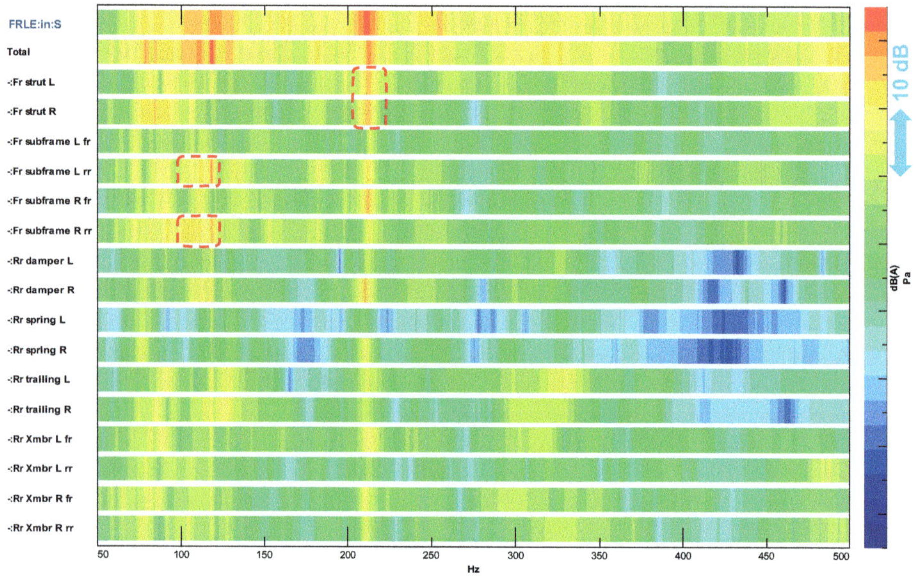


**Fig. 21.6.** Contribution map showing suspension path contributions (3 DoFs per connection point combined). The critical suspension paths are indicated for target responses at booming (around 110 Hz) and tire cavity mode (around 210 Hz).

As an example, the evaluation of the contribution distribution shows dominant contributions from the front strut connections at the second tire cavity mode frequency, around 210 Hz. In the frequency range of booming, around 110 Hz, dominant contributions are indicated from the front subframe (rear) connections.

### 21.5.2 Evaluation of Component Modifications

In the virtual stage, a multitude of assembly variants can be quickly evaluated. From the identified critical suspension paths, through which the body receiver structure is excited, a targeted resolution can be efficiently developed to improve the NVH performance of the virtual assembly. The resolution development focuses on the critical paths, in this example the front strut and the rear connections of the front subframe. A parametric study is set up to evaluate the sensitivity of the virtual assembly to component modifications. This sensitivity analysis provides directions for improving the road noise performance.

In the parametric study, different component modifications are considered: the connection point impedance of the receiver part, the connection point impedance of the active parts, and the suspension bushing dynamic stiffness. In this study, pragmatic variations by scaling from the nominal values are evaluated to investigate the assembly sensitivity. Among the evaluated variations there are the body receiver impedance, bushing dynamic stiffness, and chassis impedance, with variations of $\pm$ 50%, $\pm$ 20%, and $\pm$ 10% from the nominal (0%) values.

**Use-Case 1: Front Strut Contribution**

In the first use-case, the parametric study focuses on the connections of the front strut, resulting in dominant contributions at the second tire cavity mode.

Evaluation of the effects of the component modification, as shown in Fig. 21.7, indicates changes of the body impedance at the front strut connections have no effect on the predicted noise levels. Whereas, reducing the front strut insulator dynamic stiffness with 50%, would result in a target noise level improvement of up to −0.9 dB(A).

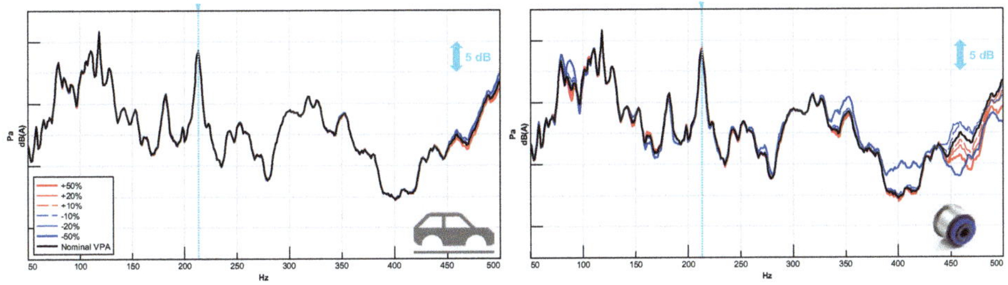


**Fig. 21.7.** Evaluation of component modifications of the front strut on cabin pressure calculated through VPA: (left) variations in body-to-front-strut connection impedance; (right) variations in bushing stiffness values.

**Use-Case 2: Front Subframe Rear Contribution.** In the second use-case, the parametric study focuses on the rear connections of the front subframe, targeting the booming phenomena around 110 Hz. To reduce front subframe contributions exciting the body receiver, in practice, modifications can be made at the impedance of the receiver parts and active parts, as well as the bushing stiffness. The results obtained in the sensitivity analysis for the aforementioned component modifications show that the body receiver connection point impedance affects the road noise performance, optimizing bushing dynamic stiffness values will improve performance, and the chassis connection point impedance (and transfer) are also relevant for the vehicle road noise performance. The results of the parametric studies can provide meaningful directions for improving the road noise performance. This approach enables early detection and solution development in the virtual stage, demonstrating the possibility of identifying issues at the mass production stage during the preliminary phase (Fig. 21.8).

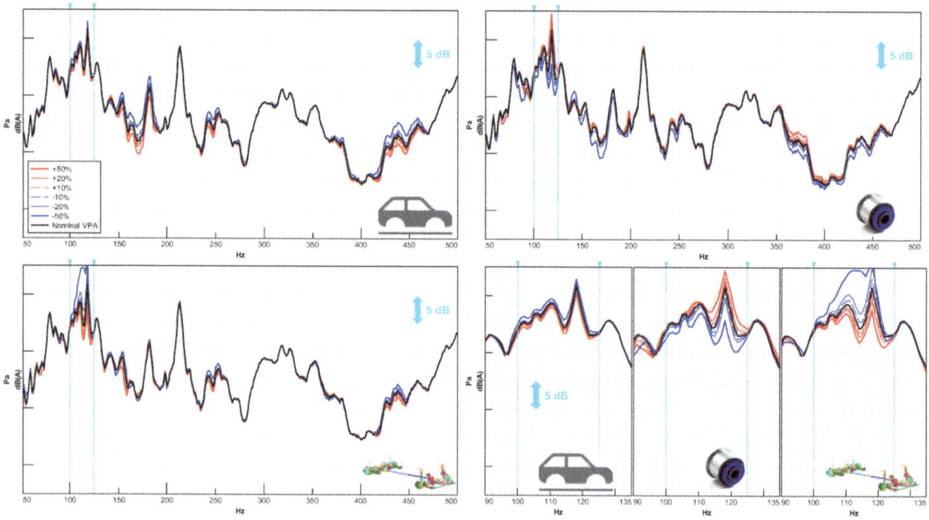


**Fig. 21.8.** Evaluation of component modifications at the front sub frame on cabin pressure calculated through VPA: (upper left) variations in front sub frame to body connection impedance; (upper right) variations in bushing stiffness values; (lower left) variations in chassis (front sub frame) connection impedance; (lower right) a close-up of the variations focusing on the booming phenomena around 110 Hz.

## 21.6 Conclusion

This paper presents an innovative hybrid approach for virtual prototyping that combines experimental and numerical techniques to characterize and optimize road noise performance. The methodology integrates test-based components with simulation-based suspension models in a unified Virtual Prototype Assembly framework, enabling comprehensive noise assessment without physical prototypes.

Results demonstrate significant advantages of FBS decoupling for accurate tire model extraction under operational conditions. While direct measurement and decoupling approaches provide similar results up to 230 Hz, the decoupling technique captures critical high-frequency dynamics more accurately by isolating true component behavior from test setup influences. This accuracy improvement is increasingly relevant as NVH analysis expands toward higher frequencies. The multi-level

TPA implementation allows detailed identification of critical transmission paths from wheel excitations to interior noise. Case studies focusing on the front strut and front subframe connections demonstrated how this approach can pinpoint specific areas requiring optimization. Parametric sensitivity studies revealed that targeted modifications to component parameters, particularly bushing dynamic stiffness, can yield significant improvements in road noise performance, providing engineers with quantitative guidance for component modifications.

This methodology enables automotive manufacturers to efficiently evaluate multiple assembly variants during early development stages, particularly valuable as vehicle electrification emphasizes the importance of road noise characterization. Future research should focus on extending the hybrid methodology to incorporate additional nonlinear phenomena and wider ranges of operational conditions. By combining the strengths of experimental and numerical domains, this methodology contributes to the evolution of vehicle development processes toward more efficient, data-driven frameworks.

**Acknowledgements.** This work has been developed in collaboration with Hyundai Motor Europe Technical Center and Hyundai Motor Company. Their interest to use the proposed Virtual Prototype Assembly techniques to enhance performance prediction capabilities, the joint efforts and technical discussions are gratefully acknowledged. An additional acknowledgement is given to the Flemish Agency for Innovation and Entrepreneurship (VLAIO) for the financial support of this study in the frame of the "ADRENALINE" research project.

## References

1. Kindt, P.: Structure-Borne Tyre/Road Noise due to Road Surface Discontinuities. PhD thesis, KU Leuven, Leuven (2009).
2. Van der Auweraer, H., Janssens, K.: A Source-Transfer-Receiver Approach to NVH Engineering of Hybrid/Electric Vehicles. SAE Technical Paper, SAE International (2012).
3. Allen, M.S., Rixen, D., Tiso, P., et al.: Substructuring in Engineering Dynamics—Emerging Numerical and Experimental Techniques. Springer, Cham (2018).
4. Gonzalez Diaz, C., Kindt, P., et al. J.: Dynamic behaviour of a rolling tyre: Experimental and numerical analyses. Journal of Sound and Vibration 364, 147–164 (2016).
5. Minervini, D., Stella, A.: Experimental Frequency-Based Decoupling for Tire Characterization. In: Proceedings of the ISMA International Conference on Noise and Vibration Engineering, pp. 1–14. KU Leuven, Leuven (2022).
6. Minervini, D., Park, S., Dirickx, T., Geluk, T.: FBS Decoupling at Suspension Level for Road Noise Applications. SAE International Journal of Advances and Current Practices in Mobility 5(2), 937–951 (2023).
7. Voormeeren, S.N., Rixen, D.J.: A Dual Approach to Substructure Decoupling Techniques. In: Proceedings of the 28th International Modal Analysis Conference, pp. 601–609. Springer, Orlando (2010).
8. Wagner, P., Bianciardi, F., Corbeels, C., Hülsmann, A.: High frequency source
9. characterization of an e-motor using component-based TPA. In: Proceedings of the SIA Automotive NVH Comfort conference (2021).
10. Ortega Almirón, J., Bianciardi, F., Corbeels, P., Pieroni, N., et al.: Vehicle road noise prediction using component-based transfer path analysis from tire test-rig measurements on a rolling tire. Journal of Sound and Vibration 523, 116694 (2022).

# Methodology for Road Noise Optimization Using Stiffness Injection for Mount Modification in Combination with a Physical Noise and Vibration Simulator

Eric Pasma[1] and Graham Franks[2(✉)]

[1] Vibes Technology B.V., Burgwal 8, 2611GJ Delft, Niederlande
epasma@vibestechnology.com
[2] VI-grade UK ltd, Station Lane Millbrook, Bedford MK43 7RB, UK
graham.franks@vi-grade.com

**Abstract.** Whilst the process of characterising and optimising automotive design through TPA has been the subject of many technical papers, combining these advanced TPA methodologies with the ability for decision makers to experience the results of these processes in a multi attribute noise and vibration simulator is a relatively new development.

Subjective assessment using Driver-in-Loop simulators to experience data has been shown to improve the quality of vehicles and reduce development time and uncertainty. Efficient development processes require a seamless interface from detailed objective data to subjective evaluations suitable for high level decision makers. In the context of perceived vehicle noise and vibration, the need for a bridge between complex data and realistic subjective evaluation of tactile response is most compelling. A suite of VI-grade noise and vibration (NVH) simulators have been developed to meet this challenge.

This paper will present the results of a component-based TPA using virtual point techniques to calculate the forces at the wheel hub virtual centres performed on a Hyundai vehicle using VIBES software solutions. A methodology will be presented to investigate the sensitivity to changes in suspension bushing stiffness and the stiffness injection via a substructuring approach used to demonstrate a change in a particular bush stiffness which was then validated through vehicle measurement and assessed on the multi attribute NVH simulator.

**Keywords:** Component based-TPA · Frequency-Based Substructuring · Stiffness injection · NVH simulator

## 22.1 Introduction

Whilst the process of characterising and optimising automotive design through TPA has been the subject of many technical papers, combining these advanced TPA methodologies with the ability for decision makers to experience the results of


A. Heintzel (Ed.): Automotive Acoustics Conference 2025, Proceedings, pp. 316–334, 2026. https://doi.org/10.1007/978-3-658-51249-1_22

these processes in a multi attribute noise and vibration simulator is a relatively new development.

Subjective assessment using Driver-in-Loop simulators to experience data has been shown to improve the quality of vehicles and reduce development time and uncertainty. Efficient development processes require a seamless interface from detailed objective data to subjective evaluations suitable for high level decision makers. In the context of perceived vehicle noise and vibration, the need for a bridge between complex data and realistic subjective evaluation of tactile response is most compelling. A suite of VI-grade noise and vibration (NVH) simulators have been developed to meet this challenge.

This paper will present the results of blocked force TPA using virtual point techniques to calculate the forces at the wheel hub virtual centres performed on a Hyundai vehicle using Vibes software tools. The TPA methodology will be used to investigate the sensitivity to changes in suspension bushing stiffness and the stiffness injection via a substructuring approach used to demonstrate a change in a particular bush stiffness which was then validated through vehicle measurement and assessed on the multi attribute NVH simulator.

A blocked force source characterization and component transfer path analysis was conducted for chassis input paths at constant speed operating conditions on several road surfaces. A virtual point transformation was applied to acquire the excitation force in a known coordinate frame to ensure compatibility across all source-receiver assemblies, estimating 6 DOF (translational and rotational) forces at each hub/knuckle as virtual point blocked forces.

The TPA process followed in this paper allows sub-structuring of the receiver system into subsystems such as suspension arms, bushings or subframes. The TPA was performed for multiple interior sound receivers and tactile vibration at the Driver's seat-rail. A seat transmissibility approach was applied to transfer TPA seat rail vibration estimates to seat skin vibration for subsequent replay by the NVH Simulator. Each of these path/receiver contributions may then be individually assessed and modified during replay of the model on the NVH Simulator.

The stiffness injection method was used to virtually increase or decrease the stiffness of the calculated subsystems by changes to a bush which demonstrated a high sensitivity to change, resulting in virtual prototypes with demonstrable changes in tactile vibration and sound compared to the reference vehicle. This allowed detailed subjective evaluation of the effect of the virtually modelled bush stiffness change, which was further validated by measurement of a modified vehicle.

The remainder of this paper is structured as follows: The TPA sub-structuring utilizing the stiffness injection method as applied to the subject Hyundai vehicle is described and test based validation of one of the study conditions (stiffness change to bushing 1) is presented. A second case study (stiffness change to bushing 2) is also presented.

The multi attribute simulator hardware is described in brief detail. The challenges involved in the simulation of driver interface vibration are discussed and the need for a link between simulated seat-rail vibration data and occupant touch point vibration is established. A seat transmissibility calculation is described and applied to the TPA estimated seat rail data. A complete NVH Simulator demonstration model generated from the supplied TPA data with surrogate test data covering the full audible frequency range is presented and discussed.

## 22.2 Theoretic Framework for Vehicle NVH Optimization

### 22.2.1 Frequency-Based Substructuring

FBS enables the assembly of system-level frequency response models from individual subsystem models by FRFs (Frequency Response Functions) [16]. Each component, such as the body or suspension, is characterized independently, and their dynamic interactions are coupled virtually via interface points. This modular approach simplifies what-if analyses by avoiding full system re-testing after each change.

$$\mathbf{u}_3^{\mathrm{cTPA}} = \mathbf{Y}_{32}^{\mathrm{B}} \left(\mathbf{Y}_{22}^{\mathrm{A}} + \mathbf{Y}_{22}^{\mathrm{B}}\right)^{-1} \mathbf{f}_2^{\mathrm{bl}} \;\; \mathbf{Y}_{22}^{\mathrm{A}} = \mathbf{T}_{2\mathrm{u}} \mathbf{Y}_{\mathrm{uf}}^{\mathrm{A}} \mathbf{T}_{\mathrm{f2}}^{\mathrm{T}}$$

To ensure a common interface between subsystems, the virtual point transformation (VPT) is used to acquire six-degree-of-freedom motion at a single defined point or node [18]. In FBS, the VPT enables accurate and physically realistic dynamic coupling of the subcomponents.

### 22.2.2 Component-based Transfer Path Analysis

Component-based Transfer Path Analysis (CTPA) is a technique used to isolate and examine how vibration or noise propagates from a source component into a receiving structure. It relies on describing the excitation—usually in the form of blocked forces—at the interfaces between the source and receiver. These forces can then be applied to various virtual vehicle models, enabling accurate interior noise prediction without repeated physical tests. Critically, blocked forces are independent of receiver dynamics and thus represent only the source component dynamics in isolation, allowing their reuse across multiple design variants.

Various methods exist to obtain these blocked forces [17]. In this study, blocked forces are computed at interface points (e.g., knuckles) using the *in-situ* method. This approach relies on an inverse calculation using measured FRFs on the full vehicle and measured operational responses when driving the vehicle on the road. Using the VPT, the forces and moments are expressed in the center of the knuckle, in six degrees of freedom.

$$\mathbf{f}_2^{bl} = \mathbf{Y}_{42}^{\mathrm{AB}^+} \mathbf{u}_4 \;\; \mathbf{Y}_{42}^{\mathrm{AB}} = \mathbf{Y}_{4\mathrm{f}}^{\mathrm{AB}} \;\; \mathbf{T}_{\mathrm{f2}}^{\mathrm{T}}$$

Predictions to the target points (i.e. seat rail responses, microphones at the driver and passenger ears positions ) are computed using component TPA. This is done by applying the blocked forces to the FRF/NTF from knuckle to the target points.

$$\mathbf{u}_3^{\mathrm{cTPA}} = \mathbf{Y}_{32}^{\mathrm{AB}} \, \mathbf{f}_2^{bl}$$

### 22.2.3 Stiffness Injection

Stiffness Injection is a special case of FBS that inserts virtual stiffness changes into the system's transfer path and hence on the full vehicle [19]. By modifying the dynamic stiffness of specific bushings (e.g., subframe mounts), engineers can simulate how different stiffness values alter the vibration transfer to the cabin—without any actual physical prototyping.

$$\mathbf{u}_3^{\mathrm{cTPA}} = \mathbf{Y}_{32}^{\mathrm{B}} \left(\mathbf{Y}_{22}^{\mathrm{A}} + \mathbf{Y}_{22}^{\mathrm{B}} + \mathbf{Y}_{22}^{\mathrm{SI}}\right)^{-1} \mathbf{f}_2^{\mathrm{bl}} \quad \mathbf{Z}^{SI} = \mathbf{Z}^{oring} + \mathbf{Z}^{mod}$$

Stiffness injection is a method for virtually modifying mount characteristics within a measured vehicle assembly to assess their impact on NVH performance. The process begins by measuring dynamic responses on the fully assembled system, focusing on the bushings of interest. Dynamic motion is captured on both the active side (e.g., suspension) and the passive side (e.g., vehicle body), and Virtual Point Transformation (VPT) is used to convert these measurements into rigid-body motion at defined virtual points. Additionally, responses at target locations—such as microphones at the driver's ear positions—are recorded to serve as performance indicators. With the full system behavior implicitly captured (denoted as system ABC), a virtual spring and/or damper is coupled in parallel to the real mount with the nominal system. If the nominal stiffness of the mount is known, a relative change can be introduced—either increased or decreased—to simulate alternative configurations. This allows for fast and repeatable predictions of interior noise changes without requiring new physical tests (Fig. 22.1).

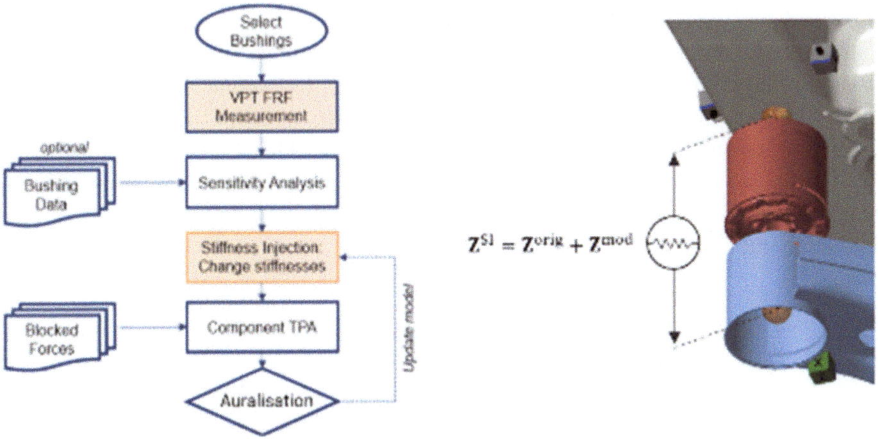


**Fig. 22.1.** On the left, schematic depiction of the vehicle NVH optimization approach using stiffness injection and component-based TPA. On the right, a picture of SI and how virtual mount modifications are applied.

## 22.3 Application Case: Front-Axle Mount Modification

The stiffness injection method was applied to a Hyundai Elantra to virtually evaluate the effect of modifying two subframe mounts on the front axle. The goal was to assess how structural changes at the mount level influence interior noise, in particular road-induced noise at the driver's ear. The workflow was carried out in five systematic steps (Fig. 22.2).

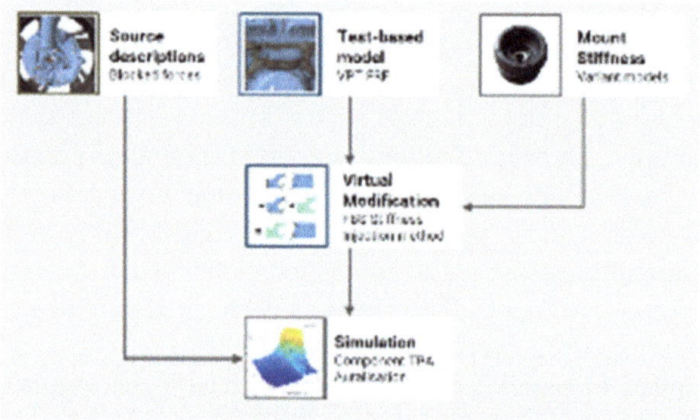


**Fig. 22.2.** Schematic representation of the approach for road noise optimization

### 22.3.1 Blocked Force Identification

Blocked forces were identified at the knuckle, during a constant-speed drive over a rough road surface at a proving ground. These computed blocked forces represent the excitation input and are invariant to changes in the vehicle's receiving structure.

The on-board component TPA synthesis of the individual front and rear axle contributions, as well as their combined effect, shows excellent correlation with the validation measurement, as can be seen in Fig. 22.4. This high level of agreement confirms the accuracy of the calculated blocked forces, thereby validating their use for subsequent simulation tasks in when performing stiffness injection predictions (Figs. 22.3 and 22.4).

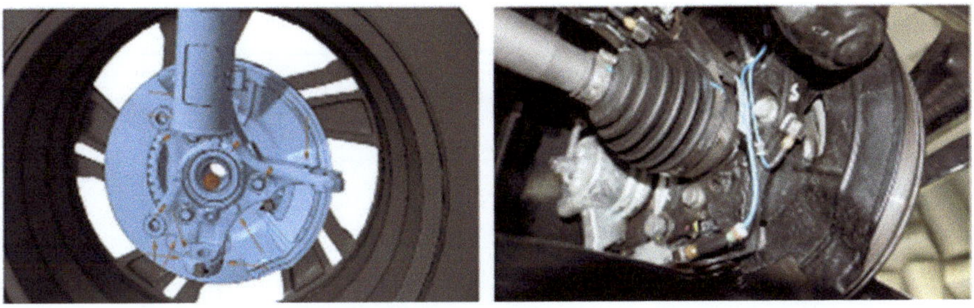

**Fig. 22.3.** The design of experiment in DIRAC, the sensor positions on the knuckle, the excitations for the blocked forces in the virtual point on the knuckle. Of the right, the actual

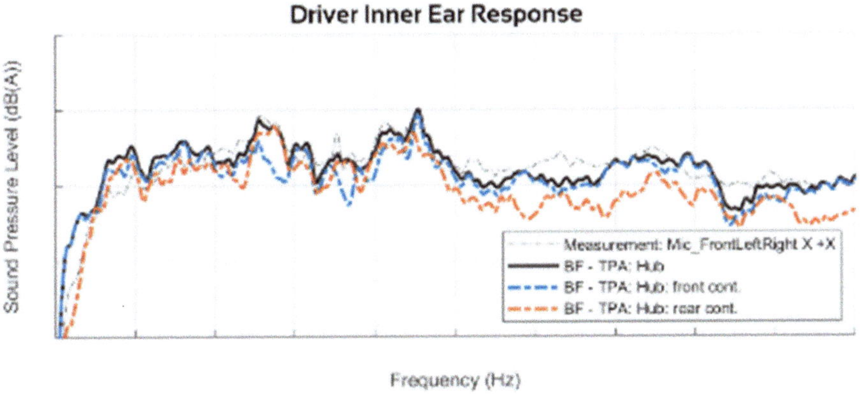


**Fig. 22.4.** The component TPA synthesis of the front and rear axle contributions separately and combined, compared with a validation measurement.

## 22.3.2 Test-Based Modelling in DIRAC

Subsequently, a test-based model of the vehicle's dynamic behavior at the mounts of interest was acquired using the DIRAC methodology. Frequency Response Functions (FRFs) were measured acquired between virtual points located on the active and passive sides of the two front-axle mounts. This is done by, per virtual point, apply 3 triaxial accelerometers and 9–12 excitation positions, all spatially well distributed for accurate VPT estimation. In the same model, the two 6 DoF excitations on the knuckles, and the target response points in cabin are included (Fig. 22.5).

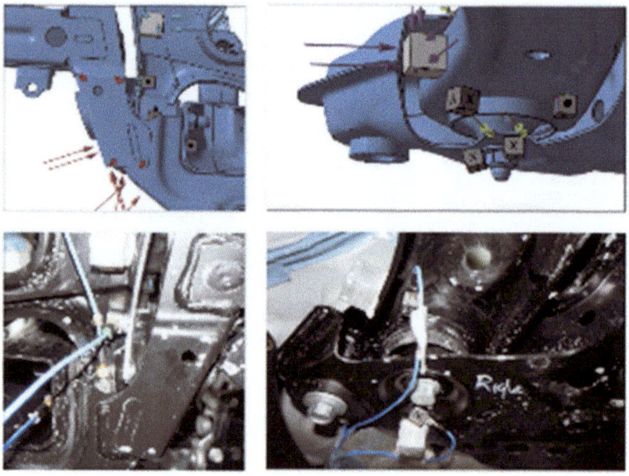

**Fig. 22.5.** VPT FRF measurement in DIRAC on the active and passive side of the mount.

### 22.3.3 Virtual Modification Using Stiffness Injection

Between the active and passive side virtual points, the mount dynamics were then virtually modified using stiffness injection. Specifically, an extreme case was considered to test the robustness of the method: one of the mounts was virtually rotated by 90 degrees, effectively swapping its stiffness characteristics in the x and y directions. This resulted in a change in directional stiffness by approximately one order of magnitude.

With the stiffness injection model established, it is now possible to compute the Sound Pressure Level (SPL) sensitivity with respect to a unitary change in mount stiffness. This sensitivity analysis enables the quantification of the acoustic impact resulting from variations in mount dynamic behavior. For a more targeted assessment, the frequency spectrum of interest is divided into three distinct ranges, corresponding to characteristic vehicle interior noise phenomena: the booming , cavity and the rumble frequency range.

The results can be seen in Fig. 22.6. What can be observed is that the sensitivity on driver's ear is that there's not one clear trend indicating how stiffness changes will improve or deteriorate the NVH performance. These detailed sensitivity studies give therefore clear insight into the impact of proposed changes, therefore, this approach is well suited for balancing of attributes.

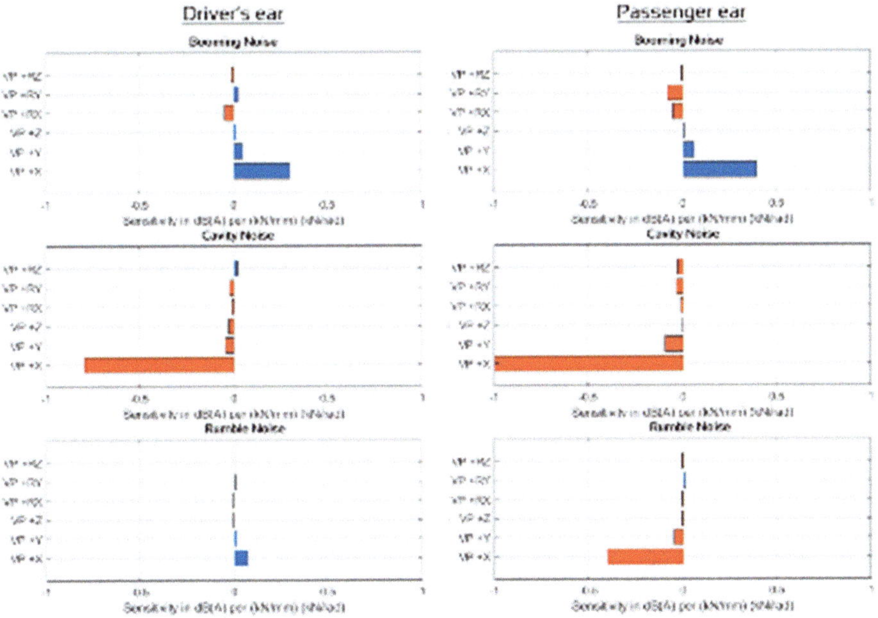


**Fig. 22.6.** Sound Pressure Level sensitivity with respect to an unit mount stiffness change. the frequency band of interest is split up into three ranges: booming, cavity and rumble noise.

### 22.3.4 Response Prediction using Component TPA

A component-based Transfer Path Analysis (TPA) was performed to predict the driver's ear Sound Pressure Level resulting from an extreme modification of the front-axle mount. In this analysis, the previously validated blocked forces—applied at the knuckle—were used as excitation. The prediction was then compared to a validation measurement obtained by physically implementing the same mount modification on the vehicle.

The correlation between the predicted and measured interior noise levels confirms the validity of the approach. Due to the modification, a sharp resonance is introduced in the vehicle, clearly observable in the cabin. This change, can be predicted accurately with the presented approach as can be seen in Fig. 22.7.

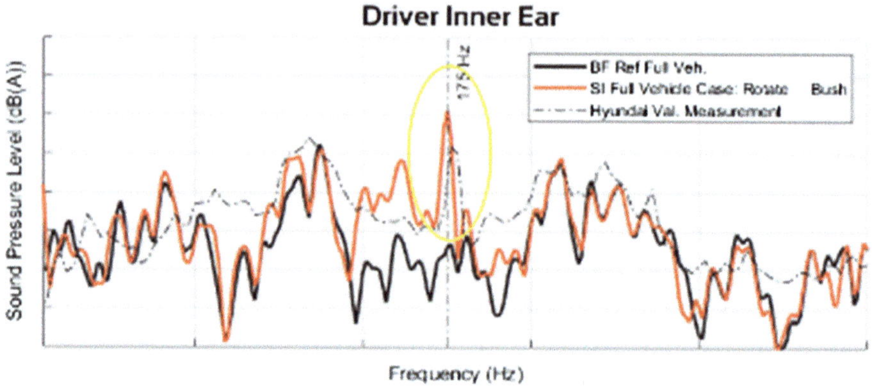


**Fig. 22.7.** Component TPA prediction on driver's ear for the extreme mount modification. the results are compared with the validation measurement, where the mount was physically changed on the vehicle.

## 22.4 TPA Data Replay on the Compact NVH Simulator

In the example presented here we will show how in a development proof-of-concept application we can import Vibes generated blocked force at the hub and combine it with the assembled hub to receiver transfer functions, to provide time domain estimates of the interior noise and tactile vibration for replay in the Compact NVH Simulator from VI-grade. The study will demonstrate the results from the stiffness variation simulation at bushing 2.

The application of noise, vibration and harshness (NVH) driver-in-the-loop simulation technology, to address the growing need to remove physical prototype testing from the engineering development process, has been widely discussed in previous publications [1–4]. The ability to evaluate test or CAE simulation results in a back-to-back, interactive driver-in-the-loop evaluation scenario offers subjective evaluation advantages over real vehicle testing, by removing the often significant time and cost required to physically introduce an engineering change in a vehicle and make it available for evaluation [5–7].

The range of frequencies of vibration perceived by a seated vehicle occupant caused by road inputs is often described as primary ride (0.8Hz – 5Hz) and secondary ride (5Hz to 20Hz) [11]. There are significant sensitivities at lower frequencies, including typically, the coupled system resonance of the seated person on the flexible seat cushion at around 5Hz [12, Fig. 22.9]. Simulated event replay of the lower frequency region is already well covered by either single axis (or multi axis) vertical displacement simulators or Stewart platform (hexapod) type simulators which deliver controlled whole vehicle vibration in six axes of motion [13]. The VI-grade Compact NVH simulator offers a unique approach in that its focus is on delivering calibrated noise and vibration from around 8–120Hz, including the region where noise and vibration interact as 'harshness', for seven tactile vibration interface degrees of freedom (DOFs). The key feature of the compact NVH simulator is its ability to independently calibrate and control these seven interface DOFs within a driver-in-the-loop simulation where the driver can 'free-drive' the vehicle. This allows for the accurate multi axis simulation of both measured and CAE simulated data, rapid switching between 'modelled' vehicles, and the independent control of individual channels with the means to modify any channel using frequency domain filters. Whilst the omission of frequencies below 8Hz changes somewhat the perception of the higher frequency vibration being presented, it is relatively easy for even novice users to quickly become accustomed to the absence of lower frequency motion and draw meaningful subjective impressions from the presented stimuli.

In this case, the presentation of multi-dimensional vibration stimuli to an assessor in a driver-in-the-loop scenario has been a key enabler in the desire to draw meaningful evaluation of CAE automotive ride predictions. The VI-grade Compact NVH Simulator was developed precisely to meet this need. The Compact NVH simulator is designed to offer the benefit of providing high fidelity individual channel vibration reproduction without the significant associated complexity of adding the sub-8Hz high displacement reproduction, which, whilst feasible, will add significant additional cost to the system. Having the ability to independently control the amplitude and phase of each vibration channel is a key feature of the NVH compact simulator enabling the interaction between vibration at different touch points to be investigated, the importance of which has been established by Jang et al [14].

### 22.4.1 Compact NVH Simulator Hardware

The VI-grade Compact NVH Simulator has been developed to meet the requirement for a NVH development tool that can provide realistic replay of either measured or simulated vibration. The Compact NVH Simulator (Fig. 22.8) provides accurate, calibrated, multi-axial vibration at all driver touchpoints, along with synchronised binaural sound and visuals enabling evaluation of the complete NVH experience. The reproduction of each of these excitation channels is essential to deliver a meaningful replay environment for the driver.

The Compact NVH Simulator provides calibrated vibration replay (from 8–120 Hz) at the following driver contact points as shown in Fig. 22.9:

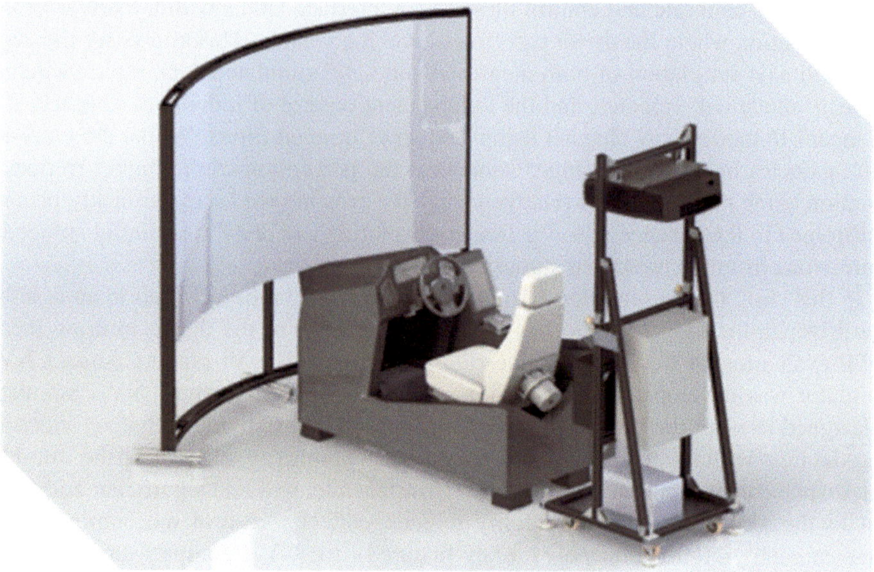

**Fig. 22.8.** Compact NVH Simulator driver-in-the-loop hardware

- Seat base (vertical)
- Seat base (lateral)
- Seat backrest (fore-aft, normal to frame)
- Floor (vertical)
- Floor (lateral)
- Steering system (vertical, normal to column)

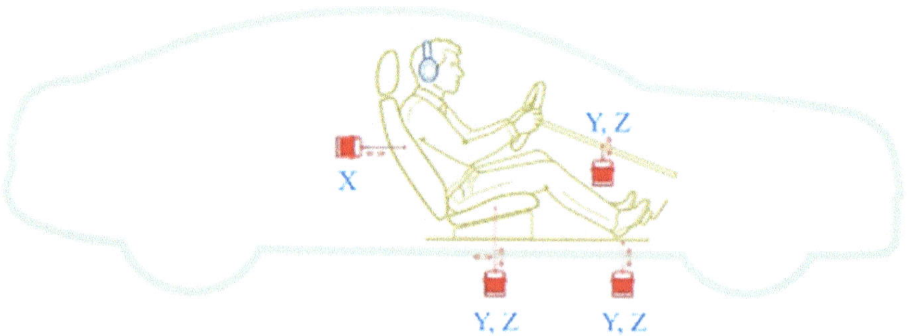


**Fig. 22.9.** Compact NVH Simulator vibration channels

Steering system (lateral)

In order to replay data representing existing physical properties, the input data should be measured directly at the driver contact points on the subject vehicle using a combination of seat and floor accelerometer pads (accelerometers embedded in rubber pads and placed on the seat and floor) and further accelerometers at hand contact points on the steering wheel rim.

The vibration replayed on the simulator at any of the interface contact points will be a function of the target drive signal provided by the simulator software, the actuation system response and the influence of the seated occupant. The Compact NVH Simulator has an embedded response calibration system that ensures the reproduced vibration is replayed at a calibrated level for any given driver, accounting for the influence that their anthropomorphic characteristics have on the replay system. This is achieved by calculating a per-channel measurement equalization (EQ) for the whole system for each occupant (or range of occupants) which is applied in real time during replay by pre-application of finite impulse response (FIR) filters. The system has been designed with largely decoupled axes of vibration which enables good results using this simplified control system, an example of which is shown in Fig. 22.10.

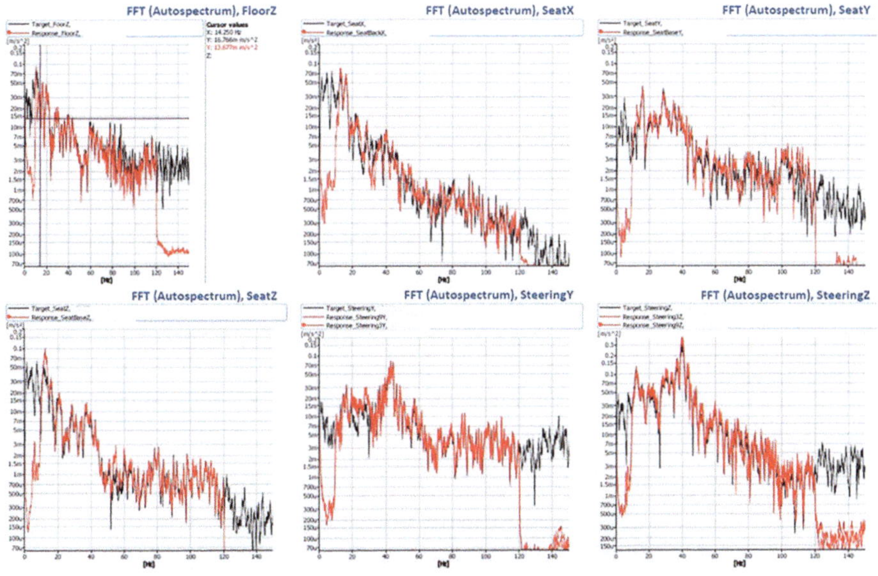


**Fig. 22.10** Compact NVH Actual vs. Target Vibration

### 22.4.2 Data Preparation For Replay on the Compact NVH Simulator

**Structure-borne Noise and Vibration Estimates from TPA.** The Vibes TPA process produced output in the form of hub blocked force at 6 Degrees of Freedom (DOF) and transfer functions from the 6 DOF hub force to Driver binaural and vibration response points.

Data was supplied as FFT vs. time blocked force estimates for the LH front hub for two cases: 60 and 120kmh for a 10 second period. Since the supplied data was stochastic in nature, a linear average of the complex spectra was obtained for each of the 6 DOFs. The source strength (hub blocked force) data was then converted to a time domain equivalent by means of convolving the complex force spectra with a unity broadband random time signal in the VI-NVHSim CAE Auditioner software. The appropriate vehicle sensitivity transfer functions are then assigned to deliver the required contact point vibration and binaural sound with a further convolution step (Fig. 22.11).

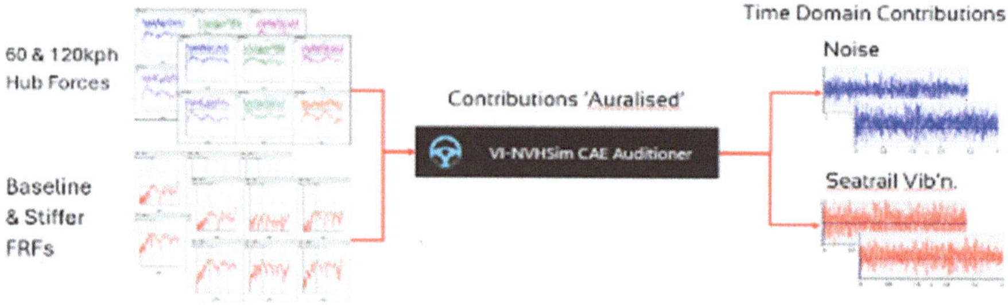


**Fig. 22.11.** Generation of TPA Path Contributions in Time Domain

As is often the case with either simulation or measured laboratory transfer functions, the seat response vibration was characterized in the TPA by transfer functions measured from the suspension hub to the seat-rail. To meet the requirement for seat contact point vibration replay in the Compact NVH Simulator it was necessary to apply some transformation from seat-rail to seat contact point vibration. A seat transmissibility matrix (T-matrix) approach [15] was therefore used.

The transmissibility matrix is derived from operating measurements made on the occupied seat of a vehicle being driven over representative road surfaces. Cross spectral matrices between the indicators (seat rail locations) and receivers (seat contact points) and between the indicators themselves are obtained from the measured data (Fig. 22.12).

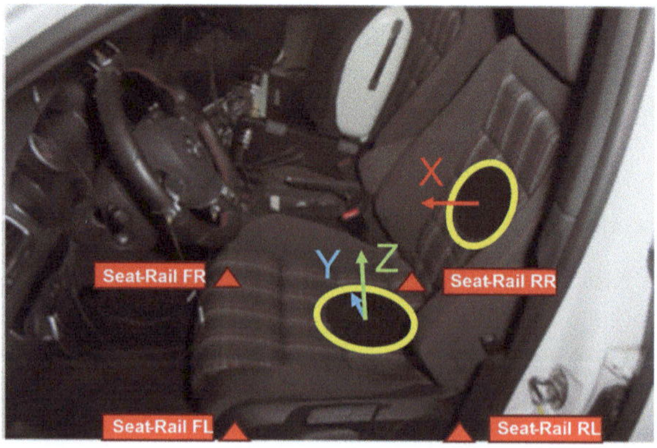


**Fig. 22.12.** Compact NVH Simulator vibration channels

The T-Matrix relationship is then given by:

$$\left[\widehat{T}_{ir,k}\right] = [C_{ir}][C_{ii}]^{-1}.$$

The T-matrix may then be applied to the seat-rail vibration estimates from the TPA output (frequency or time domain by means of T-matrix filter convolution with seat rail time histories) to estimate an equivalent contact point vibrations:

$$\left\{\widehat{G}_r\right\} = \left[T_{ir,k}\right]\{G_i\},$$

The (SB) contributions at the 9 response point receiver locations may then be calculated from the 6 blocked force hub DOFs for the LH hub input, summarized as shown below in Fig. 22.13:

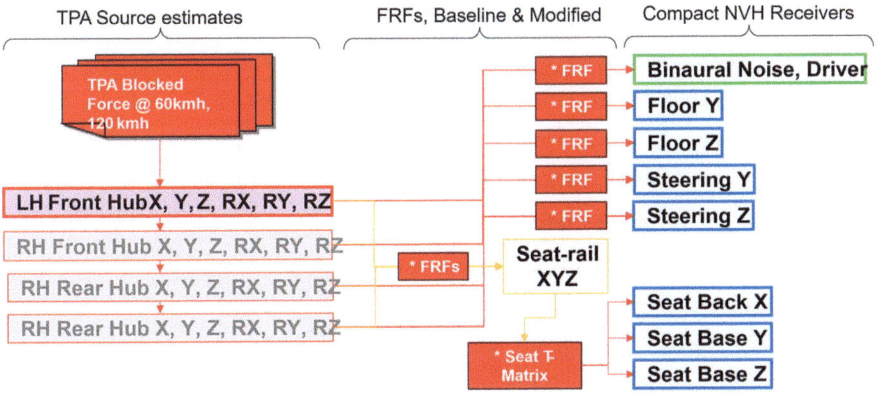


**Fig. 22.13.** Compact NVH Simulator SB Noise Estimate

The above process was performed to provide comparable SB contributions for the baseline condition and for the modified FRF assembly where the stiffness injection process was used to simulate the increased stiffness at bush 2.

**Contextual Noise for NVH Simulator.** The above described process yields SB noise contributions at the Driver's ear and vibration at the tactile point locations. The tactile point vibrations are adequately described by SB vibrations alone. The driver binaural noise however will be also heavily influenced by higher frequency tyre airborne (AB) noise (from tyre tread/road surface interactions) and wind noise. The following process of Operational Path Analysis (OPA) describes how contextual higher frequency noise may be estimated for the subject vehicle.

The subject vehicle should be instrumented according to Fig. 22.14 below with reference accelerometers at each hub and reference microphones at the front and rear of each tyre contact patch.

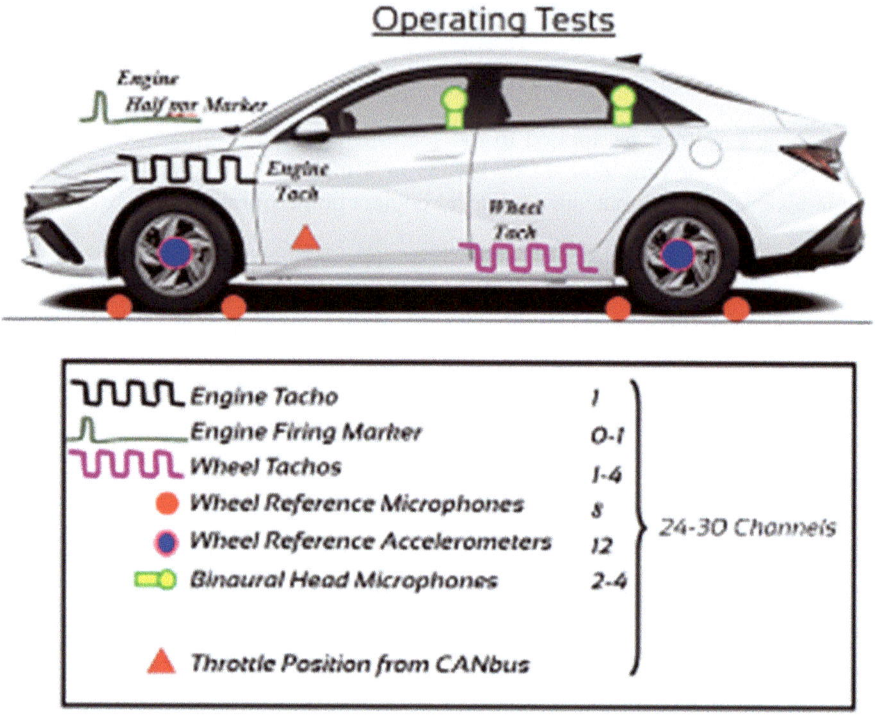


**Fig. 22.14.** Operational Path Analysis Measurement Set

Measurements are then taken on the subject vehicle, on the same surfaces and at the same speeds for which the TPA was performed. Any harmonic data is removed form the recorded signals using the tacho references and narrow envelope filters, leaving only broadband random signals. Operating Path Analysis (OPA) is then applied using a transmissibility approach to determine contributions of reference signals to response signals. Input time histories for reference and response signals are processed using short-time Fourier transforms (STFT) to create the response and reference calculation matrices:

$$T_{est} = [response, stft(f,t)]\ [reference, stft(f,t)]^{-1}$$

The contribution matrix $X_c$ is calculated using the estimated transmissibility matrix $T_{est}$ and the final result is converted to the time-domain using inverse short-time Fourier transforms for use in the NVH Simulator Model.

$$X_c = [reference, stft(f,t)]\ [T_{est}]$$

A contribution is calculated for each of the 20 references (12 hub accelerometers and 8 tyre contact patch microphones) and a wind noise estimate obtained from the measured total minus the sum of contributions as summarized in Fig. 23.15.

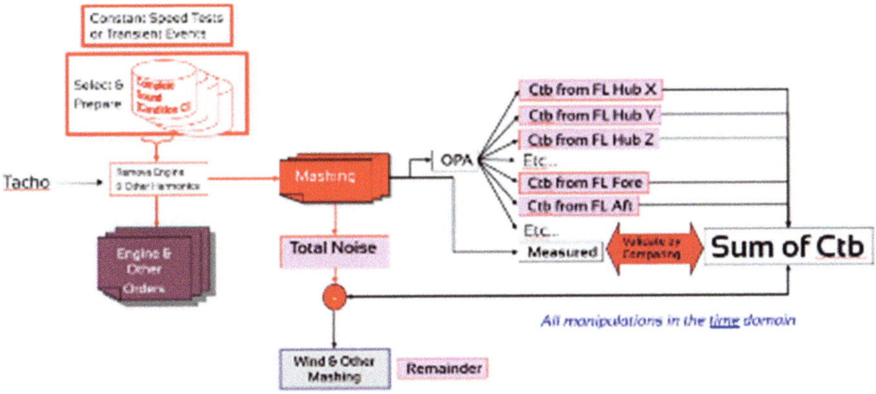


**Fig. 22.15.** Operational Path Analysis

Finally, the SB contributions from the OPA may be discarded and replaced with the more detailed TPA contributions for modelled sub-systems. In this case the LH front hub TPA contribution was raised by 3dB (to approximate the effect of adding a RH front hub) and used to represent the SB noise from the front suspension. The SB noise from the rear suspension, the AB noise from the tyres and the wind noise estimate from the OPA were then added in the simulator model to create a simulation of the complete audio range as summarised in Fig. 22.16.

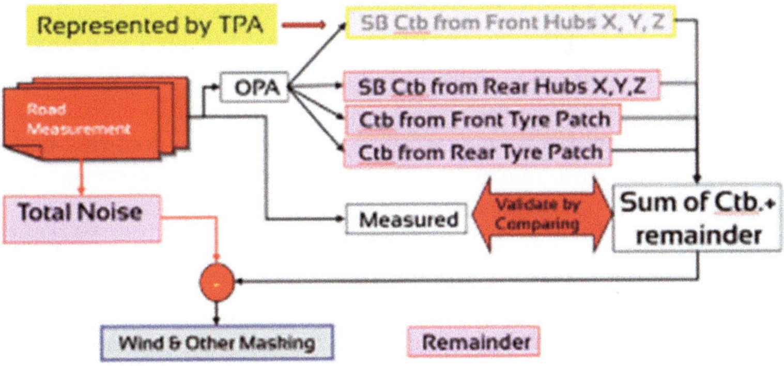


**Fig. 22.16.** Interior Noise Model Represented in NVH Simulator

**Demonstration Model and Subjective Evaluation of Bush Change.** The joint Hyundai and Vibes TPA project discussed in section 2 of this paper was conducted some time before VI-grade's involvement. It was therefore not possible to obtain a full corresponding dataset, measured on the same vehicle, to support the OPA and seat transmissibility process. In the absence of this supporting data it was decided to demonstrate the processes described in section 3 by constructing a Compact NVH

Simulator model using the Vibes calculated LH front hub TPA path contributions along with OPA and seat transmissibility data from a similar surrogate vehicle.

Subjective evaluation of this model yielded some interesting results:

- The 60kph Driver's binaural noise showed only very subtle changes as a result of the bush change, whilst the accompanying seat and floor vibration was noticeably worse for the case with the stiffened bush.
- The 120kph Driver's binaural noise showed noticeable degradation as a result of the bush stiffness increase, once again the accompanying seat and floor vibration was noticeably worse for the case with the stiffened bush.
- In general the perception of vibration change was greater than the perceived change in noise for both speeds.
- It was noted that if the assessment had been based on noise alone, focussing on the 60kmh test condition, then only small changes would have been noticed; adding vibration assessment to the evaluation allowed for a much clearer conclusion.

## 22.5 Conclusions

This paper applies a Frequency-Based Substructuring based method —the Stiffness Injection Method—to model and optimize of front-axle mounts in a full vehicle. The method enables structural modification without physically altering the vehicle. In-situ blocked force characterization at the wheel hubs was performed to independent loads. These invariant blocked forces were used as inputs in a component-based Transfer Path Analysis to predict interior noise levels.

Mount stiffness sensitivity was evaluated across key frequency regions: booming, cavity, and rumble noise. An extreme virtual mount modification was simulated, and the predicted interior noise was validated against measurements from a physically modified vehicle.

The strong agreement between simulation and experiment confirms the accuracy and practicality of the proposed approach for virtual NVH assessment and mount optimization.

An approach for a creating a multi modal noise and vibration simulation environment is presented to assess the differences between the two designs. The demonstration simulator model allowed a clear subjective judgement to be made between the two modelled bush designs to back up and clarify the interpretation of the available objective data. The comparison of such datasets in the compact NVH Simulator allows one to appreciate the important differences of complex multi-channel frequency domain data that is simply not possible with objective data alone.

Being able to experience the design changes in this way is an extremely powerful addition to the TPA process and brings a simple and accurate assessment of the results to NVH experts and non-experts alike. The whole process presents a robust, thorough and yet very straightforward means of assessing design changes in a contextually meaningful environment.

## References


1. Williams, R., Henderson, F., Allman-Ward, M., Dunne, G. et al., "Using an Interactive NVH Simulator for Target Setting and Concept Evaluation in a New Vehicle Programme," SAE Technical Paper 2005-01-2479, 2005, https://doi.org/10.4271/2005-01-2479.
2. Allman-Ward, M., Venor, J., Williams, R., Cockrill, M. et al., "The Interactive NVH Simulator as a Practical Engineering Tool," SAE Technical Paper 2003-01-1505, 2003, https://doi.org/10.4271/2003-01-1505.
3. Bogema, D., Adriano, G., Hodgkins, J., Allman-Ward, M. et al., "Utilizing a Hybrid Engineering Approach for NVH Drive Evaluations in Virtual Prototypes," SAE Technical Paper 2022-01-0980 (2022), https://doi.org/10.4271/2022-01-0980
4. Bogema, D., Newton, G., Stickler, M., Hocking, C. et al., "Idle Vibration Analysis and Evaluation Utilizing a Full-Vehicle NVH Simulator," SAE Technical Paper 2015-01-2334, 2015, https://doi.org/10.4271/2015-01-2334
5. Quinn, D., Speed-Andrews, P., Allman-Ward, M., and Heinz, T., "How Advances in On-Road NVH Simulator Technology Have Enabled Firm Targets for Delivery at the Concept Phase," SAE Technical Paper 2009-01-2178, 2009, https://doi.org/10.4271/2009-01-2178.
6. Bogema, D., "The Benefits of Experiencing NVH and CAE Data from Pre-concept to Production" , SIA Le Mans 2016.
7. Bogema, D., Guan, J.,Miskin, A., Jean Francois, A., Platten, M., "Integrating CAE, Test and Experience to Optimize NVH Design and Development", SIA Le Mans 2018
8. Fothergill D., Syred F., Kennings P., 'The use of system modelling techniques to filter test measurements and drive a physical vehicle NVH simulator', 7th Altair CAE Technology Conference, Heritage Motor Centre, Gaydon, UK, 2011.
9. Kennings, P., Senapati, U., Fothergill, D., and Syred, F., "A Novel Use of Acoustic and Vibration Simulation Techniques to Develop Better Ride Comfort for a Luxury Cabriolet Car," SAE Technical Paper 2013-01-1956, 2013, https://doi.org/10.4271/2013-01-1956.
10. Kennings, P., Layfield, J., Tarabra, M., Fothergil, D., Syred, F.,Franks., G, "Developing Powertrain Mounting Systems in the Virtual Engineering World Using a Full Vehicle NVH Simulator", Internoise 2014
11. Cieslak, MP, Kanarachos, S, Diels, C, Blundell, M, Baxendale, A & Burnett, M 2020, 'Accurate ride comfort estimation combining accelerometer measurements, anthropometric data and neural networks', Neural Computing and Applications, vol. 32, no. 12, pp. 8747–8762. https://doi.org/10.1007/s00521-019-04351-1
12. Toward, M. G. R. & Griffin, M. J., "The transmission of vertical vibration through seats: influence of the characteristics of the human body". 19 Dec 2011 In : Journal of Sound and Vibration. 330, 26, p. 65266543.
13. Xue, H., Previati, G., Gobbi, M., and Mastinu, G., "Research and Development on Noise, Vibration, and Harshness of Road Vehicles Using Driving Simulators—A Review," SAE Int. J. Veh. Dyn., Stab., and NVH 7(4):555-577, 2023, https://doi.org/10.4271/10-07-04-0035.
14. Jang, H.-K. and Griffin, J.M., "The Effect of Phase of Differential Vertical Vibration at the Seat and Feet on Discomfort," J. Sound Vib. 223, no. 5 (1999): 785-794.
15. Franks, G., Application of a Seat Transmissibility Approach to Experience Measured or Predicted Seat-Rail Vibration in a Multi-Attribute Simulator, 13th International Styrian Noise, Vibration & Harshness Congress: The European Automotive Noise Conference. https://doi.org/10.4271/2024-01-2962

16. De Klerk, D, Rixen, D, Voormeeren, S, "General Framework for Dynamic Substructuring: History, Review, and Classification of Techniques", May 2008 AIAA Journal 46(5):1169-1181 https://doi.org/10.2514/1.33274
17. Van der Seijs, M, De Klerk, D, Rixen, D, General framework for transfer path analysis: History, theory and classification of techniques, August 2015, Mechanical Systems and Signal Processing 68-69:217-244 https://doi.org/10.1016/j.ymssp.2015.08.004
18. Van der Seijs, M, Van den Bosch, D, De Klerk, D, Rixen, D, An improved methodology for the virtual point transformation of measured frequency response functions in dynamic substructuring, June 2013, Conference: 4th International Conference on Computational Methods in Structural Dynamics and Earthquake Engineering, https://doi.org/10.7712/120113.4816.C1539
19. Gagliardini, L, De Klerk, D, Stiffness Injection: A Tool for Vehicle NVH Performance Optimization, June 2022, Conference: 12th International Styrian Noise, Vibration & Harshness Congress: The European Automotive Noise Conference, https://doi.org/10.4271/2022-01-0976

# Deep Learning Surrogate Models for Automotive Acoustics Predictions

F. Hesse[2], A. Mustafa[1](✉), D. Spoljaric[2], V. Rao[2], S. Jeronimo[2], and F. Mayot[3]

[1] Dassault Systèmes, St.-Martin-Straße 82, 81541 Munich, Germany
afaq.mustafa@3ds.com
[2] Dassault Systèmes, Meitnerstraße 8, 70563 Stuttgart, Germany
{faron.hesse,Varun.RAO.intern,
svetlana.jeronimo}@3ds.com
[3] MAHLE GmbH, Pragstraße 26-46, 70376 Stuttgart, Germany
florian.mayot@mahle.com

**Abstract.** Engineers do not solely focus on the energy efficiency of the car, van or truck. A holistic approach to road vehicle development considers passenger comfort as well. And one key part of passenger comfort is pleasant acoustic characteristics of the transportation vehicle. Two noise sources in the road vehicle are the engine bay/battery pack cooling fan, as well as the heating, ventilation and air conditioning (HVAC) unit. For both noise sources, geometry optimization of the component in question for low noise must be carried out. In the past, this was done in the digital domain only using scale-resolving partial differential equation based continuum mechanics fluid solvers that, for good fidelity, are run on high-performance-computing clusters with appropriately high mesh resolution. Doing so, requires extensive compute resources and time. With the advent of more powerful computational hardware, supervised machine learning algorithms that learn from collected/historical computational aeroacoustic data can now be used to carry out accelerated optimization studies for noise mitigation. This paper presents two such optimization studies. In the first instance, multi-layer- perceptron networks are used to predict fan performance and acoustic behavior. The second case uses a combined convolutional neural network and multi-layer-perceptron architecture to determine the noise characteristics of HVAC ducts.

**Keywords:** Computational Aeroacoustics · Machine Learning · Fan Acoustics · HVAC Acoustics


A. Heintzel (Ed.): Automotive Acoustics Conference 2025, Proceedings, pp. 335–348, 2026. https://doi.org/10.1007/978-3-658-51249-1_23

## 23.1 Introduction

In recent years, the term digital twin has become increasingly widespread. It has found application in essentially all industries, from transportation & mobility to aerospace & defense and even to life science & healthcare. In short, a digital twin is an in-silico representation of a system, physical object, or process that reflects in real-time its real-world counterpart using data input from IoT devices, sensors, and more to predict object performance, thereby assisting in optimization, decision-making, and troubleshooting. To get closer to true digital twins, traditional computer-aided-engineering (CAE) workflows that employ finite volume, finite element, lattice Boltzmann, and more methods are now being augmented with machine learning (ML) techniques to reduce the high compute time and resources associated with large design-of-experiment (DOE) optimization studies. This work focuses on the application of the lattice Boltzmann method (LBM) as a computational aeroacoustic (CAA) toolkit whose derived simulation data is fed into supervised ML models to predict both performance and acoustic behavior of critical noise generating parts found inside road vehicles. The LBM code used is Dassault Systèmes' (3DS) PowerFLOW® solver, while the ML algorithms are written in Python. These tools are utilized on two key noise sources: an engine bay/battery pack cooling fan from MAHLE GmbH, and general HVAC geometries from 3DS. The remainder of this work is organized as follows: Sect. 23.2 describes the methodology adopted, Sect. 23.3 reviews the results, Sect. 23.4 provides a conclusion, and Sect. 23.5 discusses potential directions for future work.

## 23.2 Methodology

As two use-cases are explored, the methodology section is divided into two parts: Sect. 23.2.1 describes the workflow for the MAHLE GmbH fan noise study, while Sect. 23.2.2 illustrates the process for the 3DS HVAC duct noise study.

### 23.2.1 MAHLE GmbH Fan Noise Study

In the MAHLE GmbH fan noise study, the first part of the CAE workflow augmented by ML is to gather the raw LBM simulation data. This is done by running 40 cases of a morphed MAHLE GmbH fan positioned in a downstream chamber set-up as illustrated in Fig. 23.1. The following boundary conditions are critical for this study:

- The "wall" (a.k.a no-slip) condition is stipulated at the fan, chamber, shroud, motor stator, motor rotor and simulation domain floor.

- At all further simulation domain sides, the "static pressure—free flow direction" condition is applied.
- To obtain the desired mass flow rate, at the chamber's outlet a mass flow condition of 1kg/s is imposed.

Furthermore, it is important to note that the fan is surrounded by a moving/rotating mesh. This means that the fan and mesh in its immediate vicinity physically rotate and the rotation speed is set at 2550rpm. Also, the flow domain's initial pressure is set at 101325Pa and a temperature of 80°C, thereby replicating the warm conditions of a vehicle engine bay. Moving on, the DOE space is depicted in Fig. 23.2. Five morphs are implemented: "EndofBlade_height", "leadingedge", "stagger_mitte", "traileredgehub", and "traileredgering". For the training of the ML models 34 simulations (marked in red) are used, for validation during training 3 simulations (marked in green) are employed and, finally, for blind testing 3 simulations (marked in blue) are utilized. Furthermore, as part of the MAHLE GmbH fan noise study, 4 simple feed-forward deep neural networks are designed. These networks are also called multi-layer-perceptrons (MLPs).

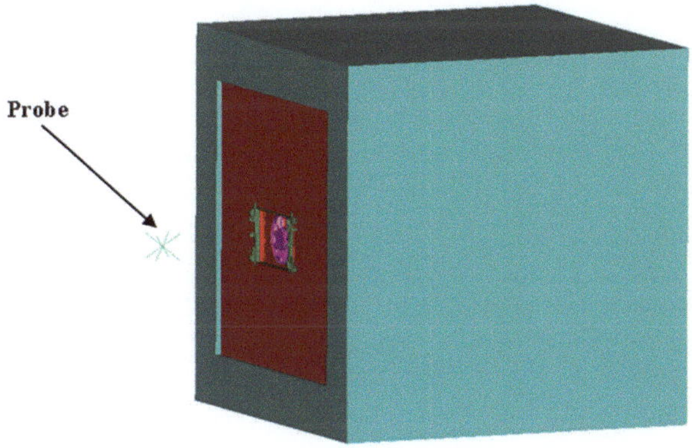


**Fig. 23.1.** Downstream chamber set-up with the morphed MAHLE GmbH fan located at the inlet

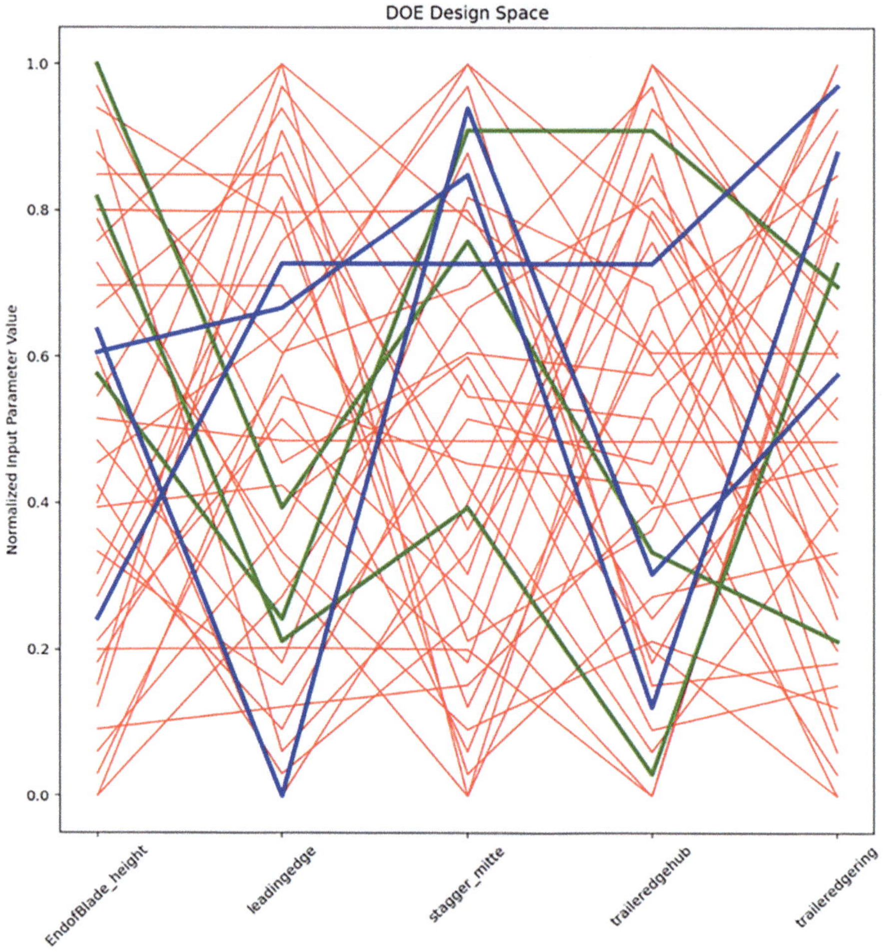


**Fig. 23.2.** DOE input morph parameter visualization

Four different MLPs are trained in order to predict 4 different sets of quantities, as outlined below:

- Fan surface static pressure is predicted using an MLP with "3D" input data where the mesh coordinates, mesh normals and the morph parameters are read in, as depicted in Fig. 23.3.
- Fluid volume slice static pressure and velocity magnitude are predicted using "3D" input data where mesh coordinates and morph parameters serve as MLP input values.
- Fan performance parameters, torque and delta pressure, are computed using a "1D" MLP where only fan morph parameters are read-in.

- Fan acoustics, in the form of sound pressure level (SPL) and overall sound pressure level (OASPL), are calculated using a "1D" MLP where only fan morph parameters are fed-in. The raw data for acoustics is derived from the pressure trace measured at the probe shown in Fig. 23.1. It is located 1m upstream of the chamber along the fan's centerline.

To evaluate the predictive capabilities of all four MLPs, contour plot comparisons to the true PowerFLOW® LBM data are made, reference line plot analysis with R-squared (R2) values is done, and uncertainty quantification using a Gaussian curve analysis is performed.

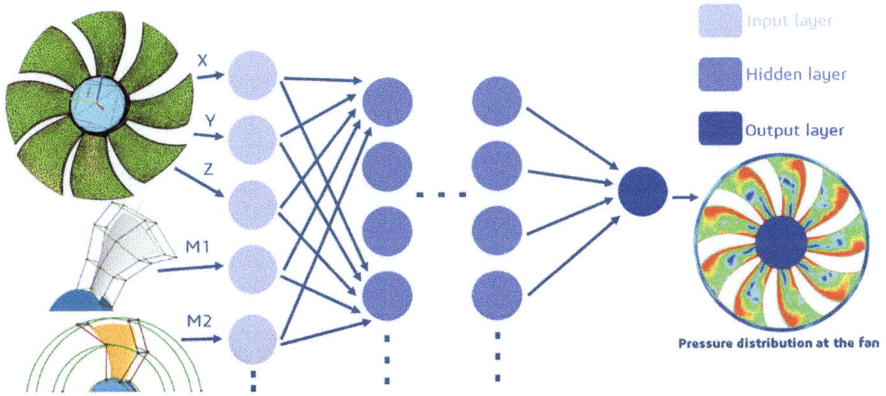


**Fig. 23.3.** Sample "3D" MLP for prediction of fan surface static pressure

### 23.2.2 3DS HVAC Duct Noise Study

In the 3DS HVAC duct noise study, the first step in the CAE workflow coupled to ML is to gather the raw LBM simulation data as well. As a baseline, two duct geometries are used in this study: (1) the EV12 rectangular duct shown in Fig. 23.4a, and (2) the EV12 circular duct depicted in Fig. 23.4b. In both simulation cases, the following boundary conditions are applied:

- The "wall" (a.k.a no-slip) condition is stipulated at the HVAC duct walls.
- The "static pressure—free flow direction" condition is set at all simulation domain boundaries.
- A mass flow condition of 0.11667kg/s is placed at the HVAC duct inlets.

The flow domain's initial pressure and temperature are set to 101325Pa and 20°C, respectively. Both geometries are morphed with lattices that shift the ducts in space and dilate the ducts. A total of 47 cases for the EV12 rectangular duct are simulated, while 5 cases of the EV12 circular duct are simulated. The EV12 rectangular duct data is used to train the subsequently described ML model(s) with a split of 75% for pure training and 25% for validation during training. Meanwhile, the EV12 circular duct data is used to test how well the ML model(s) generalize.

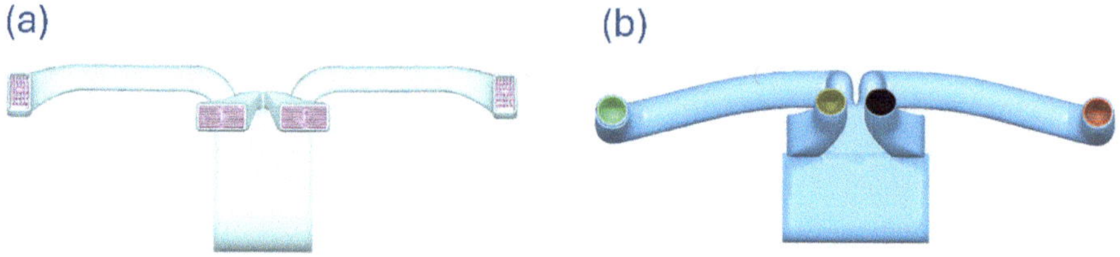


**Fig. 23.4.** **a** The EV12 rectangular duct and **b** the EV12 circular duct

The neural network employed for this study is a combined convolutional neural network (CNN) and multi-layer-perceptron ML model (Fig. 23.6). CNNs are used because the input to the neural network is a series of time-averaged fluid volume slices for velocity magnitude and pressure, as depicted in Fig. 23.5 for the EV12 rectangular duct (Fig. 23.5a) and EV12 circular duct (Fig. 23.5b).

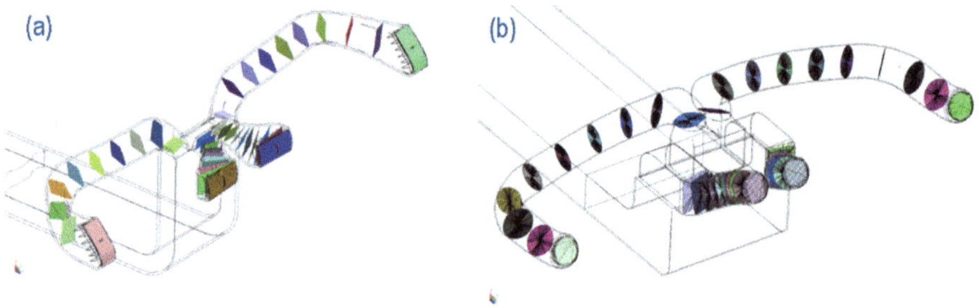


**Fig. 23.5.** CNN input fluid volume slices for **a** the EV12 rectangular duct and **b** the EV12 circular duct

The fluid volume slices per duct are stacked and then fed into a corresponding CNN; for this reason there are four CNNs, one per duct (Fig. 23.6). One calls each CNN network a 3D encoder, as it encodes the information found in the stacked fluid volume slices by successively reducing the dimensions of the input matrix (i.e. down-sampling) while increasing the number of channels (i.e. the depth). In the final CNN layer, the encoded information per duct is stacked into a 1D vector and concatenated with the data from the other ducts (Fig. 23.6). This concatenated 1D vector is then fed into either of four MLPs to predict the SPL at either of four microphones that are located downstream of the HVAC duct outlets. To assess the performance of the combined CNN-MLP neural network architecture, the true PowerFLOW® SPL plots are compared with the ML predictions, mean-square-error (MSE) is computed, weighted mean absolute percentage error (WMAPE) is determined, and R2 is evaluated too.

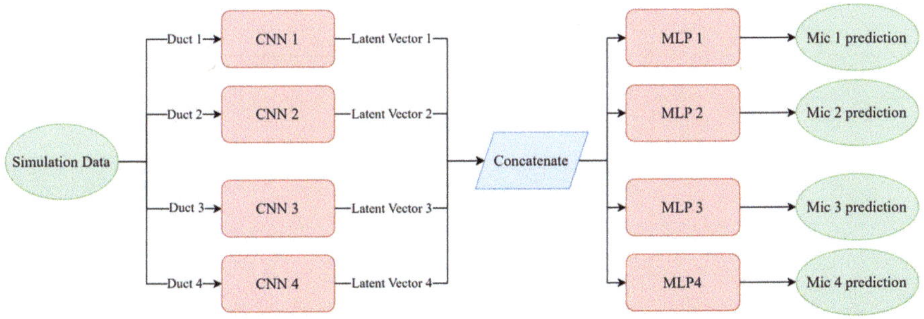


**Fig. 23.6.** The combined CNN-MLP neural network architecture for acoustics prediction

## 23.3 Results

The results section is divided into two parts: Sect. 23.3.1 describes the results for the MAHLE GmbH fan noise study, while Sect. 23.3.2 describes the results for the 3DS HVAC duct noise study.

### 23.3.1 MAHLE GmbH Fan Noise Study

For the MAHLE GmbH fan noise study, the first set of results analyzed come from the "3D" based MLP that predicts the fan surface static pressure. In total, the training time of the neural network on a single GPU is about 2 hours and inference for one case is a couple of minutes. Figure 23.7 shows the static pressure field on the MAHLE GmbH fan for the validation and the test cases. Only these cases are shown, as the ML model neuron weights are not adjusted using this data. In other words, the MLP's training is blind to these cases and the minimum of the neural network loss function is not found using this data. Therefore, these cases are perfect to assess whether the neural network generalizes well to unseen data. There are some subtle differences in the pressure distribution between the PowerFLOW® result and the ML result. However, the global trends are well captured. To better quantify the discrepancy, a reference line plot analysis (Fig. 23.8a) and Gaussian curve analysis (Fig. 23.8b) is performed. The values plotted are the integrated fan surface pressure. As can be seen in the reference line plot analysis (Fig. 23.8a), the relative error is less than 0.1% and almost all points lie on the y=x line, indicating that the ML model's predicative capability is very good. In fact, the R2 value is essentially 1.0, which means that the ML model is picking up the variance in the data set very well. Moving to the Gaussian curve plot for uncertainty quantification, one can see that with 90% confidence, the ML predicted integrated fan surface static pressure will be within −0.5Pa to about 5.5Pa of the true integrated fan surface static pressure. Compared to an absolute value between 25000Pa and 27000Pa, this is a very small delta.

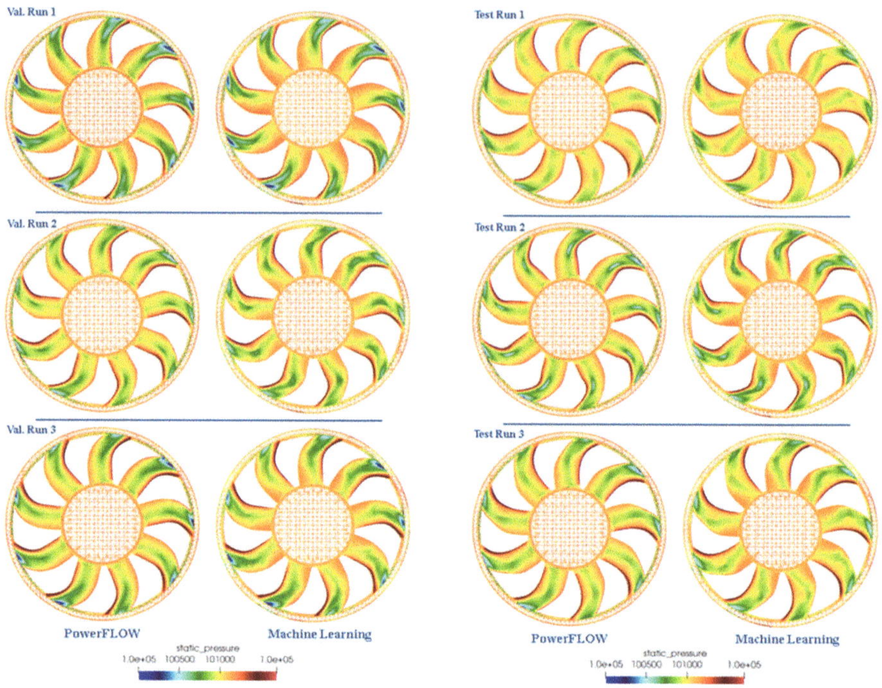


**Fig. 23.7.** Fan surface static pressure distribution for PowerFLOW® and ML. Validation and test cases are shown

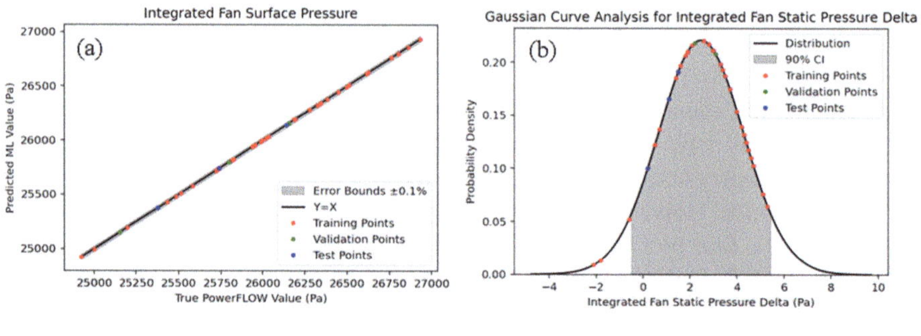


**Fig. 23.8.** **a** Reference line plot analysis and **b** Gaussian curve analysis for fan surface integrated static pressure

Next, the predicative capability of the "3D" MLP that computes the fluid volume slice static pressure and velocity magnitude is assessed. In this instance, training takes about 4 minutes on a single GPU, while inference takes a few minutes too. Given the size of this model and the speed at which training is completed, one can conclude that the MLP learns the patterns in the flow that it is able to capture quite quickly. It may, therefore, be that complex fluid patterns are not learned and this will become clear subsequently. Figure 23.9 compares the contours for static pressure and velocity magnitude between PowerFLOW® and ML for the test cases only; the patterns are the same for the validation cases. Evidently, the MLP only learns

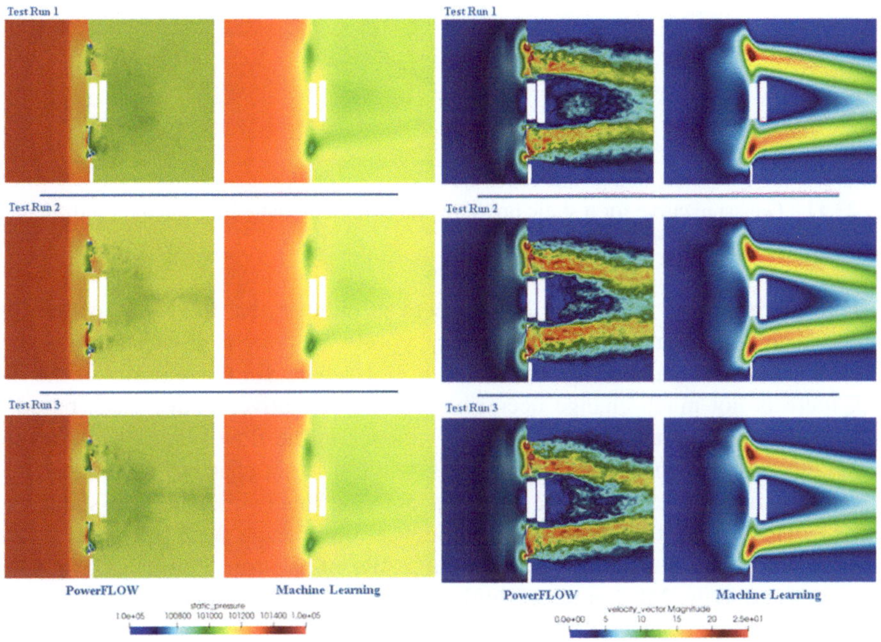


**Fig. 23.9.** Static pressure and velocity magnitude for the test cases using PowerFLOW® and ML

the average of the various flow patterns, as there is almost no discernable difference in the ML data between the different test cases. This is strange, as the PowerFLOW® data used is already averaged flow data. To better quantify the possible discrepancy between the PowerFLOW® data and the ML results, reference line plots (Fig. 23.10) and Gaussian curves (Fig. 23.11) for both average static pressure and velocity magnitude are looked at.

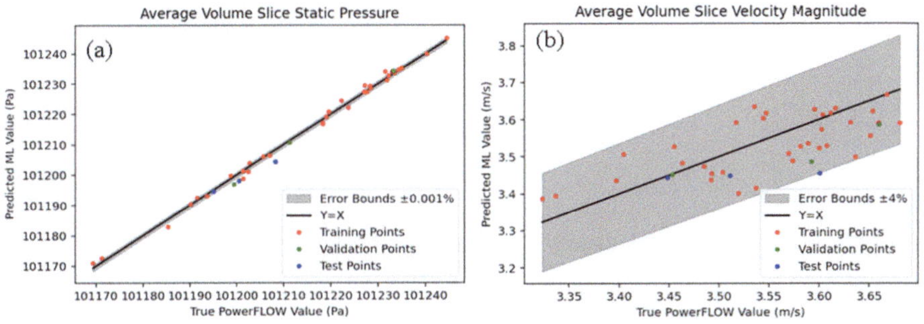


**Fig. 23.10.** Reference line plot for **a** static pressure and **b** velocity magnitude on the fluid slice

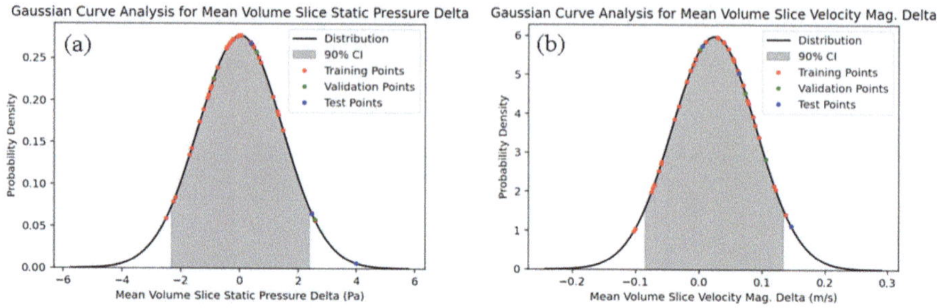


**Fig. 23.11.** Gaussian curve for **a** static pressure and **b** velocity magnitude on the fluid slice

The reference line plots (Fig. 23.10) seem to indicate that the MLP for the fluid volume slice prediction is not performing as poorly as thought. However, this can be misleading, as taking an average of a contour plot, where, for the majority of the flow field, there is not much change occurring, can be deceptive. In any case, based on analysis of the contour plot's average, the variance in the flow field is well captured for static pressure with an R2 value of 0.99, while it is not captured quite as well for the velocity magnitude, whose R2 value lies at 0.36. Meanwhile, the Gaussian curves (Fig. 23.11) illustrate that, with 90% confidence, on average the MLP predicts fluid static pressure values that lie within about 2.4Pa of the true PowerFLOW® value, while on average the MLP predicts fluid velocity magnitude data that is within about 0.1m/s of the true PowerFLOW® value.

Additionally, the "1D" MLP that determines the fan performance parameters, torque and delta pressure, is now analyzed. For this MLP, training and inference only takes a few minutes, even on a single CPU, as merely the morph parameters are fed into the network as input. The reference line plot (Fig. 23.12a) and the Gaussian curve uncertainty analysis (Fig. 23.12b) for torque show that the MLP performs quite well for this quantity.

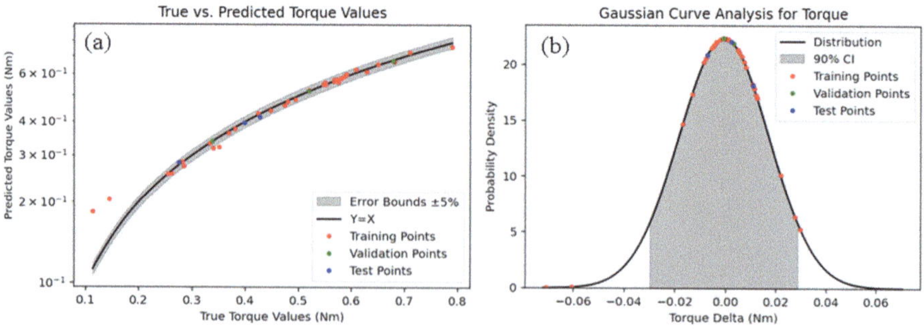


**Fig. 23.12.** **a** Reference line plot and **b** Gaussian curve for torque

R2 for torque resides at 0.99, indicating that, despite the two poorly predicted training points (see Fig. 23.12a), the MLP captures the variance inherent in the torque dataset quite well. The Gaussian curve (Fig. 23.12b) shows that, with 90% confidence, the MLP predicts torque values that are within 0.03Nm of the true PowerFLOW® value, which is very good. Having looked at torque, delta pressure is now analyzed: Fig. 23.13a is the reference line plot and Fig. 23.13b is the Gaussian curve for delta pressure. It may be seen that the MLP predictions for delta pressure are satisfactory too: its R2 value lies at 0.99 and, with 90% confidence, delta pressure values are obtained that lie within about 5Pa of the true PowerFLOW® value.

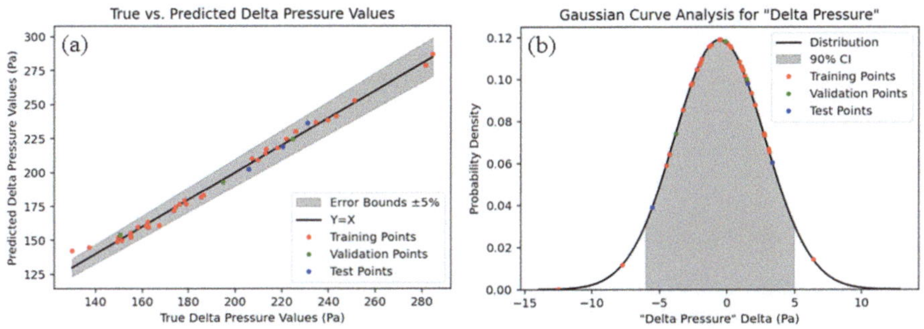


**Fig. 23.13.** **a** Reference line plot and **b** Gaussian curve for delta pressure

Finally, the "1D" MLP for acoustics predictions is assessed. Just like with the "1D" MLP for fan performance computation, the training and inference for this MLP only takes a couple of minutes on a single CPU, as only the morph parameters are fed as inputs to the neural network. Figure 23.14a depicts the reference line plot for SPL and OASPL values, while Fig. 23.14b portrays the Gaussian curve uncertainty quantification for these values. It becomes clear that the "1D" MLP computes acoustic values that lie within about 5% of the true PowerFLOW® value for an R2 value of about 0.93, which means most of the variance in the dataset is captured. The Gaussian curve shows that, with 90% confidence, acoustic values that are within 2dBA of their true PowerFLOW® value can be computed with the MLP.

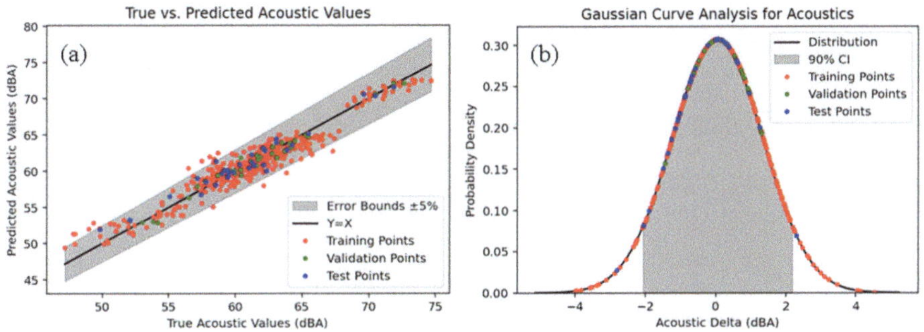


**Fig. 23.14.** **a** Reference line plot and **b** Gaussian curve for acoustics

As a final assessment of this "1D" MLP's acoustic predictive capability, the SPL plots for the validation and test cases are directly compared in Fig. 23.15. Although the curves do not lie 100% on top of each other, the trends among the six cases compare well between PowerFLOW® and ML. This is seen in the OASPL values computed too: the ML derived OASPL values are all within 1dBA of the true PowerFLOW® value.

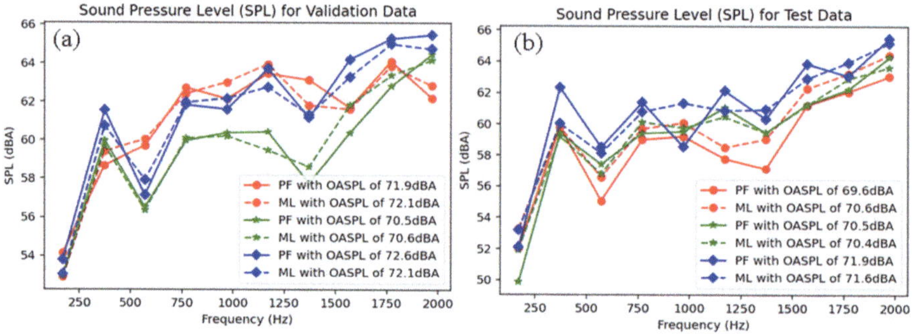


**Fig. 23.15.** PowerFLOW® and ML computed SPL plots for **a** the validation and **b** the test cases

### 23.3.2 3DS HVAC Duct Noise Study

For the 3DS HVAC duct noise study, three CNN-MLP neural networks are trained. The first is called "BaselineVel" and uses the CNN-MLP architecture with only flow velocity as input. "LargerVel" uses a CNN-MLP network with a larger feature extraction capacity per CNN and more hidden layers in each MLP. Meanwhile, "LargerVelPress" uses the previously augmented CNN-MLP architecture, and uses both flow velocity and pressure as input on the slices. Table 23.1 contains the MSE, WMAPE and R2 value for the validation dataset, which comes from the EV12 rectangular duct cases (Fig. 23.4a), and the test dataset, which is made up of EV12 circular duct cases (Fig. 23.4b).

**Table 23.1.** MSE, WMAPE and R2 for HVAC duct validation and test cases

| | Validation Data (EV12 Rect. Duct) | | | Testing Data (EV12 Circ. Duct) | | |
|---|---|---|---|---|---|---|
| Model | MSE (dB) | WMAPE (%) | R2 | MSE (dB) | WMAPE (%) | R2 |
| BaselineVel | 1.52 | 3.84 | 0.980 | 163 | 42.2 | −4.25 |
| LargerVel | 1.33 | 3.57 | 0.983 | 254 | 56.4 | −7.16 |
| LargerVelPress | 1.53 | 4.17 | 0.980 | 219 | 51.5 | −6.07 |

Looking at Table 23.1, one may conclude that using a larger/more complex CNN-MLP network and/or having more input variables does not significantly nor conclusively improve the predictive capabilities of the ML model in question. However, Table 23.1 does very obviously show that all three models perform very poorly for the EV12 circular duct test cases, unlike for the EV12 rectangular duct validation cases for which the R2 value is close to 1.0, indicating very good model performance. This becomes even more clear when SPL plots are subsequently looked at. Such behavior of the CNN-MLP networks should come as no surprise though. The EV12 circular duct test cases represent set-ups for which the CNN-MLP neural network must extrapolate. By nature, these models only behave well when interpolating into the design space. In order to be able to extrapolate, other types of neural networks must be used, as detailed in the section on future work. Figure 23.16 depicts a sample SPL plot for a validation case (Fig. 23.16a) and for a test case (Fig. 23.16b). The trends reflect the performance measures contained in Table 23.1: for the validation set-up, the SPL plots computed using the CNN-MLP models correlate well with the true PowerFLOW® curve, while for the test set-up, the SPL plots derived from the CNN-MLP models are entirely offset from the true PowerFLOW® curve.

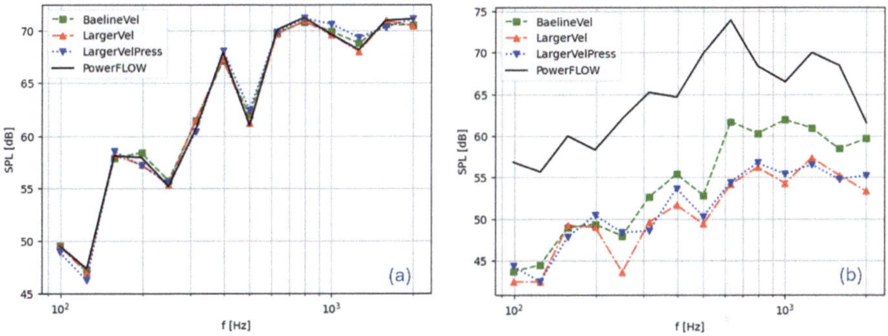


**Fig. 23.16.** Sample SPL curves for **a** one validation case and **b** one test case

## 23.4 Conclusion

Two aeroacoustic cases are presented: a MAHLE GmbH fan noise investigation and a 3DS HVAC duct noise study. In both instances, the 3DS lattice Boltzmann solver PowerFLOW® is used to collect compute and time intensive DOE simulation data. This data is then used to train ML models with which faster performance and acoustic predictions can be made. For the MAHLE GmbH fan noise study, MLPs are trained that predict fan surface static pressure, static pressure and velocity magnitude on a fluid volume slice, fan torque and delta pressure, as well as SPL and OASPL. The R2 value for all situations is above 0.9, indicating that the MLPs are capturing the variance in the dataset well. Only the predictions for the fluid volume slice are unsatisfactory; the mean flow structures are captured, while the subtleties in each test case are not resolved by the MLP. Meanwhile, for the 3DS HVAC duct noise study, a combined CNN-MLP neural network model is used to predict the SPL at four microphone locations using averaged fluid volume slices as input. The CNN-MLP performs well on the EV12 rectangular duct whose simulation data is used to train the model; R2 values for these cases are consistently above 0.9. However, when using the trained CNN-MLP on an entirely unseen HVAC duct, the EV12 circular model, SPL predictions are very poor. This does not come as a surprise though, as the CNN-MLP should not perform well when extrapolating.

## 23.5 Future Work

There are several items that can be worked on as part of future work. First, in the case of the MAHLE fan noise investigation, the MLP predictions for the fluid volume slice should be improved so that more subtleties in the flow fields are captured. Additionally, for the HVAC duct noise study, transversal/horizontal time-averaged fluid slices could be fed into the CNN-MLP model, thereby allowing information near the registers of the duct, where significant noise is generated, to be seen by the ML models. Next, graph neural networks (GNNs), which are mesh/graph-based methods, can be investigated to allow non-parametric based ML DOE investigations to be performed. Instead of feeding in the morph parameters to the neural network, only the mesh of the geometry in question is used as input to the ML model. Furthermore, physics-informed neural networks (PINNs), where there is both a data and partial differential equation based loss function, should be tried. The literature says that they perform quite well when extrapolating. Finally, generative artificial intelligence methods could be investigated. In this approach, given certain geometry and physics constraints, one may be able to specify a desired performance and acoustic behavior, and then generate a fan or HVAC duct that fulfills those criteria.

# The Relevance of Artificial Heads in Test Based and Numerical Scenarios

Haiko Brücher[1](✉), Martin Abele[2], Maximilian Kentgens[1], Tim Kamper[1], and Lukas Henseler[2]

[1] HEAD Acoustics GmbH, Herzogenrath, Germany
{haiko.bruecher,maximilian.kentgens, tim.kamper}@head-acoustics.com
[2] Dr. Ing. H.C. F. Porsche Aktiengesellschaft, Weissach, Germany
{martin.abele,lukas.henseler}@porsche.de

**Abstract.** As building physical prototypes for testing purposes is cost- and time-intensive, the role of simulation becomes more and more important in the development process in almost any industry branch. As part of this trend, the simulation of Noise, Vibration, and Harshness (NVH) problems is also becoming more prevalent. Apart from that, the application of artificial heads for the test-based investigation of NVH problems has been established for many years.

This work investigates the influence of an artificial head on the sound field inside of a vehicle cabin – especially at the driver's ears positions – by application of FE-based simulation (finite element method, FEM). Thereby it assesses the relevance of the use of artificial heads in test based and numerical scenarios. Using the artificial head, two effects can be observed which resemble the effects of a human driver on the perception of acoustics inside the car cabin: directivity, which is mainly characterized by the head, ear and torso geometry as well as a modification of the cabin room modes.

A numerical model of the artificial head is built and validated on behalf of measurements under free field conditions. It is then integrated into a fully equipped car model and the resulting simulation output is compared to measurements in a matching test car. As one of many possible real application scenarios from the automotive industry, a psychoacoustic analysis of simulated and measured data is performed to show the capabilities of binaural simulation results.

**Keywords:** HRTF Simulation · Binaural Rendering · Modal Participation Factor

## 24.1 Introduction

Numerical simulations gain increasingly more importance to speed up the development process and to cut down the cost for physical prototypes all over different industries and products. However, even though the number of measurements performed


A. Heintzel (Ed.): Automotive Acoustics Conference 2025, Proceedings, pp. 349–365, 2026. https://doi.org/10.1007/978-3-658-51249-1_24

during product development may decrease, it is mandatory to correlate and compound numerical simulations with measurements using real existing physical prototypes.

For the test-based evaluation of NVH properties binaural artificial Head And Torso Simulator (HATS, [1]) recordings are crucial components to capture and analyze the sound experience. By replicating the acoustic properties of the human body, HATS allow for application-oriented measurements on the one hand, and provide a realistic listening experience on the other hand. To better match numerical simulations with physical measurements it is helpful to provide binaural signals from the numerical simulation as well. Thus, it is recommended to generate a numerical model of an artificial head and incorporate it into the numerical model of the car cavity [2]. This way it is possible to consider diffraction effects of head and shoulder as well as the artificial head's influence on the sound field on the vehicle's interior cavity [3].

In this paper the numerical model of the virtual head as well as the model of the car cabin and their validation are discussed. Finally, the influence of the virtual artificial head on the recorded sound signals at the ear positions is shown and discussed.

Using a novel approach, this influence is separated into two parts to separately consider the diffraction and shadowing effects of head and shoulder and the change in the sound field in the car cavity caused by the presence of the HATS.

## 24.2 Overview of Present Methodology

### 24.2.1 Simulation Model Description and Validation

**Artificial head and torso model** For both the measurements and the numerical model, a HEAD acoustics HMS IV HATS was used. The numerical model is based on CAD data of the measurement system.

The model was built in MSC Actran [4] using the Finite Element (FE) method. Since it should only take airborne sound into account and no structural dynamics should be considered, the closed surface model is sufficient; only the outer surfaces of the artificial head are considered. The model comprises approximately 76000 triangular elements with an average edge length of 3 mm. The element size is non-uniform as smaller features near the auricle and the external auditory canal require a much finer element size than flat surfaces such as head and shoulder. Like the real HMS IV, the numerical model is modular and includes simplified geometries of the shoulder, head and pinna. Meanwhile, the ear canal is excluded from consideration. The study focuses on frequency range analyses up to 8 kHz. The final FE mesh is depicted in Fig. 24.1.

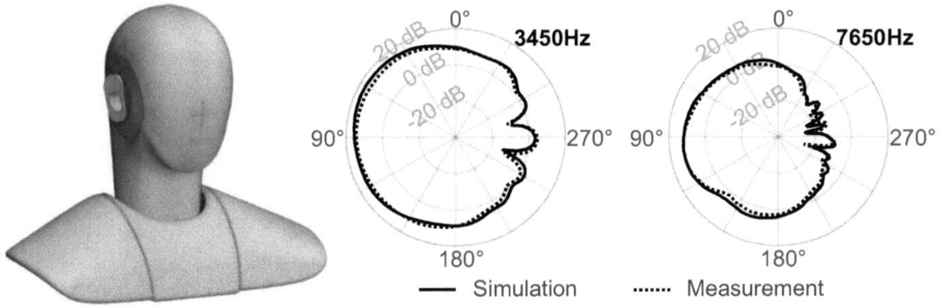


**Fig. 24.1.** Finite element model of the HMS IV HATS (left). Measured and simulated directivity for the left ear in the horizontal plane at 3450 Hz (center), and 7650 Hz (right).

**Validation of virtual artificial head model** The accuracy of the numerical model is evaluated by comparing Head Related Transfer Functions (HRTF) from measurement and simulation for different sound source positions around the artificial head. The corresponding measurements were conducted in the anechoic chamber at RWTH Aachen university. Figure 24.1 shows the measured and simulated HATS directivity for the left ear at 3450 Hz and 7650 Hz in the horizontal plane. Overall, good agreement is found across the 50 to 8000 Hz range [3].

**Preparation for the use in the vehicle model** To use the virtual artificial head in the full vehicle model and compare it with test data, several changes and adjustments are required. Like in the measurement, the artificial head is extended by a box representing the human torso, see Fig. 24.2. Since the upper frequency limit of the full vehicle is about 1 kHz, the simulation performance can be improved by increasing the mesh width of the virtual artificial head. This mesh adaptation requires some geometric simplifications compared to the previous model such as closing the collar and approximating the ear radii polygonally. The resulting model including the torso box consists of 20360 elements.

For validation, Fig. 24.2 compares the HATS-HRTF of the right ear for source directions 0° (front) and 270° (ipsilateral) from the measurement with a high-resolution (A) and the reduced (B) simulation model. Both simulation models generally agree well with the measurement. Below 600 Hz, A and B provide nearly identical results. Above this frequency, acceptable deviations occur due to the larger elements.

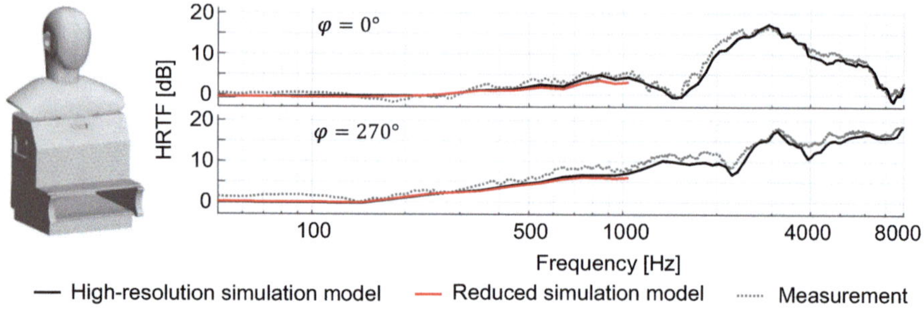


**Fig. 24.2.** Reduced FE model of the HATS (with attached torso box) and HRTFs of the right ear for the azimuth angles 0° (front) and 270° (ipsilateral). Comparison of measurement, high-resolution, and reduced simulation model.

### 24.2.2 Method and Model Description for Modes Investigation

In a first study [2] the influence of the virtual artificial head on a simplified cabin cavity of a Porsche 911 Carrera was investigated within MSC Actran [4], see Fig. 24.3. Seats are included, but details like the steering wheel or handles are omitted. The cavity is modeled with a general absorption coefficient to simplify the model. Thus, the simulation results are expected to show fundamental effects rather than exact real-world matches.

Several volume velocity sources and microphone receiver positions are distributed throughout the cabin; Fig. 24.3 shows one exemplary source position in the driver's footwell, depicted by a yellow sphere. Microphones are placed close to their corresponding sources, and sound pressure is also output at the driver's ear positions for both empty and equipped cavities (with the HATS on the driver's seat). Volume velocity sources are defined as unit excitations, providing frequency response functions (FRFs) without further postprocessing. Room modes are calculated to investigate changes in interior acoustics and diffraction effects caused by the HATS. Both FRFs and room modes are simulated for empty and equipped cavities, with selected matching room modes compared. All simulation outputs have an upper limit of 1 kHz.

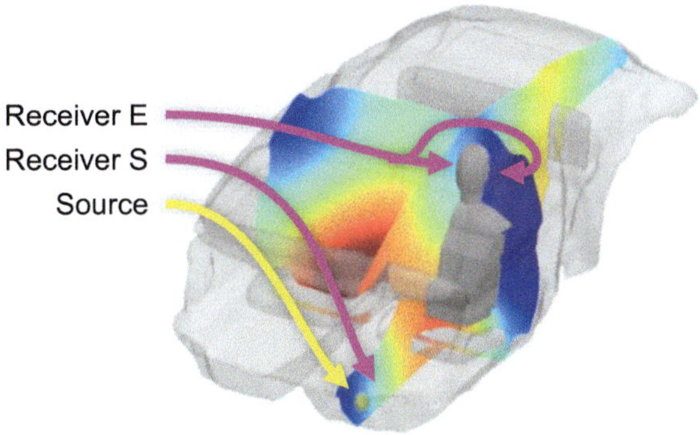


**Fig. 24.3.** Simplified cabin cavity of Porsche 911 Carrera. Cutting planes as well as investigated source and receiver positions.

**Analysis of simulation results and modes investigation.** Exemplarily, the aforementioned source position in the driver's footwell is evaluated. FRFs at the relevant receiver positions are presented for the empty cavity as well as the one equipped with the HATS in Fig. 24.4: driver's left ear position (top), driver's right ear position (center), close to volume velocity source, implying the input impedance (bottom).

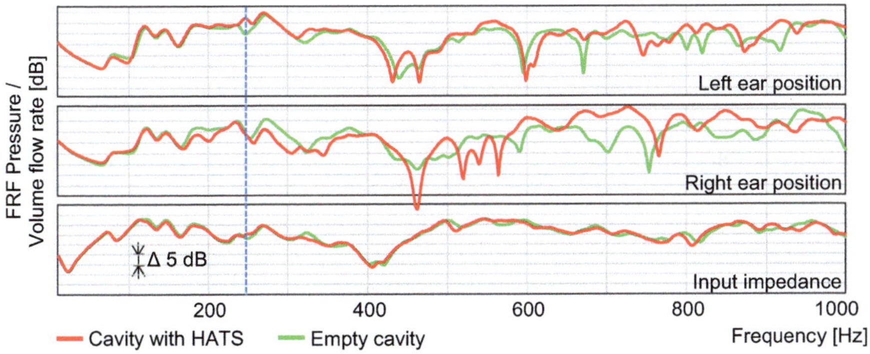


**Fig. 24.4.** Comparison of simulated frequency response functions in empty and equipped cavity with a volume velocity source in the driver's footwell and receivers at left ear (top) and right ear (center) position, as well as close to the source, resulting in the input impedance (bottom). The blue dashed line indicates the frequency to be investigated in detail.

Small differences between empty and equipped cavity can be observed at frequencies around 200 Hz for both ear positions, whereas severe deviations, ranging from 650 to 750 Hz and 820 to 900 Hz, appear at the right ear position. The input impedance shows only small deviations over the whole considered frequency range, which means the acoustic field in the driver's footwell is only slightly affected by the presence of the HATS.

Motivated by the low level of complexity, the main contributing room mode at the first significant deviation at the left ear position (~250 Hz, indicated by the blue dashed line in Fig. 24.4) is investigated. It is visualized in Fig. 24.5. Blue and red areas both show maxima which have a phase shift of 180°. Green areas indicate regions that are not affected by this particular mode at all. Due to the HATS presence, the mode is slightly shifted downwards in frequency, from 254 to 247 Hz. Its global characteristics remain, but it is deformed locally, resulting in a raised amplitude at the left ear. The amplitude at the right ear does not change at all.

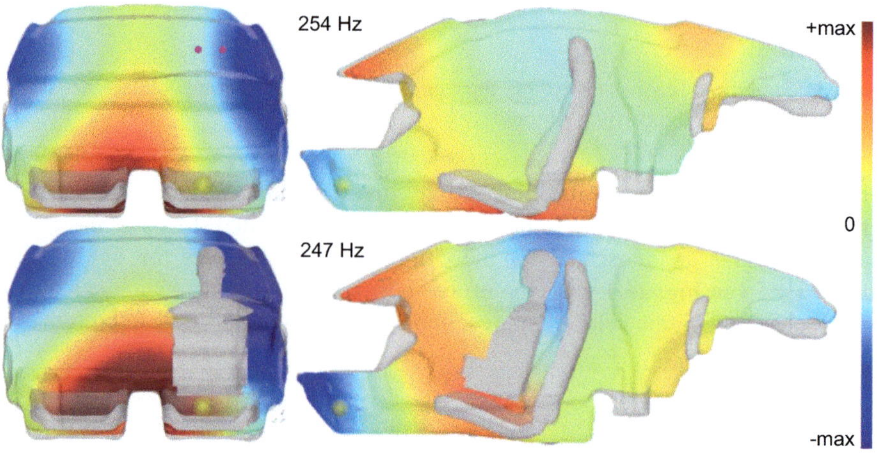


**Fig. 24.5.** Comparison of related room mode for empty (top) and equipped (bottom) cavity. Cutting planes: lateral (left) and longitudinal (right).

In the longitudinal cut, the slightly raised input impedance amplitude can be seen easily, as the blue area in the footwell gains in darkness and expanses in its spatial distribution. In [2], a Porsche Taycan cavity with a higher interior volume is investigated as well, showing the same effects described above.

### 24.2.3 Psychoacoustic Investigation in the Full Vehicle Model

A more advanced full vehicle simulation model of the Porsche 911 Carrera is introduced in [3]. To validate the virtual artificial head in the full vehicle model, various Acoustic Transfer Functions (ATFs) are both derived from simulation as well as measured on an acoustic test bench. The $ATF(f)$ represents the ratio of the sound pressure $p(f)$ to the source strength $Q(f)$ and it is denoted in the frequency domain with frequency $f$:

$$ATF(f) = p(f)/Q(f). \tag{24.1}$$

The ATFs are determined using both a single microphone and a HATS under several load cases, with measurements taken at specific seat and reference positions. The third octave band comparison between the single microphone and the dummy head signals at the driver's position, depicted for one exemplary load case in Fig. 24.6 (left), illustrates the behaviour of simulation and measurement for the full car with and without HATS. It shows that the sound pressure levels (SPL) are matching well over frequency in measurement and simulation. Between 200 Hz and 400 Hz, the artificial head's left ear has a higher SPL, while above 400 Hz, the right ear has the highest level. In [3], a more extensive validation is presented.

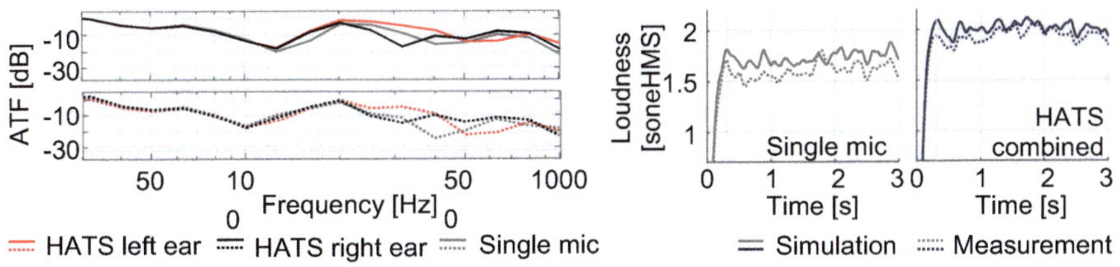


**Fig. 24.6.** Left: Comparison of the sound pressure at the single microphone to the HATS ears for simulation (top) and measurement (bottom) in the third-octave band, for one loadcase. Right: Loudness rating according to ECMA 418-2; comparison of simulation and measurement for the single microphone and combined for the artificial head.

Using the FE-based HATS in the full vehicle model allows for psychoacoustic evaluations of simulation results, e.g. loudness, translating human perception of noise loudness into a linear scale. The psychoacoustic parameter, based on the Sottek Hearing Model [5] and standardized in ECMA-418-2 [6], necessitates post-processing of artificial head measurements from experiments or simulations to ensure an equalized microphone signal (excluding diffraction effects). Given the ambiguous sound field within a vehicle cabin (neither an ideal free field nor an ideal diffuse field), the ID (Independent of Direction) equalizer is selected. To compute loudness in accordance with the standard, a three-second white noise time signal is convolved with the transfer functions derived from both simulation and measurement. Through binaural combination, the loudness calculation for the artificial head generates a single curve, which can be compared to the loudness curve obtained from a single microphone, depicted in Fig. 24.6, right. As per the standard, the loudness curve becomes meaningful starting at approximately 0.3 s. Analysis of the loudness reveals a significant difference between the empty and equipped cavity for the simulation as well as for the measurement. This indicates an additional impact of the HATS on the sound field in the cavity beyond the diffraction and shadowing effect.

## 24.3 Separation of Acoustic Effects

The above-described methodology of modes investigation lacks the separation of two effects occurring on the output signal when measurements or simulations with a HATS in a cavity are performed: on the one hand, there is the HATS directivity, caused by diffraction, shadowing, and reflections, as well as ear canal resonances. On the other hand, a change of cavity acoustics results from the presence of the HATS. This chapter addresses two approaches that help to investigate both effects more precisely: binaural rendering based on simulating a non-intrusive open microphone array and the consideration of modal participation factors (MPFs).

### 24.3.1 Binaural Rendering of Virtual Microphone Array

In order to separate the acoustic effects directly related to the HATS from the effects that are caused by the acoustic interaction of the HATS with the cavity, i.e., the acoustic coupling of the measuring object and the measurement apparatus, a trick is applied. Firstly, the cavity is simulated without HATS but with an open microphone array at the position of the HATS. The open microphone array is a set of purely virtual receivers that have no impact on the sound field. From the microphone array signals, a virtual binaural HATS signal can be rendered with the acoustic properties as if a HATS was placed in the cavity, however without the acoustic interaction with the cavity. Hereby, the rendering approach is adopted from microphone array theory and spatial audio signal processing. However, while similar binaural rendering techniques are applied in multimedia applications, where a perceptually plausible binaural rendering is often sufficient, we here aim at accurate binaural reproduction both in terms of phase and magnitude fidelity.

**Virtual microphone array design** As microphone array, a dual-radius open sphere design is used, which is well-studied in the context of spherical beamforming and plane-wave decomposition [7]. A total of 916 receivers are uniformly distributed on two acoustically transparent concentric spherical shells with diameters of ~ 19 cm and ~ 23 cm, further called virtual double sphere array, and placed in the cavity at the acoustic center of the HATS. The number of microphones is chosen by rule of thumb [8, 9] such that accurate least squares binaural rendering is achieved over a wide range of the human audible frequency range. Note that the number of microphones is much higher than necessary for the frequency range ultimately considered in the scope of this paper.

**Binaural rendering filter design** The idea of the binaural rendering is to convert the microphone array signal

$$\boldsymbol{x}(f) = \left[x_1(f), \ldots, x_{916}(f)\right]^{\mathrm{T}} \tag{24.2}$$

into an estimate $\widehat{\boldsymbol{y}}(f)$ of the binaural signal $\boldsymbol{y}(f)$,

$$\boldsymbol{y}(f) = \left[y_{\text{left}}(f), y_{\text{right}}(f)\right]^{\mathrm{T}}. \tag{24.3}$$

Here, signals are denoted in the frequency domain with frequency $f$. The conversion is realized using a $(2 \times 916)$-dimensional filter matrix $\boldsymbol{D}(f)$ multiplied with the microphone array signal

$$\widehat{\boldsymbol{y}}(f) = \boldsymbol{D}(f) \cdot \boldsymbol{x}(f) \approx \boldsymbol{y}(f). \tag{24.4}$$

Each row of the filter matrix $\boldsymbol{D}(f)$ can be interpreted as a specially designed beam-forming filter, one for each ear, that exhibits a directivity and phase response pattern that corresponds to the respective ear of the HATS.

The filter matrix design is based on the HRTFs of the HATS $\boldsymbol{h}_{\text{HATS}}(f,\theta,\phi)$ for a full sphere of 5000 directions $(\theta,\phi)$, which are obtained using a numerical free-field simulation based on the previously described HATS model (Fig. 24.2). For the same directions, the free-field transfer functions of the virtual array $\boldsymbol{h}_{\text{array}}(f,\theta,\phi)$ are calculated analytically. The matrix $\boldsymbol{D}(f)$ is then obtained by minimizing the squared error between the effective virtual HRTFs $\boldsymbol{D}(f)\cdot\boldsymbol{h}_{\text{array}}(f,\theta,\phi)$ and the actual HRTFs $\mathbf{h}_{\text{HATS}}(f,\theta,\phi)$ independently for each frequency $f$, i.e., solving the convex optimization problem

$$\boldsymbol{D}(f) = \underset{\boldsymbol{D}(f)}{\arg\min} \sum\nolimits_{\theta,\phi} \|\boldsymbol{D}(f)\cdot\boldsymbol{h}_{\text{array}}(f,\theta,\phi) - \mathbf{h}_{\text{HATS}}(f,\theta,\phi)\|^2. \tag{24.5}$$

In practice, regularization is necessary to stabilize the optimization problem particularly at low frequencies.

For validation of the binaural rendering, we analyze the free field performance of the resulting filter design by comparing the effective virtual HRTFs (resulting from the binaural rendering) with the actual HRTFs of the HATS simulation model. Figure 24.7 shows directivity results in the horizontal plane for three frequencies. The results agree well over the entire frequency range examined in the horizontal plane with a mean magnitude error of 0.3 dB, while the maximum difference of 1.9 dB appears at 750 Hz at a small angular range around $\varphi = 224°$, thus in the contralateral direction.

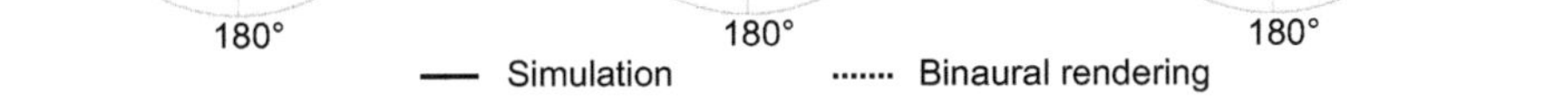


**Fig. 24.7.** Comparison of simulation and binaural rendering. Directivity in the horizontal plane for the left ear at 100 Hz (left), 750 Hz (center) and 1000 Hz (right).

**Binaural synthesis in the car cavity** The previously described binaural rendering approach facilitates investigating the impact of the interaction of HATS and cavity on the binaural signals. Given that they exceed the margin of error of the binaural rendering, differences between the recorded or rather simulated HATS signal from the cavity and binaural rendering of the virtual double sphere array in the cavity can be attributed to the effect of acoustic coupling of HATS and cavity. These differences reflect acoustic properties that are altered due to the pure presence of the HATS.

The virtual double sphere array used for the binaural rendering is designed to capture diffraction and shadowing effects approximately within the human listening range. Thereby it is highly overdetermined in terms of the investigated frequency range of below 1 kHz. A reduction of the number of receivers or the utilization of existing microphone arrays is not investigated within this paper and could be the subject of future studies.

### 24.3.2 Investigation of Modal Participation Factors

Modal participation factors (MPFs) are important quantities in the finite element method (FEM) that describe the contributions of the individual modes (mode shapes) to a particular response of a system. They are often used in vibro-acoustic analysis to understand how much each mode contributes to the overall sound pressure level of a transfer function. The contribution depends on the modal base of the system as well as on the specific position of excitation and measurement (input-output-conditions). A comprehensive overview of different definitions for MPFs and differing investigation methods is given in [10]. Here, the MPF represents the relative involvement of each mode in the total sound pressure at one certain position and frequency. Participation factors of more than 100% may occur for individual modes due to cancelation effects in the interaction with other modes. The MPF-weighted phase-aware sum of all modes will always result in a total of 100%.

## 24.4 Enhanced Analysis of Simulation Results

### 24.4.1 Full Vehicle Model (Porsche Macan)

In this contribution the car cavity of the new electric Macan is investigated (Fig. 24.8). In contrast to the previously analyzed 911 Carrera [3], this is a much larger cavity with completely different dimensions. Analogous to [3] various ATFs in the vehicle interior are calculated, with additional monopole source positions (load cases): loudspeakers in doors, driver's footwell, overhead control unit, headrest driver and headrest co-driver.

The ATFs are determined using an FE model consisting of a structural domain (primarily metal and plastic components), the interior cavity domain and the acoustically effective insulating components, the so-called trim domain (Fig. 24.9). The domains are coupled with each other so that their feedback effects on each other are considered in the simulation. With this FE model, the frequency range from 20 Hz to 1000 Hz can be investigated.

The previously mentioned load cases are applied on two simulation configurations with their respective sound pressure output positions: (i) HATS in cavity with output at the ear positions and (ii) empty cavity with output at the ear positions as well as at all 900 nodes on the acoustically transparent double sphere that will then serve as input for the binaural rendering algorithm.

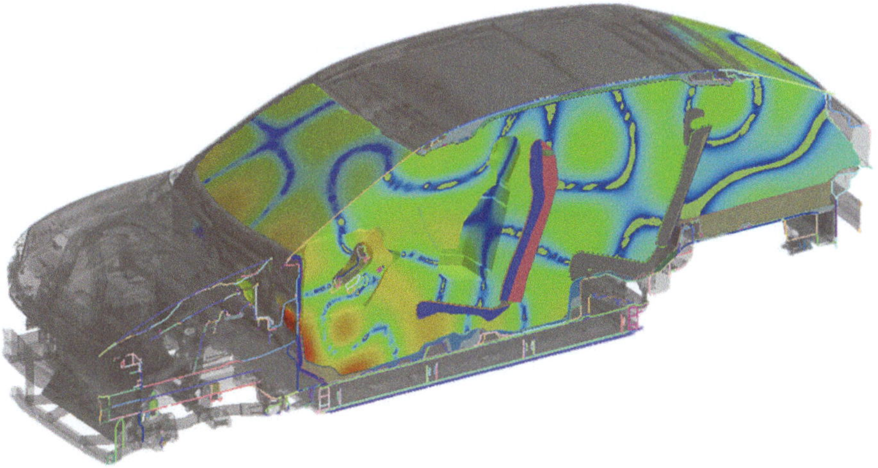

**Fig. 24.8.** Longitudinal cut of investigated full-car model of Porsche Macan with exemplary sound pressure distribution in the car cavity.

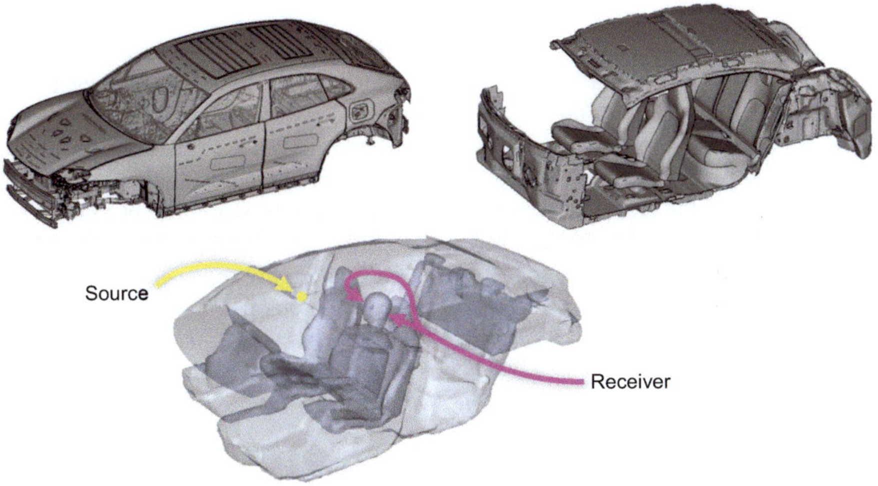


**Fig. 24.9.** Representation of the simulation model: structural domain (top left), trim domain (top right), interior acoustic domain with investigated source and receiver positions (bottom).

### 24.4.2 Detailed Analysis of the Simulation Results

From the multitude of simulated data, one example load case is presented here to demonstrate the application of the proposed methodology. The source and measurement positions that are used for the detailed investigation are shown in Fig. 24.9 (bottom). Observing the source position at the overhead control unit, three FRFs at the driver's left and right ear positions are plotted in Fig. 24.10, considering the following three scenarios: (a) HATS-equipped cavity, (b) empty cavity and (c) binaural rendering based on empty cavity, where the underlying virtual microphone array has no impact on the empty cavity acoustics.

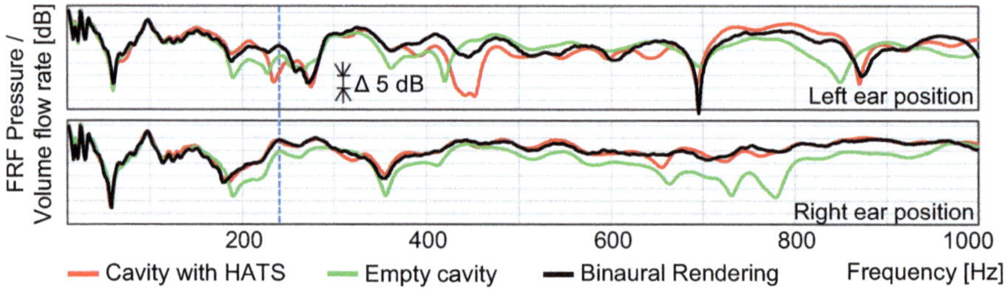


**Fig. 24.10.** Frequency response functions from source at the overhead control unit to the driver's left (top) and right (bottom) ear positions, considering three scenarios: (a) HATS-equipped cavity, (b) empty cavity and (c) Binaural rendering based on empty cavity. The blue dashed line indicates the frequency to be investigated in detail.

Significant differences between HATS-equipped (a) and empty cavity (b) can be observed starting from about 200 Hz. At first glance, the binaurally rendered FRF (c) shows good correlation with the HATS equipped cavity over a wide frequency range. Within these ranges, the differences between (a) and (b) are the result of the artificial head's directivity and the sound field seems to be unaffected by the presence of the artificial head. For the right ear, which is oriented ipsilateral with respect to the source position and does not directly face reflecting surfaces, the binaurally rendered FRF shows very good agreement with (a) over almost the whole investigated frequency range; in contrast, at the left ear, being oriented contralateral against the source position and facing the window as a reflective surface, high deviation hot spots between HATS-equipped cavity and binaural rendered FRF appear (247 Hz, 450 Hz, 650 Hz). For this reason, the focus in the following paragraphs is on the sound pressure distribution at 247 Hz.

Considering the MPFs for the left ear position, shown in Fig. 24.11, one mode stands out significantly for the empty cavity, whereas, when inserting the HATS, an almost equal participation of many modes can be observed, fluctuating with respect to the sign. The previously most contributing room mode 23 (Fig. 24.12) is locally modified by the presence of the HATS and loses its relevance for the contribution on the left ear position. The remaining contributing modes basically cancel out each other, finally resulting in a low sound pressure level at the left ear position at that certain frequency.

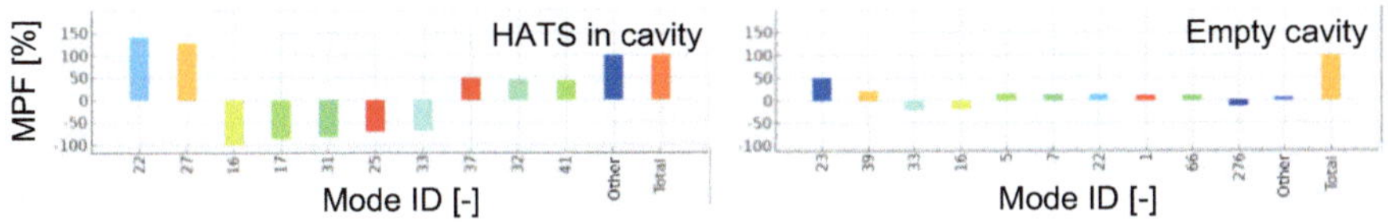


**Fig. 24.11.** Modal participation factors for the left ear position at 247 Hz for the configurations HATS in cavity (left) and empty cavity (right).

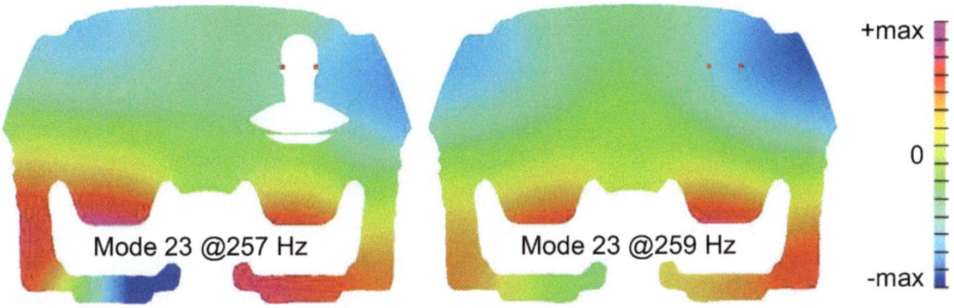


**Fig. 24.12.** Room mode 23 in lateral cutting plane for equipped (left) and empty cavity (right).

When focusing on the MPFs at the right ear position (Fig. 24.13), the clear dominance of one single mode becomes evident: mode 20 for the case of the HATS in the cavity, and mode 21 for the empty cavity. The contour plots in Fig. 24.14 suggest that they are the same mode with slightly altered characteristics. Additionally, the use of the HATS results in a little downward shift in frequency from 250 to 246 Hz. The turquoise coloration around the right ear position remains almost the same. The greenish coloration at the left ear position indicates a low magnitude; this explains why the considered room modes with the index 20/21 do not appear in the MPF distributions in Fig. 24.11 for both the empty and the equipped cavity.

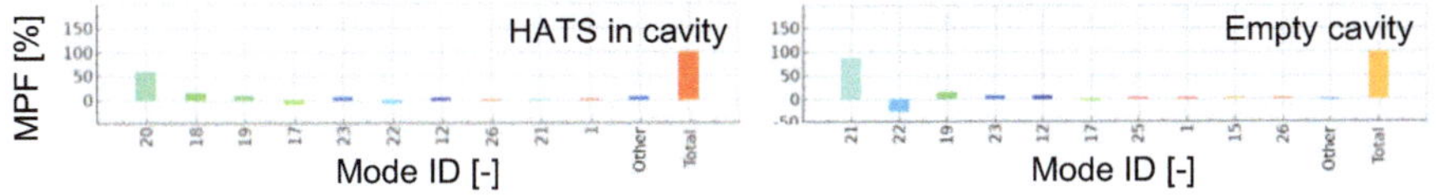


**Fig. 24.13.** Modal participation factors for the right ear position at 247 Hz for the configurations HATS in cavity (left) and empty cavity (right).

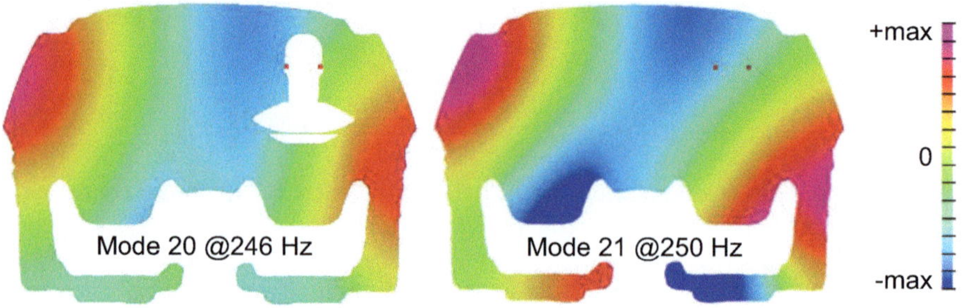


**Fig. 24.14.** Room modes 20/21 in lateral cutting plane for equipped (left) and empty cavity (right).

Figure 24.15 shows the overall sound pressure level at 247 Hz on a vertical cutting plane of the car cavity, for the considered load case (unit excitation with monopole source at the overhead control unit). It is the MPF-weighted superposition of all room modes. The plot clearly shows the asymmetric modification of the pressure distribution due to the use of the HATS. The dip at the left ear position and the elevation at the right ear position in the HATS-equipped cavity, already visible in the FRF in Fig. 24.10, can thus be understood. However, examining the contour plot alone is not sufficient to distinguish the effect of diffraction and shadowing from the change in the acoustic characteristics of the room. Since the binaural rendering filter was designed only for the two ear positions, no analogous plot can be derived that would allow a graphical visualization of the binaural rendered signal in the vertical plane.

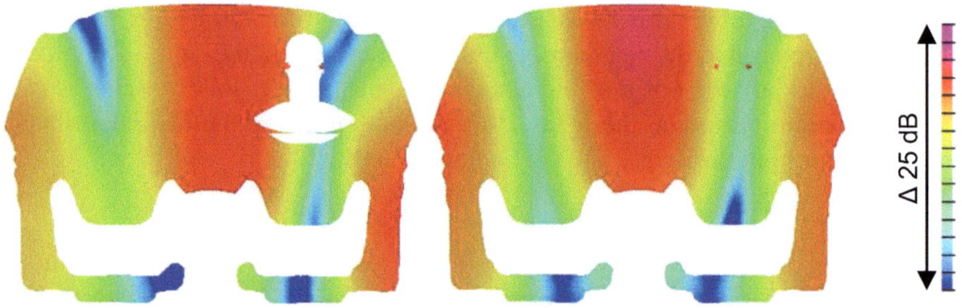


**Fig. 24.15.** Sound pressure level at 247 Hz in lateral cutting plane induced by unit excitation with a monopole source at the overhead control unit for equipped (left) and empty cavity (right).

Other load cases comprise deviations in broader frequency ranges where changes in the acoustic field characteristics are dominating instead of HATS directivity.

The investigation of additional load cases reveals deviations of more than 5 dB between HATS-equipped cavity and empty cavity, respectively binaural rendering, over larger frequency ranges. Hence in these ranges the changes in the acoustic field dominate over the directivity-induced effect. Figure 24.16 illustrates two example FRFs to emphasize this observation. The upper diagram shows the left ear position with the source on the left at the headrest, and the lower one shows the right ear position with the source in the driver's footwell. Both FRFs contain the aforementioned deviations over wide frequency ranges (670 to 800 Hz and 600 to 900 Hz, respectively).

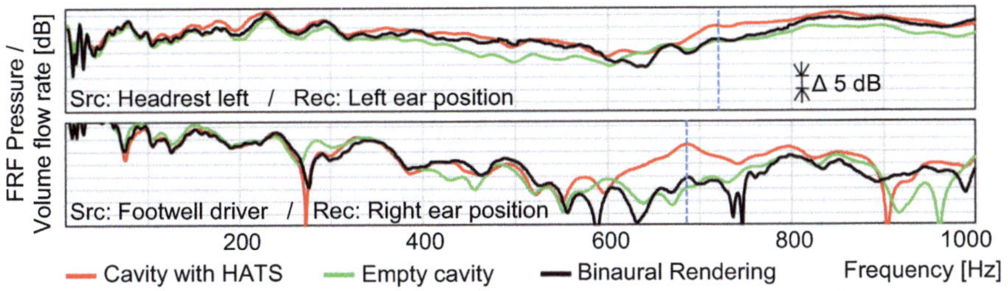


**Fig. 24.16.** Frequency response functions of additional loadcases, considering three scenarios: (a) HATS-equipped cavity, (b) empty cavity and (c) Binaural rendering based on empty cavity. The blue dashed lines indicate the frequencies to be investigated in detail.

Examining the MPFs for individual frequencies inside of these ranges (depicted in Fig. 24.17 and Fig. 24.18) reveals an even participation of many modes, whose individual consideration does not provide any insights. However, in the last case considered, it becomes clear for the empty cavity that the alternating contributions lead to an overall lower level. The response is massively altered by the presence of the HATS.

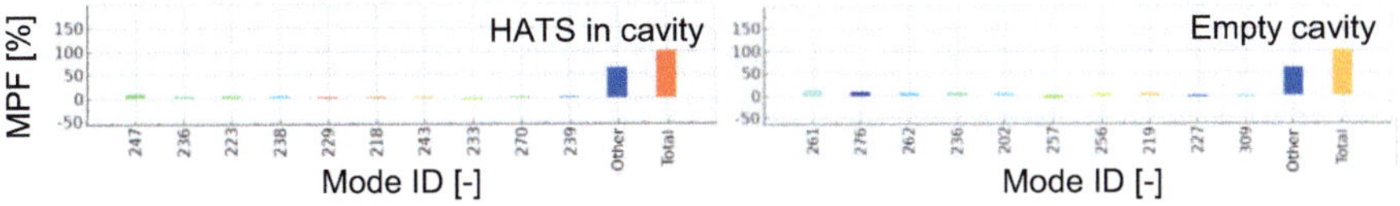


**Fig. 24.17.** Modal participation factors for left ear pos. at 720 Hz, source at driver's headrest left.

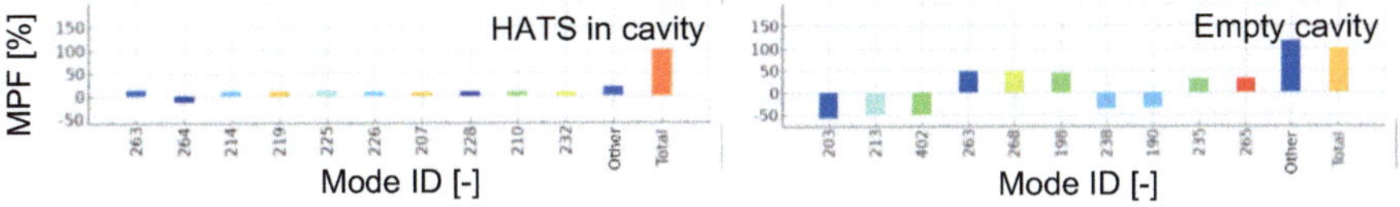


**Fig. 24.18.** Modal participation factors for right ear pos. at 688 Hz, source at driver's footwell.

## 24.5 Summary

Artificial heads are an important tool for aurally accurate investigations of NVH performance in many industries. In recent years, there has been the target to decrease the number of physical prototypes during product development. Thus, it is no longer possible to perform acoustic measurements at all stages of development anymore. Acoustic performance should still be considered as continuously as possible, though. This makes the introduction of a virtual artificial head into simulation scenarios a logical consequence. However, for correctly capturing binaural data apart from free field applications, the simple definition of a directivity- and phase-dependent simulation output will not serve as sufficient substitute, as it cannot account for all aspects that would be covered by a simulation model based on the geometry of a real artificial head. A dummy head is not only used for aurally accurate subjective evaluation of cabin acoustics, but also because the dummy head itself interacts with the vehicle acoustics in a manner comparable to a human. The relevance of the influence of the dummy head on the cabin acoustics is elaborated in this paper.

The article validates an FE model of an artificial head using head-related transfer functions up to 8 kHz. This model realistically considers diffraction and shadowing effects, serving as a surrogate for a human passenger. It ensures comparability between simulations and measurements. The model is extended by a torso box for vehicle applications and reduced by means of geometric resolution and small features to match the upper frequency limit of a full vehicle simulation model of about 1 kHz and reduce computational costs. The simulated transfer functions inside the car cavity show satisfactory agreement with the measurement and even allow for psychoacoustic evaluations, which is demonstrated at the example of loudness investigation.

With binaural rendering based on an acoustically transparent virtual double sphere receiver array, a tool is introduced to enable distinction of two major effects occurring at a dummy head's ear positions in terms of sound pressure level changes when a Head And Torso Simulator (HATS) is placed in a car cavity. One effect is the HATS directivity, mainly caused by diffraction and shadowing. The other effect is a changing cavities' room acoustic characteristics induced by the pure presence of the HATS.

An example investigation of one load case for the empty car cavity as well as for the HATS inside of the car cabin, examines differences in the simulated sound pressure level at the ear positions. The application of binaural rendering in the empty car cavity model reveals that these differences can be explained by the directivity induced by diffraction and shading over a wide, but by no means uninterrupted frequency range; section-wise deviations of more than 10 dB can be observed. Considering the modal participation factors, it can be observed that at the ranges where the binaural rendering is unable to predict the correct sound pressure at the ear positions (that occurs when the HATS is placed in the cavity), a multitude of room modes are involved, leading to cancellation or amplification effects.

**Conclusion** Aurally accurate simulation models, which can only be reached by geometrically replicating the critical parts of the human body like head, pinnae, shoulders and torso, enhance the result quality, allowing for subjective evaluation of realistic auditory experiences early in product development and optimization, with a focus on human perception. Simulated acoustic scenarios can be implemented in various contexts, providing decision-makers with a broad array of tools to make critical decisions early in the development process. Applications range from binaural sound examples and the NVH desktop simulator to full vehicle simulators, such as the Porsche NVH-Lab or the mobile NVH simulator within the actual vehicle.

## References

1. ITU-T P.58: Head and torso simulator for telephonometry. 6th edition (2021).
2. Brücher, H., Wegerhoff, M., Beljan, D., Kamper, T.: Investigations of the influence of an artificial head on acoustic characteristics of vehicle cabins based on FE simulation results. In: Proceedings of the German Annual Conference on Acoustics, Deutsche Gesellschaft für Akustik, Hamburg (2023).
3. Wegerhoff, M., Brücher, H., Henseler, L., Abele, M.: NVH Optimization with a Virtual Artificial Head. ATZ Worldwide 125, 18–25 (2023).
4. Hexagon AB, Actran (Version 2023) [Software].
5. Sottek, R.: Modelle zur Signalverarbeitung im menschlichen Gehör, Dissertation, RWTH Aachen, Aachen (1993).
6. ECMA-418-2: Psychoacoustic Metrics for ITT Equipment – Part 2: Models based on human perception. 2nd edition (2022)
7. Balmages, I., Rafaely, B.: Open-Sphere Designs for Spherical Microphone Arrays. In: IEEE Transactions on Audio, Speech, and Language Processing, vol. 15, no. 2, pp. 727–732 (2007).
8. Ward, D. B., Abhayapala, T. D.: Reproduction of a plane-wave sound field using an array of loudspeakers. In: IEEE Transactions on Speech and Audio Processing, vol. 9, no. 6, pp. 697–707, (2001).
9. Hold, C., Meyer-Kahlen, N., Pulkki, V.: Magnitude-Least-Squares Binaural Ambisonic Rendering with Phase Continuation. In: Proceedings of the German Annual Conference on Acoustics, Deutsche Gesellschaft für Akustik, Hamburg (2023).
10. Guzman Nieto, M., ElSayed, M., Walch, D.: Modal participation factors and their potential applications in aerospace: a review. In: Proceedings of The Canadian Society for Mechanical Engineering. CSME International Congress, Toronto (2018).

# Coupling Acoustic and Structural Modes via Non-Conforming Interfaces

Manuel Gutbrod[1](✉), Christoph Gabriel, and Florian Toth[2]

[1] BMW AG, Munich, Germany
manuel.gutbrod@bmw.de
[2] Institute for Mechanics and Mechatronics, TU Wien, Vienna, Austria
florian.toth@tuwien.ac.at

**Abstract.** Interior acoustics play a crucial role in the increasing driving comfort requirements of electric vehicles. Numerical analysis is a cost-effective approach to compare design variants and improve acoustic perception, but directly simulating the coupled vibro-acoustic system of a whole vehicle is computationally prohibitive. This work presents a numerical analysis method to predict low-frequency interior noise in a structure due to exterior excitation, enabling practical industrial applications.

The approach couples modal models of the mechanical structure and acoustic fluid on non-conforming interfaces. This allows independent combinations of mechanical and acoustic modes. The modal coupling significantly reducing computation time compared to full-order models. Coupling terms on the interface are calculated and inserted into the modal system matrices.

For demonstrating the method on a simple example three approaches are compared: semi-analytical with analytical ansatz-functions, numerical using finite-element ansatz-functions, and an artificial measurement approach. These comparisons determine the required coupling interface resolution on non-conforming grids. The results demonstrate the numerical efficiency and flexibility of the modal coupling approach, paving the way for industrial applications involving whole vehicle models for improving interior acoustics.

**Keywords:** Modal coupled field problem · Non-conforming interface · Low-frequency acoustics

## 25.1 Introduction

Interior acoustics play an important role in the continuously increasing driving comfort requirements in particular in electric vehicles. To compare different design variants to improve the acoustic perception of passengers, numerical analyses are cheaper than experiments on prototypes. However, the direct numerical simulation of the coupled vibro-acoustic system from a whole vehicle and the vehicle interior is not possible with a high level of detail in a practical time. We use model order reduction, i.e.


A. Heintzel (Ed.): Automotive Acoustics Conference 2025, Proceedings, pp. 366–379, 2026. https://doi.org/10.1007/978-3-658-51249-1_25

we use acoustic and mechanical modes and couple them on non-conforming interfaces. The presented numerical analysis method is intended to pave the way for practical applications. By applying the method to a structure, the low-frequency interior noise caused by excitation on the outside of the structure is to be predicted. The modal coupling of vibro-acoustic modes in the low-frequency range can be carried out as a superposition of the uncoupled modes due to the low mode density. The use of non-conforming interfaces is advantageous for coupling any data from mechanical and acoustic modes.

In our approach based on the work of Engelmann et al. [1], modal models for the mechanical structure and the acoustic fluid are extracted from either measurement or finite element models. These uncoupled descriptions are coupled on a non-conforming interface to obtain a dynamic model of the whole system, i.e. the whole system behavior is described by a superposition of the uncoupled modes. As a result of modal reduction, it is not necessary to solve for thousands or millions of degrees of freedom, but only to solve for the number of modes considered, which significantly reduces computation time. For vibro-acoustic coupling, the coupling terms on the coupling surface must be calculated and inserted into the modal system matrices. To make independent combinations of mechanical and acoustic modes possible, the coupling terms are calculated on non-conforming interfaces. A more detailed description of the single and coupled finite element systems can be found in e.g. [2].

This article starts with a recall of the most important equations for modal coupling of an acoustic and mechanical system. The method is demonstrated on a simple example consisting of a cuboid acoustic air box bounded by a flexible plate. We compare the results of a semi-analytical, a numerical and an artificial measurement approach. For the semi-analytical approach, we use analytical ansatz-functions and integrate analytically. In the numerical approach ansatz functions of a finite-element model are used. And in the third approach, the modal models are obtained from an artificial measurement. The last two approaches use numerical integration. After a brief description of the analytical approach and the resulting expressions, the assumed dimensions, material parameters and boundary conditions are given. A subset of mode shapes for the acoustic and mechanical systems are then shown. From the comparison of the three approaches we answer the questions, what resolution on the coupling interface is required to calculate the coupling terms with sufficient accuracy. We demonstrate the numerical performance of the modal approach in comparison to a full numerical system.

## 25.2 Modal Coupling of Acoustic and Mechanical Modes

Based on the governing equations for the acoustic and mechanical domains, the finite element formulations for the full numerical system with the coupling conditions at the coupling interface were derived in [1]. For sake of completeness, we recall the most important equations. The acoustic system description in matrix notation

$$\left(-\omega^2 \mathbf{M}_\mathrm{a} + \mathbf{K}_\mathrm{a}\right)\widehat{\mathbf{p}}_\mathrm{a} = -i\omega\widehat{\mathbf{q}}_\mathrm{b} \tag{25.1}$$

must be solved for the acoustic pressure $\widehat{\mathbf{p}}_\mathrm{a}$ as algebraic degrees of freedom. The dynamic stiffness matrix contains the angular frequency $\omega$ and the acoustic mass and stiffness matrices $\mathbf{M}_\mathrm{a}$ and $\mathbf{K}_\mathrm{a}$, that have been divided by the ambient density. By assuming no excitation, the equation also includes the coupling condition to the mechanical system, which is the acoustic boundary source $\widehat{\mathbf{q}}_\mathrm{b}$ resulting from mechanical boundary displacement. Similarly, the mechanical system description must be solved for the mechanical displacements $\widehat{\mathbf{u}}_\mathrm{s}$ as

$$\left(-\omega^2\mathbf{M}_\mathrm{s} + \mathbf{K}_\mathrm{s}\right)\widehat{\mathbf{u}}_\mathrm{s} = \widehat{\mathbf{f}}_\mathrm{s} + \widehat{\mathbf{f}}_\mathrm{b}, \tag{25.2}$$

while $\mathbf{M}_\mathrm{s}$ and $\mathbf{K}_\mathrm{s}$ are the mechanical mass and stiffness matrices. The right-hand side vectors are the discrete volume force $\widehat{\mathbf{f}}_\mathrm{s}$ and the mechanical boundary tractions $\widehat{\mathbf{f}}_\mathrm{b}$ resulting from acoustic pressure. On the coupling interface, the continuity of momentum and displacement must apply, which can be divided into a kinematic and dynamic coupling condition. Since $\widehat{\mathbf{q}}_\mathrm{b}$ is proportional to $\widehat{\mathbf{u}}_\mathrm{s}$ and $\widehat{\mathbf{f}}_\mathrm{b}$ is proportional to $\widehat{\mathbf{p}}_\mathrm{a}$, the coupling matrices $\mathbf{H}_\mathrm{a}$ and $\mathbf{H}_\mathrm{s}$ can be obtained. The full system can be written in matrix notation

$$\begin{bmatrix} -\omega^2\mathbf{M}_\mathrm{a} + \mathbf{K}_\mathrm{a} & -\omega^2\mathbf{H}_\mathrm{a} \\ \mathbf{H}_\mathrm{s} & -\omega^2\mathbf{M}_\mathrm{s} + \mathbf{K}_\mathrm{s} \end{bmatrix} \begin{bmatrix} \widehat{\mathbf{p}}_\mathrm{a} \\ \widehat{\mathbf{u}}_\mathrm{s} \end{bmatrix} = \begin{bmatrix} 0 \\ \widehat{\mathbf{f}}_\mathrm{s} \end{bmatrix}. \tag{25.3}$$

A more detailed description of the acoustic, mechanical and coupled systems can be found in e.g. the book of Kaltenbacher [2] or the doctoral thesis of Davidsson [3].

In modal coupling, the coupled system is not described by its physical degrees of freedom, but by superposition of a set of uncoupled modes of the subsystems [4]. Considering $n$ acoustic and $m$ mechanical modes, the physical degrees of freedom $\widehat{\mathbf{p}}_\mathrm{a}$ and $\widehat{\mathbf{u}}_\mathrm{s}$ are expressed by the mode shape matrices $\mathbf{\Psi}_\mathrm{a}$ and $\mathbf{\Psi}_\mathrm{s}$ and modal coordinates $\mathbf{x}_\mathrm{a}$ and $\mathbf{x}_\mathrm{s}$ as

$$\widehat{\mathbf{p}}_\mathrm{a} = \sum\nolimits_{i=1}^{n} \Psi_{\mathrm{a}\,i} x_{\mathrm{a}\,i} = \mathbf{\Psi}_\mathrm{a}\mathbf{x}_\mathrm{a} \text{ and } \widehat{\mathbf{u}}_\mathrm{s} = \sum\nolimits_{j=1}^{m} \Psi_{\mathrm{s}\,j} x_{\mathrm{s}\,j} = \mathbf{\Psi}_\mathrm{s}\mathbf{x}_\mathrm{s}. \tag{25.4}$$

The modal descriptions of the subsystems can be obtained numerically by solving the eigenvalue problem for a given mass and stiffness matrix $\mathbf{M}_r$ and $\mathbf{K}_r$

$$(\mathbf{K}_r - \lambda_r\mathbf{M}_r)\mathbf{\Psi}_r = 0 \text{ with } r \in [\mathrm{a}, \mathrm{s}], \tag{25.5}$$

where $\lambda_r$ is a set of eigenvalues and $\mathbf{\Psi}_r$ are the corresponding eigenvectors. For vibro-acoustic systems, the acoustic and mechanical modes considered are collected column-wise in mode shape matrices $\mathbf{\Psi}_\mathrm{a}$ and $\mathbf{\Psi}_\mathrm{s}$ and multiplied from right and left to the mass, stiffness and coupling matrix blocks of the coupled full system. As a result of modal reduction, for example, the modal acoustic mass matrix $\widetilde{\mathbf{M}}_\mathrm{a}$ as

$$\widetilde{\mathbf{M}}_\mathrm{a} = \mathbf{\Psi}_\mathrm{a}^T\mathbf{M}_\mathrm{a}\mathbf{\Psi}_\mathrm{a} \tag{25.6}$$

and the modal coupling block $\widetilde{\mathbf{H}}_\mathrm{a}$ as

$$\widetilde{\mathbf{H}}_\mathrm{a} = \mathbf{\Psi}_\mathrm{a}^T\mathbf{H}_\mathrm{a}\mathbf{\Psi}_\mathrm{s} \tag{25.7}$$

are obtained. Because no acoustic volume velocity source is assumed, only the mechanical right-hand side excitation vector must be transformed into the modal subspace by pre-multiplication with the corresponding mode shape matrix. The modal mechanical right-hand side vector $\widetilde{\mathbf{f}}_\mathrm{s}$ as

$$\widetilde{\mathbf{f}}_\mathrm{s} = \mathbf{\Psi}_\mathrm{s}^T \widehat{\mathbf{f}}_\mathrm{s} \tag{25.8}$$

contains the modal excitation forces. The resulting modal system of equations reads

$$\left( -\omega^2 \begin{bmatrix} \widetilde{\mathbf{M}}_\mathrm{a} & \widetilde{\mathbf{H}}_\mathrm{a} \\ \mathbf{0} & \widetilde{\mathbf{M}}_\mathrm{s} \end{bmatrix} + \begin{bmatrix} \widetilde{\mathbf{K}}_\mathrm{a} & \mathbf{0} \\ \widetilde{\mathbf{H}}_\mathrm{s} & \widetilde{\mathbf{K}}_\mathrm{s} \end{bmatrix} \right) \begin{bmatrix} \mathbf{x}_\mathrm{a} \\ \mathbf{x}_\mathrm{s} \end{bmatrix} = \begin{bmatrix} \mathbf{0} \\ \widetilde{\mathbf{f}}_\mathrm{s} \end{bmatrix} \tag{25.9}$$

for calculating the system response in the modal coordinates $\mathbf{x}_\mathrm{a}$ and $\mathbf{x}_\mathrm{s}$. If the eigenvectors for each mode are correctly scaled, the system can be written in mass-normalized form [3], i.e. $\widetilde{\mathbf{M}}_\mathrm{a}$ and $\widetilde{\mathbf{M}}_\mathrm{s}$ are identity matrices and the eigenvalues of the modes considered are on the main diagonal of $\widetilde{\mathbf{K}}_\mathrm{a}$ and $\widetilde{\mathbf{K}}_\mathrm{s}$. The system response in the physical degrees of freedom can be obtained by using the modal ansatz in Eq. (25.4).

## 25.3 Analytical Approach for an Air Box Coupled to a Flexible Plate

We design the example such that analytical solutions for the eigenmodes of the individual problems exist. For the acoustics, we assume a cuboid air cavity with sound hard walls. For the structure, we consider a simply supported rectangular plate. The general setup is shown in Fig. 25.1 and a more detailed description can be found in Sect. 25.4. The modes of the single systems are then used as generalized coordinates for model order reduction of the coupled system equations, which are treated numerically by a Galerkin-type approach. For the modal system written in matrix notation, we derive the expressions of the entries in the matrices to obtain this semi-analytical approach.

### 25.3.1 System Description and Analytical Eigenvectors

The governing equation for the acoustics is the Helmholtz equation in the weak form

$$-\int_\Omega p' \frac{\omega^2}{c^2} p \,\mathrm{d}\Omega + \int_\Omega \nabla p' \cdot \nabla p \,\mathrm{d}\Omega - \int_\Gamma p' \rho_\mathrm{L} \omega^2 \mathbf{u} \cdot \mathbf{n} \,\mathrm{d}\Gamma = \mathbf{0}, \tag{25.10}$$

where $c$ is the speed of sound, $\rho_\mathrm{L}$ the density of air, $p$ the acoustic pressure, $p'$ the corresponding test function and $\mathbf{u} \cdot \mathbf{n}$ the mechanical displacement normal to the coupling surface. The acoustic eigenvector $V_{\mathrm{p}i}$ can be derived for a cuboid volume with perfectly smooth and rigid walls as

$$V_{\mathrm{p}\,i}(x, y, z) = \cos\left(\frac{\kappa_\mathrm{x} \pi x}{L_\mathrm{x}}\right) \cos\left(\frac{\kappa_\mathrm{y} \pi y}{L_\mathrm{y}}\right) \cos\left(\frac{\kappa_\mathrm{z} \pi z}{L_\mathrm{z}}\right). \tag{25.11}$$

The acoustic volume is oriented in the coordinates $x$, $y$ and $z$ and has the extents 0 to $L_{\mathrm{x}}$, $L_{\mathrm{y}}$ and $L_{\mathrm{z}}$, respectively. The analytical ansatz function of the $i$-th acoustic mode is described by a combination of the parameters $\kappa_{\mathrm{x}}$, $\kappa_{\mathrm{y}}$ and $\kappa_{\mathrm{z}}$ with positive integer values 0, 1, 2 and so on. The corresponding eigenvalues $\lambda_{\mathrm{p}\,i}$ are computed by

$$\lambda_{\mathrm{p}\,i} = c^2\left(\left(\frac{\kappa_{\mathrm{x}}\pi}{L_{\mathrm{x}}}\right)^2 + \left(\frac{\kappa_{\mathrm{y}}\pi}{L_{\mathrm{y}}}\right)^2 + \left(\frac{\kappa_{\mathrm{z}}\pi}{L_{\mathrm{z}}}\right)^2\right). \tag{25.12}$$

Following standard plate theory in [5] and [6] the governing equation for a homogeneous Kirchhoff plate in the weak form reads

$$\begin{aligned} -\int_{\Gamma} w'\omega^2\frac{\rho_{\mathrm{P}}h}{B}w\mathrm{d}\Gamma + \int_{\Gamma} w'\frac{\partial^4 w}{\partial y^4}\mathrm{d}\Gamma + \int_{\Gamma} w' 2\frac{\partial^4 w}{\partial y^2\partial z^2}\mathrm{d}\Gamma + \int_{\Gamma} w'\frac{\partial^4 w}{\partial z^4}\mathrm{d}\Gamma \\ + \int_{\Gamma} w'\frac{1}{B}p\mathrm{d}\Gamma = \int_{\Gamma} w'\frac{1}{B}f_{\mathrm{s}}\mathrm{d}\Gamma, \end{aligned} \tag{25.13}$$

where $w$ is the transverse displacement, $w\prime$ the corresponding test function, $\rho_{\mathrm{P}}$ the density of the plate, $h$ the plate thickness and $B$ the flexural rigidity, which is defined by

$$B = \frac{Eh^3}{12(1-\nu^2)}. \tag{25.14}$$

Equation (25.13) considers the acoustic pressure $p$ on the coupling interface at $x = 0$ to the acoustic volume and a volume force $f_s$ as an excitation on the other plate side. The plate is oriented in the coordinates $y$ and $z$ and has the same extents as the acoustic volume $L_{\mathrm{y}}$ and $L_{\mathrm{z}}$, respectively. For the boundary condition simply supported on all sides of the plate the expression for the eigenmodes $V_{\mathrm{w}\,j}$ was derived in literature as

$$V_{\mathrm{w}\,j}(y,z) = \sin\left(\frac{k_{\mathrm{y}}\pi y}{L_{\mathrm{y}}}\right)\sin\left(\frac{k_{\mathrm{z}}\pi z}{L_{\mathrm{z}}}\right), \tag{25.15}$$

where $k_{\mathrm{y}}$ and $k_{\mathrm{z}}$ are the mode order parameters with positive integer values 1, 2, 3 and so on for the $j$-th mode. The corresponding eigenvalues $\lambda_{\mathrm{w}\,j}$ are computed by

$$\lambda_{\mathrm{w}\,j} = \frac{B}{\rho_{\mathrm{P}}h}\left(\left(\frac{k_{\mathrm{y}}\pi}{L_{\mathrm{y}}}\right)^2 + \left(\frac{k_{\mathrm{z}}\pi}{L_{\mathrm{z}}}\right)^2\right)^2. \tag{25.16}$$

### 25.3.2 Modal Reduction and Galerkin-Type Approach

For modal reduction of the acoustic pressure degrees of freedom, we choose a modal basis with our analytical eigenmodes similarly to Eq. (4)

$$p(x,y,z,\omega) = \sum_{i=1}^{N} V_{\mathrm{p}\,i}(x,y,z)p_i(\omega) \tag{25.17}$$

and analogously for the test function $p\prime$. In the same way we do this for the mechanical displacement degrees of freedom as

$$w(y,z,\omega)=\sum\nolimits_{j=1}^{M}V_{\mathrm{w}\,j}(y,z)w_j(\omega) \tag{25.18}$$

and also for the test function $w\prime$.

After inserting the modal bases in Eqs. (25.10) and (25.13) and applying a Galerkin-type approach, the following $N$ equations are obtained for acoustics, for example

$$-\int\limits_{\Omega}V_{\mathrm{p}\,i}\frac{\omega^2}{c^2}V_{\mathrm{p}\,i}p_i\,\mathrm{d}\Omega+\int\limits_{\Omega}\nabla V_{\mathrm{p}\,i}\cdot\nabla V_{\mathrm{p}\,i}p_i\,\mathrm{d}\Omega-\int\limits_{\Gamma}V_{\mathrm{p}\,i}\rho_{\mathrm{L}}\omega^2V_{\mathrm{w}\,j}w_j\,\mathrm{d}\Gamma=0. \tag{25.19}$$

The single terms in (19) together with the mechanical equation terms can be transformed into a system description analogous to Eq. (25.9) in matrix notation, where the modal degrees of freedom – the modal coordinates $p_i$ and $w_j$ – are separated.

The entries of the modal system matrices are written on the corresponding places according to the $i$- and $j$-indices. The non-zero entries on the diagonal of $\underset{\sim}{\mathbf{M}}_{\mathrm{a}}$ read

$$m_{\mathrm{a}\,i,i}=\int_{\Omega}V_{\mathrm{p}\,i}(x,y,z)\frac{1}{c^2}V_{\mathrm{p}\,i}(x,y,z)d\Omega=\frac{L_{\mathrm{x}}L_{\mathrm{y}}L_{\mathrm{z}}}{8c^2}, \tag{25.20}$$

and depend only on the volume dimensions and the speed of sound for all modes. The diagonal elements of $\underset{\sim}{\mathbf{K}}_{\mathrm{a}}$ for each mode $i$

$$k_{\mathrm{a}\,i,i}=\int_{\Omega}\nabla V_{\mathrm{p}\,i}(x,y,z)\cdot\nabla V_{\mathrm{p}\,i}(x,y,z)d\Omega=\pi^2\frac{L_{\mathrm{x}}^2L_{\mathrm{y}}^2\kappa_{\mathrm{z}}^2+L_{\mathrm{x}}^2L_{\mathrm{z}}^2\kappa_{\mathrm{y}}^2+L_{\mathrm{y}}^2L_{\mathrm{z}}^2\kappa_{\mathrm{x}}^2}{8L_{\mathrm{x}}L_{\mathrm{y}}L_{\mathrm{z}}} \tag{25.21}$$

are the only non-zero entries in the matrix and depend on the dimensions and the mode order parameters $\kappa_{\mathrm{x}}$, $\kappa_{\mathrm{y}}$ and $\kappa_{\mathrm{z}}$ for acoustics.

The non-zero entries on the diagonal of $\underset{\sim}{\mathbf{M}}_{\mathrm{s}}$ result in

$$m_{\mathrm{s}\,j,j}=\int_{\Gamma}V_{\mathrm{w}\,j}(y,z)\frac{\rho_{\mathrm{P}}h}{B}V_{\mathrm{w}\,j}(y,z)d\Gamma=\frac{L_{\mathrm{y}}L_{\mathrm{z}}\rho_{\mathrm{P}}h}{4B} \tag{25.22}$$

and only depend on the plate dimensions, density and flexural rigidity for all modes. The diagonal elements of $\underset{\sim}{\mathbf{K}}_{\mathrm{s}}$ for each mode $j$ simplify to

$$\begin{aligned}k_{\mathrm{s}\,j,j}=&\int\limits_{\Gamma}V_{\mathrm{w}\,j}(y,z)\frac{\partial^4V_{\mathrm{w}\,j}(y,z)}{\partial y^4}d\Gamma\\&+\int\limits_{\Gamma}2V_{\mathrm{w}\,j}(y,z)\frac{\partial^4V_{\mathrm{w}\,j}(y,z)}{\partial y^2\partial z^2}d\Gamma\\&+\int\limits_{\Gamma}V_{\mathrm{w}\,j}(y,z)\frac{\partial^4V_{\mathrm{w}\,j}(y,z)}{\partial z^4}d\Gamma\\=&\frac{\pi^4}{4}\left(\frac{\left(L_{\mathrm{y}}^2k_{\mathrm{z}}^2+L_{\mathrm{z}}^2k_{\mathrm{y}}^2\right)^2}{L_{\mathrm{y}}^3L_{\mathrm{z}}^3}\right)\end{aligned} \tag{25.23}$$

and only depend on the width and height of the plate and the mode order parameters $k_y$ and $k_z$ for mechanics.

The coupling matrix blocks contain zero entries if the modes are orthogonal to one another in this mode combination of acoustic and mechanical modes. The non-zero entries are obtained by integrating the expressions via the coupling interface $\Gamma$. The entries for the acoustic coupling matrix $\mathbf{H}_a$ for the $i$-th acoustic and $j$-th mechanical mode

$$h_{a\,i,j} = \int_\Gamma V_{p\,i}(x,y,z)\rho_L V_{w\,j}(y,z)d\Gamma\Big|_{x=0} = \frac{L_y L_z k_y k_z \rho_L\left((-1)^{(\kappa_y+k_y)}-1\right)\left((-1)^{(\kappa_z+k_z)}-1\right)}{\pi^2\left(\kappa_y^2-k_y^2\right)\left(\kappa_z^2-k_z^2\right)} \tag{25.24}$$

and the entries for the mechanical coupling matrix $\tilde{\mathbf{H}}_s$ for the same modes $j$ and $i$

$$h_{s\,j,i} = \int_\Gamma V_{w\,j}(y,z)\frac{1}{B}V_{p\,i}(x,y,z)d\Gamma\Big|_{x=0} = \frac{L_y L_z k_y k_z\left((-1)^{(\kappa_y+k_y)}-1\right)\left((-1)^{(\kappa_z+k_z)}-1\right)}{B\pi^2\left(\kappa_y^2-k_y^2\right)\left(\kappa_z^2-k_z^2\right)} \tag{25.25}$$

are proportional by the factor $\rho_L B$.

## 25.4 Description of the Simple Example

This example consists of a cuboid air volume and a flexible steel plate as a coupled system of an acoustic and a mechanical domain. The origin of the coordinate system is chosen at a lower corner of both the air box and the plate, as both ansatz functions are defined for a domain with the limits of zero and the dimensions. The model of a Kirchhoff plate is only two-dimensional and describes the transverse displacements of the midplane. To avoid double/ multiple eigenvalues in acoustics, the geometric lengths must not be integer multiples of one another and therefore the selected air box has a length in the $x$ direction of $L_x = 0.85$ m, a width in the $y$ direction of $L_y = 0.41$ m and a height in the $z$ direction of $L_z = 0.71$ m. The flexible plate has the same geometric dimensions in the $y$ and $z$ directions as the air volume and a thickness $h$ of 2 mm. The coupling interface is located at $x = 0$. The geometrical setup can be seen in Fig. 25.1, showing the non-conforming interface with 5 continuum elements for the finite element simulation for air and 12 for the plate in $y$ direction. The following simulation runs are named according to this element count.

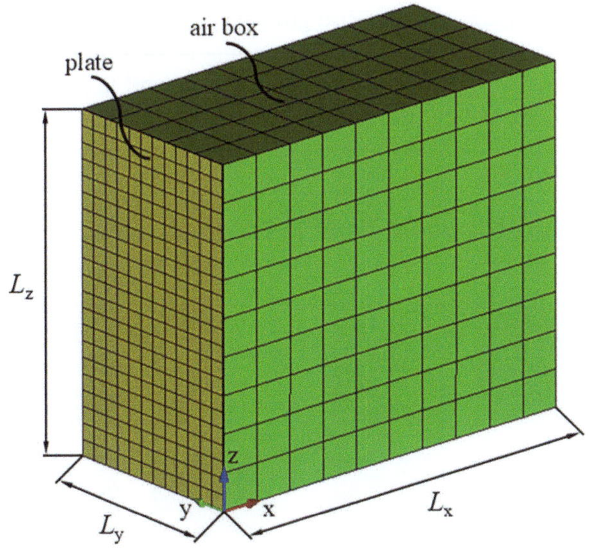


**Fig. 25.1.** Geometry of the simple example

For the air, acoustic material parameters are assumed, so the speed of sound $c$ is set to 343 m/s and the density of the air $\rho_L$ is set to 1.204 kg/m$^3$. The mechanical material parameters of the steel plate are density $\rho_P$ at 7850 kg/m$^3$, Young's modulus $E$ at 210 GPa and Poisson number $\nu$ at 0.3.

Rigid walls are assumed for the acoustic system, which corresponds to the natural boundary condition sound hard. The simply supported boundary condition is applied to the plate, i.e. all translational degrees of freedom are fixed at the edges of the midplane.

## 25.5 Modes of the Uncoupled Systems

The dimensions $L_x$, $L_y$ and $L_z$ were used to visualize the acoustic and mechanical mode shapes. The eigenvectors are normalized to a range of values between $-1$ and $+1$ and are represented by a diverging color scheme with ten levels. The corresponding eigenfrequencies can be computed from the positive eigenvalues by

$$f_{\text{eig,a}\,i} = \frac{1}{2\pi}\sqrt{\lambda_{a\,i}}\, and\, f_{\text{eig,s}j} = \frac{1}{2\pi}\sqrt{\lambda_{sj}} \tag{25.26}$$

### 25.5.1 Acoustic Modes of the Air Box

The three side surfaces of the sound pressure field calculated by (11) are shown, with the top view at the top, the front view at the bottom left and the side view at the bottom right. The first two acoustic modes under 300 Hz are shown in Fig. 25.2 with their corresponding eigenfrequencies. The first one has a constant pressure in every $y$

-$z$ plane and shows a sinusoidal pressure gradient of half a wavelength in $x$ direction with an eigenfrequency of 201.9Hz. The describing mode order parameters are $\kappa_x$, $\kappa_y$ and $\kappa_z$ with the values 1, 0 and 0, respectively. The second acoustic mode at 241.7Hz can be described by the mode order parameters $\kappa_x$, $\kappa_y$ and $\kappa_z$ 0, 0 and 1 and has a sinusoidal gradient in $z$ direction of half a wavelength.

### 25.5.2 Mechanical Modes of the Plate

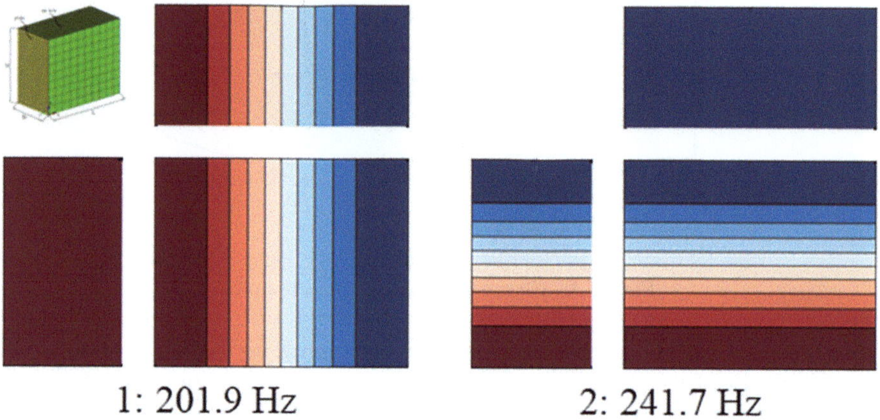


**Fig. 25.2.** Collection of first two acoustic modes with corresponding eigenfrequencies

The first ten plate modes under 300 Hz are visualized in Fig. 25.3 and the corresponding eigenfrequencies are written underneath. The mode order parameters $k_y$ and $k_z$ count the number of maxima along each coordinate dimension, e.g. mechanical mode 1 has parameters 1 and 1 and mode 3 has parameters 1 and 3.

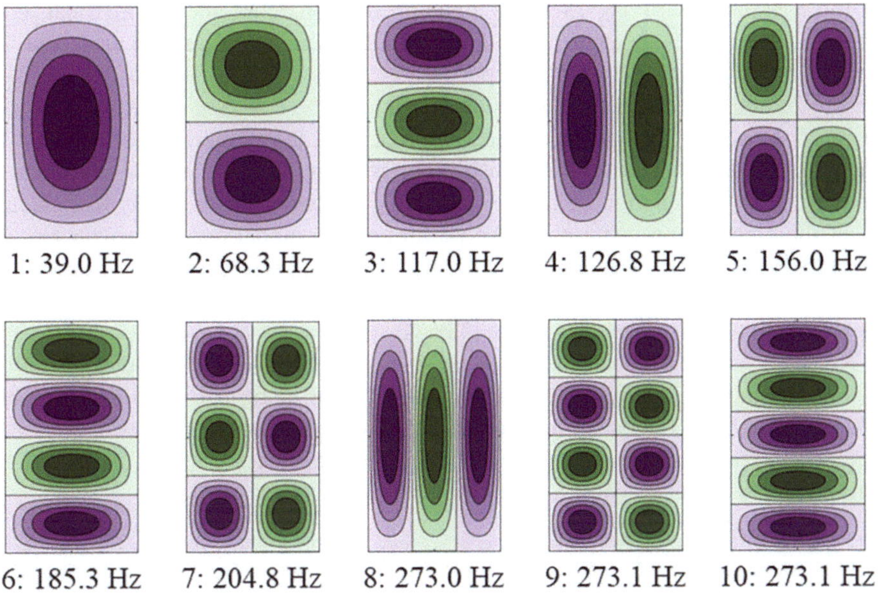


**Fig. 25.3.** Collection of first ten mechanical modes with corresponding eigenfrequencies

## 25.6 Results

An eigenvalue analysis of the modal coupled system again led to essentially the same eigenmodes and quite similar eigenfrequencies. We can confirm for our demonstrated example the statement of Engelmann et al. [1] that the acoustic and mechanical system show only weak coupling.

In the following sections we show the system responses due to harmonic excitation with a uniform pressure $p_{\text{excite}}$ of 1Pa for our three different approaches, namely analytical, numerical and synthetic measurement data approach. Then we investigate the accuracy for different plate resolutions and a refinement of the acoustic domain.

### 25.6.1 Comparison of the Analytical, Numerical and Synthetic Measurement Data Approach

For the analytical approach the right-hand-side forcing vector entries need to be calculated by analogously integrating the expression of the weak form over the coupling interface. Also, for the right-hand-side excitation the same mechanical mode order parameters $k_y$ and $k_z$ per each mechanical mode $j$ are used. The entries of $f_{sj}$ result in

$$f_{sj} = \int_{\Gamma} V_{wj}(y,z) \frac{1}{B} p_{\text{excite}} d\Gamma = \frac{4 L_y L_z p_{\text{excite}} \sin\left(\frac{k_y \pi}{2}\right)^2 \sin\left(\frac{k_z \pi}{2}\right)^2}{B k_y k_z \pi^2}. \tag{25.27}$$

For the comparison of the different undamped approaches the geometrical setup from Fig. 25.1 with 5 air elements and 12 plate elements in $y$ direction was used. We obtain good agreement between the approaches. The system response $x$ displacement between 20 and 300 Hz in the middle of the plate in the coupling surface can be seen in Fig. 25.4, where (a) is the analytical approach. In the numeric cases (b) and (c) quadratic elements were used for acoustics and mechanics, while in case (d) quadratic elements for the simulation of the air and linear elements for the analytical calculation of the eigenvectors were used. We performed the modal reduction with numerical (c) and analytical (d) eigenvectors to show, that this resolution would need to be refined in both numerical cases (b) and (c), whereas for the synthetic measurement data approach (d) this resolution of the plate is sufficient.

We observe three resonance peaks at 39 Hz, 117 Hz and 273 Hz, that are dominated by the corresponding mechanical modes of the plate in Fig. 25.3. The fourth peak around 203 Hz stands for an acoustically dominated mode, because there are no symmetrically excitable modes of the plate at this frequency. Due to the symmetric excitation, only the first acoustic mode from Fig. 25.2 can be excited.

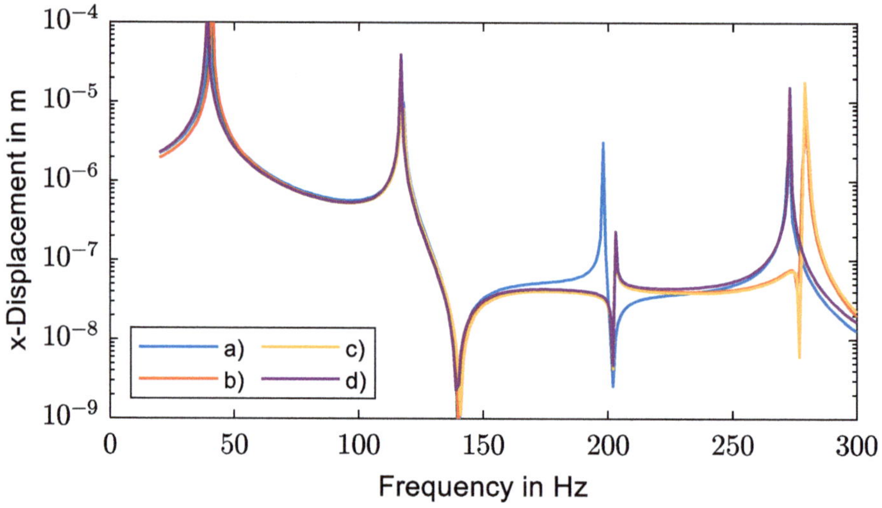


**Fig. 25.4.** Comparison of x-displacements from the four approaches: a) analytical, b) numerical, c) modal reduction by using numerical eigenvectors and d) modal reduction by using analytical eigenvectors (synthetic measurement data approach)

### 25.6.2 Investigation of the Required Resolution in the Coupling Interface

To investigate the change in accuracy, we observed a set of different mesh resolutions applying a non-conforming interface for mainly the plate but also the air box. For acoustics, we chose an element size of five quadratic elements in $y$ direction and calculated the eigenmodes up to 300 Hz with very good agreement to the analytical solution. For the plate, we calculated the system response for linear element sizes between four and twenty-seven elements in $y$ direction. In order to examine the remaining potential accuracy by refining the acoustic mesh to ten elements in the width direction, the plate meshes with twelve and fourteen elements in the $y$ direction are used.

In Fig. 25.5 the mechanical part of the system responses due to a uniform pressure excitation on the outside of the plate for the synthetic measurement data approach can be seen for different non-conforming mesh resolutions of the air and plate. We observe for all configurations three resonances at approximately 39 Hz, 117Hz and 273Hz, where the first, third and eighth mechanical modes from Fig. 25.3 were identified. Since these frequencies match the mechanical eigenfrequencies, we can confirm the weak coupling of the system. In addition, we see a resonance at around 203 Hz, with the first acoustic mode dominating here. The other eigenmode could not be excited because a symmetrical excitation was applied. Similar results were obtained for the finer resolutions of the plate. A refinement of the air volume has not shown any improvement in accuracy, so we conclude a low requirement for acoustic resolution in our case.

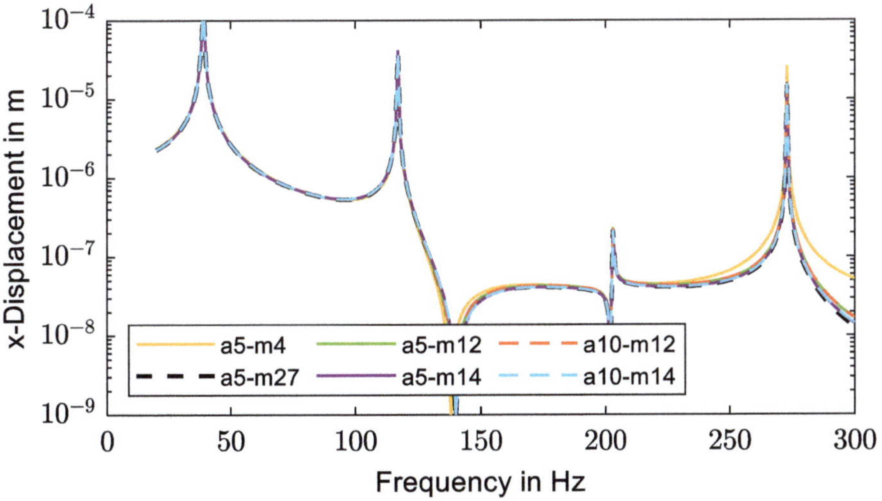


**Fig. 25.5.** Comparison of different resolutions for acoustics (a) and mechanics (m)

A convergence study is carried out to quantify descriptive accuracy between our different configurations of the simple example. Mesh resolution was used as a quality criterion in the calculation of modal coupling terms. We looked at the relative error in calculating the modal coupling terms and use as a reference a fine mesh with an element size of fifteen elements in $y$ direction with a conforming interface and modal reduction by numerical eigenvectors. Most coupling terms are numerically zero because the associated mode combinations of acoustics and mechanics are orthogonal. These coupling entries do not have to be considered. For our symmetrical excitation, we only considered the three coupling terms of the major resonance peaks and plotted the relative error with respect to our chosen reference over the number of nodes of the plate in Fig. 25.6. In addition, we plot the mean of the relative errors of the three coupling terms.

Our results show a linear decrease in the relative error in the double-logarithmic plot with convergence against a numerical error. Because of a higher resolution per wavelength, a lower error is detected at lower modes. A general statement about the required resolution cannot be made at this point, as the accuracy depends on the application.

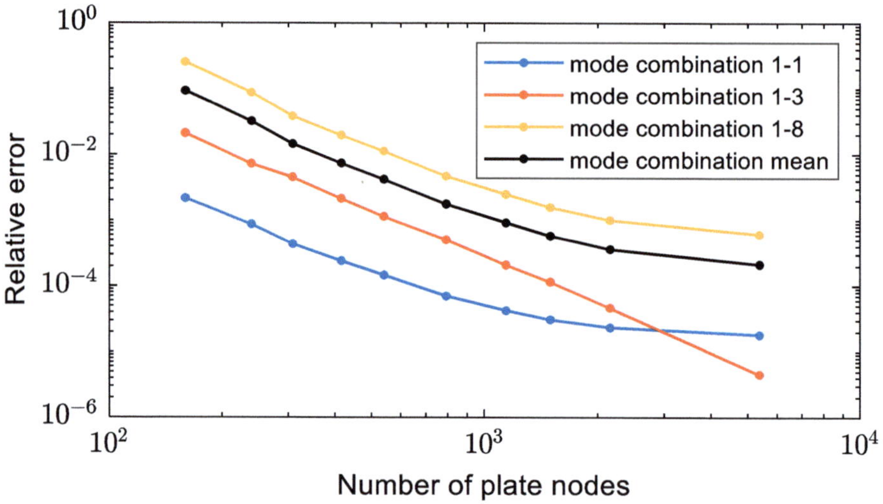


**Fig. 25.6.** Relative error of coupling terms for relevant mode combinations

### 25.6.3 Computation Performance Compared to Fully Numerical Simulation

Using the modal approach, a system with the algebraic degrees of freedom of acoustic pressure and mechanical displacement is described as a superposition of uncoupled modes. In our finest mesh configuration, for example, with an element size of five quadratic elements for the acoustics and twenty-seven linear elements for the plate in the $y$ direction, the full system has approximately 20,000 degrees of freedom under the boundary conditions used. Considering 2 acoustic and 10 mechanical modes below 300 Hz, we were able to reduce the modal system matrices to dimensions of $12 \times 12$, which significantly reduces the computation time.

## 25.7 Conclusion

In this article, we demonstrated an approach for computing low-frequency vibro-acoustic system responses based on modal descriptions of the single field uncoupled modes of the acoustic and mechanical domain. We presented a semi-analytical approach for a simple example using the analytical solutions of the individual problems as a modal basis. Our synthetic measurement data approach for applying modal reduction showed good quality of results, while being more efficient.

## References

1. Engelmann, R., Toth, F., Gabriel, C., Kaltenbacher, M.: An approach for modal coupling based on experimental and computed modes using non-conforming grids. Journal of Sound and Vibration 534, 117041 (2022).
2. Kaltenbacher, M.: Numerical Simulation of Mechatronic Sensors and Actuators. 3rd edn. Springer Berlin, Heidelberg (2015).
3. Davidsson, P.: Structure-acoustic analysis; finite element modelling and reduction methods. (Doctoral thesis), Lund University, Division of Structural Mechanics, Box 118; 221 00 Lund (2004).
4. Allen, M. S., Rixen, D., van der Seijs, M., Tiso, P., Abrahamsson, T., Mayes, R. L.: Substructuring in Engineering Dynamics. 1st edn. Springer Cham, Cham (2019).
5. Leissa, A. W.: Vibrations of plates. NASA-SP-160, Washington, D.C. (1969).
6. Hake, E., Meskouris, K.: Statik der Flächentragwerke. 2nd edn. Springer Berlin, Heidelberg (2007).

# NVH Digital Prototype for Trucks: Prediction of Customer Feature Values

Oleg Danilov[1](✉), Santhakumar Aruchamy[2], Guillaume Estienne[1], Anders Hedlund[3], Otto Hjalmarsson[3], Anders Lindström[3], and Vijay Ramanathan[2]

[1] Volvo GTT, St Priest, France
{oleg.danilov,guillaume.estienne}@volvo.com
[2] Volvo Group India Private Limited, Bengaluru, India
{santhakumar.aruchamy,vijay.ramanathan}@volvo.com
[3] Volvo GTT, Dept. BF74131, Gothenburg, Sweden
{Anders.Hedlund,otto.hjalmarsson,
anders.ali.lindstrom}@volvo.com

**Abstract.** Numerical simulation of NVH features ave been an integrated part of car and truck development process for many years. Verification of different requirements is usually done numerically for components and modules. This could include characteristics like natural frequency, stiffness, input impedance, as well as more integral characteristics like wall transmission loss or NTF (noise transfer function). Different tools are used for verifications: FE simulations for low and medium frequency range & SEA for high frequencies. Further step of NVH simulation is a Digital Prototype. In this case, the sub-feature indicators felt by driver (sound pressure at ears level and vibration on seatbase and steering wheel) are calculated. The noise & vibrations felt by drivers are a sum of different contributions, like booming noise from ICE, aerodynamic noise, air compressor noise in case of BEV or ICE vehicles, and different other contributors defined by customer operational conditions. That is why, the building of NVH Digital Prototype requires creation of different numerical models for each noise contribution. Each model includes source definition, propagation path from source to cabin, and in-cab propagation model. Simulations normally done with using hybrid approach: MBD source simulation and FE propagation in low frequency range, measured Sound Power Level and BE / SEA propagation model for high frequency airborne noise. Excitation forces obtained by block force method are used mainly in case of noise & vibration sources for BEV application. Several examples of sub-feature simulations are presented and discussed.

**Keywords:** VAP · Virtual acoustic prototype · Complete vehicle model · Structure-borne noise


A. Heintzel (Ed.): Automotive Acoustics Conference 2025, Proceedings, pp. 380–393, 2026. https://doi.org/10.1007/978-3-658-51249-1_26

## 26.1 Introduction

The numerical simulation of NVH features is an integrated part of car and truck development process since many years [1]. We can separate this work into two main areas: system and complete vehicle ones. In the case of trucks, the system approach is very widely used, we have really big systems like chassis or cabin. Otherwise only complete truck models allow us to predict of customer feature values. FE prediction of vibration values on complete truck developed in Volvo in $90^{th}$. In the beginning of 2000, the FE prediction of vibration levels on complete vehicles was integrated in the development process of Volvo [2, 3]. In the case of NVH sub-features, the road induced vibrations were mainly evaluated with quite good confidence level [4].

VAP (vehicle acoustic prototyping) or NHV digital prototype we consider as a numerical prediction of noise and vibration felt by driver. A model of feature felt on complete vehicle is needed for this. The numerical prediction of structure-borne and air-borne noises are frequently based on hybrid approach: one part of model, like input excitation forced are determined experimentally, and a model of car body is built numerically. In case of air-borne noise, hybrid way also includes an experimental determination of source strength.

## 26.2 Complete Vehicle Features Prediction: General Approaches

### 26.2.1 Structure-Borne Noise

On one side, the vehicle development process includes verification of NVH requirements on components, like cabin and chassis elements. In low frequency range, this is a numerical verification on system level, where the criteria could be a target curve of FRF or natural frequencies. These FE models built for verification work are the main support for complete vehicle NVH models.

On other side, the sub-feature model for complete vehicle requires noise and vibration sources and connections elements. The main improvement of structure-borne noise simulations was obtained with improvement of methodology of vibration source characterization. Usage of blocked force method allows us to predict SPL with good precision [5]. The typical VAP model for structure born noise can be presented as it is shown in Fig. 26.1:

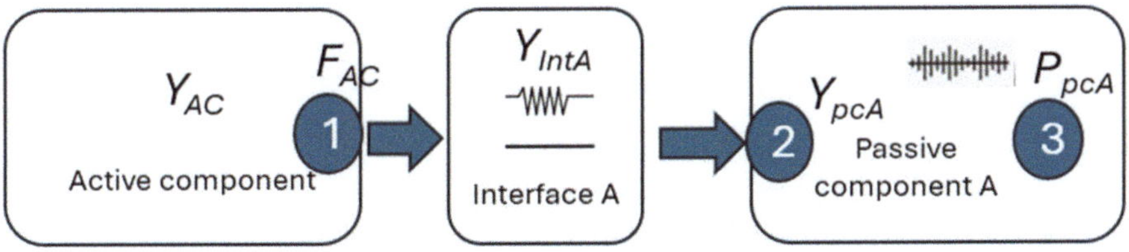


**Fig. 26.1.** A typical VAP model for structure born noise.

The equation to define sound pressure will be:

$$\{P_3\} = \left[Y_{pcA23}\right]\left[Y_{ac11} + Y_{PCA22} + Y_{IntA12}\right]^{-1}\left[Y_{ac11}\right]\left[F_{ac11}\right] \quad (26.1)$$

In the case of truck, where the transfer paths include engine suspension, chassis structure, cabin suspension with different elements, and cabin, the next schema will appear (see Fig. 26.2).

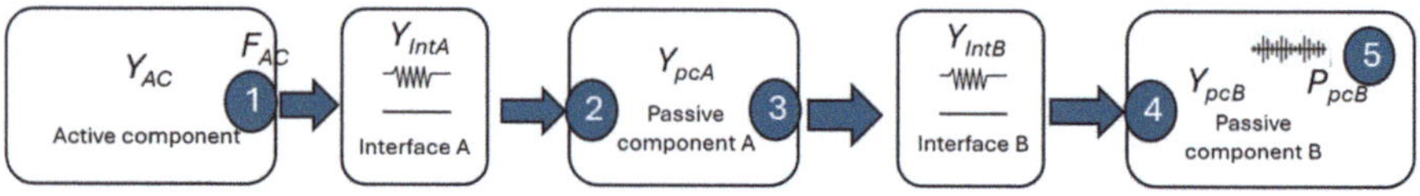


**Fig. 26.2.** FRF based sub-structuring for ICE truck.

This leads to this equation with many more elements

$$\{P_5\} = \left[Y_{pcB45}\right]\left[Y_{pcB44} + Y_{PCA33} + Y_{IntB34}\right]^{-1}\left[Y_{pcA23}\right] \left[Y_{ac11} + Y_{PCA22} + Y_{IntA12}\right]^{-1}\left[Y_{ac11}\right]\left[F_{ac11}\right] \quad (26.2)$$

where Active Component is engine, Interface A is an engine suspension, Passive Component A is chassis, Interface B is an engine suspension, and Passive Component B is a cabin. $Y$ is component transfer function, $F_{ac}$ is internal forces of active component or blocked forces. All these transfer matrices should be validated by tests and can be obtained by tests or by simulation. One knows that the matrix inversion is an important source of error.

In current case for ICE vehicles, we do not use substructure approach for this type of simulation, but one-shot calculation on complete vehicle FE model (see Fig. 26.3).

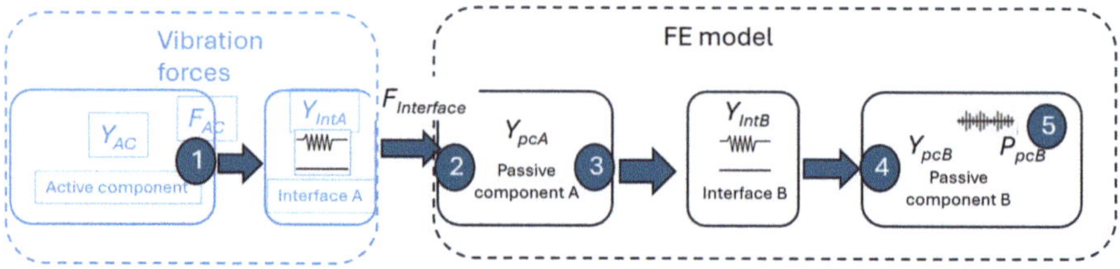


**Fig. 26.3.** FE approach for structure-borne noise on ICE truck.

Even the one-shot simulation, the main elements of model like cabin or chassis should be verified separately and on complete vehicle model.

### 26.2.2 Air-Borne Noise

Like structure-borne noise, complete vehicle models for air-borne noise are created as well. The verification of transmission loss of cabin panels requires SEA models of cabin.

The prediction of air-borne noise contribution requires source modeling, transfer path and receiver (Fig. 26.4).

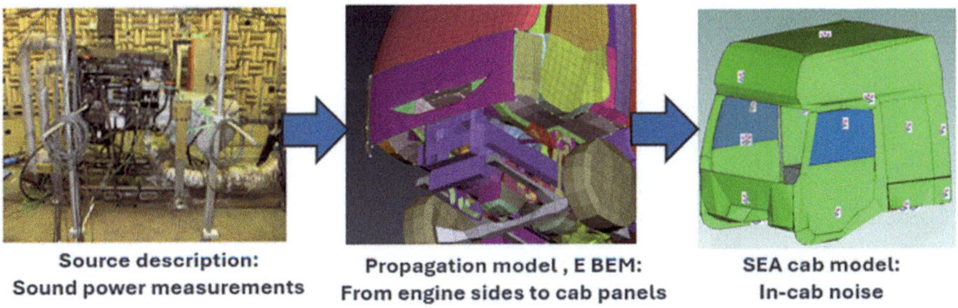


**Fig. 26.4.** Simulation of air-borne noise contribution in high / mid frequency range.

Procedure of simulation described in [6]. As a source, the sound power measured in engine test cell is used. An Energy Boundary Element Method [7] applied to calculate acoustic transfer from engine sides to cabin, and SEA model of cabin. These types of complete vehicle simulations allow us to predict the contribution of air-borne engine noise to interior noise.

## 26.3 Examples of Application

### 26.3.1 Structure Born Noise: ICE Booming Noise Model

One example of application ICE vehicle booming noise. This type of noise could be very annoying. Main source is engine induced vibration on 3$^{rd}$ engine order, and in function of operational condition can cover 25- 100 Hz frequency range.

As described above, structure born noise is calculated in "one-shot" by FE model, including cabin, cabin suspension, chassis and cabin models. The interface forces applied to chassis on engine suspension are used. We use two types of force: measured and simulated by means of MBD software.

One can see an overview of complete vehicle model for booming noise (Fig. 26.5).

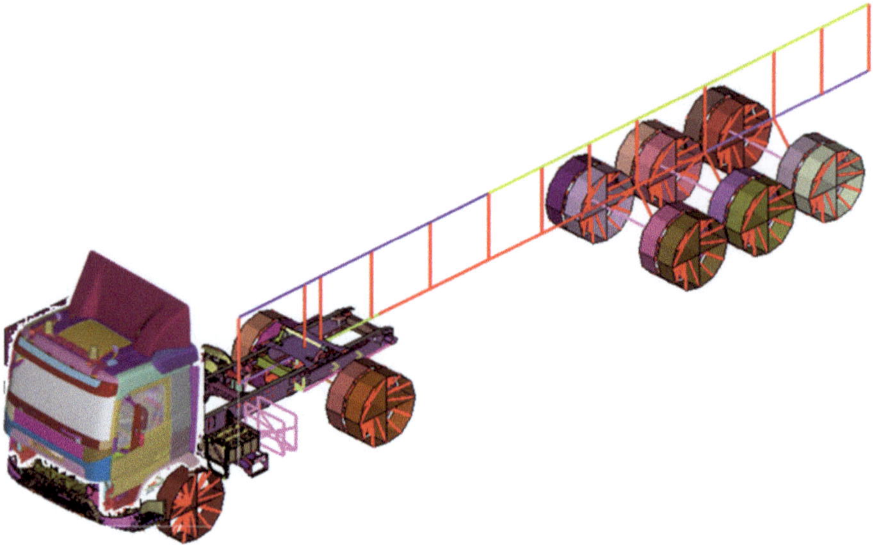

**Fig. 26.5.** ICE booming noise: FE model overview, force determination by MBD simu.

Firstly, let's look at numerical approaches for forces. The ISO 21955:2021 [8] describes procedure for vibration force determination on bench and transfer these forces to the body. We built a numerical case $\boldsymbol{F}_{bench} = \boldsymbol{F}_{vehicle}$. The MBD model includes a chassis super-element in four connection points with engine suspension (12 DOF), see Fig. 26.6

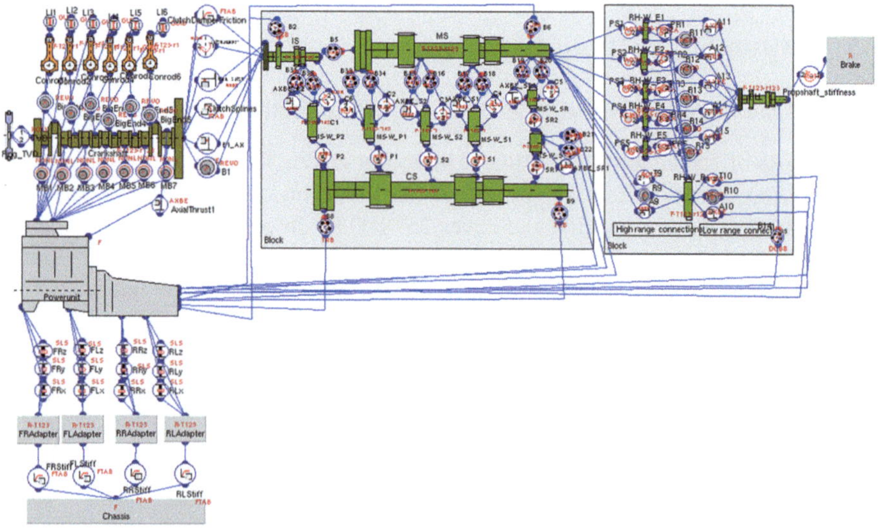


**Fig. 26.6.** ICE booming, force determination by MBD simulations.

An important challenge is finding the engine mounts dynamic stiffness, which corresponds to evaluated operational condition. The pre-load of engine torque added to one of engine weight gives a static displacement of mount. It could be a compression as well as an extension of mount. A dynamic stiffness should be found for this point from force /displacement curve.

The FE chassis model is a "classical" NVH model, we have quite few welding points and more screw connection. Chassis suspension, even if it is not an element of transfer path, is also included.

An important challenge is modeling cabin suspension. The suspension should allow tilting the cabin, so it has complex kinematics, which brings a multiplication of transfer paths. Suspension includes dampers, different bushings, and mechanical and air springs. All these elements have non-linear behavior and linearization for each operational condition is needed (Fig. 26.7).

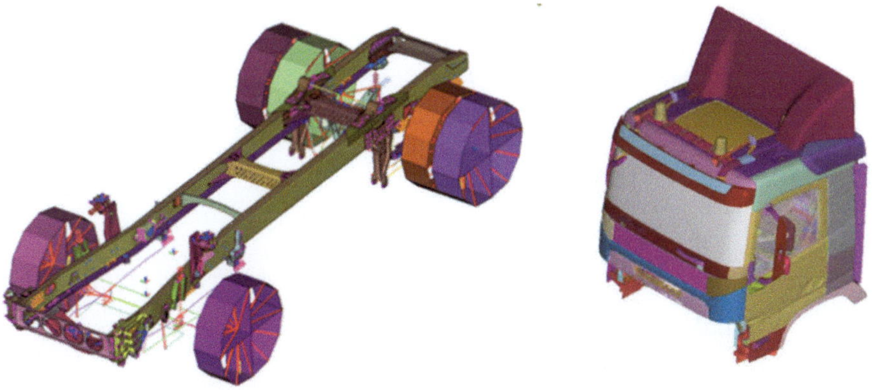

**Fig. 26.7.** ICE booming noise, chassis and cabin models overview.

The comparison test versus simulation is generally used for verification of numerical models. In this case, the two levels of verification: sub-structure alone and complete model.

The next Fig. 26.8 shows the comparison of measured and simulated forces and vibrations on the chassis side. This allows us to see the reliability of the MBD engine model. The forces were measured by dynamic stiffness method, same dynamic stiffness of mount as for MBD simulations. Measurements were done on complete vehicle: a special bench reproducing a chassis stiffness is not available.

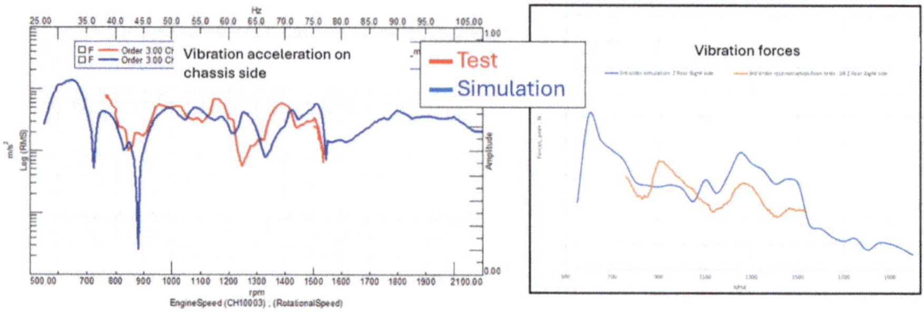


**Fig. 26.8.** ICE booming, MBD engine model verification, measured and simulated vibration acceleration and forces.

An example of verification on a sub-system alone is a cabin. The conference paper F. Acher and [9] describes a cabin model with integrated trims and carpets. Verifications were done on trimmed cabin by FRF measurements, the cabin alone installed on air suspension, see Fig. 26.9

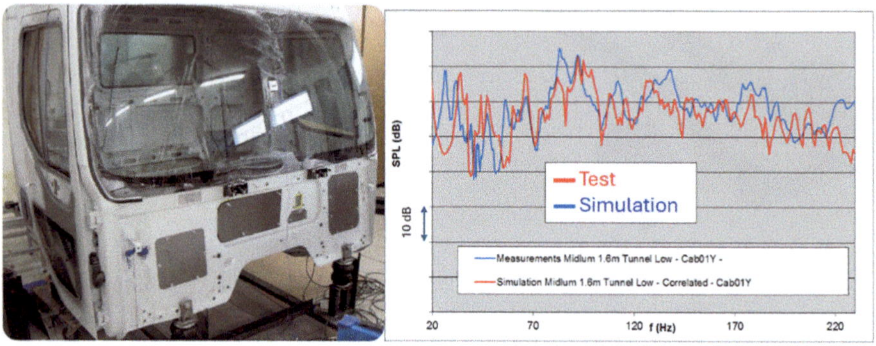


**Fig. 26.9.** ICE booming noise, cabin model verifications overview.

In case of chassis, the verifications were done on complete vehicle, but both FRF measurements and tests in operational conditions (Fig. 26.10).

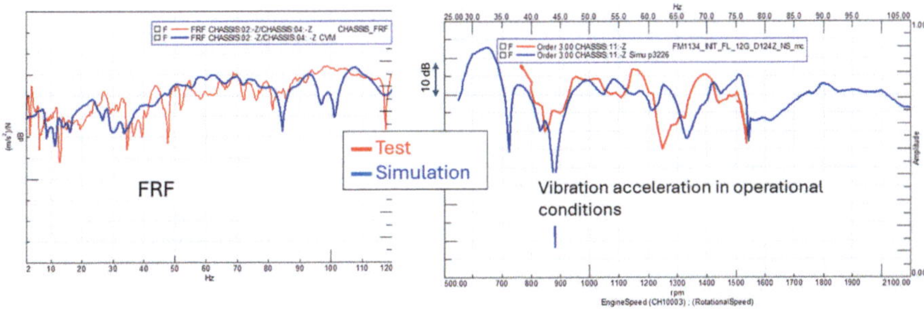


**Fig. 26.10.** ICE booming noise, chassis model verifications overview.

Results on feature point are shown on the next Fig. 26.11, sound pressure were measured and simulated on the level of driver ears. The simulation / tests comparison is shown for two different trucks. The same engine model was used for both trucks, but the chassis and especially cabins were different.

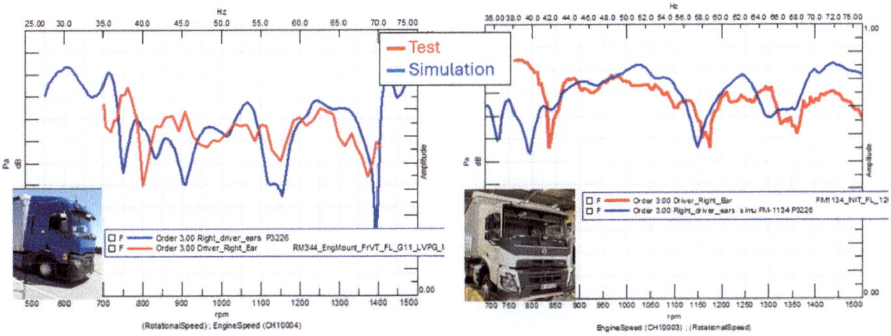


**Fig. 26.11.** ICE booming noise: simulated and measured sound pressure in control point.

This comparison shows a good confidence level of simulations.

### 26.3.2 BEV Application Interesting Example on Perceived Quality

The second example of structure-borne noise simulation is linked with BEV. In the case of different BEV accessories, like air compressors or steering pumps are ensured by separate electric motors. And it could be a source of air-borne and structure-borne noise.

Next Fig. 26.12 shows a photo of an installation of electric compressor on truck.

In this case we have a specific type of structure-borne noise, which is radiated by chassis structure and perceived mainly in exterior of the truck. This sub-feature is placed in two domains: noise and perceived quality (exterior non-regulatory noise). It could be audible at the exterior of a truck when it works in idling conditions. This gives a perception of the truck quality.

**Fig. 26.12.** BEV air compressor installed on chassis.

The air compressor connected with electric motor is installed on rigid support, which connected to one bracket through rubber mounts. This bracket is fixed on a chassis. Vibration blocked forces were measured at the level of rigid connection compressor/motor and support. The next Fig. 26.13 shows a scheme of installation.

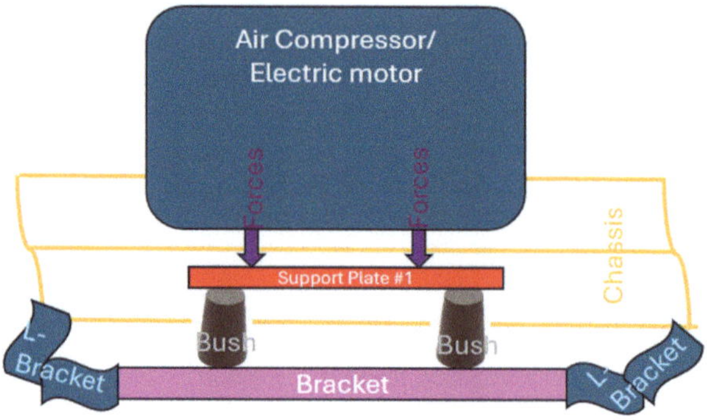


**Fig. 26.13.** BEV air compressor installed on chassis: scheme principal.

The radiated noise is simulated by one-shot calculation. The block of motor + compressor is presented as mass /inertia. The main risk for this case is noise radiation by the bracket developed for this version of installation. An overview of numerical model is shown on Fig. 26.14. An optimization of bracket was done taken into account number of mounts, bushing stiffness, plate & bracket thickness, mounting direction, material, design of the plate etc.

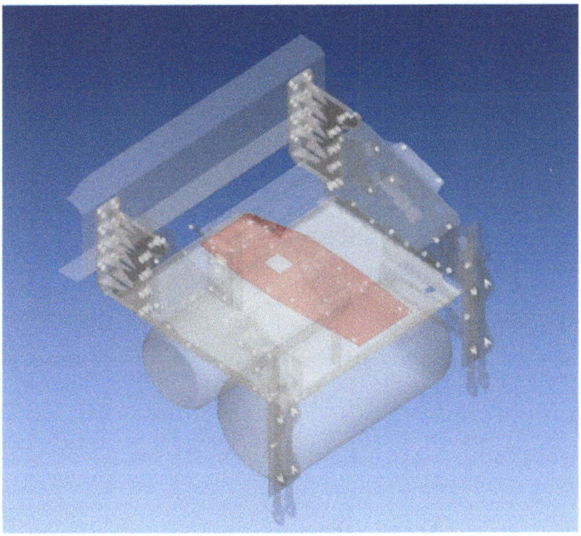

**Fig. 26.14.** BEV air compressor installed on chassis: model overview.

The criteria are radiated power and SPL at an estimated point at 1 m. The next Fig. 26.15 presents input data for simulation (measured blocked forces) and simulation results sound pressure in feature point.

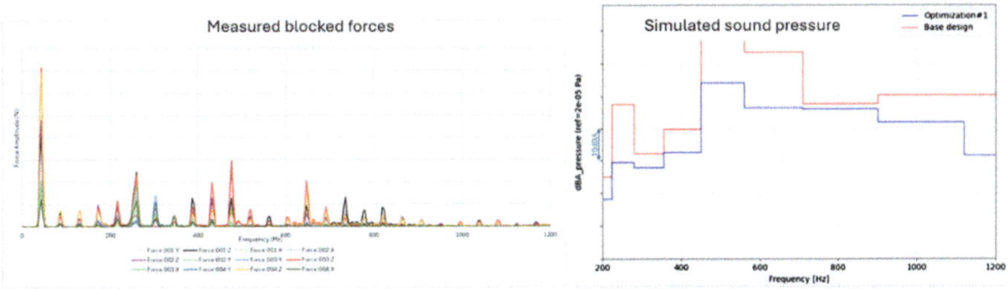


**Fig. 26.15.** BEV, air compressor installed on chassis: measured blocked forces and simulated sound pressure.

With this method we can design the attachment plate based on the NVH quality requirement. We can decide the thickness, design of the plate & most importantly alignment/position of the plate based on the radiated noise from the plate.

### 26.3.3 Acoustic Comfort in Resting Condition

Acoustic comfort in resting and sleeping conditions is one of the more valuable comfort features. An important part of this feature is interior noise when exterior acoustic excitation disturbs driver. It happens in truck parking, if another truck passes close, or a refrigerated trailer is parked very close, or connected to our truck.

We know that the most disturbing noise source in sleeping and resting conditions comes from the refrigeration unit. As well as being very loud, the noise is intermittent (the units start and stop to maintain the trailer at the desired temperature), making it extremely disturbing. Two truck configurations are considered: in the first, the road tractor is coupled to the refrigerated trailer, and in the second, the trailer is positioned longitudinally to the truck. The numerical model presented here is the worst case, when the refrigerated trailer is coupled with tested truck.

The workflow for this case is shown on Fig. 26.16.

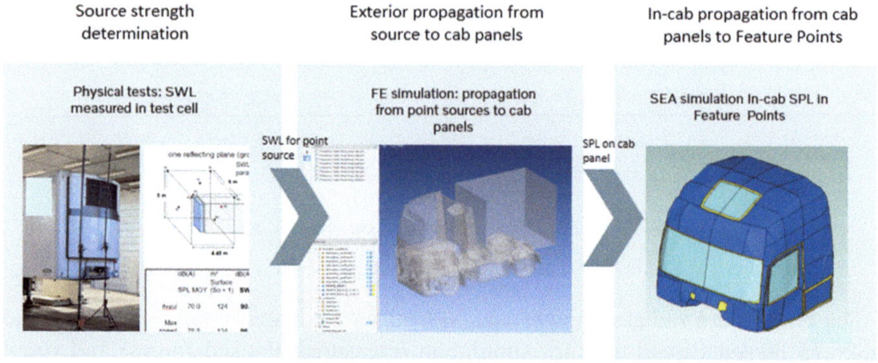


**Fig. 26.16.** Resting operational condition: simulation procedure.

Main difference with former approach for ICE noise contribution (Fig. 26.4), a FE model is used for propagation model from source to cab panels. The frequency range, covered by this simulation, achieved a 1/3 octave band of 3150 Hz. The source sound power was measured and presented in model as a monopole. The cab is always presented by SEA, this limits our model in low frequency range.

As explained in [5], the SEA cab model could be hybridized: cabin transmission loss can be simulated or measured. In our case the measured values of cabin transmission loss are used.

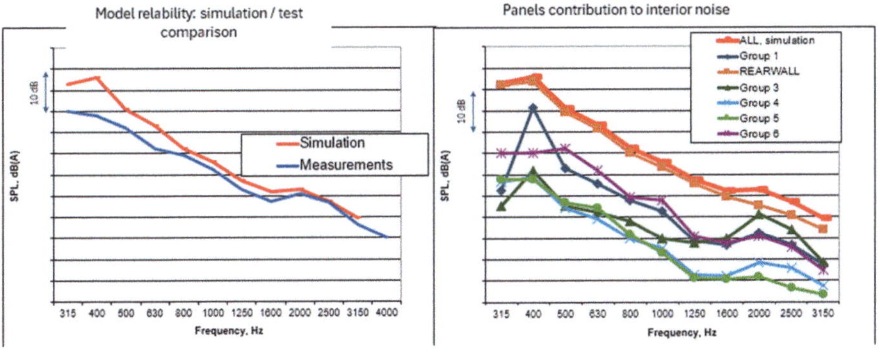


**Fig. 26.17.** Resting condition: measured and simulated results.

Figure 26.17 shows comparison tests / simulation and an example of panel contribution ranking. Results show good reliability of this model. This simulation gives us a sound pressure level heard by a customer. The main handicap, disadvantage of this approach, is an impossibility to cover low frequency, this is linked with using a SEA model.

## 26.4 Discussion, Further Steps, Tendencies

These three examples of VAP showed a possibility to predict acoustic behavior of truck with good precision. In all cases the final customer feature is simulated. In two cases we use a kind of "one shot" simulation approach, and not a matrix multiplication, which is a usual case for structure born noise VAP. As it is discussed in chapter 2, the matrix multiplication with matrix inversion could be a source of error. At the same time, we cannot say the "VAP matrix multiplication" will not be used more. The huge FE model of complete truck is challenging to solve in a higher frequency range. This high frequency limit covers ICE vehicles, but the BEV structure-borne noise sources can be placed much higher in frequency and a hybrid approach with measured transfer matrix for chassis seems to be an optimal solution.

Main objective of prediction is verification of technical solutions, and FE is best updated to build a detailed model. This is true in both cases: structure- and air-borne noises. So, FE approach will progressively replace BEM and SEA with moving in higher and higher frequency range.

The main advantage of complete vehicle models is the possibility to evaluate a technical solution from final customer feature: if driver will feel an impact of this modification. Does it mean that evaluation on system level will disappear? Our opinion: not yet. The system model is normally a one brick in complete vehicle model, and simplified models will always be present.

Complete vehicle model building is huge work with the implication of specialists from different departments, different development sites. We need support from suppliers. Input data like Biot characteristics for porous material or dynamic stiffness for bushings are needed. Also, it is important to obtain precise data, verified on system level.

The complete vehicle simulations give the same values as we obtain by test: we reproduce the same operational conditions. So, verification by simulation will take more and more place, progressively replacing physical verifications. This is a tendency. Could it completely replace tests? Taken into account that there are some sub-features, which are not covered by simulations, with an acceptable confidence level, there will always be a need for physical tests. The verification tests remain but will be progressively moved in later and later part of projects. The number of physical tests used to feed and verify simulation models will increase.

**Acknowledgement.** Authors would like to express gratitude to colleagues from Volvo GTT for their help and useful discussions: Michel Chanal, Melen Le Moigne, Kavya L, Ney Siqueira, Frederic Wullens, Laetitia Idir, Markus Andmarsjö, Clement Coppel, Fabien Acher.

## Definitions, Acronyms, Abbreviations

| | |
|---|---|
| BE | boundary elements |
| BEV | battery electric vehicle; |
| FE | finite elements |
| FRF | frequency response function |
| ICE | internal combustion engine |
| MBD | multi body dynamics |
| NTF | noise transfer function |
| SEA | statistical energy analysis |
| SPL | sound pressure level |
| VAP | virtual acoustic prototype |

## References

1. Kohler J. Möller J.-S.: Improvement of CAE Prediction Quality of NVH Full Vehicle Models by Applying Hybrid Methods, In SIA NVH Conference, 2014, Le Mans, France
2. Johansson I., Gustavsson M.: FE-based Vehicle Analysis of Heavy Trucks, Part 1: Methods and Models, Proc. of the 2nd MSC worldwide Automotive Conference, Dearborn, USA, 2000.
3. Öijer F.: FE-based Vehicle Analysis of Heavy Trucks, Part II: Prediction of Force Histories for Fatigue Life Analysis Proc. of the 2nd MSC worldwide Automotive Conference, Dearborn, USA, 2000.
4. Sjögren, S., Lindström, A., and Sundell, P.: Structured Feature Analysis in Heavy Truck Platform Projects, SAE Technical Paper 2009-01-1226, 2009, https://doi.org/10.4271/2009-01-1226.
5. Moorhouse A., Elliott A., Meggitt J.: Virtual Acoustic Prototyping - A story of four decades, In Inter Noise, Glasgow, 2022

6. Danilov O., Andre F., Tufano AR..: CAE methods for prediction of airborn noise in truck cabin, In Inter Noise, Innsbruck, 2013
7. Thivant M., Bouvet P., Cloix A., Blairon N., Braguy C., Peyroux C.: Boundary Element Energy Method for the acoustic prediction of external and interior vehicle noise – Application on a Truck, SAE Technical Paper 2010-01-1406.
8. ISO 21955:2021 "Vehicles - Experimental method for transposition of dynamic forces generated by an active component from a test bench to a vehicle", published 2021-08
9. Acher F., Sandier C., Gerges Y., Ege K.: Simulation of Truck Cabin Vibro-acoustic Response in Low Frequency Range including Effect of the Acoustic Treatments; Comparison with Experiment, SIA NVH Conference, Le Mans France , 2016

# Heuristic Approach on NVH Design of e-Motor Drives

Léon Gavric(✉) and Wilfrid Raguenet

Stellantis Auto SAS, Technical Center Carrières Sous Poissy, Carrières Sous Poissy, France
{leon.gavric,wilfrid.raguenet}@stellantis.com

**Abstract.** This paper presents a heuristic approach to the Noise, Vibration, and Harshness (NVH) design of Electric Drive Systems (EDS), emphasizing the integration of fundamental design principles from the early stages of development. The proposed methodology focuses on simple yet effective strategies to address key NVH challenges, including stator deformation, various excitation sources such as magnetic field interactions of poles, torque ripple, and cogging effects.

The design process begins with a comprehensive consideration of these parameters, ensuring that each factor is accounted for from the outset. By doing so, even the initial design of the EDS is inherently robust, minimizing the need for extensive modifications later in the development cycle. This proactive approach not only enhances the overall NVH performance but also streamlines the design process, reducing time and cost. Only at the end of design process, Finite Element Method (FEM) numerical tools are used for final optimizations and to avoid resonant effects.

The results of this heuristic design approach demonstrate significant improvements in NVH performance, highlighting the importance of early-stage parameter consideration and the strategically appropriate use of numerical optimization tools. This paper provides valuable insights for engineers and designers in the field of electric drive systems, offering a practical framework for achieving superior NVH characteristics through thoughtful and systematic design practices.

**Keywords:** Electric vehicles · Electric powertrain · Design · Engineering approach

## 27.1 Battery Electric Vehicles—BEV

### 27.1.1 Technology, Driving Experience and NVH

The automotive industry is undergoing a significant transformation with the increasing adoption of Battery Electric Vehicles (BEVs). This shift is driven by the need for sustainable transportation solutions that reduce environmental impact and improve


A. Heintzel (Ed.): Automotive Acoustics Conference 2025, Proceedings, pp. 394–403, 2026. https://doi.org/10.1007/978-3-658-51249-1_27

energy efficiency. BEVs, which utilize electricity stored in battery packs to power electric motors, offer a compelling alternative to traditional Internal Combustion Engine Vehicles (ICEVs). This introduction chapter provides a comparative analysis of BEVs and ICEVs, focusing on their technological differences and Noise, Vibration, and Harshness (NVH) behavior, setting the stage for a detailed exploration of electric motor design. BEVs and ICEVs differ fundamentally in their propulsion systems and supporting technologies:

- **Battery electric vehicles (BEVs):**
  - *Electric motor as an alternative to IC engine:* BEVs are powered by electric motors that convert electrical energy from the battery pack into mechanical energy to drive the wheels.
  - *Power Electronics (Controller) instead of Transmission System:* These components manage the power flow between the battery and the electric motor, ensuring efficient operation.
  - *Battery Pack as a replacement for Fuel/Air streams:* The energy storage system that supplies electricity to the motor, which can be recharged using grid electricity from a wall socket or dedicated charging unit.

### 27.1.2 Driving Experience and NVH Comfort

The driving experience in BEVs is shaped by its propulsion technologies:

- **Fun to drive:** BEVs often feature one-pedal driving and smooth power delivery, making them easy and enjoyable to drive.
- **Instant torque:** Electric motors deliver near-instant torque, resulting in rapid acceleration and a smooth driving experience at any duty point. This contrasts with ICEVs, where torque increases with engine revolutions (RPM)

Noise, Vibration, and Harshness (NVH) characteristics are critical factors influencing the driving experience. BEVs and ICEVs exhibit distinct NVH behaviors:

- The noise of the electric propulsion system is considerably lower, which results in a noticeable improvement of NVH comfort.
- The road noise and vibration excitation, which occurs at all vehicle speeds, remain unchanged.
- Wind noise which usually dominates at speeds over 100 km/h is independent of the propulsion system.

Completely new and "smoother" soundscape inside of the vehicle, but:

- Sound quality decreased due to whining of e-powertrain
- Lack of adequate acoustic feedback
  Remark: Residual noises, specifically auxiliary sounds (e.g., fans, compressors) remain perceptible.

This comparative analysis highlights the technological and experiential differences between BEVs and ICEVs, providing a foundation for the subsequent chapters that will delve into the design and optimization of electric motors for BEVs.

## 27.2 Electric Drive Module—EDM

### 27.2.1 Electric vs Internal Combustion Propulsion

This chapter is aimed to provide a comprehensive comparison between electric powertrains (e-powertrains) and internal combustion engine (ICE) powertrains, focusing on their characteristics and the implications for Noise, Vibration, and Harshness (NVH). Electric powertrains exhibit several distinct characteristics when compared to ICE powertrains:

- **Lower mass and moment of inertia:** Electric powertrains have a lower mass and moment of inertia, resulting in reduced excitation from road-induced vibrations.
- **High torque at low motor speeds:** They deliver high torque even at low motor speeds, which affects driveline behaviors such as "tip-in" and "let-off."
- **Low torque irregularity:** The torque ripple in electric motors is generally minimal, typically 5 to 10% of the overall torque.
- **High rotational speeds:** Electric motors can achieve rotational frequencies up to 30,000 rpm.
- **Absence of linear moving masses:** Unlike IC engines, electric motors do not have pistons or valves, eliminating unbalanced inertial forces.
- **No idle speed (noise & vibrations):** Electric powertrains do not experience noise and idle speed vibrations or shaking during stop and start operations (Fig. 27.1).

**Fig. 27.1.** Electric propulsion technology essentials: electric powertrain (EDM), reduction gears and electric motor

### 27.2.2 EDM Whining and Its Sources

This section explores the sources and design parameters that influence NVH characteristics. Whining noise in electric powertrains can originate from several components:

- **Electric Motor:**
  - The motor itself can generate whining noise.
  - The dimensions and design of the active parts (rotor and stator) play a significant role. The configuration of poles, slots and phases in the motor affects noise levels. The winding type layout and skewing can influence NVH characteristics.
- **Speed Reduction Gears:**
  - These gears can significantly contribute to the overall noise.
  - Errors in transmission and the rigidity of the axle impact NVH. The surface finish and machining of gear teeth affect noise generation. The stiffness of housings and their resonances are crucial.
- **Power and Control Electronics:**
  - The electronics involved in controlling the motor can also be a source of noise.
  - The frequency of PWM used in power electronics as well the "strategy" of torque delivery can influence NVH.

The NVH parameters not only affect the noise and vibration characteristics but also influence the efficiency, torque, and power characteristics of the powertrain. Understanding the differences between electric and IC powertrains, particularly in terms of NVH, is essential for optimizing the design and performance of modern vehicles. By addressing the sources and design parameters that influence NVH, engineers can enhance the overall driving experience and efficiency of electric powertrains.

## 27.3 Electric Motors and Automotive Applications

### 27.3.1 Permanent Magnet Synchronous Motors

Electric motors, specifically Permanent Magnet Synchronous Motors (PMSM), are widely used in electrified vehicle powertrains. The rotor of a PMSM is equipped with permanent magnets that follow a rotating magnetic field, the speed of which is proportional to the current frequency. In the simplest configuration, a single pair of poles motor, the frequency of the electric current and the rotation frequency are identical. The rotating field is generated by three magnetic fluxes produced by coils positioned at 120° angle distances. Motor torque is proportional to the phase angle between the magnetic field and the rotor, known as the load angle.

PMS motors are characterized by high power density, meaning they offer significant power relative to their weight and volume. Modern PMS motors used in automotive applications typically have 3 or 4 pole pairs.

- The magnetic field and flux are distributed according to a sinusoidal angular curve that corresponds to the number of magnetic pole pairs
- The rotation (mechanical) frequency equals the electric (current) frequency divided by the number of pole pairs

The sinusoidal current of varying frequency that powers PMS motors is generated using appropriate electronic devices and Pulse Width Modulation (PWM) techniques.

### 27.3.2 Electromagnetic Excitation of PMSM

Electromagnetic excitation in PMS motors is primarily due to the complex pattern of forces resulting from the magnetic flux and Maxwell pressure (i.e. magnetomotive force, permeance, ...). These forces change shape and intensity over time while remaining periodic. Both radial and tangential forces must be considered, as they generate radial stator displacements. Maxwell forces can be developed into double Fourier series in time and space/angle domains and analyzed as fundamental excitation components. The time-space pattern defines the resulting excitation frequency, which can be experimentally observed (Fig. 27.2).

Principal electromagnetic excitations of electric motors include:

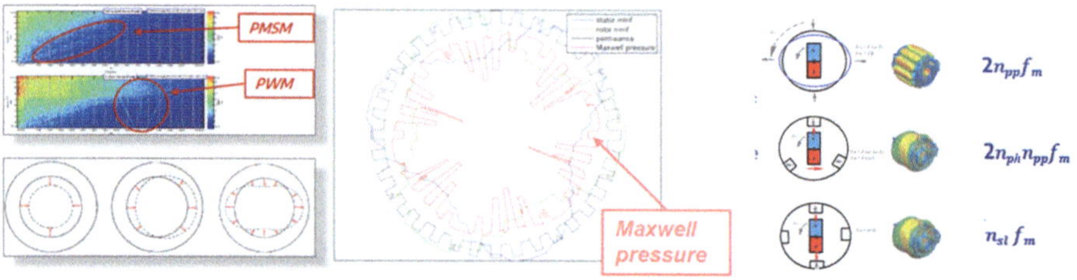


**Fig. 27.2.** Whining is the primary NVH issue for EDMs, originating from the e-motor, reduction gears, or power electronics. The e-motor whining is generated by Maxwell forces, which are distributed in a complex time-space pattern. However, several principal excitation types can be identified: "running magnetic," "phase-pole pass-by," and "slot-pole pass-by."

- The running magnetic excitation wave, which corresponds to a sinusoid with a period equal to twice the number of pole pairs and rotates at the rotor's mechanical frequency. The observed vibration frequency equals twice the electric current frequency.
- The poles-phase passing excitation, with a frequency of twice the electric current multiplied by the number of phases.
- The poles-slot passing excitation frequency is equal to the mechanical one multiplied by the number of stator slots. This excitation is more complex if is not a multiple of the number of poles (sic-not pole pairs).
- Excitation by power electronics, specifically PWM frequency, is typically around 10 kHz and may be constant or variable.

The sinusoidal current which powers PMS motors is generated using appropriate electronic devices and Pulse Width Modulation (PWM) techniques.

## 27.4 NVH Design of e-motors

### 27.4.1 Number of Poles

A high number of pole pairs in electric motors can improve Noise, Vibration, and Harshness (NVH) characteristics but require higher current and PWM frequencies, which can affect the efficiency of the e-motor. The magnetic field deforms the circumference of the stator into a polygon with number of sides equal to number of rotor poles, resulting in high deformation amplitude for a low number of pole pairs.

This deformation rotates at the rotor speed (mechanical frequency), leading to stator vibration. The observed vibration frequency equals the mechanical frequency multiplied by number of poles.

Generally, a lower number of pole pairs increases the vibration amplitude of the first harmonics and structure-borne noise.

### 27.4.2 Number of Slots

The design of slots influences magnetic permeance and is crucial for stator noise and vibration. The magnetic field deforms the stator circumference into a polygon, causing high deformation amplitude (quasi-static excitation).

- A smart choice of number of slots is a multiple of number of poles. This results in force distribution exciting higher order circumferential modes, with lowest equal to number of poles (6th, 8th, 10th and higher circumferential modes, i.e., with higher resonant frequency than the breathing mode).
- On the contrary, an inappropriate choice of slot and pole numbers can create low-order spatial distribution of radial forces, which excite low-frequency stator modes (1st, 2nd, 3rd, and 4th circumferential modes, i.e., lower than the breathing mode).

This can create significant acoustic radiation of the e-motors and affect NVH behavior of e-motor.

### 27.4.3 Interaction of Slots and Poles and Multiplication of Excitations

The design of slots also influences magnetic permeance and is critical for stator noise and vibration. The following.

- The highest force amplitude is observed when the pole is close to the slot (marked with a red circle O),
- The lowest force amplitude is observed when the pole is between two slots (marked with a blue circle o) (Fig. 27.3).

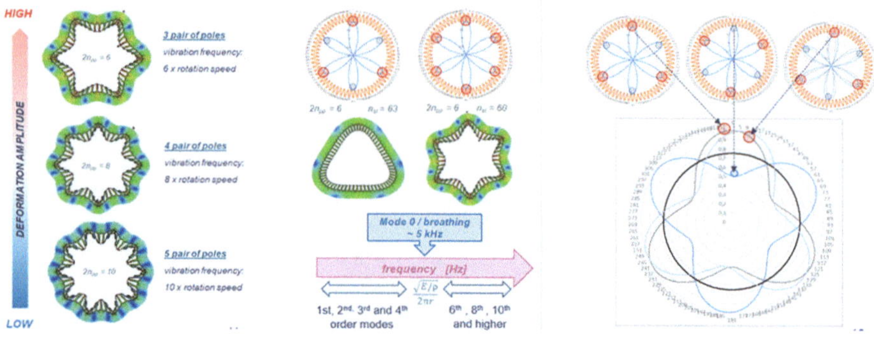


**Fig. 27.3.** The number of poles in an e-motor significantly influences its vibration deformation pattern and the resulting noise. Generally, a higher number of poles is advantageous for NVH. For any given number of poles, the slot number must be carefully selected to prevent the excitation of low-order circumferential modes caused by pole-slot interaction.

As the e-motor runs, it passes through these two steps, generating excitation efforts. Excitation forces and pressures fluctuate based on the proximity of the rotor pole and the stator slot, creating a vibrating and rotating pattern. These forces are distributed in a sinusoidal arrangement corresponding to the number of pole pairs.

This force distribution simultaneously:

- vibrates with the pass-by slot frequency in the rotor coordinates
- rotates with the rotor speed, resulting in stator vibration frequency.

The superposition of these phenomena results in excitation at two separate frequencies, resulting in rather complex vibrational phenomenon observed on the stator:

- $f_1 = (n_{sl} + n_{pp})\, f_m$
- $f_2 = (n_{sl} - n_{pp})\, f_m$

involving both, number of slots, number of poles as well the mechanical frequency of rotation.

### 27.4.4 Frequency Analysis

The analysis of excitation frequencies in electric motors involves examining the "rpm" and "frequency" patterns of the first orders of various excitations, including running magnetic excitation, pole-phase pass-by excitation, and pole-slot pass-by excitation (with two curves). If deemed potentially significant, the second order of each curve can also be included. The analysis can be divided into two frequency domains: structure-borne transfer predominant and airborne transfer predominant. This division helps identify critical excitation modes.

Additionally, the same diagram can be divided into two rpm domains: the constant torque domain (low rpm, high excitation of some excitations) and the constant power domain (flux weakening at higher rpms and diminishing of some excitations). This approach allows for the observation of areas of "quasi-static" response and model response. The strategy is to avoid resonances and adapt the excitation to minimize response. Several parameters must be considered in this analysis:

- Transfer parameters:
  - Structure-borne transfer: < 2.0 kHz
  - Airborne noise: > 2.0 kHz
- Excitation/response parameters:
  - Flux weakening: > nominal speed (as a rule ~ 50% of Max rpm)
  - Resonances stator yoke: > 3.0 kHz
- Perception/human parameters:
  - 3dB A weighting: < 0.5 kHz
  - Effect of "beating/modulation" (Fig. 27.4)

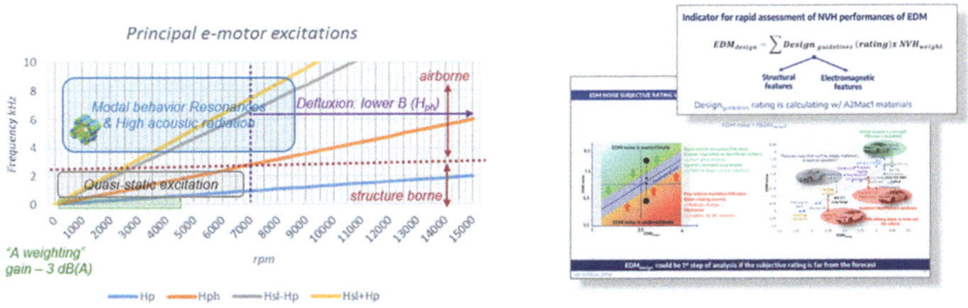


**Fig. 27.4.** The initial vibro-acoustic study involves analyzing the excitation and potential response of the stator in relation to the frequency-rpm diagram. Key inputs include significant excitation orders, modal frequencies, nominal rpm flux weakening, and frequency domains corresponding to vehicle airborne and structure-borne transfer. Subsequently, more advanced tools can be employed, leveraging the weighting of NVH levers. This process can be automated using machine learning or other AI techniques.

### 27.4.5 Pre-Design Methodologies Based on Benchmark Data

The pre-design methodology/tool provides an order of magnitude for the Electric Drive Motor (EDM) noise. This methodology is based on NVH weighting of technological content and is well correlated with subjective ratings of competitors.

It is useful as the first step of NVH analysis and allows for fast responses to "what if..." questions during development. This process may be automated using a deep learning AI approach.

## 27.5 Integration of e-motors in BEVs

### 27.5.1 Insulation of EDMs

In the realm of Battery Electric Vehicles (BEVs), the integration of Electric Drive Motors (EDMs) presents unique challenges and opportunities. One critical aspect is the insulation of EDMs to manage airborne noise.

Acoustic shielding proves effective for high frequencies, typically above 1.5 kHz. To achieve this, complete encapsulation of the e-motor, reduction gear housing, and electronic device covers is necessary. This encapsulation involves multiple parts to ensure efficient shielding.

### 27.5.2 Isolation of EDMs

Isolation of EDMs is another vital consideration. Functionally, the modal frequencies of the e-powertrain on elastic mounts should be higher to minimize the response to road excitation. However, these mounts must also be soft enough to attenuate structure-borne whining excitation.

Platform requirements vary based on the type of vehicle. For multi-energy platforms, a modified pendulum system is often chosen as it allows the integration of Internal Combustion Engines (ICE) as well. In contrast, native electric vehicles benefit from dedicated mount systems. Rear EDMs typically require double isolation if an isolated subframe is accounted for. On the other hand the front EDMs may use single isolation, while the overall architecture is either "pendulum type" or "horizontal with mounts fixed to cradle" (Fig. 27.5).

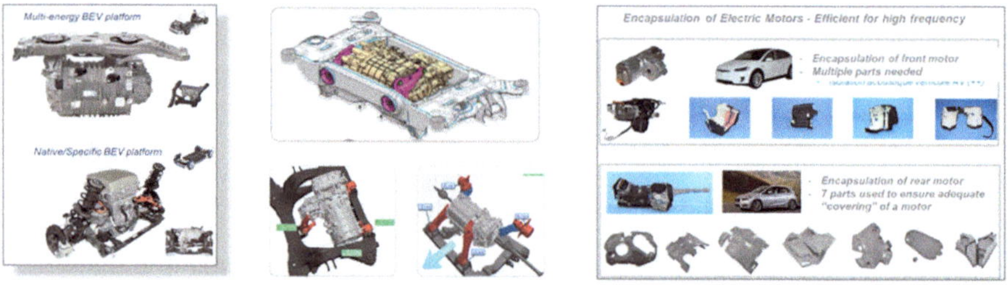


**Fig. 27.5.** A critical aspect of managing airborne noise in Electric Drive Motors (EDMs) is insulation, with acoustic shielding proving effective for high frequencies, typically above 1.5 kHz. Equally important is the isolation of EDMs. Mastering both insulation and isolation is essential when integrating EDMs into Battery Electric Vehicles (BEVs) to effectively address airborne and structure-borne noise transfer.

### 27.5.3 Electric Auxiliary Systems and Accessories

Innovative electrified accessories are required for BEVs due to the absence of an ICE, which typically drives accessories. These accessories are crucial because their noise significantly impacts the perceived noise comfort and performance of BEVs.

Without the masking effect of an ICE, device and system noises become more pronounced. Electrically driven accessories introduce new NVH (Noise, Vibration, and Harshness) challenges, including supplementary whining noises and whistling sounds. Therefore, NVH specifications for system and device noise must be strengthened to account for the relative quietness of BEVs.

For citations of references, we prefer the use of square brackets and consecutive numbers. Citations using labels or the author/year convention are also acceptable. The following bibliography provides a sample reference list with entries for journal articles [1], an LNCS chapter [2], a book [3], proceedings without editors [4], as well as a URL [5].

## 27.6 Concluding Remarks

### 27.6.1 Why to Use Heuristic NVH Approach?

Battery Electric Vehicles (BEVs) exhibit distinct Noise, Vibration, and Harshness (NVH) characteristics compared to traditional Internal Combustion Engine (ICE) vehicles. The noise generated by Electric Drive Motors (EDMs) is significantly lower than that of ICEs, leading to a marked improvement in NVH comfort.

However, the e-motor can become a principal noise source if not properly designed. This can result in sound quality issues, particularly the production of annoying whining noises. Addressing these issues early in the concept phase is crucial, even when detailed design and geometry are not yet defined. The NVH design levers strongly interact with other performance aspects of the EDM, necessitating early trade-offs.

The principles of NVH heuristic pre-design are straightforward: reduce excitation, optimize responses, and avoid resonances. Once a fundamentally performant concept is established, Finite Element Method (FEM) and test optimization of the product can be conducted efficiently, even by less experienced engineers.

## Reference

1. Leon GAVRIC: NVH refinement issues for BEV; 5th International ATZ Conference on Vehicle Acoustics—9 and 10 July 2019 | Zurich, Rueschlikon, Switzerland

# Author Index





A. Heintzel (Ed.): Automotive Acoustics Conference 2025, Proceedings, pp. 404–406, 2026.
https://doi.org/10.1007/978-3-658-51249-1

MIX
Papier aus verantwortungsvollen Quellen
Paper from responsible sources
FSC® C105338

If you have any concerns about our products,
you can contact us on
**ProductSafety@springernature.com**

In case Publisher is established outside the EU,
the EU authorized representative is:
**Springer Nature Customer Service Center GmbH**
**Europaplatz 3, 69115 Heidelberg, Germany**

Printed by Libri Plureos GmbH
in Hamburg, Germany